T0310168

TRANSPORT PHENOMENA IN MICROFLUIDIC SYSTEMS

TRANSPORT PHENOMENA IN MICROFLUIDIC SYSTEMS

Pradipta Kumar Panigrahi

Indian Institute of Technology, Kanpur, India

This edition first published 2016

Registered office
John Wiley & Sons Singapore Pte Ltd, 1 Fusionopolis Walk, #07-01 Solaris South Tower, Singapore 138628

For details of our global editorial offices, for customer services and for information about how to apply for permission to reuse the copyright material in this book please see our website at www.wiley.com.

Library of Congress Cataloging-in-Publication Data applied for.

A catalogue record for this book is available from the British Library.

ISBN: 9781118298411

Set in 10/12pt, TimesLTStd by SPi Global, Chennai, India.

Printed and bound in Singapore by Markono Print Media Pte Ltd

1 2016

Contents

About the Author

Dr Pradipta Kumar Panigrahi is the N.C. Nigam Chair Professor and Head of the Mechanical Engineering Department at IIT Kanpur, India. He was previously the Head of Photonics Science and Engineering Program and Center for Lasers and Photonics at IIT Kanpur. Dr Panigrahi received an MS in Mechanical Engineering, MS in System Science, and PhD in Mechanical Engineering from Louisiana State University, USA. His research focuses on optical methods in thermal sciences at both the macro and micro scales, microfluidics, heat transfer, and flow control. He has authored over 65 refereed journal papers, 62 conference papers, 3 popular articles, 6 book chapters, 2 Springer conference proceedings, and 2 Springer monographs. He received the Humboldt Research Fellowship, Germany; BOYSCAST Fellowship, Japan; and AICTE Career Award and Swarnajayanti Fellowship, India.

Preface

Several advances on microscale devices and systems have taken place in the past few decades. These devices have taken advantage of low cost and superior performance for the augmentation in transport processes because of their small scale. However, there is a limited understanding of physical processes in these devices. More experimental and simulation studies are essential for further improvement and development of these microsystems. Therefore, I decided to pursue research on the emerging field of microfluidics and heat transfer. My first interest was to extend my prior expertise on experimental techniques for macroscale systems to microsystems. While initiating research on this topic, I also proposed an optional course at IIT Kanpur to expose the students to this new exciting research area. I searched for a textbook on this topic but could not find a single book that satisfies all the requirements of my course proposal. Therefore, I had to refer to many reference books for the preparation of my class notes. This book is the result of several revisions of my class notes.

This book is intended as an optional course for senior undergraduate and graduate level students of various engineering and science disciplines owing to the interdisciplinary nature of the subject. It introduces different transport processes related to microdevices. The purpose of this book is to prepare students with the fundamentals and tools needed to model and analyze different microsystems. It may also serve as a reference book for microsystem designers and researchers.

The primary objective of this book is to provide a detailed overview of this subject. All aspects of transport processes relevant to microsystems, that is, mass transfer, momentum transfer, energy transfer, charge transfer, surface tension-driven flow, magnetofluidics, microscale conduction, and microscale convection, have been discussed. It is also felt that a student needs to be exposed to various microfabrication capabilities in order to appreciate the scope and significance of microscale transport phenomena. Therefore, a brief introduction to microfabrication technology has also been included in one of the chapters. Characterization of microscale transport processes is essential for validation of different simulation models and for testing of prototypes. Therefore, experimental techniques for the characterization of microscale transport processes have also been included as a separate chapter. Sensors and actuators form an integral part of both macrosystems and microsystems for the optimization of their performance. Therefore, different microsensors and actuators are also included as a chapter for highlighting the potential applications of microsystems. A micro heat pipe involving several complexities of microscale transport processes is discussed at the end of the book as one of the practical examples. Several other examples of microscale devices and systems are also included in this book depending on the importance of the specific transport process for that device.

Rapid development in microsystem technology has taken place in the past few years. It is not possible to include all the developments in an introductory textbook. Online or ancillary teaching materials need to be used by the instructor for exposing the students to several recent developments in this field.

This work owes a great deal to several published literature on microfluidics and heat transfer. I have used examples and problems from these published works while developing my course notes for the class. As I did not keep the record of all references in my early years of teaching, I have tried to eliminate most of these materials as much as I knew. However, I would like to express regret if few of them have been unintentionally included. Finally, I would appreciate receiving suggestions from readers in improving the contents of the book and the online supplementary/ancillary material.

Pradipta Kumar Panigrahi
IIT Kanpur, India

Acknowledgement

I would like to acknowledge with gratitude the initial support by the Centre for Development of Technical Education (CDTE) of IIT Kanpur for initiation of the book writing proposal. I also thank the Ph.D. students, Tapan, Balakrishna and Archana for offering miscellaneous help during final preparation of the text book. A special note of appreciation is due to Alok for preparation of Figures and Manoj for handling many of the secretarial details.

Finally, I would like to dedicate this work to my parents; Mahendra and Saraswati, in-laws; Birendra and Binodini, brothers; Jyoti and Prafulla, sisters; Shanti and Trupti, daughters; Prapti and Pragya, and wife, Mamata for their love, encouragement and support.

List of Figures

List of Tables

1

Introduction

This chapter introduces the terminology related to microfluidics and its practical applications. The historical perspective of this emerging discipline is introduced first. Subsequently, different natural systems with microscale structure and transport phenomena are discussed and correlated to the microfluidic systems. Various practical examples of microfluidic devices are presented to illustrate the importance of studying transport processes in microscale. Scaling laws are used to demonstrate different flow physics of small-scale devices.

1.1 History

The year 1959 is considered as the beginning of microtechnologies and nanotechnologies. In December 1959, R.P. Feynman gave a visionary speech during the American Physical Society meeting at Caltech entitled "*There is plenty of room at the bottom*".

The beginning of the speech was as follows:

"I would like to describe a field, in which little has been done, but in which an enormous can be done in principle. This field is not quite the same as the others in that it will not tell us much of fundamental physics (in sense of "what are the strange particles?") but it is more like solid-state physics in the sense that it might tell us much of great interest about the strange phenomena that occur in complex situations. Furthermore, a point that is most important is that it would have an enormous number of technical applications."

Some other excerpt of his speech is:

"How many times when you are working on something frustratingly tiny like your wife's wrist watch, have you said to yourself, 'If I could only train an ant to do this!' What I would like to suggest is the possibility of training an ant to train a mite to do this. What are the possibilities of small but movable machines? They may or may not be useful, but they surely would be fun to make."

Feynman's suggestion didn't remain in fantasy world. The first microbeam was fabricated in 1982 and the first microspring was fabricated in 1988. Microfluidics emerged in the beginning of the 1980s and has been used in the development of ink-jet printheads, DNA

Transport Phenomena in Microfluidic Systems, First Edition. Pradipta Kumar Panigrahi.

Companion Website: www.wiley.com/go/panigrahi/microfluidic

chips, lab-on-a-chip technology, micropropulsion, and microthermal technologies. In 1995, the word IBM was spelled out using only few atoms. *Microfluidics* is a field associated with flows that are constrained to small geometries, where the characteristic dimensions are of the order of few hundred microns. It deals with the behavior, precise control, and manipulation of flows at submillimeter scale. It is a multidisciplinary field intersecting engineering, physics, chemistry, microtechnology, and biotechnology, with practical applications to the design of systems in which small volumes of fluids are used.

1.2 Definition

A fundamental question initially arises about the definition of microfluidics. The basic meaning of this terminology is the flow at small scales. The primary advantage is the utilization of breakdown phenomena in scaling laws for new effects and better performance. Hence, the importance is not the size of surrounding instrumentation and the material of the device but the *space where the fluid is processed*. The minimization of the entire system may be beneficial but is not a *requirement* of a microfluidic system. The key issue of microfluidics is the microscopic quantity of fluid in which small-scale causes change in fluid behavior.

There are different points of view regarding device size and fluid quantity for the definition of microfluidic device. The microelectromechanical systems (MEMS) terminology indicates that the device size *must be smaller than 1 mm*. Electrical and mechanical engineers are interested to work on microfluidics because of their fabrication capabilities using microtechnology. Their idea is to shrink the device size and thus define microfluidics in terms of size to take advantage of the new effects and better performance. The objective is to shrink down the pathway of the chemicals. Another preferred way to define microfluidics is based on *fluid quantities*. Figure 1.1(a,b) shows the size and volume characteristics of different microsystems.

Nanodevices are of size less than 1 μm. Human hair is between 1 μm and 1 mm, and has similar size as microsystem. Microneedle, micropumps, microanalysis system, and microreactor are best defined based on the volume of fluid handled. Microanalysis system handles fluid volume more than 1 μl. Microneedle handles fluid volume between 1 pl and 1 μl. Microreactors handle fluid volume between 1 pl and 1 ml.

1.3 Analogy of Microfluidics with Computing Technology

Many days/hours of computing are required to perform numerical simulation for weather forecasting and various computational fluid dynamics applications. The development of *parallel architecture* in modern computers has contributed significantly to speed up the computational speed for these applications. Similar to parallel computer, microfluidics can revolutionalize chemical screening power. Compared to *manual and bench-top experiments*, microfluidics can allow pharmaceutical industry to screen combinatorial libraries at high throughput. A microfluidic assay can have several hundreds to *several thousands* parallel processes in comparison to *few hundreds* parallel processor of parallel computer. This high-performance capability is important for DNA-based diagnostics in pharmaceutical and healthcare applications.

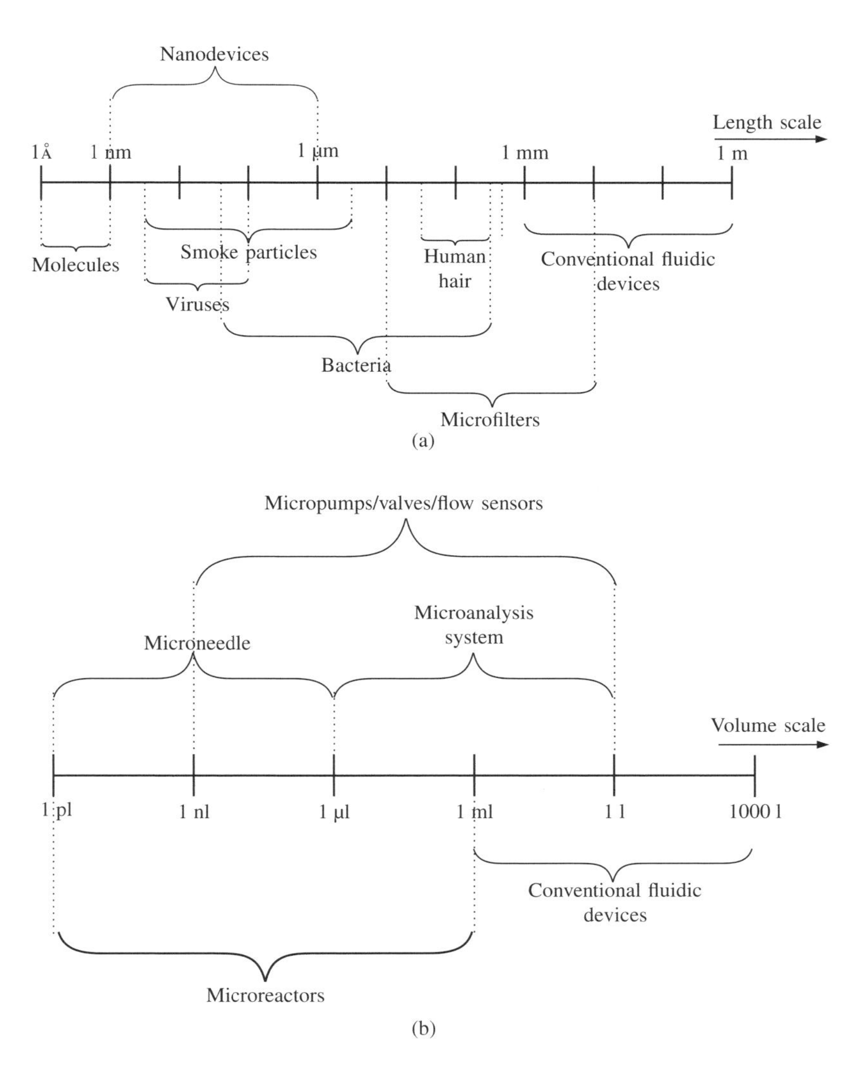

Figure 1.1 (a) Size characteristics and (b) volume characteristics of different microsystems

1.4 Interdisciplinary Aspects of Microfluidics

Owing to the rapid development of microfabrication technologies, it is now possible to miniaturize mechanical, fluidic, electromechanical, and thermal systems to micrometer sizes for various applications. This new development led to the creation of a new discipline known as microfluidics. *Microfluidics* is defined as the study of flows, which can be simple or complex, mono- or multiphasic circulating in artificial microsystems, that is, systems fabricated using new technologies, namely, etching, photolithography, and microimpression.

Research on microfluidics has become a truly multidisciplinary field representing almost all traditional engineering and scientific disciplines. Initially, microfluidics developed as a part of MEMS technology using the available infrastructure and technology of *microelectronics*. Electrical and mechanical engineers are interested in contributing to the fabrication technology related to microfluidics. Fluid mechanics researchers are interested to study the *new phenomena* in fluid flows. The flow physics in microfluidic devices is governed by a transitional regime between the continuum and molecular dominated regimes. There are a new class of fluid measurement tools for microscale flows using in situ microinstruments in addition to new analytical and computational models. Microfluidic tools allow the *life scientists* and *chemists* to explore new effects that are not possible in traditional devices.

1.4.1 Microfluidics in Nature

Microfluidics is not limited to "systems made by man." Nature also produces micrometric systems with impressive characteristics having controlled circulation of fluids.

One example is tree. The question is: how can a tree bring water and nutrients to the leaves? Nature used a complex network of capillaries to achieve this (see Figure 1.2). The trunk of the

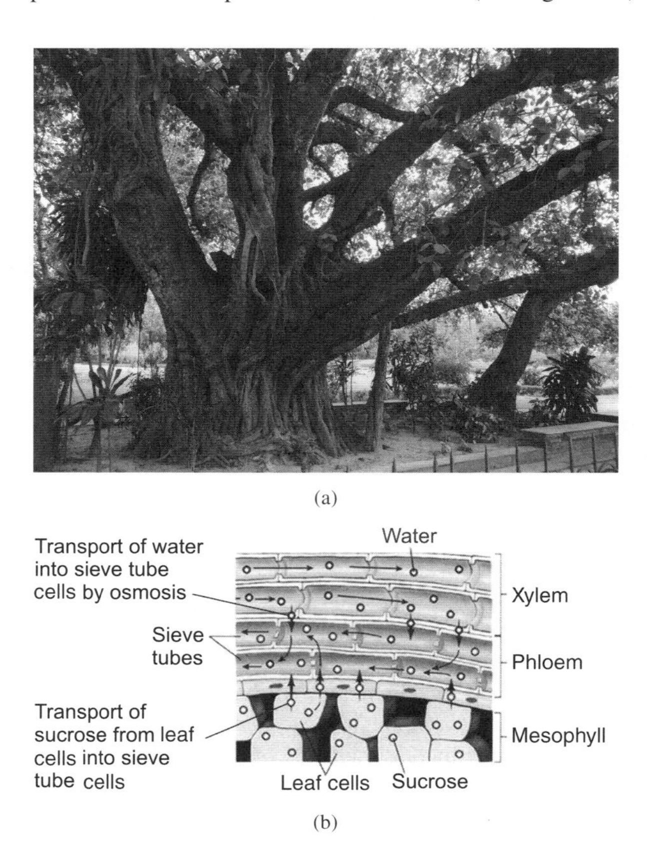

Figure 1.2 (a) A banyan tree and (b) schematic of transport process inside the capillary network

Figure 1.3 A picture of spider web having more strength than that of steel

tree has microcapillaries of a hundred *micrometers* size and leaves have *microchannels* of size equal to *several tens* of *nanometers*. Air on leaf surfaces causes water to evaporate creating a pull in the water column. There is transport of water into the cells of the root by osmosis. There is simultaneous active transport of sucrose from leaf cells into the cells of the root or stem. The supply of food carrying liquids, that is, sap, is *homogeneous* despite the complexity of the network. The pressure drop in the complex network is significant, that is, several tens of bars, implying that the sap is subjected to *negative pressure*. The hydrodynamic of this system involves consideration of deformability of the channels under the effect of pressure.

Spider web is another example of micrometric flows appearing in nature (Figure 1.3). Spider produces a long silken thread by synthesis of protein in a gland mixed with a solution. The silken thread has *exceptional mechanical properties*. Individual threads of spider web look fragile. However, they are extremely strong, that is, more than that of steel. The spider silks may be useful for human applications such as making medical sutures and high-performance ropes or used as filling in bulletproof vests. There are many examples of microfluidic systems existing in nature. Man-made systems are far from being able to compete with the natural systems.

1.4.2 Unit Systems in Small Scales

The development of various microsystems also requires new definition of unit system to describe these devices. Volume is associated with exponent equal to 3 of the length scale of a system that is, $\forall \sim l^3$. If length scale reduces from centimeters to micrometers, the volume decreases by 12 orders of magnitude. Therefore, micro world requires appropriate units for describing small quantities. Table 1.1 shows different units for description of microsystems.

A microfluidic system commonly contains volume of about 10 to few hundred nanoliters. A *biological cell* having a typical size of 10 μm diameter encloses a volume of 4 pl. Recently, it has been shown that miniaturized electrochemical method can detect the presence of about *zeptomole of DNA*.

Table 1.1 Unit system used to describe microsystems

Unit name	Power	Unit name	Power
milli	10^{-3}	atto	10^{-18}
micro	10^{-6}	zepto	10^{-21}
nano	10^{-9}	yocto	10^{-24}
pico	10^{-12}		
femto	10^{-15}		

1.5 Overall Benefits of Microdevices

Micron-sized devices are becoming more prevalent in both commercial and research applications. The growing number of publications and patents in the area of microfluidics is constantly showing the versatility of this technology. The introduction of commercially successful devices is bringing microfluidics out of research and development into a variety of applications. Unlike microelectronics, where the current emphasis is on reducing the size of transistors, microfluidics is focusing on making more complex systems of channels with more sophisticated fluid-handling capabilities, rather than reducing the size of the channels. The following are the various benefits of microfluidic devices.

- Ability to manipulate and detect small volumes
- Low consumption of reagents
- Less human error
- Higher repeatability
- Quick system response
- Reduction of power consumption
- Parallel devices and faster processes leading to high throughput
- High rate of heat transfer in heat exchanger applications
- Safety, reliability, portability, and user-friendly devices.

One important advantage of the microfluidic devices is that miniaturized components and processes use smaller volumes of fluid, thus leading to reduced reagent consumption. This decreases costs and permits small quantities of precious samples to be stretched further (e.g., divided up into a much larger number of screening assays). Quantities of waste products are also reduced. The low thermal mass and large surface-to-volume ratio of small components facilitate rapid heat transfer, enabling quick temperature changes and precise temperature control. In exothermic reactions, this feature can help to eliminate the buildup of heat or "hot spots" that could otherwise lead to undesired side reactions. The large surface-to-volume ratio is also an advantage in processes involving support-bound catalysts or enzymes, and in solid-phase synthesis. Microfluidic devices sometimes enable tasks to be accomplished in entirely new ways. For example, fluid temperature can be rapidly cycled by moving the fluid among chip regions with different temperatures rather than heating and cooling the fluid in place. The laminar nature of fluid flow in microchannels permits new methods for performing solvent exchange, filtering, and two-phase reactions.

Many microfluidic technologies permit the construction of devices containing multiple components with different functionalities. An integrated chip could perform significant biological or chemical processing from beginning to end, for example, the sampling, preprocessing, and measurement involved in an assay. This is the kind of vision that led to the terms "lab-on-a-chip" (LOC) and "micrototal analysis system (μ-TAS)". Performing all fluid-handling operations within a single chip saves time, reduces risk of sample loss or contamination, and can eliminate the need for bulky, expensive laboratory robots. Furthermore, operation of microfluidic devices can be fully automated, thus increasing throughput, improving ease of use, improving repeatability, and reducing the element of human error. Automation is also useful in applications requiring remote operation, such as devices performing continuous monitoring of chemical or environmental processes in inaccessible locations.

1.5.1 Importance of Flow through Microchannels

Nominally, microchannels can be defined as channels whose dimensions are less than 1 mm and greater than 1 μm. For channels of size greater than 1 mm, the flow exhibits behavior similar to that of macroscopic flows. For channels of size lesser than 1 μm, the flow is better characterized as nanoscopic. Most microchannels used today in various microdevices fall into the range of 30–300 μm. Microchannels are fabricated from glass, polymers, silicon, and metals by using various processes, that is, surface micromachining, bulk micromachining, molding, embossing, and conventional machining with microcutters. The advantages of microchannels are due to their high surface-to-volume ratio and their small volumes. The large surface-to-volume ratio leads to higher heat and mass transfer rates, making microdevices as excellent tools for compact heat exchangers. Microchannels are also used to transport biological materials such as (in order of size) proteins, DNA, cells, and embryos or to transport chemical samples and analytes. One successful example of the application of microchannels is in the area of bio-microelectromechanical systems (Bio-MEMS) for biological and chemical analyses. The primary advantages of microscale devices in these applications are good match with the scale of biological structures and potential for placing multiple functions for chemical analysis on a small area, that is, the concept of a "*chemistry laboratory on a chip*". Flow rate in biological and chemical microdevices are usually much slower than those in heat transfer and chemical reactor microdevices.

1.5.2 Multiphase Microfluidics

In multiphase microfluidic systems, interfacial forces dominate over inertia and gravity. The first application is the generation of particles using fluid–fluid interfaces. Instabilities can occur at the interface between two fluids in motion because of the difference in shear velocities, which causes waves to propagate at the interface. This can lead to droplet breakup with correct geometric configuration. The second application is using gas–liquid interface for the enhancement of heat and mass transfer. Air bubbles inside a liquid-filled microfluidic channel elongate into plugs. These plugs are surrounded by thin liquid film, which causes the bubbles to move faster than the liquid creating a recirculating wave behind the bubble. This can be used to enhance the mixing and heat transfer inside the channel. The third application is using solid–liquid interface for control of boiling. Single bubbles form and depart from the wall for nucleate pool boiling. The frequency and size of the bubbles can be influenced by the surface wet ability. By patterning surfaces with wetting and nonwetting regions, the growth of bubbles can be controlled to enhance the heat transfer.

1.5.3 Microfluidic Applications

In microfluidics, small volumes of solvent, sample, and reagents are moved through microchannels embedded in a chip. Current applications of microfluidics include detection and control of chemical reactions, sample preparation, various sensors (flow sensors, pressure sensors, and chemical sensors), actuators (microvalves, micropumps, and microfluidic amplifiers) for proper manipulation, control of flow in various microdevices as LOC and μ-TAS (total analysis system) for biological diagnostics, such as DNA analysis by means of PCR (polymerase chain reaction), and so on. The large surface-to-volume ratio makes microdevices as excellent tools for compact heat exchangers and fuel cells, which provide a powerful platform for thermal management of high power density microprocessors and cellular phones. The printing of text is achieved by a well-synchronized cooperation between a precisely manufactured set of micronozzles and a myriad of microelectronic circuits of an ink-jet printer.

In early days of microcomputing, computers were accessible to a limited part of the population. However, they are today commonly encountered in almost every sphere of life. It is expected that products based on microfluidics will be an integral part of our lives in the near future. The current trend in microfluidics is toward the development of integrated devices, which may transform our world in a manner similar to that of microelectronics.

A number of microfluidic applications in biology, analytical biochemistry, and chemistry have grown as a range of new components and techniques have been developed and implemented for delivering, mixing, pumping, and storing fluids in microfluidic channels. Many companies have ventured into developing microscale devices for chemical and biological analyses. Some of the applications of microfluidics are discussed in the following sections.

1.5.3.1 Lab-on-a-Chip

Lab-on-a-chip (LOC) is a term for devices that integrate (multiple) laboratory functions on a single chip of only millimeters to a few square centimeters in size. These devices are capable of handling extremely small fluid volumes down to less than picoliters. LOC devices are a subset of MEMS devices and often indicated as μ-TAS. Microfluidics is a broader term that also describes mechanical flow control devices such as pumps and valves or sensors such as flow meters and viscometers. In other words, "lab-on-a-chip" indicates generally the scaling of single or multiple lab processes down to chip format, whereas "μ-TAS" is dedicated to the integration of the total sequence of lab processes to perform chemical analysis. The following processes can be performed with LOC:

- Real-time PCR; detect bacteria, viruses, and cancers.
- Biochemical assays.

- Immunoassay; detect bacteria, viruses, and cancers based on antigen–antibody reactions.
- Dielectrophoresis-based detection of cancer cells and bacteria.
- Blood sample preparation; can crack cells to extract DNA.
- Cellular lab-on-a-chip for single-cell analysis.
- Ion channel screening.

1.5.3.2 Microreactors

Microreactors are used to synthesize material more effectively than batch technologies. Microreactors are devices in which chemical reactions take place in a confinement with typical lateral dimensions less than 1 mm. The main feature of microstructured reactors is the high surface area-to-volume ratio in comparison to the conventional chemical reactors. Heat transfer coefficient in microchannel reactor is significantly higher than that of traditional heat exchangers. The high heat-exchanging efficiency allows to carry out reactions under isothermal conditions. In addition to heat transport, mass transport is also improved considerably in microstructured reactors. Process parameters such as pressure, temperature, residence time, and flow rate are more easily controlled in this reactor. Owing to enhanced mass transfer capability and controlled thermodynamics, microreactors can be effectively used to synthesize material.

1.5.3.3 Microdiluters

Microfluidic diluters ('Microdiluters') are systems in which solutions or liquid reagents are carried through a series of controlled dilutions and then used in assays. These diluters perform some of the functions of multiple well plate assays, but use smaller quantities of reagents, and are less labor-intensive. Microdiluters that perform multiple cycles of dilution are the microfluidic version of a multiwell plate. Two fluids are repeatedly split at a series of nodes, combined with neighboring streams, and mixed in a microdiluter system (Dertinger *et al.*, 2001).

1.5.3.4 Microarrays

Microfluidic systems consisting of crossed sets of microchannels provide a way of studying the interaction of a large number of molecules with proteins or cells in a combinatorial layout. A microfluidic chemostat with an intricate system of plumbing is used for growing and studying bacteria (Balagadde *et al.*, 2005). Ismagilov *et al.* (2001) described a combinatorial tool based on two layers of polydimethylsiloxane (PDMS) microfluidic channels bonded together at right angles to each other and separated by a thin membrane. The membrane permits chemical contact between the two layers of channels, while keeping small particles from crossing the two streams.

1.5.3.5 Biomedical Industry

Disposable blood pressure transducers have become an integral part of various biomedical devices, that is, respirators, lung capacity meters, medical process monitoring, kidney dialysis equipments, and so on. Microchannels provide a convenient mechanism for treatment of cells – or portions of cells – with soluble reagents. Takayama *et al.* (2001) exposed selected regions of mammalian cells to fluids containing fluorescent dyes by positioning cells at the interface of two different streams of fluids flowing at low Reynolds number.

1.5.3.6 Micro Heat Exchangers

Micro heat exchanger technology exploits enhanced heat transfer resulting from structurally constraining streams to flow in microchannels, which reduces resistance to transferring heat. Fluid flowing through the channels on a plate may evaporate or condense, and heat is transferred. Micro heat exchangers have been demonstrated with high convective heat transfer coefficients (10,000–35,000 W/m^2-°C, or about one order of magnitude higher than that typically seen in conventional heat exchanger) with low-pressure drops. This technology has high potential for mass production and can be used in a variety of applications, such as automobiles, commercial and residential heating/cooling, manufacturing, and electronics cooling.

1.5.4 Consumer Products

Various microdevices, that is, MEMS products, have also found applications in various consumer products, that is, sport shoes with automatic cushioning control, digital tire pressure gauges, smart toys, washers with water-level controls, and so on.

1.6 Microscopic Scales for Liquids and Gases

One of the basic scales of gas and liquid systems is the *size of the molecules*. The size of simple molecule is of the order of few angstroms. A second important scale is the *average distance between* molecules. Let us consider the intermolecular distance of air. The volume occupied by a single air molecule can be written as

$$l^3 = \forall / N \tag{1.1}$$

where $\forall$ is the volume of air containing N molecules. The length scale l can be represented as average intermolecular distance. The ideal gas law from kinetic theory of gases is

$$P\forall = nRT = NK_BT \tag{1.2}$$

where n is the number of moles; R, the universal gas constant = 8.314 J/mol-K; K_B, the Boltzmann constant = 1.38066×10^{-23} J/K = R/N_A; and N_A, the Avogadro's number = 6.0221×10^{23}/mol. Using the gas law, we have

$$l = \left(\frac{\forall}{N}\right)^{\frac{1}{3}} = \left(\frac{K_BT}{P}\right)^{\frac{1}{3}} \tag{1.3}$$

For $P = 10^5$ Pa, $T = 300$ K, $K_B = 1.3807 \times 10^{-23}$ J/K, $l = 3.5$ nm.

Table 1.2 Typical values of mean free path of different gases at normal conditions

Gas name	λ (nm)
Helium	177
Argon	64
Nitrogen	60
Air	61

The typical intermolecular distance of liquids is 0.3 nm, which is 10 times smaller than that of air. For liquids, the intermolecular distance is *comparable* to the size of the molecule. For gases, the intermolecular distance is much larger than the *size of the* molecule.

The *mean free path*, λ is another fundamental scale necessary for describing the dynamical properties of gases. Mean free path is defined as the average distance traveled by the molecule between two successive collisions. Kinetic theory of gases establishes the following two expressions for the mean free path of gases.

$$\lambda = \frac{1}{\sqrt{2}\pi \rho a^2} = \frac{K_B T}{\sqrt{2}\pi P a^2} \tag{1.4}$$

where ρ is the density, that is, the number of molecules per unit volume, T is the temperature, K_B is the Boltzmann constant, P is the pressure, and a is the size of the molecule. At normal condition, the typical values of mean free path for different gases are given in Table 1.2. The above equation indicates that mean free path increases with an increase in temperature and decreases with an increase in pressure.

1.7 Physics at Micrometric Scale

One question arises while dealing with microscale systems, that is, is there anything extraordinary happening at this scale?

In a simple liquid, molecular sizes and intermolecular distances are of the order of nanometers, which is much smaller than the dimensions of any ordinary microsystem. In a *cubic micrometer* space, there are about 10^{12} atoms of tetradecane, which is sufficiently large to disregard the identity of atoms. The *thermodynamic fluctuations* can be neglected justifying the application of macrometer approach. In an *interface*, the ranges of intermolecular forces are no larger than 30 nm, which is smaller than the size of the micrometric systems. The above facts do not support any requirement of special attention to microscale systems. However, there are several microsystems, where the microscopic description must be amended. For *gas flows in microchannels*, the boundary condition need to be modified with an expression for the mean free path of the gas, which is not traditionally present in regular hydrodynamics problems. Boundary conditions for liquid flows in microsystems are not same as that in macrosystems because of various reasons, which are described in the next chapter. Similarly, surface tension forces, electrical forces, and magnetic forces are predominant in some microsystems, necessitating special treatment compared to conventional macrosystems. Other examples are large molecules, that is, DNA or high-molecular-weight proteins that must be *individually treated* in a microsystem.

1.7.1 Macromolecules

Macromolecules are molecules containing a large number (few thousands or millions) of atoms, that is, proteins, DNA, polymers, and so on. Figure 1.4 represents three biological molecules of different sizes showing the difference in size between molecules and large macromolecules. The macromolecules *adopt different* shapes, that is, sheets, helices, and so on, depending on the *solvent type* and *temperature* in which they are immersed. Three geometric measurements are used to characterize the macromolecules.

Contour length (R_c) is the length of the macromolecules measured along its backbone. For example, a polymer having N monomers with separation distance between each of them equal to l, the contour length, R_c is given by

$$R_c = Nl \tag{1.5}$$

Contour length of a DNA molecule is the arc length of the backbone contour, that is, the distance we would travel if we move along the curved backbone from one end of the molecule to the other. *Radius of gyration* (R_g) is the average distance between the extremities of a frame around the macromolecules in a folded form. The radius of gyration of a DNA molecule is the statistical measure of the linear distances between different points on the DNA backbone. The following relationship represents the radius of gyration of a macromolecule as a function of number of atom, N making the molecule.

$$R_g \sim N^{\frac{1}{2}} \tag{1.6}$$

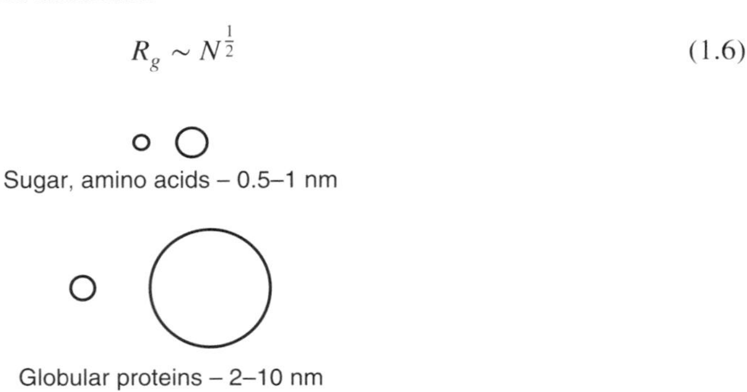

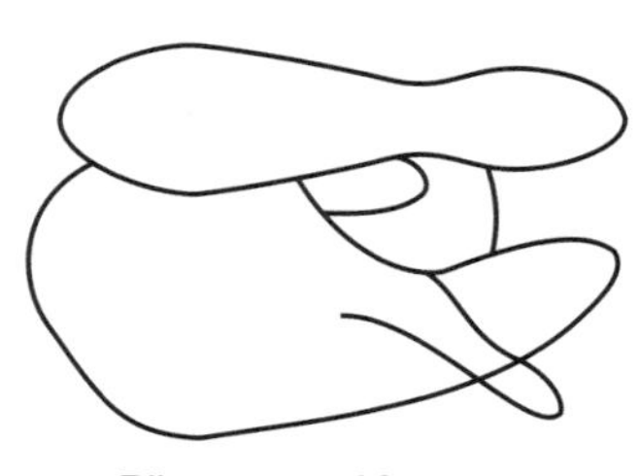

Figure 1.4 Three molecules of different sizes

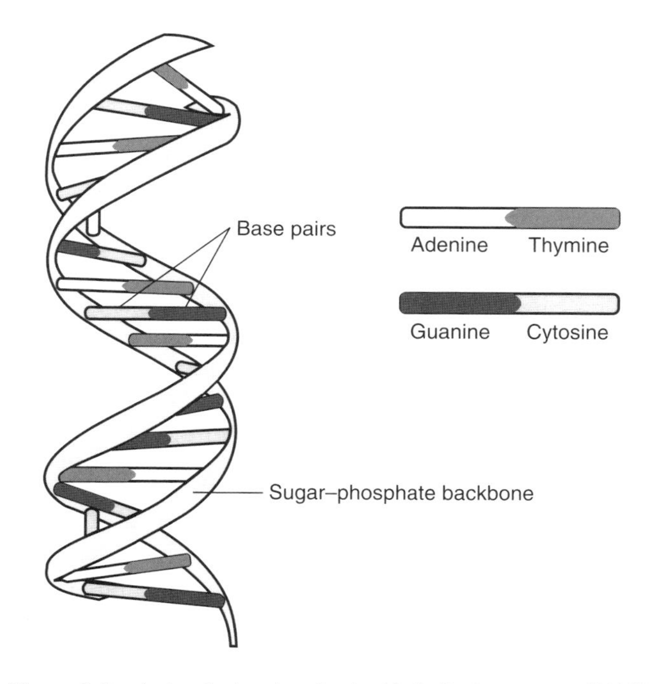

Figure 1.5 A sketch showing the double helical structure of DNA

Table 1.3 Comparison of contour length and radius of gyration for two different molecules (see Figure 1.5)

Molecules type	Contour length	Radius of gyration
DNA of a human chromosome	5 cm	1 μm
DNA of the bacteria *Escherichia coli*	1.5 mm	100 nm

Table 1.3 compares the size measure of two macromolecules. It shows the order of magnitude difference between the contour length and radius of gyration of a molecule.

1.8 Scaling Laws

Scaling laws are simple and are generally deduced at the macroscopic level. However, they can be used to understand the behavior of the micro and nano worlds. Scaling laws illustrate the fact that shrinking a body leads to not only size reduction but also different modifications of physical effects. Scaling laws are useful to analyze the *equilibrium points* in miniaturized systems, that is, the transformation in behavior when ordinary macrometric systems are miniaturized. Different physical quantities are related to the size l of the system. For an object having similar dimensions in order of magnitude in three spatial dimensions the length scale, l represents the order of magnitude of the object size. For *anisotropic object* (length larger than width and height) l, should be the scale controlling all the dimensions of the system, that is, change in l should lead to corresponding change in other dimensions of the system maintaining same aspect ratios. Mass (M) and volume ($\forall$) can be related as

$$M \sim \forall \sim l^3 \tag{1.7}$$

The relationship between physical quantity under consideration and the object size, l may or may not be influenced by other physical quantities depending on the situations. Let us consider an example of fluid flow circulating through a microchannel at a fixed pressure difference. For a laminar flow, inside a tube of radius a, we have the flow rate Q as

$$Q = \frac{\pi a^4 \Delta P}{8\mu L} \tag{1.8}$$

Thus, the order of magnitude of velocity inside the capillary can be written as

$$U \sim \frac{a^2 \Delta P}{\mu L} \tag{1.9}$$

where U is the average velocity, a is the transverse dimension, L is the channel length, μ is the fluid viscosity, and ΔP is the pressure difference applied along the channel length. For a fixed ΔP, a, and μ, we can write

$$U \sim l^{-1} \tag{1.10}$$

We note that for this situation, the average velocity is inversely proportional to the length of the channel. Similarly, keeping ΔP, μ, and L constant, we observe different relationships between the average velocity and the transverse dimension of the channel.

1.8.1 Application of Scaling Law to Natural System

Let us look into the small animals in nature. Animals in general eat food *proportional to their weight*, which is the source of chemical energy. We can assume the generation of thermal energy (Q_G) to scale as

$$Q_G \sim l^3 \tag{1.11}$$

The animals transfer heat toward outside. Assuming the heat transfer to take place through a layer of thickness δ and surface area A, the heat transferred per unit time, Q_R, by the animal is

$$Q_R = \frac{KA}{\delta}\Delta T \tag{1.12}$$

where K is the thermal conductivity of the surroundings and ΔT is the temperature difference between the animal body and the exterior. For constant value of K and ΔT, we can write the scaling law for heat transfer as

$$Q_R \sim l \tag{1.13}$$

For smaller organism (l is small) from equations (1.11) and (1.13), we have $Q_R > Q_G$. Therefore, it is necessary to compensate for thermal losses by eating food and to maintain the *internal temperature level*. An example is *pygmy shrew* (second smallest mammal), which must eat at regular intervals to maintain its internal temperature at a fixed value for survival. It is like a live *jet engine, consuming* fuel and converting it to energy quickly. Therefore, small animals are not warm blooded. In contrast, larger animals have difficult time to get rid of the thermal energy. The *whale*, the largest mammal in earth, uses its blood circulation to transfer heat toward exterior. The arterial wall properties of whale are considerably different from other animals for enhancing the heat transfer. When *hunters* kill the whale, the blood circulation stops leading to the cessation of the heat exchange. This causes internal tissues to get cooked and the hunter prefers to cook the meat this way.

Another application of scaling law is *insects walking* on water surface. Gravitational force is associated with l^3, and *surface tension force* is associated with l. Thus, at small scales, the surface tension force dominates in comparison to the gravitational forces. The gravitational force that tends to immerse the foot is compensated by the capillary force.

Note: One may think that in micro world, *automobile accidents* would not take place by collisions but by drops of water abandoned in the road. One may also be surprised not to have many machines driven by surface tension till now.

1.8.2 Scaling Laws in Microsystems

The dynamics of a microsystem depend on the relevant forces, that is, volume forces (gravity and inertia) and surface forces (viscous effect). The ratio of these two forces based on the scaling law is given by

$$\frac{\text{Surface forces}}{\text{Volume forces}} \sim \frac{l^2}{l^3} = l^{-1} \tag{1.14}$$

As l becomes smaller, the ratio approaches infinity. Hence, volume forces, which are prominent in real-life system, are unimportant in microfluidic system. The surface forces become dominant, and our intuition may not be valid at all situations in case of microsystem. Some examples are provided in the following sections to clarify this point.

1.8.2.1 Parallel-Plate Capacitor

The capacitance C of a parallel-plate capacitor, with plate area equal to A, voltage difference between the two plates V, and separation distance d, is given by

$$C = \frac{\varepsilon A}{d} \tag{1.15}$$

where ϵ is the permittivity of the medium. The electrostatic energy stored in the parallel-plate capacitor is

$$W_e = \frac{1}{2}CV^2 = \frac{\varepsilon A V^2}{2d} \tag{1.16}$$

Force used to displace one of the plates leads to a gradient of the energy given as

$$F = -\frac{dW_e}{dZ} = \frac{\varepsilon A V^2}{2d^2} \tag{1.17}$$

where Z is the axis perpendicular to the plate.

Using $V = Ed$ where E is the electric field, we have

$$F = \frac{\varepsilon A E^2}{2} \sim \varepsilon E^2 l^2 \tag{1.18}$$

Hence, for constant electric field E and fixed environment, we have

$$\text{Electrostatic force} \sim l^2 \tag{1.19}$$

Note: (1) Gravitational and inertia forces vary as l^3. The above equation indicates that the electrostatic force is dominant over gravitational and inertia forces for microsystems. Therefore, electric field can be used to exert *rapid acceleration* in microsystems. The weak inertia allows detection of sudden impacts involving automobiles and the microsensor using parallel-plate capacitor, which can be used as an *excellent accelerometer* for air bag.

(2) Miniaturization also helps in *preventing electric discharge* inside the parallel plates of the sensor. The sudden imposition of electric field between two electrodes can lead to the formation of sparks because of *ionization of gas molecules* situated between the two electrodes. A large number of electrons are liberated and electric current flows between the two electrodes. There is a simultaneous luminous emission known as *electric arc*. Generally, under normal conditions, the breakdown electric strength for air is of the order of 30 kV/cm. In miniaturized systems, much higher electric field can be produced without the generation of an electric arc.

Figure 1.6 shows the electric field breakdown voltage of air as a function of the pressure–separation distance product. It shows that at large values of pressure–distance product, the breakdown electric field strength is approximately equal to 30 kV/cm. At small values of this product, the breakdown electric field strength increases significantly. This phenomenon is due to the *rarefaction effect*. When the mean free path becomes comparable to

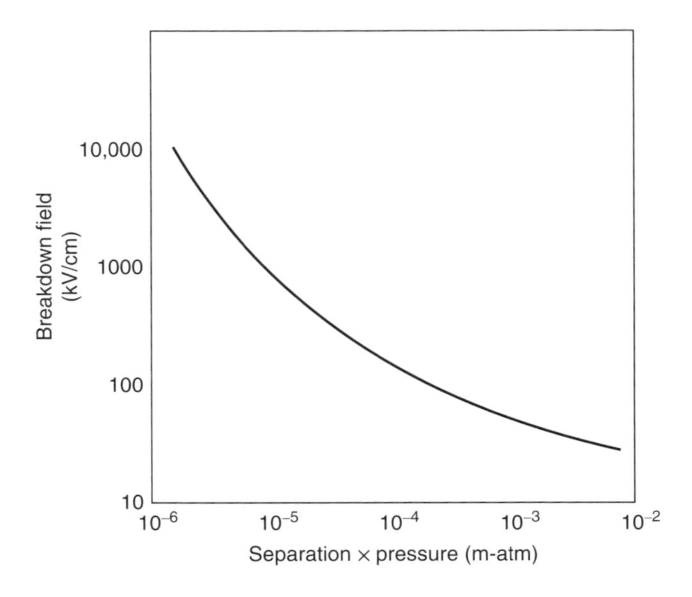

Figure 1.6 Electric field strength for dielectric breakdown of a parallel-plate capacitor in air

the distance between the electrodes, the majority of collision takes place between the gas and the electrode surface and not within the gas. This inhibits the formation of electric discharge in the gas.

1.8.2.2 Thermal Inertia of Chemical Reactor

The temporal heat flux ($q(t)$) associated with a variation of temperature dT during time period dt is given as

$$q(t) = \rho C_P \frac{dT}{dt} \tag{1.20}$$

where ρ is the density and C_P is the specific heat at constant pressure. Suppose that the heat flux is attained by an external heat source of higher temperature by conduction. The heat flux will have a corresponding heat transfer given as

$$q(t) = -K\nabla^2 T \tag{1.21}$$

Thus, combining these two equations, the time constant for this process can be written as

$$\tau \sim \frac{\rho C_P l^2}{K} \sim \frac{l^2}{\alpha} \tag{1.22}$$

where α is the thermal diffusivity of the material. This expression shows that *miniaturization lead to lower time constant*. Hence, miniaturized system subjected to a sharp temperature change can reach thermal equilibrium faster when compared to the macrosystems. This characteristic is very useful for *chemical sciences and engineering* applications. Many reactions of chemical processes are dependent on precise control of thermal conditions. Improper thermal control leads to development of *unwanted parasite reactions*. Miniaturization improves the *selectivity* of the process output.

1.8.2.3 Microscale Heat Exchanger

The heat produced by chemical reaction in a microcombustor (Q_V) can be expressed as

$$Q_V \sim l^3 \tag{1.23}$$

The heat transfer from the combustor (Q_R) can be estimated based on the Fourier law as

$$Q_R = \frac{KA\Delta T}{l} \tag{1.24}$$

where A is the area, ΔT is the temperature difference, and K is the thermal conductivity. We can write the following scaling law for conductive heat loss.

$$Q_R \sim Kl\Delta T \sim l \tag{1.25}$$

The above scaling law shows that the exponent for volumetric heat generation is higher than the heat evacuated by conduction indicating that $Q_R > Q_V$. Thus, *micro heat exchangers* are very effective in miniaturized systems for transport of combustion energy. One can correlate this to a microscale power plant where the role of boiler for transferring heat from combustion becomes very efficient due to its small scale.

1.8.2.4 Microdroplets

Droplets are present in many engineering systems, that is, fuel injection system, LOC system, and so on. The evaporation of a drop of diameter d at time t is governed by D^2 law, that is,

$$d^2 = d_0^2 - \beta t \tag{1.26}$$

where d_0 is the initial diameter and β is a constant independent of the drop size.

The time τ for the drop to disappear, that is, $d = 0$, is given by

$$\tau = \frac{d_0^2}{\beta} \sim l^2 \tag{1.27}$$

The above-mentioned scaling law indicates rapid evaporation of droplets in microsystems because of smaller time constants in comparison to the larger droplets. This can be a disadvantage for LOC applications involving the transport of a small amount of liquids because of the loss of liquid by evaporation during transport. However, it can be an advantage in microcombustor applications, where the evaporation of droplets takes place rapidly.

1.8.2.5 Miniaturized Resonator

Micromechanical resonator has emerged as a promising component for rapidly developing *telecommunication systems*. These resonators also have a wide range of *sensing applications*. High-frequency resonators can use *capacitive or piezoelectric transduction*. To understand the importance of miniaturization, let us consider a cantilever beam. The natural frequency of the beam is given by the relation:

$$f \approx \frac{hC}{2\pi L^2} \tag{1.28}$$

where h is the thickness of the beam, L is its length, and C is the speed of sound in the beam. Thus, the scaling law can be expressed as

$$f \simeq l^{-1} \tag{1.29}$$

This expression indicates that resonance frequency of the system increases as its size reduces. A silicon beam of 1 μm thick and 3 μm long where $C \simeq 7470$ m/s has a resonant frequency equal to $f \simeq 132$ MHz. This value is well suited for radiofrequency applications. However, a larger silicon beam of 1 mm thick and 3 mm long will have resonant frequency equal to 132 kHz.

1.8.3 Scaling Laws Limitation

Scaling laws are simple. However, the conclusion based on scaling is not always directly applicable to microsystem. Let us consider an example of mixing in microsystems. The question is whether mixing time can be reduced in microfluidic system by agitating the fluids. This can be answered using the mixing timescale by convection and diffusion mechanisms. We can write the *hydrodynamic transport time* as

$$\tau_a \sim \frac{l}{U} \sim l^0 \tag{1.30}$$

where l is the length scale and U is the speed.

The *molecular diffusion time* is

$$\tau_D \sim \frac{l^2}{D} \tag{1.31}$$

where D is the diffusion coefficient. For constant diffusivity coefficient D, we can write

$$\tau_D \sim l^2 \tag{1.32}$$

Comparing the equations (1.30) and (1.32) for a microsystem, we have

$$\tau_a \gg \tau_D \tag{1.33}$$

This expression indicates that it is *useless to agitate* the fluid in order to enhance the mixing process.

We can also use the complete expression instead of scaling laws as

$$\frac{\tau_D}{\tau_a} = \frac{Ul}{D} = \text{Pe} \;\; \text{(Peclet number)} \tag{1.34}$$

Let us take an example of mixing fluorescein with water ($D = 3 \times 10^{-6}$ cm^2/s). For a microsystem with $l \sim 100$ μm and $U \sim 30$ μm/s, the Peclet number is equal to 10. Thus, $\tau_D > \tau_a$, which is contrary to equation 1.33 . This indicates that *diffusion phenomenon is much slower* than hydrodynamic transport phenomenon, which is contrary to what was suggested based on the scaling analysis. Hence, the scaling laws cannot be used blindly. It provides an estimate of the process, which need to be verified from the exact analysis.

1.9 Shrinking of Human Beings

A famous movie named "The Fantastic Voyage" by Richard Fleischer in 1966 used the concept of shrinking human beings. In this movie, a submarine and its human occupants are shrunk to the size of a few hundred micrometers. They are then injected into the body of a Czech scientist for repairing some brain problem. Based on the movie, the question is what would be our shape to survive under such conditions? Are we going to have the same concept of *good-looking* boys and girls?

Let us consider shrinking the body size by a factor of 10^3. The weight of the body will decrease by a factor of 10^9. In order to support this weight, the diameter of the bone has to be reduced to a comparable factor. The weight of the body should be balanced by the maximum stress limit (σ) of the bone as

$$\sigma \sim \frac{\text{mg}}{A} \quad \text{or} \quad \sigma A \sim \text{mg} \tag{1.35}$$

For a given material, σ is constant. Therefore, the diameter d of the supporting material, that is, bone, can be estimated based on the above-mentioned equation as

$$d^2 \sim L^3 \quad \text{or} \quad d \sim L^{3/2} \tag{1.36}$$

Therefore, the diameter of the bone has to decrease by a factor upto $10^{9/2} \simeq 3.16 \times 10^4$. Thus, the diameter of the legs should be comparatively much smaller than the size. For such a small size of the foot, the adhesion force of the feet on the ground would be more important than the gravitational force. Human beings will adhere to the ceiling *similar to insects*. Therefore, the dimensions of the feet would have to be similar to small points in order to move on the ground. This will lead to stability problem. Human beings will require at least *four legs* instead of walking on two legs.

The other critical issue is the vision. The minimum diameter of the irradiated zone is given by $L = 2\lambda/\pi\text{NA}$ where NA is the numerical aperture. Numerical aperture (NA) is a dimensionless number, which is used to characterize the range of angles over which an optical system can accept or emit light. It is defined as $\text{NA} = n\sin\theta$, where n is the index of refraction of the working medium and θ is the half angle of the maximum cone of light that can enter or exit the lens. In case of monochromatic light such as laser beam, θ is considered as the divergence of the beam and $\text{NA} \simeq \frac{\lambda}{\pi w_0}$, where $2w_0$ is the diameter of the beam at its narrowest spot. Hence, the wavelength of illumination $\lambda \sim L$, that is, the same order of magnitude as the diameter of the lens. Hence, the eye would have to look at the wavelength of emitted light of 10^{-3} times that of visible light, that is, X-rays. We know that the absorption length of X-rays in living matter is of the order of few centimeters compared to few micrometers for visible light. Thus, the thickness of the eye should be larger than the length of the body.

Problems

1.1 Based on the scale analysis, estimate the ratio of surface force to volume force for a reservoir of 1 μm^3 volume with that of 1 mm^3 volume.

1.2 Compare the ratio of electrostatic force to inertia force of an automobile air bag using an accelerometer based on parallel-plate capacitor with separation distance of 1 mm between the two plates with that of separation distance of 10 μm.

1.3 A 1-mm-diameter methanol droplet takes 1 min for complete evaporation at atmospheric condition. What will be the time taken for a 1-μm-diameter methanol droplet for complete evaporation at same conditions based on the scaling analysis?

1.4 If the mean free path of air at $P = 10^5$ Pa, $T = 300$ K is equal to 61 nm, what will be the mean free path if the temperature of air is raised to 600 K?

1.5 What is the natural frequency of a microcantilever (thickness: 10 μm and length: 30 μm) used in a piezoelectric transducer? The speed of sound in the beam is equal to 8000 m/s.

References

Balagadde FK, You LC, Hansen CL, and Arnold FH 2005 Quake SR: long-term monitoring of bacteria undergoing programmed population control in a microchemostat. *Science*, **309**, pp. 137–140.

Dertinger SKW, Chiu DT, Jeon NL, and Whitesides GM 2001 Generation of gradients having complex shapes using microfluidic networks. *Anal. Chem.*, **73**, pp. 1240–1246.

Ismagilov RF, Ng JMK, Kenis PJA, and Whitesides GM 2001 Microfluidic arrays of fluid-fluid diffusional contacts as detection elements and combinatorial tools. *Anal. Chem.*, **73**, pp. 5207–5213.
Takayama S, Ostuni E, LeDuc P, Naruse K, Ingber DE, and Whitesides GM 2001 Subcellular positioning of small molecules. *Nature*, **411**, pp. 1016.

Supplemental Reading

Gad-el-Hak M 2000 *Flow Control, Passive, Active and Reactive Flow Management*. Cambridge University Press.
Nguyen NT and Wereley ST 2006 *Fundamentals and Applications of Microfluidics*. Artech House, Inc.
Tabeling P 2005 *Introduction to Microfluidics*. Oxford University Press.

2

Channel Flow

2.1 Introduction

Microchannels are one of the common components of most of the microfluidic devices. Therefore, appropriate analysis tools are required for designing microchannels. Contrary to traditional channels, microchannels cannot be easily fabricated to a desired shape or a cross-section. The geometrical shape of the microchannels depends on the fabrication methodology. This chapter presents the analytical relationships for microchannel flow with different geometrical shapes and arrangements. Electrical analogy has been proposed as one of the convenient tools for the design of microchannel network. The influence of viscous dissipation and compliance of the channel wall on the modeling of microchannel flow are also presented.

2.2 Hydraulic Resistance

The pressure-driven, steady-state flow of an incompressible fluid through a straight channel known as Poiseuille flow provides the basic fundamentals about the flow through channels relevant to liquid handling in a lab-on-a-chip (LOC) system. A constant pressure drop ΔP results in a constant flow rate Q. This result is summarized as the Hagen–Poiseuille law:

$$\Delta P = R_{\text{hyd}} Q \tag{2.1}$$

Here, the proportionality factor R_{hyd} is known as the hydraulic resistance. For a pipe (radius a and length L) flow of fluid with viscosity μ, the hydraulic resistance is expressed as

$$R_{\text{hyd}} = \frac{8\mu L}{\pi a^4} \tag{2.2}$$

The Hagen–Poiseuille law equation (2.1) is completely analogous to Ohm's law, $\Delta V = RI$, where the electrical current I through a wire is related to the electrical resistance R of the wire and the electrical potential drop ΔV along the wire. The flow rate Q is analogous to current; the pressure drop is analogous to electrical potential drop ΔV; and the hydraulic resistance is

Transport Phenomena in Microfluidic Systems, First Edition. Pradipta Kumar Panigrahi.

Companion Website: www.wiley.com/go/panigrahi/microfluidic

analogous to electrical resistance R. The SI units used in the Hagen–Poiseuille law are

$$[Q] = \frac{\text{m}^3}{\text{s}}, \quad [\Delta P] = \text{Pa} = \frac{\text{N}}{\text{m}^2}, \quad [R_{\text{hyd}}] = \frac{[\text{Pa-s}]}{\text{m}^3} \tag{2.3}$$

The concept of hydraulic resistance is central in characterizing and designing microfluidic channels in LOC systems. This chapter presents both fundamental and applied aspects of hydraulic resistance in relation to microfluidic applications.

2.3 Two Connected Straight Channels

Most of the microfluidic devices consist of microchannel of different dimensions interconnected with each other. In the case of two straight channels of different dimensions connected to form one long channel, the translation invariance, that is, fully developed flow assumption, is broken. Hence, the expressions for the ideal Poiseuille flow no longer apply. However, it is expected that the ideal description is approximately correct if the Reynolds number (*Re*) of the flow is sufficiently small. This is because a very small value of *Re* corresponds to a vanishing small contribution from the nonlinear term $(\vec{V} \cdot \vec{\nabla})\vec{V}$ in the Navier–Stokes equation (N–S equation), a term that is strictly zero in ideal Poiseuille flows due to translation invariance.

The influence of the Reynolds number on the velocity field of a backward-facing step is illustrated in Figure 2.1. The backward-facing step flow can be considered equivalent to two infinite parallel-plate channels with heights h_1 and h_2, and hydraulic resistances R_1 and R_2 are joined in a series coupling. At low Reynolds number $Re = \rho V_0 h_1/\mu \simeq 0.01$ (Figure 2.1(a)), the transition from a perfect Poiseuille flow in R_1 is smooth. One can assume the validity of Hagen–Poiseuille flow to the individual channels, that is, channel 1 and channel 2. At high Reynolds number $Re = \rho V_0 h_1/\mu \simeq 100$ (Figure 2.1(b)), a convection roll forms in the entrance region of R_2. This is a simple example of how it is a fair approximation to assume

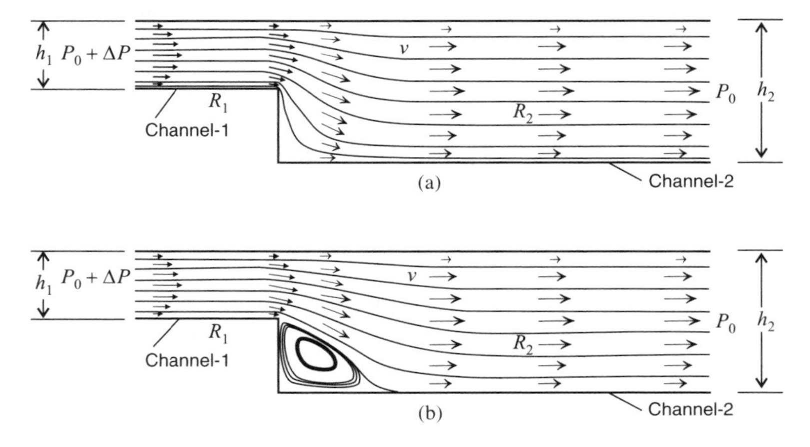

Figure 2.1 Flow patterns as a function of Reynolds number (*Re*) of a backward facing step: (a) $Re = 0.01$ and (b) $Re = 100$. The flow is analogous to two hydraulic resistors, R_1 and R_2, connected in series for flow inside channel 1 and channel 2 of varying cross-sections

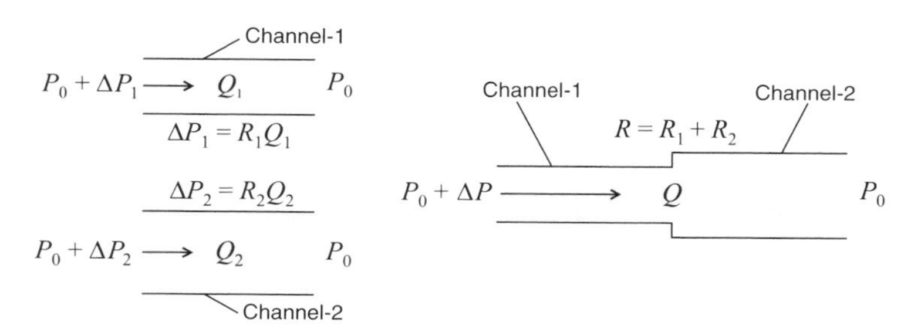

Figure 2.2 The series coupling of two channels (channel-1 and channel-2) with hydraulic resistances R_1 and R_2

ideal Poiseuille flows in individual parts of a microfluidic network at low Reynolds numbers, whereas the approximation is dubious at high Reynolds numbers.

Note that in microfluidics, the Reynolds number $Re = \rho V_0 L_0 / \mu$ tends to be low because of the small length scales L_0 involved and low velocities. Thus, Hagen–Poiseuille law can be applied beyond the ideal situation on various channel geometries.

2.3.1 Straight Channels in Series

Consider the series coupling of two hydraulic resistors as shown in Figure 2.2. If we assume the validity of the Hagen–Poiseuille law for each of the resistors after they are connected, then using the additivity of the pressure drop along the series coupling is straightforward to show the law of additivity of hydraulic resistors in a series coupling results in equivalent resistance R as

$$R = R_1 + R_2 \tag{2.4}$$

It should be noted that the additivity law is only valid for low Reynolds numbers and for long and narrow channels.

2.3.2 Straight Channels in Parallel

Consider the parallel coupling of two hydraulic resistors as shown in Figure 2.3. If we assume the validity of the Hagen–Poiseuille law for each of the resistors after they are connected, then using the conservation of flow rate, that is, $Q = Q_1 + Q_2$ in the parallel coupling, it is straightforward to show the law of additivity of inverse hydraulic resistances in a parallel coupling,

$$R = \left(\frac{1}{R_1} + \frac{1}{R_2} \right)^{-1} = \frac{R_1 R_2}{R_1 + R_2} \tag{2.5}$$

Similar to the above-mentioned section, this inverse-additive law is only valid for low Reynolds numbers and for long and narrow channels far apart. If N identical hydraulic resistors R_1 are connected in parallel, we can show by induction that

$$R(N) = \frac{R_1}{N}$$

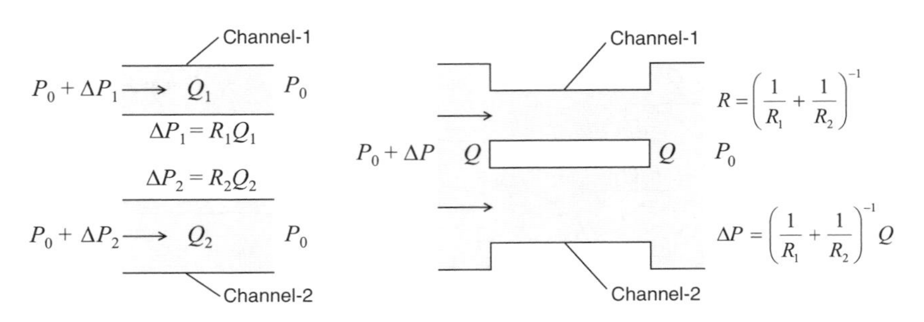

Figure 2.3 The parallel coupling of two channels (channel-1 and channel-2) with hydraulic resistances R_1 and R_2

2.4 Equivalent Circuit Theory

Given the complete analogy between the Hagen–Poiseuille law and Ohm's law for series and parallel couplings of resistors, it is an obvious advantage to apply the well-known methods from electric circuit theory to microfluidic networks on LOC systems. For a given microfluidic network, one can draw the equivalent electric network. Channels with hydraulic resistances R_{hyd} become resistors, flow rates Q become currents, and pumps delivering pressure differences ΔP become batteries. One can then apply Kirchhoff's laws using the following assumptions:

1. The flow rates entering a given point equals the flow rates leaving the point.
2. The total pressure difference (including pumps) in a closed loop of the circuit is zero.

Volume in fluidics is equivalent to electrical charge, and mechanical elasticity of the walls of the channels (known as compliance defined as change of volume due to change in pressure) is equivalent to electrical capacitance. More details on analogy related to channel with compliant wall will be discussed in a later section. Also the inertia of fluids has an electrical analog, namely, electrical inductance.

In Figure 2.4(a), sketch of a electro-osmotic pump has been shown. The microfluidic network of the pump is not simple. Series and parallel channel networks are used to increase the flow output and to develop high pressure. One single stage of micropump contains two main parts: the collection of 10 parallel narrow channels, and 1 wide single-channel. One can calculate the basic hydraulic resistance R_{hyd} of each channel depending on the cross-section of the channel. Then, the equivalent circuit diagram of the micropump can be developed. The micropump is a series coupling of the hydraulic resistance, R_{stage} of each single stages in the pump. In Figure 2.4(b), R_{stage} has been broken further down into a series coupling of three hydraulic resistors. $R_{\text{hyd},1}$ is the parallel coupling of 10 identical narrow channels, $R_{\text{hyd},2}$ corresponds to the single channel connecting each stage and $R_{\text{hyd},3}$ is the hydraulic resistance corresponding to the broad channel connecting the multichannel with single-channel segments.

Thus, using equivalent circuit theory, it is possible to obtain a good estimate of the total hydraulic resistance of the microfluidic network without performing complicated numerical simulations. This is extremely helpful when designing LOC systems, and if all the involved channels are long and narrow, then the result is very accurate.

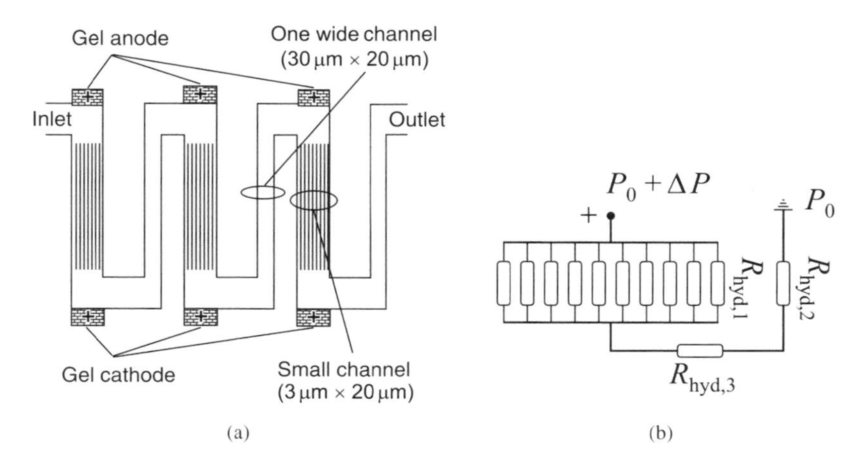

Figure 2.4 An equivalent circuit analysis of a cascade electro-osmotic micropump: (a) a sketch of the micropump, consisting of 10 narrow channels in parallel followed by a single wide channel in series, and (b) the equivalent circuit diagram for calculation of the hydraulic resistance of single stage of a micropump consisting of two wide channels and one multichannel combination of the micropump

2.5 Reynolds Number

The simplified circuit model discussed in the previous section depends on the validity of fully developed flow assumption. The present section discusses the fully developed flow criteria from nondimensionalization of the governing equation.

The proper way to verify whether the nonlinear term $(\vec{v} \cdot \vec{\nabla})\vec{v}$ in the N–S equation can be neglected is to make the equation dimensionless. This means that we express all physical quantities, such as length and velocity, in units of the characteristic scales, for example, L_0 for length and V_0 for velocity. Reynolds number appears in the nondimensionalized N–S equation and the relative importance of different terms in the can be verified for small Reynolds number assumption. We explain this in the following sections using two examples.

2.5.1 Microsystems with Only One Length Scale

Let the microsystem under consideration is characterized by only one length scale L_0 and one velocity scale V_0. One example is the flow inside two infinite parallel plates. The expressions of coordinates and velocity in terms of dimensionless coordinates and velocity can be written as

$$x = L_0 x^*, \quad v = V_0 v^* \tag{2.6}$$

where the superscript $*$ indicates a quantity without physical dimension. Once the length and velocity scales L_0 and V_0 have been fixed, the reference scales T_0 and P_0 for time and pressure are

$$t = \frac{L_0}{V_0} t^* = T_0 t^*, \qquad P = \frac{\mu V_0}{L_0} P^* = P_0 P^* \tag{2.7}$$

Note that a quantity often can be made dimensionless in more than one way. The pressure term can be nondimensionalized by using reference pressure P_0 equal to $\mu V_0 / L_0$ or ρV_0^2. The former is suitable as the scale of pressure in case of small velocities $\left(\nabla P = \mu \nabla^2 v;\ P_0 = \frac{\mu V_0}{L_0}\right)$, such as in microfluidics, whereas the latter is used at high velocities $(\nabla P = \rho(\vec{V} \cdot \vec{\nabla})\vec{V})$. Let us begin with the simplified N–S equation for constant density and viscosity:

$$\rho\left(\frac{\partial \vec{v}}{\partial t} + (\vec{v} \cdot \vec{\nabla})\vec{v}\right) = -\nabla P + \vec{\mu} \nabla^2 \vec{v}$$

By substituting equations (2.6) and (2.7) in the N–S equation and using the straightforward scaling of the derivatives, $\frac{\partial}{\partial t} = \frac{1}{T_0}\frac{\partial}{\partial t^*}$ and $\nabla = (1/L_0)\nabla^*$, we get

$$\rho\left[\frac{V_0}{T_0}\frac{\partial}{\partial t^*}v^* + \frac{V_0^2}{L_0}(v^* \cdot \nabla^*)v^*\right] = -\frac{P_0}{L_0}\nabla^* P^* + \frac{\mu V_0}{L_0^2}\nabla^{*2} v^* \tag{2.8}$$

which after simplification becomes

$$Re\left[\frac{\partial^*}{\partial t}v^* + (v^* \cdot \nabla^*)v^*\right] = -\nabla^* P^* + \nabla^{*2} v^* \tag{2.9}$$

Here, we have introduced the dimensionless number Re (Reynolds number) as

$$Re = \frac{\rho V_0 L_0}{\mu} \tag{2.10}$$

In the limit of small Reynolds number, the nonlinear N–S equation (2.9) reduces to linear form as

$$0 = -\vec{\nabla} P + \mu \nabla^2 \vec{V} \tag{2.11}$$

We clearly observe from equation (2.9) that for $Re \ll 1$ the viscous term $\nabla^{*2} v^*$ dominates, whereas in the steady state for $Re \gg 1$, the inertia term $(v^* \cdot \nabla^*)v^*$ is the most important term.

The corresponding dimensionless form of the continuity equation for incompressibility assumption, $\vec{\nabla} \cdot \vec{v} = 0$, is quite simple. Using $\vec{\nabla} = \frac{1}{L_0}\nabla^*$ and $\vec{v} = V_0 v^*$, we have

$$\nabla^* \cdot v^* = 0 \tag{2.12}$$

From this nondimensionalization analysis, we can conclude that the solutions obtained for the ideal Poiseuille flows, where the nonlinear term $(\vec{v} \cdot \vec{\nabla})\vec{v}$ is identical zero, remains approximately valid if the Reynolds number is small, $Re \ll 1$. This approximation may be valid for most of the microfluidic applications with low velocity.

2.5.2 *Microsystems with Two Length Scales*

The dimensional analysis reported in the previous section does not include the influence of channel aspect ratio or size. We try to bring the effect of channel cross-section size in this section. In comparison to the example discussed in the last section, most systems are characterized by more than one length scale, which leads to a more involved Reynolds number analysis. As an example, consider a section of length L and width w of the infinite, parallel-plate channel with height h shown in Figure 2.5. The system is translation invariant

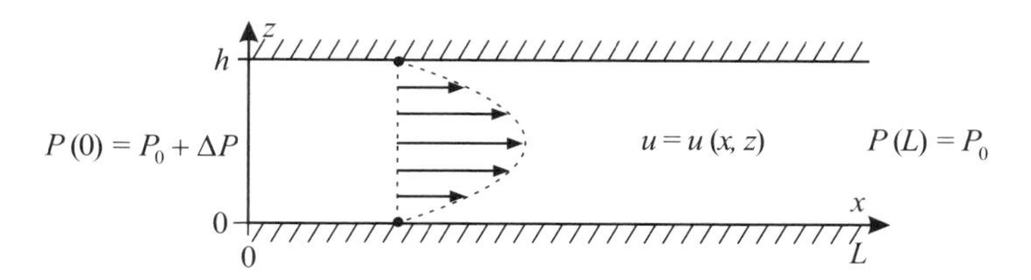

Figure 2.5 A sketch of an infinite, parallel-plate channel of height h in the x–z plane. The fluid is flowing in the x-direction because of a pressure drop ΔP over the section of length L

in the y-direction due to high value of w/h ratio so that only the x- and z-coordinates enter in the analysis that follows. Although the system as shown is also translation invariant in the x-direction, we perform the analysis as if this invariance is weakly broken rendering a nonzero vertical velocity, w. This will be applicable when the height of the channel is large (nonslender channel geometry) or short length of the channel leading to developing flow situation. The two length scales entering the problem are the length L and the height h,

$$x = Lx^*, \quad z = hz^* \tag{2.13}$$

The spatial derivatives are represented as

$$\frac{\partial}{\partial x} = \frac{1}{L}\frac{\partial}{\partial x^*} \equiv \epsilon \frac{1}{h}\frac{\partial}{\partial x^*}, \qquad \frac{\partial}{\partial z} = \frac{1}{h}\frac{\partial}{\partial z^*} \tag{2.14}$$

where we have introduced the aspect ratio ϵ defined by

$$\epsilon = \frac{h}{L} \ll 1 \tag{2.15}$$

The characteristic velocity in the x-direction is given by the mean velocity

$$V_0 = Q/(wh) \tag{2.16}$$

where Q is the flow rate through a section of width w and height h. The characteristic time T_0 is therefore given by

$$t = \frac{L}{V_0}t^* = T_0 t^* \tag{2.17}$$

The nondimensionalization expressions for the two velocity components can be written as

$$u = V_0 u^*, \qquad w = \frac{h}{T_0}w^* = \frac{h}{L}\frac{L}{T_0}w^* = \epsilon\, V_0 w^* \tag{2.18}$$

Finally, the characteristic pressure is given by the pressure drop P_0, see Table 2.3,

$$P_0 = R_{\text{hyd}} Q \simeq \frac{\mu L}{h^3 w} Q = \frac{\mu V_0 L}{h^2} \tag{2.19}$$

Here, for convenience, we have dropped the numerical factor of 12 $\left(R_{\text{hyd}} = \frac{12\mu L}{h^3 w}\right)$. We can also estimate the characteristic pressure P_0 by using

$$\nabla P = \mu \nabla^2 V \tag{2.20}$$

Thus, the reference pressure can be written using this equation as

$$P_0 = \frac{\mu V_0 L}{h^2} \tag{2.21}$$

If we follow the convention that the Reynolds number Re should contain the smallest length scale of the problem, here h, we define

$$Re = \frac{\rho V_0 h}{\mu} \tag{2.22}$$

Using the above-mentioned nondimensionalization parameter, we can rewrite the two-component (x- and z-directions) N–S equation and the continuity equation in terms of dimensionless variables as

$$\epsilon Re \left(\frac{\partial}{\partial t^*} + u^* \frac{\partial}{\partial x^*} + w^* \frac{\partial}{\partial z^*} \right) u^* = -\frac{\partial}{\partial x^*} P^* + \left(\frac{\partial^2}{\partial z^{*2}} + \epsilon^2 \frac{\partial^2}{\partial x^{*2}} \right) u^* \tag{2.23}$$

$$\epsilon^3 Re \left(\frac{\partial}{\partial t^*} + u^* \frac{\partial}{\partial x^*} + w^* \frac{\partial}{\partial z^*} \right) w^* = -\frac{\partial}{\partial z^*} P^* + \left(\epsilon^2 \frac{\partial^2}{\partial z^{*2}} + \epsilon^4 \frac{\partial^2}{\partial x^{*2}} \right) w^* \tag{2.24}$$

$$\frac{\partial}{\partial x^*} u^* + \frac{\partial}{\partial z^*} w^* = 0 \tag{2.25}$$

In the limit of small ϵ, $\epsilon \to 0$, keeping the term of first order in ϵ we have:

$$\epsilon Re \left(\frac{\partial}{\partial t^*} + u^* \frac{\partial}{\partial x^*} + w^* \frac{\partial}{\partial z^*} \right) u^* = -\frac{\partial}{\partial x^*} P^* + \frac{\partial^2}{\partial z^{*2}} u^* \tag{2.26}$$

$$0 = -\frac{\partial}{\partial z^*} P^* \tag{2.27}$$

$$\frac{\partial}{\partial x^*} u^* + \frac{\partial}{\partial z^*} w^* = 0 \tag{2.28}$$

Comparing with equation (2.9) for flow system with single length scale, we can conclude that the effective Reynolds number Re_{eff} for this two-length-scale problem is

$$Re_{\text{eff}} = \epsilon Re = \frac{\rho V_0 h}{\mu} \frac{h}{L} \tag{2.29}$$

Note that at small value of Re_{eff}, the left-hand side of equation (2.26) is equal to zero. This effective Reynolds number can therefore be arbitrarily small compared to the conventional Reynolds number ($Re = \rho V_0 h/\mu$) given a sufficiently long channel. Thus, microchannel flows easily satisfy the Poiseuille flow assumption leading to $(\vec{V} \cdot \vec{\nabla})\vec{V} = 0$ compared to the conventional channels.

2.6 Governing Equation for Arbitrary-Shaped Channel

In a Poiseuille flow, the fluid is driven through a long, straight, and rigid channel by imposing a pressure difference between the two ends of the channel. Originally, Hagen and Poiseuille studied channels with circular cross-sections. However, in microfluidics, one frequently encounters other shapes. One example, shown in Figure 2.6(a,b), is the Gaussian-like profile that results from producing microchannels by laser ablation in the surface of a piece of the polymer poly methyl methacrylate (PMMA). The heat from the laser beam cracks the PMMA into monomer, MMA, which by evaporation leaves the substrate. A whole network of microchannels can then be created by sweeping the laser beam across the substrate in a well-defined pattern. The channels are sealed by placing and bonding a polymer lid on the top of the structure.

Figure 2.7 shows the steady-state Poiseuille flow in an arbitrary cross-sectional-shaped channel. The channel flow is parallel to the x-axis and is assumed to be translation invariant in that direction, that is, fully developed in nature. The constant cross-section C in the y–z plane has boundary ∂C. A constant pressure difference ΔP is maintained over a segment of length L of the channel, that is, $P(0) = P_0 + \Delta P$ and $P(L) = P_0$. The continuity and N–S equation for

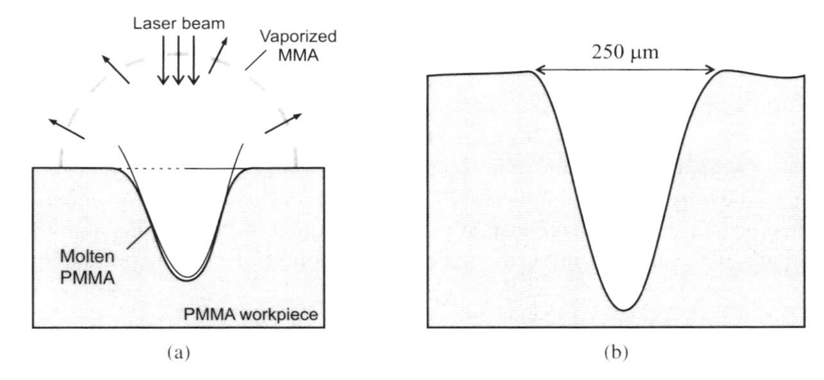

Figure 2.6 Fabrication of microchannels by laser ablation of polymer, PMMA: (a) schematic diagram of the laser beam, the laser ablated groove, including the molten PMMA and the vaporized hemispherical cloud of MMA leaving the cut-zone, (b) Scanning electron microscope (SEM) micrograph of the cross-section of an actual microchannel showing the resulting Gaussian-like profile

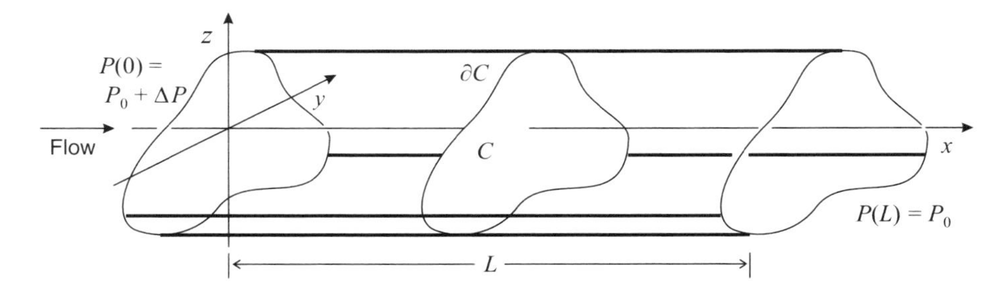

Figure 2.7 The Poiseuille flow problem in the x-direction of a channel, which has an arbitrarily shaped cross-section C in the y–z-plane. The boundary of C is denoted by ∂C. The pressure at the left end, $x = 0$, is an amount ΔP higher than that at the right end, $x = L$

isothermal incompressible flow of constant density and viscosity, respectively, are given as

$$\vec{\nabla} \cdot \vec{V} = 0 \tag{2.30}$$

$$\rho \left(\frac{\partial \vec{V}}{\partial t} + (\vec{V} \cdot \vec{\nabla})\vec{V} \right) = -\vec{\nabla} P + \mu \nabla^2 \vec{V} + \vec{F}_B \tag{2.31}$$

where F_B is the body force.

Based on the dimensional analysis of the previous section, we can assume the nonlinear term $(\vec{V} \cdot \vec{\nabla}\vec{V}) = 0$ for a high aspect ratio channel. The vanishing force in y–z surface indicates that only the x-component of velocity is nonzero. The steady-state N–S equation becomes

$$\vec{V}(r) = u_x(y, z)\vec{e}_x, \tag{2.32a}$$

$$0 = \mu \nabla^2 [u_x(y, z)\vec{e}_x] - \vec{\nabla} P \tag{2.32b}$$

Since the y- and z-component of the velocity field are zero, it follows that $\frac{\partial P}{\partial y} = 0$ and $\frac{\partial P}{\partial z} = 0$, and consequently the pressure field only depends on x, that is, $P(r) = P(x)$. Using this result, the x-component of N–S equation (2.32b) becomes

$$\mu \left[\frac{\partial^2}{\partial y^2} + \frac{\partial^2}{\partial z^2} \right] u_x(y, z) = \frac{\partial P(x)}{\partial x} \tag{2.33}$$

Here, it is seen that the left-hand side is a function of y and z while the right-hand side is a function of x. The only possible solution is thus that the two sides of the N–S equation is equal to the same constant. However, a constant pressure gradient $\frac{\partial}{\partial x}P(x)$ implies that the pressure must be a linear function of x, and using the boundary conditions for the pressure we obtain

$$P(x) = \frac{\Delta P}{L}(L - x) + P_0 \tag{2.34}$$

Thus, we obtain the second-order partial differential equation as

$$\left[\frac{\partial}{\partial y^2} + \frac{\partial}{\partial z^2} \right] u_x(y, z) = -\frac{\Delta P}{\mu L}, \quad \text{for} \quad (y, z) \in C \tag{2.35}$$

This governing equation can be solved using the no-slip boundary condition:

$$u_x(y, z) = 0, \quad \text{for} \quad (y, z) \in \partial C \tag{2.36}$$

Once the velocity field is determined, it is possible to calculate the flow rate Q defined as the fluid volume discharged by the channel per unit time.

$$Q \equiv \int_c u_x(y, z) dy\, dz \tag{2.37}$$

This is the theoretical basis for determination of Poiseuille flow without specifying the actual shape of the channel. The following sections present the pressure drop versus flow rate relation for various channel cross-sections (see Figure 2.8).

2.6.1 *Elliptic Cross-section*

For the elliptic cross-section, let the center of the ellipse be at $(y, z) = (0, 0)$. The major axis of length a and the minor axis of length b are parallel to the y-axis and z-axis, respectively, as

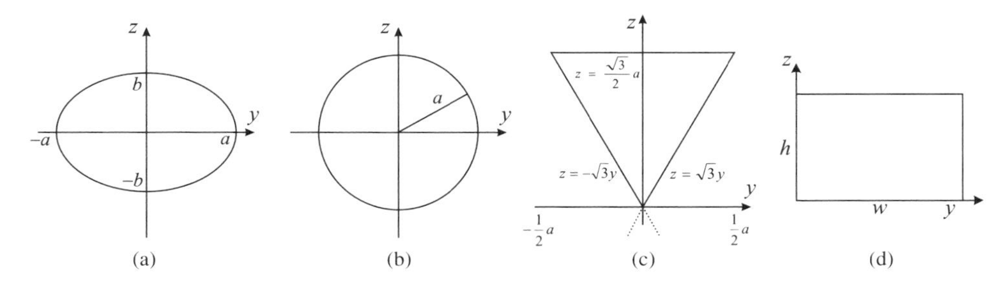

Figure 2.8 The four specific cross-sectional shapes for the Poiseuille flow problem: (a) the ellipse with major axis a and minor axis b, (b) the circle with radius a, (c) the equilateral triangle with side length a, and (d) the rectangle with height h and width w

shown in Figure 2.8(a). The boundary ∂C of ellipse is given by the expression:

$$1 - \frac{y^2}{a^2} - \frac{z^2}{b^2} = 0 \tag{2.38}$$

Let us choose a trial solution of the velocity field as

$$u_x(y, z) = v_0 \left(1 - \frac{y^2}{a^2} - \frac{z^2}{b^2}\right) \tag{2.39}$$

This trial velocity field guarantees that $u_x(y, z)$ satisfies the no-slip boundary condition (equation (2.36)). It also automatically satisfies the continuity equation. Substitution of the trial solution into the left-hand side of the N–S equation (2.35) yields

$$\left[\frac{\partial^2}{\partial y^2} + \frac{\partial^2}{\partial z^2}\right] u_x(y, z) = -2v_0 \left(\frac{1}{a^2} + \frac{1}{b^2}\right) \tag{2.40}$$

The substitution of equation (2.40) in the N–S equation (2.35) gives the constant v_0 as

$$v_0 = \frac{\Delta P}{2\mu L}\left(\frac{a^2 b^2}{a^2 + b^2}\right) \tag{2.41}$$

To calculate the flow rate Q for the elliptic channel, we need to evaluate a $2D$ integral in an elliptically shaped integration region. This is accomplished by coordinate transformation. Let (ρ, ϕ) be the polar coordinates of the unit disk, that is, the radial and azimuthal coordinates, which obey $0 \le \rho \le 1$ and $0 \le \phi \le 2\pi$, respectively. The physical coordinates (y, z) and the velocity field u can then be expressed as functions of (ρ, ϕ) as

$$y(\rho, \phi) = a\rho \cos \phi, \tag{2.42}$$

$$z(\rho, \phi) = b\rho \sin \phi, \tag{2.43}$$

The velocity field given by equation (2.39) can be written as

$$u_x(\rho, \phi) = v_0(1 - \rho^2) \tag{2.44}$$

The advantage of this coordinate transformation is that now the boundary ∂C can be expressed in terms of just one coordinate instead of two. Hence, at the boundary ∂C, we have

$$\rho = 1 \tag{2.45}$$

The (y, z) surface integral in equation (2.37) can be transformed into (ρ, ϕ) coordinates as

$$\int_c dy\,dz = \int_c \begin{vmatrix} \frac{\partial y}{\partial \rho} & \frac{\partial z}{\partial \rho} \\ \frac{\partial y}{\partial \phi} & \frac{\partial z}{\partial \phi} \end{vmatrix} d\rho\,d\phi$$

$$= \int_0^1 d\rho \int_0^{2\pi} d\phi \begin{vmatrix} +a\cos\phi & +b\sin\phi \\ -a\rho\sin\phi & +b\rho\cos\phi \end{vmatrix} = ab \int_0^{2\pi} d\phi \int_0^1 d\rho\,\rho \tag{2.46}$$

The flow rate Q for the elliptic channel is calculated as

$$Q = \int_c dy\,dz\,u_x(y,z) = ab \int_0^{2\pi} d\phi \int_0^1 d\rho\,\rho u_x(\rho,\phi) = \frac{\pi}{4}\frac{1}{\mu L}\frac{a^3b^3}{a^2+b^2}\Delta P \tag{2.47}$$

The hydraulic resistance for an elliptic channel thus can be calculated as

$$R_{\text{hyd}} = \frac{\Delta P}{Q} = \frac{4\mu L(a^2+b^2)}{\pi a^3 b^3} \tag{2.48}$$

The area of the ellipse is given by πab.

Hence, the mean velocity can be written as

$$v_m = \frac{Q}{\pi ab} = \frac{a^2b^2(\Delta P/L)}{4\mu(a^2+b^2)} \tag{2.49}$$

The hydraulic diameter of the elliptic cross-section tube can be calculated by using the expression for its perimeter. For "a" not more than three times larger than "b" an approximation for calculation of perimeter, p (within 5% of true value) is:

$$p \simeq 2\pi\sqrt{\frac{a^2+b^2}{2}} \tag{2.50}$$

Indian mathematician Ramanujan came up with a better approximate of perimeter as

$$p \simeq \pi\left[3(a+b) - \sqrt{(3a+b)(a+3b)}\right] \tag{2.51}$$

The perimeter of the ellipse can also be calculated as

$$p = 4a \int_0^{\pi/2} \sqrt{1 - K^2\sin^2\theta}\,d\theta \tag{2.52}$$

which is an elliptic integral and the eccentricity K is defined as

$$K = \frac{\sqrt{a^2-b^2}}{a} \tag{2.53}$$

Note: Eccentricity is 0 for a circle and $0 < K < 1$ for an ellipse.

The integral (equation (2.52)) can be evaluated as an infinite sum

$$p = 2\pi a\left[1 - \left(\frac{1}{2}\right)^2 K^2 - \left(\frac{1\times 3}{2\times 4}\right)^2 \frac{K^4}{3} - \left(\frac{1\times 3\times 5}{2\times 4\times 6}\right)^2 \frac{K^6}{5} \cdots\right] \tag{2.54}$$

Table 2.1 The friction factor of an elliptic tube for different cross-sections (b/a ratios)

b/a	$p/2\pi a$	$f \cdot Re$
1.0	1.0000	64.00
0.8	0.9027	64.40
0.6	0.8125	65.91
0.4	0.7325	69.18
0.2	0.6690	74.35
0.1	0.6483	76.89

The hydraulic diameter of the elliptic tube is

$$D_h = \frac{4\pi ab}{p} \tag{2.55}$$

The friction factor is

$$f = -\frac{\Delta P}{L}\frac{D_h}{\frac{1}{2}\rho v_m^2} \tag{2.56}$$

Table 2.1 shows the friction factor for different elliptic cross-sections and indicates that among all possible elliptic shapes, circular cross-section offers least resistance to the flow.

2.6.2 *Circular Cross-Section*

The circle (Figure 2.8(b)) is just the special case of the ellipse with $a = b$. Therefore, we can immediately write down the result for the velocity field and flow rate for the Poiseuille flow problem in a circular channel. From equations (2.39), (2.41), and (2.47) using $a = b$, it follows that

$$u_x(y, z) = \frac{\Delta P}{4\mu L}(a^2 - y^2 - z^2) \tag{2.57a}$$

$$Q = \frac{\pi a^4}{8\mu L}\Delta P \tag{2.57b}$$

The hydraulic resistance can be written from equation (2.48) as

$$R_{\text{hyd}} = \frac{8\mu L}{\pi a^4} \tag{2.58}$$

2.6.2.1 Alternative Approach

The same result can also be obtained by direct calculation using cylindrical coordinates (x, r, ϕ) and thus avoiding the trial solution equation (2.39). For cylindrical coordinates with the x-axis as the cylinder axis, we have

$$(x, y, z) = (x, r\cos\phi, r\sin\phi) \tag{2.59a}$$

$$\vec{e}_x = \vec{e}_x \tag{2.59b}$$

$$\vec{e}_r = +\cos\phi\vec{e}_y + \sin\phi\vec{e}_z \tag{2.59c}$$

$$\vec{e}_\phi = -\sin\phi\vec{e}_y + \cos\phi\vec{e}_z \tag{2.59d}$$

$$\nabla^2 = \frac{\partial^2}{\partial x^2} + \frac{\partial^2}{\partial r^2} + \frac{1}{r}\frac{\partial}{\partial r} + \frac{1}{r^2}\frac{\partial^2}{\partial \phi^2} \tag{2.59e}$$

The symmetry considerations reduce the velocity field to $v = v_x(r)\vec{e}_x$, so that the N–S equation (2.35) becomes an ordinary differential equation of second order as

$$\left(\frac{\partial^2}{\partial r^2} + \frac{1}{r}\frac{\partial}{\partial r}\right) v_x(r) = -\frac{\Delta P}{\mu L} \tag{2.60}$$

The solutions to this inhomogeneous equation is the sum of a general solution to the homogeneous equation, $v_x'' + v_x'/r = 0$, and one particular solution to the inhomogeneous equation. Here, $'$ indicates the derivative with respect to r. It is easy to observe that the general homogeneous solution has the linear form $v_x(r) = A + B\ln\ r$, while a particular inhomogeneous solution is $v_x(r) = -(\Delta P/4\mu L)r^2$. Using the boundary conditions $v_x(a) = 0$ and $v_x'(0) = 0$, we can arrive at

$$v_x(r,\phi) = \frac{\Delta P}{4\mu L}(a^2 - r^2) \tag{2.61}$$

The flow rate Q can be obtained by integration as

$$Q = \int_0^{2\pi} d\phi \int_0^a dr\, r\frac{\Delta P}{4\mu L}(a^2 - r^2) = \frac{\pi}{8}\frac{a^4}{\mu L}\Delta P \tag{2.62}$$

The average velocity, v_m can be calculated as

$$v_m = \frac{Q}{\pi a^2} = \frac{a^2}{8\mu}\frac{\Delta P}{L} \tag{2.63}$$

The friction factor is defined as

$$f = -\frac{\Delta P}{L}\frac{2a}{\frac{1}{2}\rho v_m^2} = \frac{64}{Re} \tag{2.64}$$

where Re is the Reynolds number equal to $\dfrac{\rho v_m 2a}{\mu}$.

2.6.3 *Equilateral Triangular Cross-section*

No analytical solution exists to Poiseuille flow problem with a general triangular cross-section. In fact, there is analytical solution only for the equilateral triangle defined in Figure 2.8(c). The domain C in the y–z-plane of the equilateral triangular channel cross-section can be thought of as the union of the three half-planes, plane 1: ($z \le \sqrt{3}/2a$), plane 2: $z \ge \sqrt{3}y$, and plane 3: $z \ge -\sqrt{3}y$. Similar to the trial solution of the elliptic channel, we can propose a trial solution by multiplying together the expression for the three straight lines defining the boundaries of the equilateral triangle as

$$u_x(y,z) = \frac{v_0}{a^3}\left(\frac{\sqrt{3}}{2}a - z\right)\left(z - \sqrt{3}y\right)\left(z + \sqrt{3}y\right) = \frac{v_0}{a^3}\left(\frac{\sqrt{3}}{2}a - z\right)(z^2 - 3y^2) \tag{2.65}$$

This trial solution satisfies the no-slip boundary condition on ∂C. The substitution of this trial solution on the Laplacian term of the governing equation (2.35) gives

$$\left[\frac{\partial^2}{\partial y^2}+\frac{\partial^2}{\partial z^2}\right]u_x(y,z)=-2\sqrt{3}\frac{v_0}{a^2} \tag{2.66}$$

The substitution of equation (2.66) in the N–S equation (2.35) yields the constant v_0 as

$$v_0=\frac{1}{2\sqrt{3}}\frac{\Delta P}{\mu L}a^2 \tag{2.67}$$

The flow rate Q can be found by integrating over y and then over z as

$$Q=2\int_0^{(\sqrt{3}/2)a}dz\int_0^{(1/\sqrt{3})z}dy\quad u_x(y,z)=\frac{4v_0}{3\sqrt{3}a^3}\int_0^{(\sqrt{3}/2)a}dz\left(\frac{\sqrt{3}}{2}a-z\right)z^3$$

$$=\frac{3}{160}v_0\quad a^2=\frac{\sqrt{3}}{320}\frac{a^4}{\mu L}\Delta P \tag{2.68}$$

Thus the hydraulic resistance for the equilateral triangle channel is

$$R_{\text{hyd}}=\frac{320\mu L}{\sqrt{3}a^4} \tag{2.69}$$

The hydraulic diameter of the equilateral triangle is

$$D_h=\frac{4A}{P}=\frac{4\times\frac{1}{2}a\times\frac{\sqrt{3}}{2}a}{3a}=\frac{a}{\sqrt{3}} \tag{2.70}$$

In terms of friction factor, it can be shown that

$$f\cdot Re=53.36 \tag{2.71}$$

2.6.4 *Rectangular Cross-section*

For LOC systems, many fabrication methods lead to microchannels having rectangular cross-section (Figure 2.8(d)). One example is the microchannel made in the polymer SU-8 by hot embossing. The SU-8 is heated up slightly above its glass transition temperature, where it gets soft, and then a hard stamp containing the negative of the desired pattern is pressed into the polymer. The stamp is removed and later a polymer lid is placed on top of the structure and bonded to make a leakage-free channel.

There is no analytical solution for the Poiseuille flow problem with a rectangular cross-section. It is only possible to find a Fourier sum representing the solution.

The governing equation for flow inside rectangular cross-section channel is given by

$$\left(\frac{\partial^2}{\partial y^2}+\frac{\partial^2}{\partial z^2}\right)u(y,z)=-\frac{\Delta P}{\mu L} \tag{2.72}$$

The boundary condition at the wall from the no-slip boundary conditions are

$$u(y,z)=0\quad \text{At}\quad y=0\quad\text{and}\quad y=w$$
$$\text{and}\quad z=0\quad\text{and}\quad z=h$$

Let us propose a trial solution for velocity distribution, $u(y, z)$ as a double sine series given by

$$u(y,z) = \sum_m \sum_n A_{mn} \sin \frac{m\pi y}{w} \sin \frac{n\pi z}{h} \tag{2.73}$$

where $m, n = 1, 2, \ldots$.

Note that this trial solution automatically satisfies all the above-mentioned no-slip boundary conditions.

Inspection of the trial solution (equation (2.73)) and governing equation (2.72) indicates that we can rewrite the governing equation (2.72) as

$$-\frac{\Delta P}{\mu L} = \sum_m \sum_n B_{mn} \sin \frac{m\pi y}{w} \sin \frac{n\pi z}{h} \tag{2.74}$$

where B_{mn} is an additional constant.

Let us define $\alpha = \frac{-\Delta P}{\mu L}$.

From the theory of Fourier series, we can write

$$\begin{aligned} B_{mn} &= \frac{4\alpha}{wh} \int_0^w \int_0^h \sin \frac{m\pi y}{w} \sin \frac{n\pi z}{h} dy\, dz \\ &= \frac{4\alpha}{wh} \left(\int_0^w \sin \frac{m\pi y}{w} dy \right) \left(\int_0^h \sin \frac{n\pi z}{h} dz \right) \\ &= \frac{4\alpha}{mn\pi^2} (1 - (-1)^m)(1 - (-1)^n) \end{aligned} \tag{2.75}$$

To find A_{mn}, the assumed $u(y, z)$ profile need to be substituted in the governing equation. We thus calculate

$$\frac{\partial^2 u}{\partial y^2} = -\sum_m \sum_n \left(\frac{m\pi}{w}\right)^2 A_{mn} \sin \frac{m\pi y}{w} \sin \frac{n\pi z}{h} \tag{2.76}$$

$$\frac{\partial^2 u}{\partial z^2} = -\sum_m \sum_n \left(\frac{n\pi}{h}\right)^2 A_{mn} \sin \frac{m\pi y}{w} \sin \frac{n\pi z}{h} \tag{2.77}$$

Substituting equations (2.76) and (2.77) in the governing equation (equation (2.72)), we get

$$A_{mn} = \frac{-B_{mn}/\pi^2}{m^2/w^2 + n^2/h^2}$$

The mean velocity through the tube can be calculated as

$$\begin{aligned} u_m &= \frac{1}{wh} \int_0^w \int_0^h u(y,z) dy\, dz \\ &= -\sum_m \sum_n \frac{4\alpha}{m^2 n^2 \pi^6} \frac{(1 - (-1)^m)^2 (1 - (-1)^n)^2}{\frac{m^2}{w^2} + \frac{n^2}{h^2}} \end{aligned} \tag{2.78}$$

Hydraulic diameter is $D_h = \frac{4A}{p} = \frac{2wh}{w+h}$.

Table 2.2 $f \cdot Re$ as a function of h/w for a rectangular cross-section channel

h/w	$f \cdot Re$
1.0	56.880
0.9	57.048
0.8	57.510
0.6	59.920
0.4	65.470
0.2	76.280
0.1	84.640

The friction factor is

$$f = \frac{\frac{-\Delta P}{L} D_h}{\frac{1}{2}\rho v_m^2} \tag{2.79}$$

$$Re = \frac{\rho v_m D_h}{\mu} \tag{2.80}$$

We can derive the friction factor relationship using equations (2.78), (2.79), and (2.80) as

$$\frac{2}{fRe} = \sum_m \sum_n \frac{1}{\pi^6} \frac{(1-(-1)^m)^2(1-(-1)^n)^2(1+h/w)^2}{m^2 n^2 \left[(h/w)^2 m^2 + n^2\right]} \tag{2.81}$$

For a square duct ($h = w$), we have

$$f \cdot Re = \frac{\pi^6}{2} \frac{1}{S} \tag{2.82}$$

where

$$S = \sum_m \sum_n \frac{1}{m^2 n^2} \frac{1}{m^2 + n^2} (1-(-1)^m)^2 (1-(-1)^n)^2 \tag{2.83}$$

For one term expansion, $m = n = 1$, we have $S = 8$ and $f \cdot Re = 60$. For $m = n = 10$, we have $f \cdot Re = 56.9$. Table 2.2 shows fRe values as a function of the aspect ratio of rectangular cross-section channel. It shows the friction loss to have minimal value for a square cross-section channel, which increases with increase in the aspect ratio.

2.6.5 *Infinite Parallel-plate Channel*

In microfluidics, the aspect ratio of a rectangular channel can often be so large that the channel is well approximated by an infinite parallel-plate configuration. Owing to large dimension of the channel in y-direction compared to that in z-direction, the y dependence drops out from equation (2.72) and the simplified governing equation is:

$$\frac{d^2}{dz^2} u(z) = -\frac{\Delta P}{\mu L} \tag{2.84}$$

The no-slip boundary conditions are given by

$$u(0) = 0 \tag{2.85}$$

$$u(h) = 0 \tag{2.86}$$

The solution of the governing equation (2.84) with the boundary conditions (equations (2.85) and (2.86)) is a simple parabola as

$$u(z) = \frac{\Delta P}{2\mu L}(h - z)z \tag{2.87}$$

The flow rate Q through a section of width w and height h can be expressed as

$$Q = \int_0^w dy \int_0^h dz \frac{\Delta P}{2\mu L}(h - z)z = \frac{h^3 w}{12\mu L}\Delta P \tag{2.88}$$

Thus, the hydraulic resistance for this configuration is

$$R_{\text{hyd}} = \frac{12\mu L}{h^3 w}$$

2.7 Summary of Hydraulic Resistance in Straight Channels

Using the results derived for the Poiseuille flow in straight channels, the hydraulic resistance R_{hyd} for six different cross-sections has been listed in Table 2.3. These results are all valid for the special case of a translation invariant (straight) channel. The vanishing of the nonlinear term $(\vec{v} \cdot \nabla)\vec{v}$ in the N–S equation has been used in the derivation. The analysis is carried out using the approximation of low dimensionless Reynolds number and high aspect ratio of microchannel.

Table 2.3 The hydraulic resistance (R_{hyd}) for straight channels with different cross-sectional shapes

Shape		R_{hyd}
Circle	a	$\frac{8}{\pi}\mu L \frac{1}{a^4}$
Ellipse	b a	$\frac{4}{\pi}\mu L \frac{1+(b/a)^2}{(b/a)^3}\frac{1}{a^4}$
Triangle	a a a	$\frac{320}{\sqrt{3}}\mu L \frac{1}{a^4}$
Two plates	h w	$12\mu L \frac{1}{h^3 w}$
Rectangle	h w	$\frac{12\mu L}{1-0.63(h/w)}\frac{1}{h^3 w}$
Square	h h h h	$\frac{28.4\mu L}{h^4}$

2.8 Viscous Dissipation of Energy

Electrical resistance leads to dissipation of electrical energy in the form of Joule heating. Similarly, hydraulic resistance leads to viscous dissipation of mechanical energy into heat by internal friction in the fluid. The role of viscous dissipation can be explained based on the schematic of transient flow behavior shown in Figure 2.9. Let an incompressible Poiseuille fluid flow takes place inside a channel at times $t < 0$. The constant Poiseuille-type velocity field is maintained by a constant over-pressure ΔP applied to the left end of the channel. The over-pressure ΔP is suddenly removed at time, $t = 0$. However, the fluid flow continues due to the inertia of the fluid. The internal viscous friction of the fluid gradually slows down the motion of the fluid, and eventually in the limit $t \to \infty$ the fluid comes to rest relative to the channel walls. As time passes, the kinetic energy of the fluid at $t = 0$ is gradually transformed into heat by the viscous friction.

Let us calculate the rate of change of the kinetic energy at any instant $t > 0$, after the over-pressure has been removed. Any influence of gravitational and electrical forces is neglected. The kinetic energy of the fluid can be expressed as an integral over the space $\forall$ occupied by the channel as

$$E_{\text{kin}} = \int_{\forall} \frac{1}{2} \rho v^2 d\forall \tag{2.89}$$

The rate of change of E_{kin} is

$$\frac{\partial}{\partial t} E_{\text{kin}} = \int_{\forall} \rho v \frac{\partial v}{\partial t} d\forall \tag{2.90}$$

The time-derivative $\frac{\partial v}{\partial t}$ can be obtained by using the N–S equation for incompressible flow as

$$\rho \left(\frac{\partial \vec{v}}{\partial t} + (\vec{v}.\vec{\nabla})\vec{v} \right) = -\vec{\nabla} p + \mu \nabla^2 \vec{v} + \rho \vec{g} \tag{2.91}$$

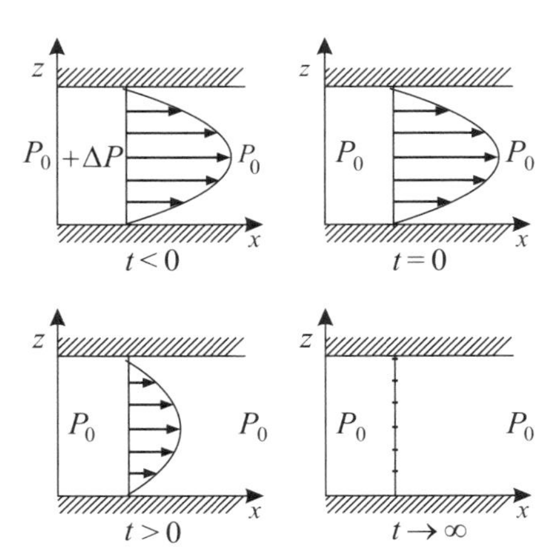

Figure 2.9 Schematic of transient fluid flow behavior inside a channel. For times $t < 0$ the flow is in steady-state, due to the over-pressure ΔP applied to the left. The flow profile is a characteristic parabola. Suddenly, the pressure ΔP is turned off at $t = 0$. However, the inertia keeps up the flow. For $t > 0$ the fluid velocity diminishes due to viscous friction, and in the limit $t \to \infty$ the fluid comes to rest relative to the channel walls

Without the pressure force and neglecting the gravitational force, we have:

$$\rho\frac{\partial\vec{v}}{\partial t} = -\rho(\vec{v}\cdot\vec{\nabla})\vec{v} + \mu\nabla^2\vec{v} \tag{2.92}$$

Substituting equation (2.92) in equation (2.90), the rate of change of the kinetic energy can be written as

$$\frac{\partial}{\partial t}E_{\text{kin}} = \int_{\forall}\left(-\rho(\vec{v}\cdot\vec{\nabla})\vec{v} - \mu\nabla^2\vec{v}\right)\vec{v}d\forall \tag{2.93}$$

Using vector identity

$$\vec{v}\cdot\vec{\nabla}f = \vec{\nabla}\cdot f\vec{v} + f\vec{\nabla}\cdot\vec{v}$$

Using incompressibility flow assumption $(\vec{\nabla}\cdot\vec{v}) = 0$ and Gauss divergence theorem:

$$\int_{\forall}\vec{\nabla}\cdot\vec{Q}d\forall = \int_{A}d\vec{A}\cdot\vec{Q}$$

We have

$$\frac{\partial}{\partial t}E_{\text{kin}} = -\int_{\partial S}\frac{1}{2}\rho v^2\vec{v}\cdot d\vec{S} - \int_{\forall}\phi d\forall \tag{2.94}$$

where ϕ is the dissipation per unit volume.

$$\phi = 2\mu\left[\left(\frac{\partial u}{\partial x}\right)^2 + \left(\frac{\partial v}{\partial y}\right)^2 + \left(\frac{\partial w}{\partial z}\right)^2 - \frac{1}{3}\left(\vec{\nabla}\cdot\vec{v}\right)^2\right]$$
$$+\mu\left[\frac{\partial v}{\partial x} + \frac{\partial u}{\partial y}\right]^2 + \mu\left[\frac{\partial w}{\partial y} + \frac{\partial v}{\partial z}\right]^2 + \mu\left[\frac{\partial u}{\partial z} + \frac{\partial w}{\partial x}\right]^2 \tag{2.95}$$

The surface, ∂s has been shown pictorially in Figure 2.10. The channel surface ∂S consists of three parts: the solid side wall ∂S_{wall}, the open inlet ∂S_1, and the open outlet ∂S_2. The contribution to the surface integral in equation (2.94) from ∂S_{wall} is zero due to the no-slip boundary condition that ensures $\vec{v} \equiv 0$ on solid walls. The two contributions from ∂S_1 and ∂S_2 exactly cancel each other due to the translation invariance of the Poiseuille flow problem. Therefore, the first term in right-hand side of equation (2.94) is equal to zero. The change in kinetic energy of the flow inside the channel is due to viscous dissipation only.

Let's assume the Poiseuille flow $\vec{v} = u(y,z)\vec{e}_x$. The simplified viscous energy dissipation in a Poiseuille flow is therefore given by the volume integral

$$\frac{\partial}{\partial t}E_{\text{kin}} = -\mu\int_{\forall}d\forall\left[\left(\frac{\partial u}{\partial y}\right)^2 + \left(\frac{\partial u}{\partial z}\right)^2\right] \tag{2.96}$$

We note that the kinetic energy is diminishing in time, $\frac{\partial}{\partial t}E_{\text{kin}} < 0$ since the integrand is always positive and the viscosity coefficient μ is positive. The viscous dissipation will be also large due to small size of the channel in the y and z directions compared to that of the macrosystems. Let W_{visc} denotes the heat generated by the viscous dissipation. Then, we have:

$$\frac{\partial}{\partial t}E_{\text{kin}} = \frac{\partial}{\partial t}W_{\text{visc}} \tag{2.97}$$

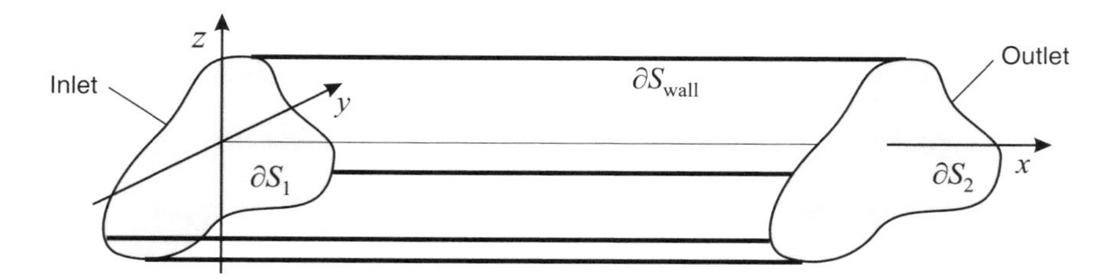

Figure 2.10 A sketch of the geometry for calculating the viscous energy dissipation in a Poiseuille flow

2.8.1 Energy Equation in Microgeometries

The relaxation of Poiseuille flow to obtain an expression for the rate of viscous dissipation of energy, $\frac{\partial}{\partial t}W_{\text{visc}}$ is presented in the previous section. In this section, let us analyze the steady-state Poiseuille flow. Here, the pressure $P(\partial S_1)$ to the left-hand side on ∂S_1 is higher than the pressure $P(\partial S_2)$ to the right-hand side on ∂S_2 given as

$$P(\partial S_1) = P(\partial S_2) + \partial P \tag{2.98}$$

For such a flow, the velocity field is constant and consequently the kinetic energy of the fluid is constant. The rate $\frac{\partial}{\partial t}W_{\text{visc}}$ of heat generation by viscous friction is balanced by the mechanical power $\frac{\partial}{\partial t}W_{\text{mech}}$ put into the fluid by the pressure force at steady-state condition, which can be expressed as

$$\frac{\partial}{\partial t}E_{\text{kin}} = \frac{\partial}{\partial t}W_{\text{mech}} - \frac{\partial}{\partial t}W_{\text{visc}} = 0 \tag{2.99}$$

For incompressible flow and constant density, the N–S equation is given by

$$\rho\left(\frac{\partial\vec{v}}{\partial t} + (\vec{v}.\vec{\nabla})\vec{v}\right) = -\vec{\nabla}P + \mu\nabla^2\vec{v} \tag{2.100}$$

Note that the left-hand side of this equation disappears due to steady-state translation invariance behavior of the Poiseuille flow. The first term in the right-hand side of equation (2.100) contributes to the mechanical energy and the second term contributes to the viscous dissipation.

We can determine the rate of change in W_{mech} by multiplying the pressure term in equation (2.100) by $\vec{v}$ and integrating over volume,

$$\frac{\partial\vec{W}_{\text{mech}}}{\partial t} = \int_{\forall}\vec{v}\cdot(-\vec{\nabla}P)d\forall = -\int_{\forall}\vec{\nabla}\cdot\vec{v}P\,d\forall = -\int_{\partial S}d\vec{A}\cdot(-\vec{v}P) \tag{2.101}$$

Note that in this simplification, we have used the Gauss divergence theorem and vector identity given as

$$\int_{\forall}\vec{\nabla}\cdot\vec{Q}\quad d\forall = \int_{A}d\vec{A}\cdot\vec{Q} \tag{2.102}$$

$$\vec{v}\cdot\vec{\nabla}f = \vec{\nabla}\cdot f\vec{v} + f\vec{\nabla}\cdot\vec{v} \tag{2.103}$$

Similar to the discussion in the previous section, the contribution from the solid walls at ∂S_{wall} is zero due to the no-slip boundary condition and only the inlet ∂S_1 and outlet ∂S_2 surface yield nonzero contributions in equation (2.101). The surface normal are opposite, $\vec{n}(\partial S_1) = -e_x = -\vec{n}(\partial S_2)$, and the pressure is constant at each end-face. So we get

$$\frac{\partial \vec{W}_{mech}}{\partial t} = P(\partial S_1)\int_{\partial S_1} dAu - P(\partial S_2)\int_{\partial S_2} dAu = \Delta P\int_{\partial S_1} dAu(y,z) = Q\Delta P \tag{2.104}$$

In this simplification, we have assumed $u(\partial S_2) = u(\partial S_1)$.

Let us consider a circular capillary of radius a. The power consumption for this geometry can be written as

$$\text{Power consumption} = Q\Delta P = \frac{(\Delta P)^2}{R_{hyd}} = \frac{\pi a^4 (\Delta P)^2}{8\mu L} \tag{2.105}$$

The second term in equation (2.100) is the viscous dissipation part. The result for the viscous dissipation of energy in steady-state Poiseuille flow can be written as

$$\frac{\partial}{\partial t} W_{visc} = \mu \int_{\forall} \left[\left(\frac{\partial u}{\partial y}\right)^2 + \left(\frac{\partial u}{\partial z}\right)^2 \right] d\forall = Q\Delta P \tag{2.106}$$

The viscous dissipation of mechanical energy results in conversion of energy to heat and therefore a rise in temperature. For a general situation with heat transfer to/from the fluid medium, the generalized energy equation can be derived from the N–S equation.

The generalized energy equation for fluid flow is expressed as

$$\rho\left(\frac{De}{Dt}\right) = -\vec{\nabla}\cdot\vec{q} - P(\vec{\nabla}\cdot\vec{v}) + \phi \tag{2.107}$$

where for a gas $e = C_V T$, C_V is the specific heat for constant volume, q is the heat flux, and ϕ is the viscous dissipation.

For a microsystem with no heat transfer and incompressible flow the first and second terms in right-hand side of equation (2.107) disappear. Therefore, the rise in temperature is possible due to viscous dissipation and small volume/mass of the microsystem.

It should be noted that in conventional flow situations, the viscous dissipation terms in the governing energy conservation equation is conveniently neglected. However, in case of microchannel flow, this may lead to significant errors. The energy equation (2.107) needs to be solved simultaneously with the momentum equation (2.100) for analysis of isothermal flow case having no heat addition. The numerical simulation using combined N–S equation and energy equation has shown influence on both friction factor and temperature inside microchannels. Figure 2.11 shows a sample simulation result for flow inside microtube with adiabatic wall.

Figure 2.11 shows that the local temperature increases along the flow direction and the temperature difference between the inlet and outlet increases as the diameter of the microtube decreases. The fluid viscosity is a function of temperature and, therefore, there will also be a change in viscosity along the tube. This will lead to a change in viscous shear force and thus the pressure distribution along the tube. Hence, the characteristics of flow in microgeometries is expected to be different in terms of friction factor compared to the conventional macrosystem. Therefore, for microchannel geometries, the energy equation needs to be used along with the momentum equation for simulation of the fluid flow.

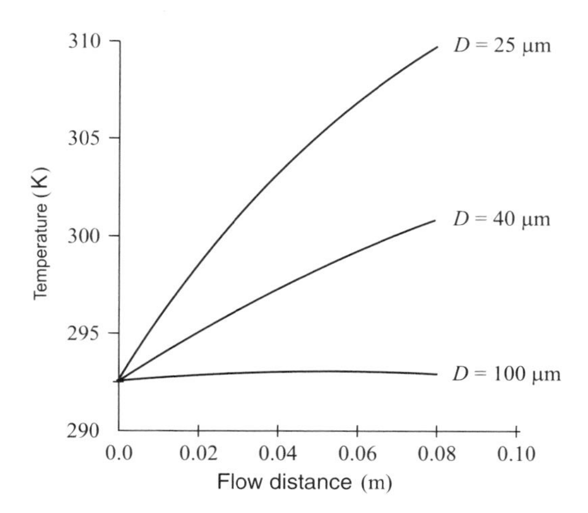

Figure 2.11 Average temperature in the flow direction of a microtube at $Re = 800$ as a function of capillary diameter (D)

2.9 Compliance

The pressure flow rate relationship inside a microchannel can be influenced by compliance properties of the system. The compliance in microsystem exists because of two reasons. One reason is one of the fluid inside the microsystem may have some amount of compressibility. The other possible reason is that the wall material is not rigid but flexible. We discussed below the effect of compliance properties in microsystems using two examples. The first example is for compressible fluid inside the channel and the second is for the flexible channel wall.

2.9.1 Compliance due to Entrapped Gas

Figure 2.12(a) shows an amount of gas trapped ahead of an advancing liquid front in a closed channel. The advancing liquid acts like a liquid piston. At atmospheric pressure, P_0 the gas fills the volume $\forall_0$. The volume $\forall$ and pressure P of the gas change as the liquid advances. Similar to electrical capacitance, $C = dq/dv$ (q is the charge, v is the voltage applied), hydraulic capacitance known as compliance is defined as

$$C_{\text{hyd}} = -\frac{d\forall}{dP} \tag{2.108}$$

Here, the negative sign is used because the volume diminishes as the pressure increases. Assuming the process to be isothermal, the ideal gas law gives: $P\forall = P_0\forall_0$. Thus, the compliance of the entrapped gas can be written as

$$C_{\text{hyd}} = \frac{P_0\forall}{P^2} \tag{2.109}$$

For a small change in pressure, $P_0 \simeq P$, we have:

$$C_{\text{hyd}} \approx \frac{\forall_0}{P_0} \tag{2.110}$$

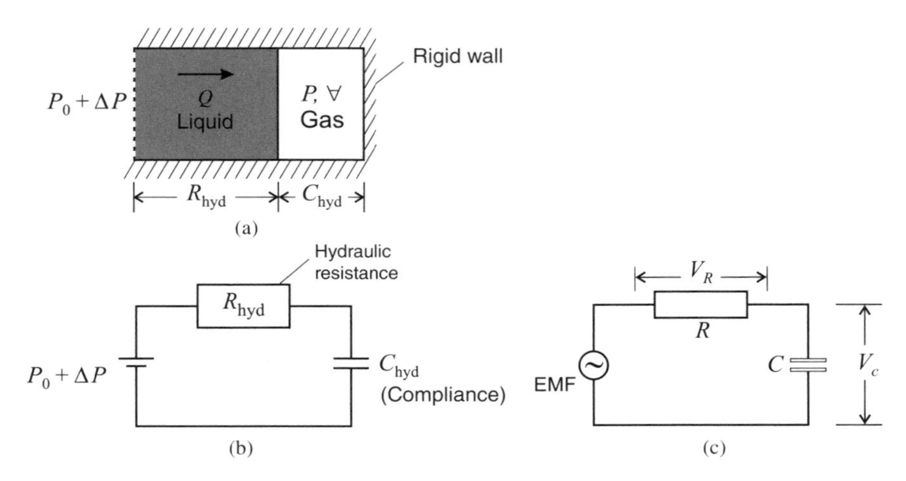

Figure 2.12 (a) A volume of gas trapped in a closed channel wall partly filled with liquid, (b) equivalent hydraulic circuit, (c) a similar series *RC* circuit

Let us assume that, pressure is everywhere at P_0 for time $t < 0$. At time $t = 0$, pressure is suddenly increased from P_0 to $P_0 + \Delta P$. The liquid advances with a flow rate $Q(t)$ and there is a change in the volume of gas given by

$$-\frac{\partial \forall}{\partial t} = Q(t) \tag{2.111}$$

Equation (2.111) is written using the incompressible property of the liquid.

From chain rule, we get

$$Q(t) = -\frac{\partial \forall}{\partial t} = -\frac{\partial \forall}{\partial P}\frac{\partial P}{\partial t} = C_{hyd}\frac{\partial P}{\partial t} \tag{2.112}$$

From Hagen–Poiseuille law and equation, we get (2.112)

$$(P_0 + \Delta P) - P(t) = R_{hyd}Q(t) = -R_{hyd}\frac{d\forall}{dt} = R_{hyd}C_{hyd}\frac{\partial P}{\partial t} \tag{2.113}$$

The solution of this equation is:

$$P(t) = P_0 + (1 - e^{-t/\tau})\Delta P \tag{2.114}$$

where $\tau = R_{hyd}C_{hyd}$ is the time constant. When time constant $\tau = 0$, the air plug instantly experiences the over-pressure ΔP. Figure 2.13 shows the typical pressure–time solution of equation (2.114). It shows that the rate of pressure rise reduces with increase in the time constant (τ) value. This indicates that the equilibrium pressure of the entrapped gas reaches at a longer time with increase in either the hydraulic resistance, R_{hyd} or capacitance, C_{hyd} value. Compliance effect of entrapped gas is higher for smaller channel size and for low pressure condition. This behavior is similar to the behavior of *RC* circuit. The charging time of the capacitor is smaller at lower value of time constant. The presence of resistance, *R* in the circuit slows down the current flow leading to increase in the charge time for changing the amount of charge in the capacitor.

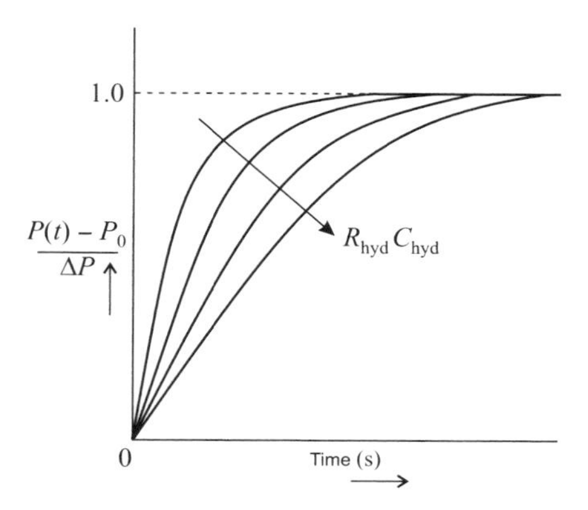

Figure 2.13 The normalized pressure–time evolution of the trapped gas inside the closed channel

Figure 2.12(b,c) shows the equivalent *RC* circuit corresponding to the gas entrapped in a closed channel with liquid. It may be noted that *RC* circuit is commonly used for blocking certain frequencies of a signal, that is, high pass filter, low pass filter, and so on.

2.9.2 *Soft-Walled Channel Flow*

Let us consider another example of compliance, that is, a soft-walled channel. The channel is filled with incompressible liquid. The increase in pressure inside the channel leads to expansion of its wall. The compliance of the channel (C_{hyd}) is related to the geometry and material properties of the channel wall (Figure 2.14(a)). Let us simplify the system assuming the channel to consist of two subchannels with hydraulic resistances $R_{\text{hyd},1}$ and $R_{\text{hyd},2}$, respectively, connected in series. The pressure (P_c) at the meeting point of two channels determines the flow due to the expansion of the channel wall.

We also assume that the pressure at the inlet be P_0 at time $t < 0$ and $P_0 + \Delta P$ for time $t > 0$. The flow rate equation between the inlet region and meeting point can be written as

$$Q_1 = \frac{P_0 + \Delta P - P_c}{R_{\text{hyd},1}} \tag{2.115}$$

The flow rate equation between the meeting point and the outlet region can be written as

$$Q_2 = \frac{P_c - P_0}{R_{\text{hyd},2}} \tag{2.116}$$

The rate of volume expansion of the channel (Q_c) is given by

$$Q_c = +\frac{\partial \forall}{\partial t} = +C_{\text{hyd}}\frac{\partial P_c}{\partial t} \qquad \text{(using chain rule)} \tag{2.117}$$

Note that here the negative sign is not used in C_{hyd} definition as the volume increases with increase in pressure contrary to the entrapped gas case discussed in the previous section.

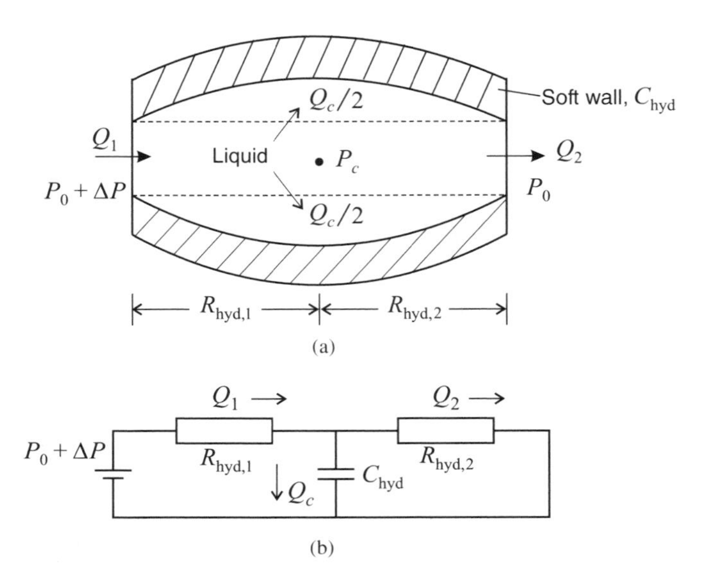

Figure 2.14 (a) Compliance due to a soft-walled channel filled with liquid, and (b) the equivalent circuit diagram corresponding to the soft-walled channel

From mass conservation, we can write

$$Q_1 = Q_2 + Q_c \tag{2.118}$$

Substituting the flow rates (equations (2.115), (2.116), and (2.117)) in equation (2.118), we have

$$\frac{\partial P_c}{\partial t} = -\left(\frac{1}{\tau_1} + \frac{1}{\tau_2}\right) P_c + \left(\frac{1}{\tau_1} + \frac{1}{\tau_2}\right) P_0 + \frac{1}{\tau_1}\Delta P \tag{2.119}$$

where, $\tau_1 = R_{hyd,1} C_{hyd}$ and $\tau_2 = R_{hyd,2} C_{hyd}$.

The solution of this differential equation is:

$$P_c(t) = P_0 + \left(1 - e^{-\left(\frac{1}{\tau_1} + \frac{1}{\tau_2}\right)t}\right)\left(\frac{\tau_2}{\tau_1 + \tau_2}\right)\Delta P \tag{2.120}$$

The equivalent circuit diagram for this case is shown in Figure 2.14(b). This case is analogous to a voltage across a capacitor charged through a voltage divider. The flow rate, Q_2 from the channel will depend on the pressure $P_c(t)$. When $P_c(t) = P_0$, equation (2.116) indicates $Q_2 = 0$, that is, no outflow from the soft-walled channel. Thus, inlet flow and outlet flow of a soft-walled channel flow may not necessarily be equal to each other.

Note that depending on the frequency of the oscillating pressure ($\Delta P(t)$), the flow rate Q_1 and Q_2 will change. In some situations, there may be no flow Q_2 and the liquid stays inside the expanding/relaxing channel.

Problems

2.1 Water (viscosity, $\mu = 1 \times 10^{-3}$ Pa-S) flows at $0.1\,\text{m}^3/\text{s}$ through two rectangular channels (height $= h$ and width $= w$) of same length but different cross-sections. The cross section of the micro channel and mini channel is given by: ($h = 100\,\mu\text{m}$, $w = 300\,\mu\text{m}$ (microchannel) and $h = 1$ mm, $w = 3$ mm (minichannel)). Calculate the ratio of viscous dissipation between the microchannel and minichannel. Discuss the important observations/conclusions for design and analysis of microdevices based on the above calculation.

Note: For rectangular channel with height, h and width w, the hydraulic resistance is given by:

$$R_{\text{hyd}} = \left(\frac{12\mu L}{1 - 0.63\left(\frac{h}{w}\right)} \right) \frac{1}{h^3 w}$$

2.2 Water flows (viscosity, $\mu = 1 \times 10^{-3}$ Pa-S) through two channels of same length ($L = 50$ mm) connected in series at a flow rate of $10^{-4}\,\text{m}^3/\text{s}$. Both channels have same width, that is, 5 mm. However, height of channel-1 is $50\,\mu\text{m}$ and channel-2 is 1 mm. Can you use the resistance concept to calculate the total pressure drop inside the channel? Explain your answer.

2.3 Figure 2.12 shows the electrical analog R–C circuit of a trapped gas in a closed channel. Derive the expression for current flow ($i(t)$) and the voltage across the capacitor $V_c(t)$ for a sudden imposition of voltage, ΔE. Compare the expression for voltage across the capacitor with that of the pressure equation (equation (2.114)) derived inside the text. Assume $\Delta E = 50$ V, $R = 100\,\text{k}\Omega$ and plot the $V_c(t)$ variation as a function of time for varying capacitance values, that is, 0.4, 0.8, and 1.0 μF. Comment on the analogous flow situation based on the plot.

2.4 For the RC circuit shown in Figure 2.14(b), derive the expression for voltage across the capacitor ($V_c(t)$) and the current flowing through the resistance $R_{\text{hyd},2}$ for a sudden imposition of potential (ΔE). Compare the voltage across the capacitor with that of the pressure equation (2.120) derived in the text. Plot $V_c(t)$ variation with time assuming $\Delta E = 50$ V, $R_1 = R_2 = 100\,\text{k}\Omega$ for various values of capacitances, that is, 0.2, 0.4, 0.8, and 1.0 μF. Comment on the analogous flow situation based on the plot.

2.5 Table 2.1 has presented the perimeter ($P/2\pi a$) and fRe for different aspect ratios of the elliptic channel. Verify these values using the relevant expressions reported in the text.

2.6 Verify that $fRe = 53.36$ for flow inside a microchannel with equilateral triangular cross-section.

2.7 Derive the expression for viscous dissipation (ϕ) in Cartesian coordinate system.

2.8 Consider two circular tubes with entrapped air of length 1 cm. One tube is of diameter equal to 1 cm and the other tube is of diameter equal to 1 mm. The length of the water plug in both the tubes is of 2 cm length. An overpressure of 10 kPa is suddenly applied to the liquid plug. Calculate the time required to experience 6.328 kPa for both the capillaries and comment on your answer.

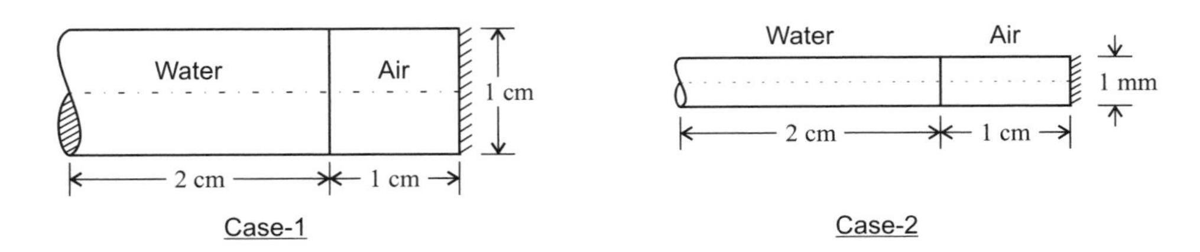

2.9 Consider Couette flow of water with upper plate of surface area equal to 1 m^2 and the velocity of the upper plate equal to 1 m/s. In one case, the separation distance between the two plates is equal to 1 cm and the other case has separation distance equal to 0.5 mm.

(a) Calculate the total viscous dissipation of energy (J/m^3-s) for both the cases.

(b) Assume that the total viscous dissipation is used to raise the temperature of the water contained inside the channel, that is, neglect the inlet and outlet energy from the fluid. What will be the average temperature of water after 1 min for both the cases?

Use: $\rho = 1000\,kg/m^3$, $C = 4.18$ J/kg-K, $\mu = 8.91 \times 10^{-4}$ Pa-s.

Supplemental Reading

Bejan A 2004 *Convection Heat Transfer*. John Wiley & Sons, Inc.

Bruus H 2007 *Theoretical Microfluidics*. Oxford University Press.

Karniadakis G, Beskok A, and Aluru N 2005 *Microflows and Nanoflows, Fundamentals and Simulation*. Springer-Verlag.

White F 2011 *Fluid Mechanics*. Tata Mcgraw Hill.

3

Transport Laws

3.1 Introduction

The Navier–Stokes equation (N–S equation) is one of the basic governing equations for the study of fluid flow related to various disciplines of engineering and sciences. It is a partial differential equation whose integration leads to the appearance of some constants. These constants need to be evaluated for exact solutions of the flow field, which are obtained by imposing suitable boundary conditions. These boundary conditions have been proposed based on physical observation or theoretical analysis. One of the important boundary conditions is the no-slip condition, which states that the velocity of the fluid at the boundary is the same as that of the boundary. Accordingly, the velocity of the fluid adjacent to the wall is zero if the boundary surface is stationary, and it is equal to the velocity of the surface if the surface is moving. This boundary condition is successful in representing a wide range of fluid flow problems. However, it has been observed that no-slip boundary condition is not valid for all situations, and there is a difference between velocity of the surface and the fluid particles near the boundary. This boundary condition is termed as the slip condition. This chapter presents the situations and reasons behind the violation of no-slip boundary condition. The modified velocity profile because of the slip flow boundary condition is discussed subsequently. Similarly, the temperature jump boundary condition, which is also not universally valid, and its influence on flow behavior are subsequently presented. The validity of continuum model in modeling fluid flows is discussed next and the use of molecular based models is presented.

3.2 Boundary Slip

Three possible velocity profiles near a solid boundary are shown in Figure 3.1. Figure 3.1(a) shows that the velocity near the stationary solid wall is equal to zero and represents the no-slip boundary condition. Figure 3.1(b) shows that the velocity near the stationary solid wall is nonzero with a relative velocity between the two, which represents the slip boundary condition. Figure 3.1(c) shows the perfect slip condition, for which there is no influence of the boundary surface on the velocity profile. The velocity when extrapolated toward the wall matches with that of the wall at some distance L_s away from it (Figure 3.1(b)), which is known as *slip length*

Transport Phenomena in Microfluidic Systems, First Edition. Pradipta Kumar Panigrahi.

Companion Website: www.wiley.com/go/panigrahi/microfluidic

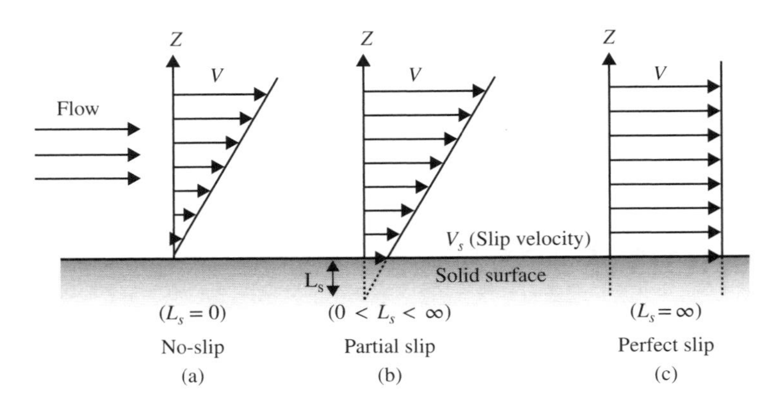

Figure 3.1 Schematic representations of the (a) no-slip, (b) partial slip, and (c) perfect slip boundary conditions. Under no-slip boundary condition, the relative velocity V_s between the fluid and the solid wall is zero at the wall. When slip occurs at the wall, V_s is finite. The extent of the slip is characterized by the slip length L_s

and is used as a measure of the slip. The slip length is a fictitious distance below the surface at which the velocity would be equal to zero if extrapolated linearly. The velocity difference between the boundary surface and the adjacent fluid particle is known as *slip velocity* V_s and is related to the velocity gradient $(\partial V/\partial y)$ near the solid boundary as

$$V_s = L_s \frac{\partial V}{\partial y}\bigg|_{\text{wall}} = \frac{L_s}{\mu}\mu\frac{\partial V}{\partial y}\bigg|_{\text{wall}} = \frac{\tau}{\frac{\mu}{L_s}} = \frac{\tau}{C_s} \tag{3.1}$$

where τ is the shear stress, C_s is the coefficient of slip, and μ is the coefficient of viscosity of the fluid. The slip length L_s is the ratio of coefficient of viscosity (μ) to the coefficient of slip (C_s). The no-slip boundary condition is equivalent to $C_s = \infty$, and the perfect slip condition is equivalent to $C_s = 0$.

The slip flow near the boundary surface can be analyzed based on the types of fluids, that is, gas, Newtonian, and non-Newtonian liquids as discussed in the following sections.

3.3 Slip Flow Boundary Condition in Gases

The slip flow in gases has been derived based on the Maxwell's kinetic theory. In gases, the concept of mean free path is well defined. The slip flow is observed when the characteristic flow length scale is of the order of the mean free path of the gas molecules. The mean free path λ depends strongly on pressure and temperature because of density variation. Knudsen number is defined as the ratio of the mean free path to the characteristic length scale, that is, $Kn = \lambda/L$. The characteristic length scale (L) can be the overall dimension of the flow geometry. The slip velocity is expressed as a function of the Knudsen number and the velocity gradient at the wall as

$$V_s = \frac{2-\sigma_v}{\sigma_v}\left(Kn\left(\frac{\partial V}{\partial n}\right)_s + Kn^2\left(\frac{\partial^2 V}{\partial n^2}\right)_s + \cdots\right) \tag{3.2}$$

where subscript s corresponds to the surface and σ_v is the momentum accommodation coefficient, which is a function of the wall and gas interaction. The slip velocity is zero when Knudsen number is small, that is, the no-slip boundary condition is valid. The slip condition is valid for large Knudsen number, that is, when $Kn \geq 0.1$.

3.3.1 Accommodation Coefficient

Figure 3.2(a) shows the possible interaction mechanism between gas molecules and solid surface. Momentum and energy transfer between the gas molecules and the surface require the specification of interactions between the *impinging gas molecules* and the *surface*, which is complicated and requires the knowledge of scattering kernels. Impingement of gas molecules on a solid surface leads to the transfer of energy in the form of heat and to the transfer of stress between the gas molecule and solid surface molecule. This transfer process is described macroscopically by two types of accommodation coefficients, namely, thermal and tangential momentum.

Let us look into some average parameter for defining gas–solid interaction from macroscopic point of view. *Tangential momentum accommodation coefficient* (σ_v) is a measure of the tangential momentum exchange of gas molecules with the solid surface defined as

$$\sigma_v = \frac{\tau_i - \tau_r}{\tau_i - \tau_w} \tag{3.3}$$

Here, τ_i and τ_r are the tangential momentums of the incoming and reflected molecules, respectively. The tangential momentum of re-emitted molecules corresponding to that of surface is termed as τ_w and is equal to 0 for stationary surface.

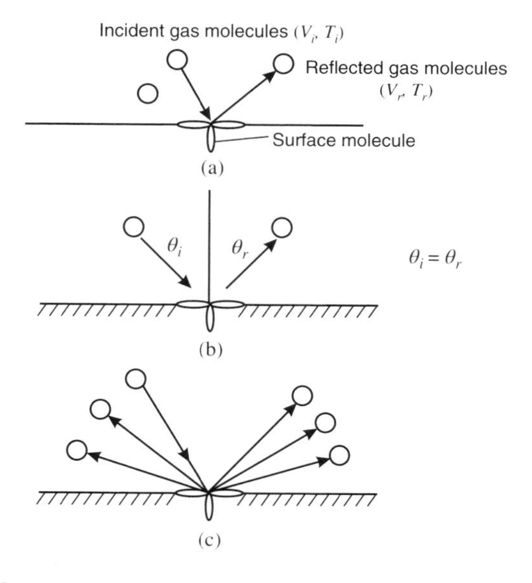

Figure 3.2 (a) Simplified view of gas–surface interaction, (b) specular reflection with incident angle θ_i same as reflected angle θ_r, (c) diffuse reflection where the gas molecules interact with the surface molecule multiple times as they are re-emitted to the gas

When $\sigma_v = 0\,(\tau_i = \tau_r)$, the interaction between molecules and surface is called *specular reflection* (Figure 3.2(b)). For specular reflection, the tangential velocity of molecules reflected from the walls is *unchanged*, but the normal velocity of the molecules is reversed due to the *normal momentum transfer* to the wall. Hence, there is no tangential momentum exchange of fluid with the wall, resulting in *zero skin friction (inviscid flow)*.

Hence, $\dfrac{\partial u_{\text{gas}}}{\partial n} \rightarrow 0$ as $\sigma_v \rightarrow 0$.

This is known as *perfect slip condition*. It may be noted that the velocity slip boundary condition for $u_{\text{gas}} - u_{\text{wall}}$ becomes obsolete, since the Euler equations require only the no-penetration boundary condition in this limit. For *diffuse reflection*, the boundary condition is $\sigma_v = 1$ (Figure 3.2(c)). Here, the molecules are reflected from the walls with zero average tangential velocity for the stationary wall case. Therefore, diffuse reflection leads to tangential momentum exchange (thus friction) of the gas with the walls.

Thermal accommodation coefficient (σ_T) is a measure of the energy exchange between gas molecules and solid surfaces defined as

$$\sigma_T = \frac{E_i - E_r}{E_i - E_w} \tag{3.4}$$

Here, E_i and E_r are the *energy fluxes* of incoming and reflected molecules *per unit time*, respectively. The energy flux of all the incoming molecules that are re-emitted with the energy flux corresponding to the surface temperature T_w is termed as E_w. For the gas, the energy flux may be defined as

$$E = C_v T \tag{3.5}$$

where C_v is the heat capacity of gas molecule. Thus, we can write equation (3.4) as

$$\sigma_T = \frac{T_i - T_r}{T_i - T_w} \tag{3.6}$$

The *perfect energy exchange* between gas molecules and solid molecules corresponds to $\sigma_T = 1$. The thermal accommodation coefficient is a measure of the fraction of heat transferred between the wall and the gas molecule. If the gas at temperature 600 K interacts with the wall at 300 K, the wall heats up and the gas molecule cools down by 300 K. Thus, the gas molecule adjacent to the solid surface satisfies the constant temperature boundary condition similar to that of the wall.

The accommodation coefficients σ_v and σ_T depend on the gas and surface temperatures and local pressure. Table 3.1 shows the typical accommodation coefficient for some gas–surface combinations. Accommodation coefficient also depends on the Knudsen number of the flow. Figure 3.3 shows the variation of accommodation coefficient with respect to the Knudsen number for nitrogen gas.

3.3.2 Slip Model Derivation

Maxwell's derivation of velocity slip and temperature jump boundary condition is based on *kinetic theory of gases*. A similar boundary condition can be derived by an approximate analysis of the motion of gas in an isothermal condition, which has been presented in this section.

Table 3.1 The accommodation coefficient for different gas–surface combinations

Gas	Surface	σ_T	σ_v
Air	Al	0.87–0.97	0.87–0.97
Air	Iron	0.87–0.96	0.87–0.93
H_2	Iron	0.31–0.55	

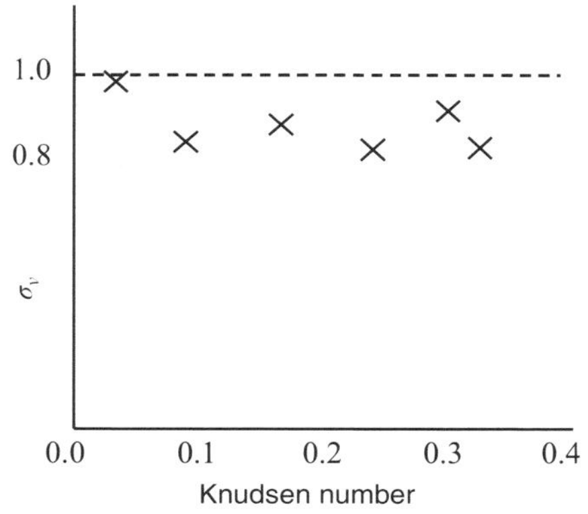

Figure 3.3 Momentum accommodation coefficient of nitrogen gas as a function of Knudsen number

Figure 3.4 shows a wall moving at a velocity u_w. A molecule is assumed to be located at a distance of mean free path (λ) from the wall. A control surface "s" near the wall is indicated as dotted line where the gas molecule has a velocity u_s. It is required to know the slip velocity $u_s - u_w$.

We can write the tangential momentum flux on a surface "s" located near the wall as equal to $n_s m \overline{v}_s u_s$. Here, n_s is the number density of molecules crossing surface "s"; m is the molecular mass; u_s is the tangential (slip) velocity on the surface; and $\overline{v}_s$ is the mean thermal speed of the molecule.

It may be noted that thermal speed $\vec{v}_s$ is a measure of temperature. It is strictly not velocity, that is, it is not a vector quantity. Rather it is simply a scalar speed and has been defined in

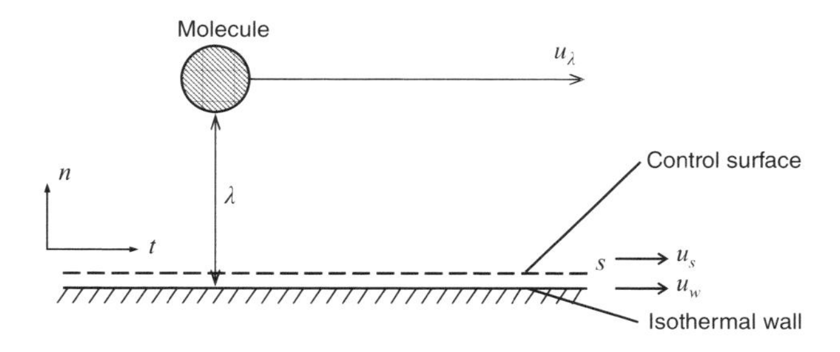

Figure 3.4 A schematic showing gas molecule interaction with the wall

different ways. According to one definition, it is defined as the root mean square of the total velocity in three dimensions. According to another definition, it is defined as the mean of the magnitude of the velocity of the atoms or molecules.

Let us assume that approximately half of the molecules passing through control surface s are coming from a layer of gas at the distance proportional to one mean free path $\left(\lambda = \frac{\mu}{p}\left(\frac{RT\pi}{2}\right)^{1/2}\right)$ away from the surface. The number of these molecules are equal to n_λ. The tangential momentum flux of these incoming molecules can be written as

$$n_\lambda m \overline{v}_\lambda u_\lambda$$

Here, $\vec{v}_\lambda$ is the thermal velocity and u_λ is the tangential velocity at distance λ.

We have assumed $n_\lambda = \frac{1}{2}n_s$.

The other half of the molecules passing through control surface s are reflected from the wall with tangential momentum flux of $n_w m \overline{v}_w u_r$ and the number of these molecules is given as

$$n_w = \frac{1}{2}n_s$$

Here, u_r is the average tangential velocity of the molecules reflected and subscript w denotes wall condition.

Let us assume that σ_v (in %) of the molecules are reflected from the *wall diffusely* (i.e., with average tangential velocity corresponding to that of the wall (u_w), and $(1-\sigma_v)$ (in %) of the molecules are reflected from the wall *specularly* (i.e., conserving their average incoming tangential velocity, u_λ). The velocity of the reflected molecules can be written as

$$u_r = (1-\sigma_v)u_\lambda + \sigma_v u_w \tag{3.7}$$

The total tangential momentum flux on the control surface s because of incoming and reflected molecules is

$$n_s m \overline{v}_s u_s = n_\lambda m \overline{v}_\lambda u_\lambda + n_w m \overline{v}_w \left[\left(1-\sigma_v\right) u_\lambda + \sigma_v u_w\right] \tag{3.8}$$

Let us assume that the temperature of the fluid and the surface *are the same*. Therefore, the mean thermal speeds are identical (i.e., $\overline{v}_s = \overline{v}_\lambda = \overline{v}_w$).

We have assumed $n_\lambda = n_w = \frac{1}{2}n_s$, which is approximately true if there is no *accumulation or condensation* of gas on the surface.

Thus, from equation (3.8) we have

$$u_s = \frac{1}{2}\left[u_\lambda + (1-\sigma_v)u_\lambda + \sigma_v u_w\right] \tag{3.9}$$

The tangential velocity u_λ can be expressed using Taylor series expression about u_s for a distance of one mean free path (λ) away from the surface or a fraction of it.

Using Taylor series expression for u_λ about u_s in equation (3.9), we obtain

$$u_s = \frac{1}{2}\left[u_s + \lambda\left(\frac{\partial u}{\partial n}\right)_s + \frac{\lambda^2}{2}\left(\frac{\partial^2 u}{\partial n^2}\right)_s + \cdots\right] \tag{3.10}$$

$$+\frac{1}{2}(1-\sigma_v)\left[u_s + \lambda\left(\frac{\partial u}{\partial n}\right)_s + \frac{\lambda^2}{2}\left(\frac{\partial^2 u}{\partial n^2}\right) + \cdots\right] + \frac{\sigma_v u_w}{2} \tag{3.11}$$

Here, the normal coordinate to the wall is denoted by n. This expression in simplified form gives the slip relation at the boundaries as

$$u_s - u_w = \frac{2-\sigma_v}{\sigma_v}\left[\lambda\left(\frac{\partial u}{\partial n}\right)_s + \frac{\lambda^2}{2}\left(\frac{\partial^2 u}{\partial n^2}\right)_s + \cdots\right] \tag{3.12}$$

After nondimensionalizing with a reference length (L) and velocity scale (U_∞), we obtain

$$U_s - U_w = \frac{2-\sigma_v}{\sigma_v}\left[Kn\left(\frac{\partial U}{\partial N}\right)_s + \frac{Kn^2}{2}\left(\frac{\partial^2 U}{\partial N^2}\right)_s + \cdots\right] \tag{3.13}$$

Note that nondimensional quantities are shown in capital letters. Here, Kn is the Knudsen number defined as λ/L.

If we neglect the higher order term in the boundary condition, we get Maxwell's slip boundary condition.

Note: The no-slip condition is valid when $Kn < 10^{-3}$. The slip flow boundary condition is valid when $10^{-3} < Kn < 10^{-1}$. For diffuse reflection, $\sigma_v = 1$, $U_s - U_w$, that is, the slip velocity is finite.

3.4 Slip Flow Boundary Condition in Liquids

Liquid slip has implications in various macroscopic applications, that is, flow through porous media, particle aggregation, liquid coating, lubrication, MEMS, and bio-MEMS applications. The movement of three-phase contact line between two immiscible fluids and solid on a substrate during the advancing or receding film motion indicates the importance of slip flow boundary condition. The visible contact angle from measurement differs from that predicted using Young–Laplace equation. The buoyancy and marangoni effect due to temperature and composition distribution is attributed to the slip flow nature of contact line movement. The conventional hydrodynamics with classical no-slip condition on the substrate generates multivalued velocity and infinite drag force near the contact line. The imposition of slip condition eliminates this viscous stress singularity. Similarly, non-Newtonian fluid flows such as polymer solutions show significant apparent slip. Molecular dynamics (MD) simulation has also confirmed the local slip near the contact line. Therefore, for liquids, the slip flow characteristics are different from that of gas and need different explanation.

The liquid slip phenomenon has been presented in the following sections. The experimental investigation quantifying the liquid slip has been discussed to begin with followed by the results from molecular simulation. Subsequently, the factors responsible for liquid slip and the possible mechanism responsible for liquid slip are discussed.

The applicability of slip flow is not well accepted to date by the academic community. One of the problems is the small length scale of slip flow regime (if present) in comparison to the length scale of the flow or system. The hydrodynamic boundary condition appears to be one of no-slip, unless the flow is examined on a length scale comparable to the slip length. Hence, very accurate techniques with high spatial resolution, capable of interfacial flow measurements, are required to detect the effects of slip. Some of the examples of slip flow in liquid are reviewed in the following section.

3.4.1 Flow Rate Measurements

The macroscopic quantity, that is, flow rate and pressure drop measurement, can be used for indirect determination of liquid slip. In this approach, a known pressure gradient ΔP is applied between the two ends of a capillary or a microchannel and the flow rate Q is measured. The flow rate for the slip flow condition is higher than that predicted from the no-slip boundary

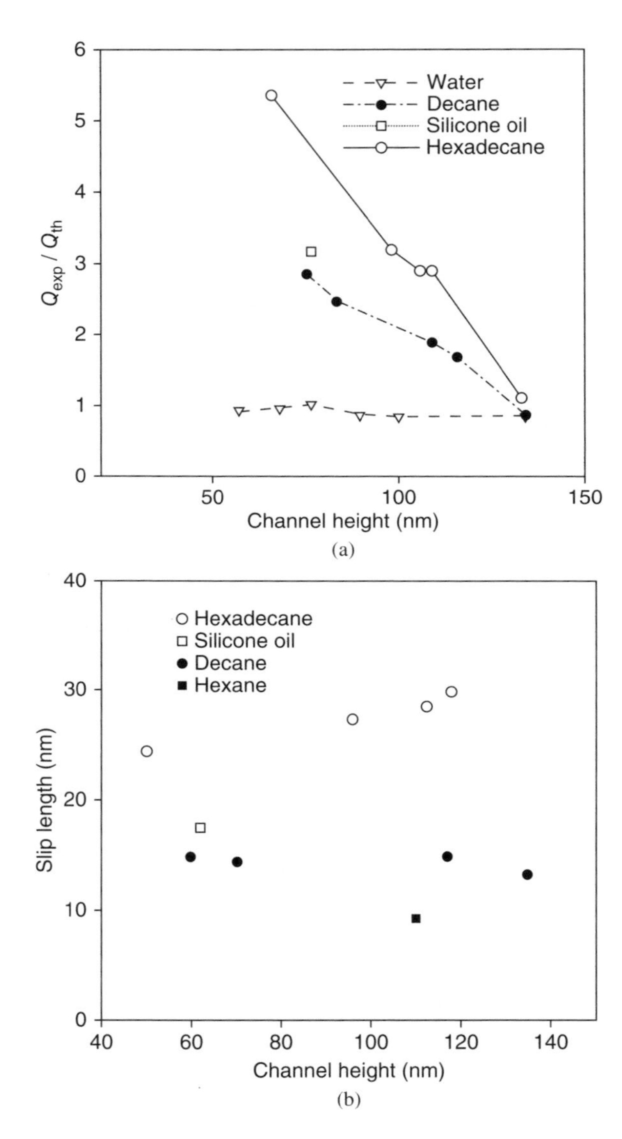

Figure 3.5 (a) The ratio of experiment flow rate (Q_{exp}) to theoretical flow rate (Q_{th}) and (b) the slip length as a function of channel height for different fluids (water, silicone oil, decane, hexane, and hexadecane)

condition. Cheng and Giordano (2002) reported the slip length data for different Newtonian fluids, that is, silicone oil, hexane, decane, and hexadecane (fluids having different molecular diameters) as a function of channel height (see Figure 3.5(a,b)). They observed the no-slip condition to be valid when water is the working fluid. However, the slip length is definite for other working fluids. The slip flow effect is insignificant for channel size greater than about 140 nm. Their study indicates the dependence of slip flow on molecular diameter of the fluid and the fluid–channel wall interaction.

In one of the experiments, Choi *et al.* (2003) presented the slip velocity and slip length of hydrophilic and hydrophobic microchannels (1 and 2 μm depth) based on the flow rate and pressure drop measurements. Sample results from their study are compiled in Figure 3.6(a,b). The flow rate for hydrophobic surface is higher than that compared to the hydrophilic surface. The corresponding slip velocity and slip length for hydrophobic surface are also higher than that of the hydrophilic surface. The slip velocity and slip length increase with shear rate.

For Poiseuille flow of fluid with viscosity μ through a narrow cylindrical channel of radius R and length L, the flow rate assuming no-slip boundary condition is

$$Q_{\mathrm{th}} = \left(\frac{\Delta P \pi R^4}{8\mu L}\right) \tag{3.14}$$

If there is slip at the walls of the channel, the Poiseuille flow velocity profile may be modified to

$$u(r) = \left(\frac{\Delta P}{4\mu L}\right)\left[(R + L_s)^2 - r^2\right] \tag{3.15}$$

The pipe radius "R" is modified to "$R + L_s$" in this expression on the basis of slip length definition (see Figure 3.1).

The flow rate using the above slip flow velocity profile (assuming $\frac{L_s}{R} \ll 1$) can be derived as

$$Q_{\mathrm{slip}} = Q_{\mathrm{th}}\left(1 + \frac{4L_s}{R}\right) \tag{3.16}$$

This equation indicates an increase in slip flow rate with a decrease in the radius of the cylindrical channel for any particular slip length of the channel wall. For the slip length $L_s = 200$ nm, the flow rate with slip (Q_{slip}) is about 6% and 23% higher than the theoretical flow rate with no-slip (Q_{th}) for capillary radius of 13.3 μm and 3.48 μm, respectively.

3.4.2 *Hydrodynamic Force Measurement*

The motion (steady or oscillatory) of a sphere toward a flat surface experiences a resistance to its motion (see Figure 3.7(a)). This resistance is due to the combined contribution from stokes drag on the sphere, the drainage force, and the drag force on the cantilever attached between the sphere and the force measuring apparatus, that is, atomic force microscopy (AFM) or surface force apparatus (SFA). Here, the substrate is driven toward a silica sphere mounted on the tip of an AFM cantilever. A piezo-electric transducer controls the drive rate and the separation distance from the surface. The deflection of the cantilever is monitored, which gives accurate measurement of surface force. The drag force on the cantilever is the main contribution to the total force at large separation distance. When the separation distance is reduced, the drag force

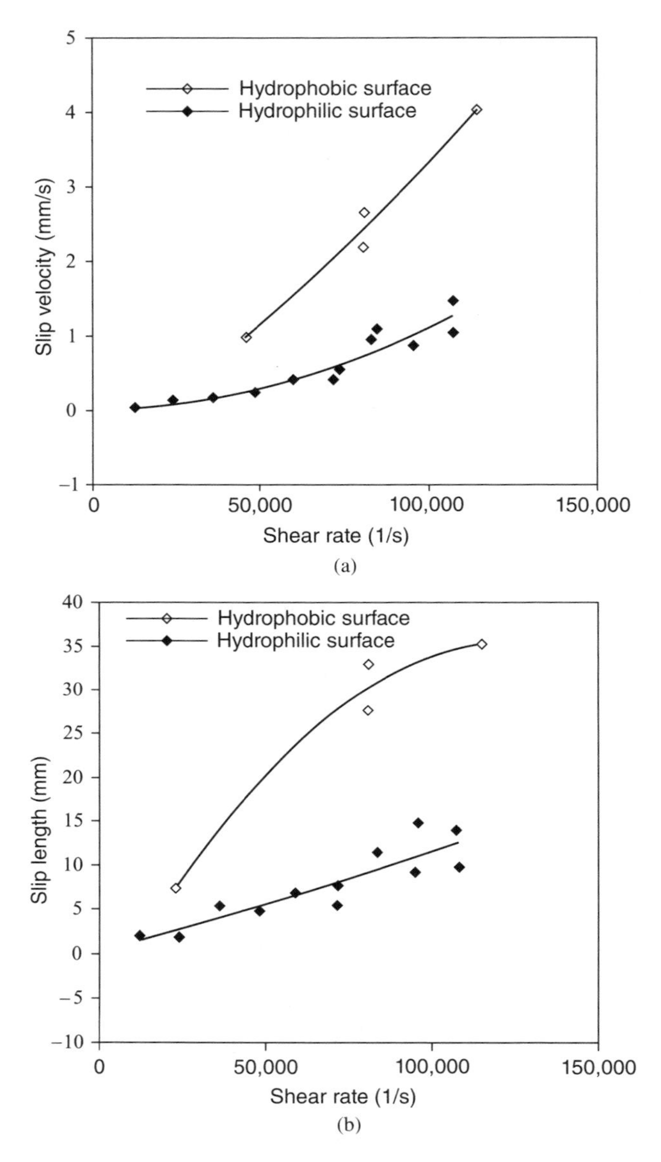

Figure 3.6 (a,b) Comparison of slip velocity and slip length between hydrophilic and hydrophobic surfaces as a function of shear rate

on the cantilever marginally increases. However, the drag force on the sphere increases rapidly compared to the cantilever because of its proximity. The measured drag force is dominated by the sphere when the separation distance is smaller than the sphere radius. As the sphere approaches the flat surface, the repulsive hydrodynamic force increases and the velocity of the

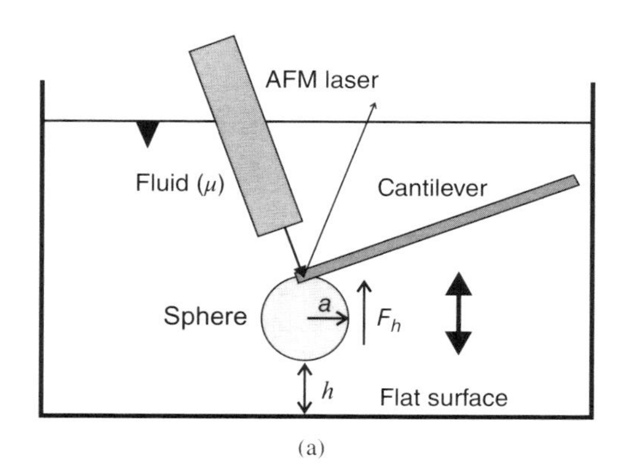

(a)

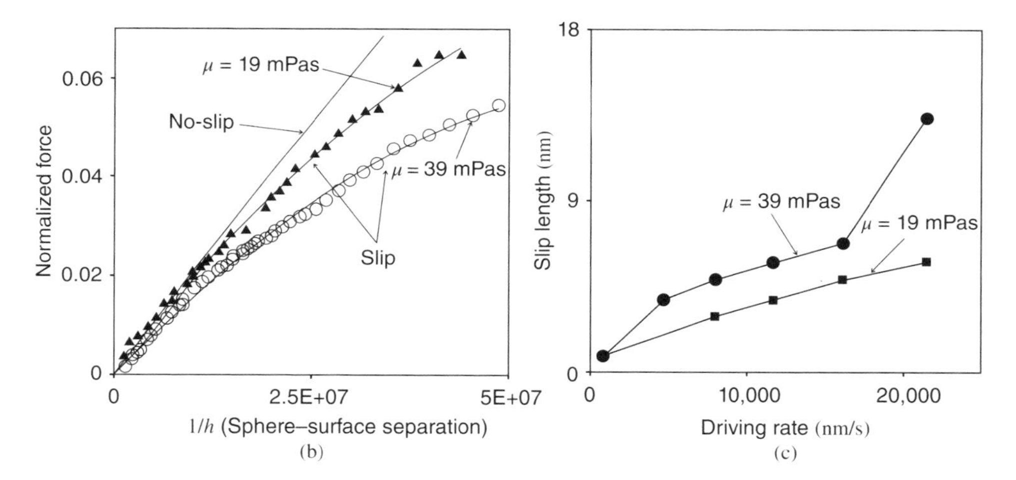

(b) (c)

Figure 3.7 (a) Schematic of the arrangement for slip length characterization, (b) the normalized hydrodynamic force versus inverse of separation for fluid of different viscosities and comparison with no-slip flow calculation, and (c) slip length as a function of viscosity and driving rate

sphere progressively decreases. The exact hydrodynamic solutions of this resistance force for a sphere of radius a, approach velocity V, viscosity μ, and closest separation distance h can be derived using creeping motion approximation. For creeping flow, advective inertia force is small compared to viscous force ($Re \ll 1$). For small surface separation h, the thickness of the liquid film separating the two surfaces is small compared to the radius of curvature of each surface; hence, both surfaces are locally considered parallel. In this lubrication regime, the approximate solution of the N–S equation for the sphere–plane geometry leads to the hydrodynamic force acting on the sphere approaching the wall as (Neto *et al.*, 2003):

$$F = f_{\text{slip}} \frac{6\pi \mu a^2 V}{h} \tag{3.17}$$

For the no-slip boundary condition, $f_{\text{slip}} = 1$. Otherwise when there is slip, $f_{\text{slip}} < 1$, we have

$$f_{\text{slip}} = \frac{h}{3L_S}\left[\left(1+\frac{h}{6L_S}\right)\ln\left(1+\frac{6L_S}{h}\right)-1\right] \tag{3.18}$$

Slip increases the rate of drainage of the fluid confined between the surfaces. For $h \gg L_s$, $f_{\text{slip}} \approx 1$ indicating that the liquid flow is unaffected by slip for surface separations much greater than the slip length. When $h \ll L_s$, $f_{\text{slip}} \simeq 0$ is given as

$$f_{\text{slip}} \simeq h\frac{\ln\,(6L_s)}{3L_s} \tag{3.19}$$

Thus, the hydrodynamic force for the slip flow case is smaller in magnitude than that for the no-slip case.

Figure 3.7(a) shows the schematic of the device for drainage force measurement. Figure 3.7(b) shows the hydrodynamic force versus inverse separation distance for fluid of different viscosity based on the measurements by Neto *et al.* (2003). The predicted drainage force based on the no-slip flow condition is also compared in Figure 3.7(b). It is clearly evident that the no-slip boundary condition is unable to describe the experimental data. The slip length obtained based on equations (3.17) and (3.18) is equal to 4 nm and 12 nm for lower and higher viscosity values, respectively. Figure 3.7(c) shows the slip length as a function of the approach velocity and fluid viscosity, indicating the slip length to be a function of both fluid types, that is, viscosity and strain rate, that is, the approach velocity.

3.4.3 *Velocity Measurements*

The micron resolution particle image velocimetry (PIV) can also be used for direct observation of slip length by measuring the velocity profile in the near-wall region. Figure 3.8(a) shows a typical μ-PIV setup for microchannel velocity measurement. The velocity measurement in hydrophilic (uncoated glass) and hydrophobic (octadecyltrichlorosilane (OTS) coating) channel has been shown in Figure 3.8(b) adapted from Tretheway and Meinhart (2002). Fluorescently dyed polystyrene particle of 300 nm diameter absorbs the green (532 nm) Nd:YAG laser light that emits red (575 nm) light. The emitted light from the particles is collected by CCD camera through the epifluorescent filter. The cross-correlation between the pair of particle images provides the velocity field information. The near-wall velocity field measurement of the microchannel (30 μm deep and 300 μm wide) shown in Figure 3.8(b) indicates different velocity profile behaviors of the hydrophilic and hydrophobic channels. The hydrophobic channel shows a shifting of the velocity profile toward higher value and a finite nonzero velocity near the bottom wall, that is, at about 450 nm from the wall surface indicating slip flow behavior.

Ou and Rothstein (2005) developed an ultrahydrophobic surface with micrometer-sized ridges (20–30 μm wide) placed 20–120 μm apart aligned in the flow direction. They demonstrated maximum 25% drag reduction for flow inside these channels. From μ-PIV measurements, they showed the existence of slip flow in the air–water interface between the ridges, while the flow over the ridges obeys the no-slip condition. They attributed the existence of slip flow to drag reduction.

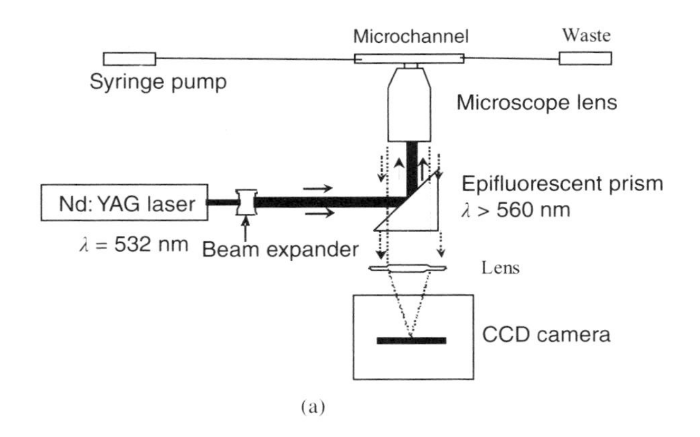

(a)

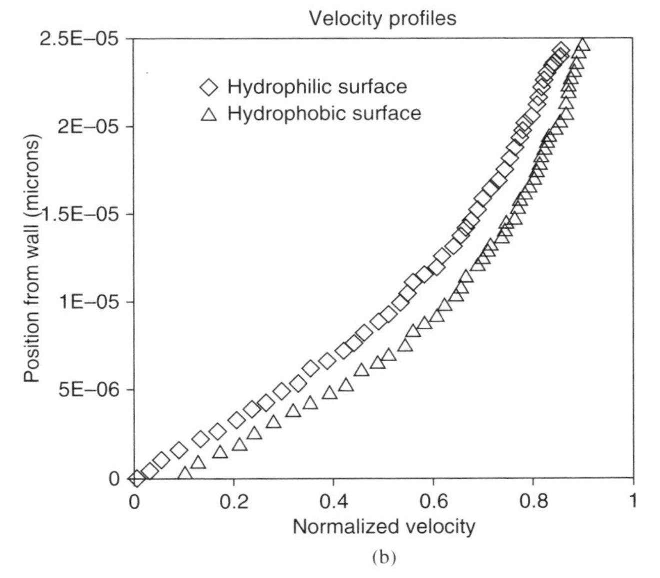

(b)

Figure 3.8 (a) Experimental arrangement for μ-PIV measurement and (b) velocity profile for hydrophilic and hydrophobic surfaces obtained from μ-PIV measurement

3.4.4 Molecular Dynamics Simulation

MD simulation is a useful tool for the study of small–scale fluid flows, where numerical integration of Newton's law of motion for particles (atoms or molecules) is carried out using

$$m_i \frac{d^2 r_i}{dt^2} = \sum_j F_{ij} \tag{3.20}$$

where m_i is the particle mass, r_i is the position of particle i, and F_{ij} is the interatomic or intermolecular force between particles i and j ($F_{ij} = -\nabla_i V_{ij}$). Lennard-Jones (L-J) two-body potential (V_{ij}) between particles is frequently used,which is given by

$$V_{ij} = \varepsilon \left[\left(\frac{\sigma}{r_{ij}} \right)^{12} - c_{ij} \left(\frac{\sigma}{r_{ij}} \right)^{6} \right] \tag{3.21}$$

where ε is an energy scale, σ is the particle size, and r_{ij} is the distance between particles i and j. The constants c_{ij} allow variation of the relative intermolecular attraction between liquids and solids, which therefore represents wetting behavior.

Barrat and Bocquet (1999) carried out the MD simulation of Couette and Poiseuille flows. In Couette flow, the upper wall is moved with a constant velocity, and in Poiseuille flow an external force drives the flow. Sample results from MD simulation are reproduced in Figure 3.9(a,b). The application of no-slip boundary condition leads to the expected linear and parabolic velocity profiles, respectively, for Couette and Poiseuille flows. However, the velocity profile obtained from MD simulation shows a sudden change of velocity in the near-wall region indicating the slip flow. The velocity profile for Couette flow away from the solid surface is linear with different slope than that of the no-slip case. The velocity for slip flow case is higher than that observed in the no-slip case for Poiseuille flow. For both Couette and Poiseuille flows, the partial slip boundary condition at the wall predicts similar bulk flow as that observed by MD simulation. Some discrepancy in the velocity profile is observed in the near-wall region.

3.4.5 Other Techniques

Sedimentation velocity and streaming potential measurements also provide indirect information about the slip length.

The *sedimentation velocity* of particles under gravity can be measured and compared with the predicted values based on no-slip and slip flow boundary conditions. The ratio of the sedimentation velocity as a function of slip length can be derived as

$$\frac{V_{\text{slip}}}{V_{\text{NS}}} = \frac{1 + 3L_s/a}{1 + 2L_s/a} \tag{3.22}$$

For small particles with radius a, the sedimentation velocity with slip (V_{slip}) for slip length L_s is larger than that with no-slip (V_{NS}). The comparison of actual velocity with the predicted velocity based on no-slip condition provides the slip length. Boehnke *et al.* (1999) observed 1 μm slip length for an experiment on silica in the atmosphere. No slip length was observed in experiment conducted under vacuum atmosphere. The slip flow observed in atmospheric condition may be attributed to the presence of dissolved gases in the liquid.

Streaming potential measurement of electrolyte flow inside a capillary under applied pressure difference depends on the slip length. Surface of the capillary acquires a net charge in contact with the electrolyte. The pressure-driven flow of the capillary creates an advection of charges resulting in surplus ions at one end of the capillary compared to the other end. If the two ends of the capillary are not short-circuited, a net steady-state potential difference termed as streaming potential develops. The streaming potential depends on the extent of slip. The ratio of streaming potential for slip (ΔE_{slip}) and no-slip (ΔE_{NS}) is given by

$$\frac{\Delta E_{\text{slip}}}{\Delta E_{\text{NS}}} = 1 + L_s k \tag{3.23}$$

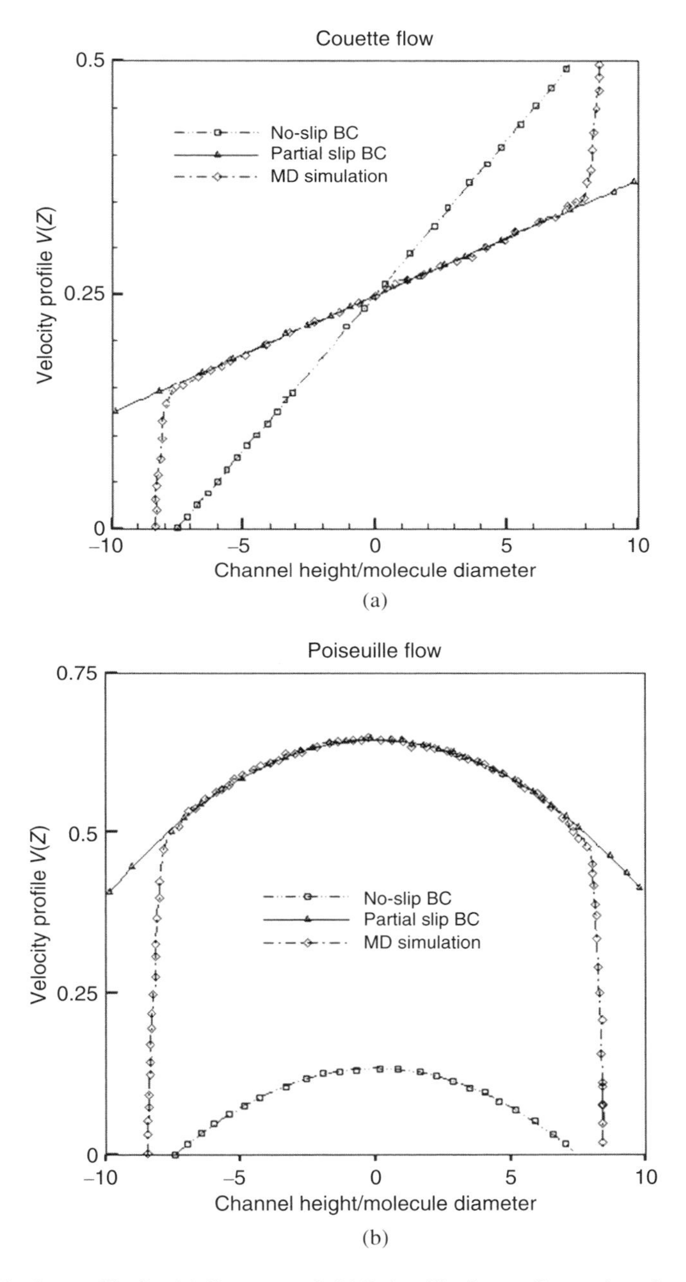

Figure 3.9 Velocity profile for (a) Couette and (b) Poiseuille flows. Comparison between molecular dynamics simulation, no-slip boundary condition, and partial slip boundary condition

where k is the Debye screening parameter, which gives the typical distance close to the surface where there is a net charge density in the liquid and $k^{-1} = (\varepsilon_r \varepsilon_0 k_B T/2e^2 n_0)^{1/2}$. Here, ε_r is the dielectric constant of the liquid; ε_0, the permittivity of the vacuum; k_B, Boltzman's constant; T, temperature; e, the electron charge; and n_0 is the number density of ions in the bulk of the solution. Churaev *et al.* (2002) have presented the streaming potential measurement for quartz and KCl solution indicating the presence of slip flow.

3.5 Physical Parameters Affecting Slip

The results reported in the last section have established the occurrence of slip flow from both experimental and simulation studies. It is important to know various physical parameters affecting slip. Some of the physical parameters affecting slip are summarized in the following sections.

3.5.1 Surface Roughness

Roughness influences the behavior at liquid–solid interfaces. Roughness induces flow around it that leads to dissipation of the mechanical energy. Therefore, there is an increase in the overall resistance to flow, and the tendency of the slip decreases due to surface roughness. The roughness also influences the dewetting behavior of the liquid. A high contact angle is indicative of a weak interaction between liquid and solid, thus causing the fluid molecules to slide across the solid. Roughness can increase the tendency to produce gas–liquid interface at the solid boundary. The surface behaves like a superhydrophobic surface and the slip tendency increases. However, systematic study of surface roughness is not easy to implement. It is difficult to produce suitable surfaces of controlled roughness. Most efforts to alter the surface roughness result in additional undesired changes to the interface properties.

In a systematic molecular study, the following observations have been made regarding the effect of surface roughness on no-slip boundary condition.

1. For microchannel flow with atomically smooth walls, the no-slip condition at the walls is valid if the global Knudsen number $Kn_g = \lambda/h \leq 0.01$, where h is the channel height and λ is the mean free path of gas.
2. For microchannel flow with atomically rough walls, the no-slip condition at the walls is valid if the local Knudsen number $Kn_l = \lambda/A$ is of the order of unity, where A is the roughness height.

3.5.2 Surface Wettability

It is generally believed that the liquid has a larger slip tendency for poorly wetted surfaces. Higher contact angle indicates weak interaction between the solid and liquid and therefore easy to overcome. Many experiments and molecular simulation results have confirmed that the level of hydrophobicity is one of the primary factors determining the level of slip.

3.5.3 *Shear Rate*

The slip length also depends on the shear rate imposed on the fluid particle. Thompson and Troian (1997) have reported the MD simulation of Couette flow at different shear rates. At lower shear rate, the velocity profile follows the no-slip boundary condition. The slip length increases with an increase in shear rate. The critical shear rate for slip is very high for simple liquids, that is, $10^{11} s^{-1}$ for water indicating that slip flow can be achieved experimentally in very small devices at very high speeds. Experiments performed with the SFA and AFM have also showed shear dependence slip in the hydrodynamic force measurements.

3.5.4 *Dissolved Gases and Bubbles*

Dissolved gases or bubbles near a solid also influence the slip flow behavior. It has been observed experimentally that the amount of slip depends on the type and quantity of dissolved gases in the fluid. From sedimentation studies it has been reported that slip is not observed in vacuum conditions while there is a clear slip when a liquid sample is in contact with air. Slip in nonwetting systems depends strongly on the environment in which the experiment is performed. Dissolved gases or nanobubbles in the near-wall region are thought to create localized defects increasing the possibility of slip.

3.5.5 *Polarity*

For electrolyte solutions and polar liquids, the amount of slip depends on the electrical properties of the liquid. The sedimentation experiments report that slip is only observed for polar liquids. Drainage force experiments report slip to increase with an increase in the dipolar moment of the liquid when liquids are polar. This phenomenon is attributed to the superlattice structure in liquid because of the dipole–dipole interactions.

3.6 Possible Liquid Slip Mechanism

The slip flow phenomena can be explained by different mechanism. The fluid slip can be described as *true* or *apparent slip*. The true slip occurs at a molecular level, where liquid molecules are effectively sliding on the solid surface. The apparent slip occurs not at the solid/fluid interface but at the fluid/fluid interface where a thin layer of liquid/gas molecules is tightly bound to the solid surface. For apparent slip, the velocity gradient close to the solid is so high that the molecules beyond the layer of liquid/gas molecules appear to slide on the surface.

The *true slip* phenomena can be attributed to the liquid–liquid and liquid–solid interactions. If the viscous friction between liquid molecules at the interface is stronger than between molecules of the liquid and molecules of the solid, then the molecules can slide on the surface. This is true for hydrophobic surfaces but might also hold for hydrophilic surfaces. If the dimensions of the liquid molecules are of comparable size as the corrugation on the solid surface, the molecules are trapped in the pits on the surface resulting in no-slip BCs. But if their size is much smaller or much larger, they can slide on the surface.

For *apparent slip*, thin gas–liquid layer with a modified viscosity and/or mobility is created near the solid surface. At room temperature and pressure, there is always some residual gas dissolved in a liquid. Critical level of shear might induce cavitations in a liquid and the generated gas bubbles might adhere to the solid surface forming a thin layer at the interface onto which the liquid can slip. The other factor can be the critical shear rate at which a microscopic surface roughness or corrugation can favor the generation of turbulent flow layer at the interface, thus modifying the viscosity of this layer with respect to the bulk even if the overall flow is laminar.

Manipulation of conditions required for generation of slip flow can have many practical applications. This can be beneficial for the development of drag reduction and mixing enhancement technologies. This is particularly important for microtechnologies and nanotechnologies, where the pressure penalty is very high and mixing is difficult due to difficulty in the generation of turbulent flow.

3.7 Thermal Creep Phenomena

In this section, the phenomena of thermal creep and its influence on the slip flow boundary condition is discussed. Let us explain the thermal creep phenomena using a simple experiment. Figure 3.10 shows the schematic of thermal creep experimental setup. The two tanks are connected with each other through an array of microchannel. The pressure relief valves are kept open, and both the tanks are initially at equilibrium with the atmosphere. Then the tanks are dipped to the fluid bath. If the continuum hypothesis is valid, the pressure will remain unchanged. If thermal creep effects are present, then pressure in the *cold reservoir will decrease* and hot reservoir will increase indicating the pumping action from the cold reservoir to the hot reservoir. It is possible to increase the *thermal creep* effects by performing the

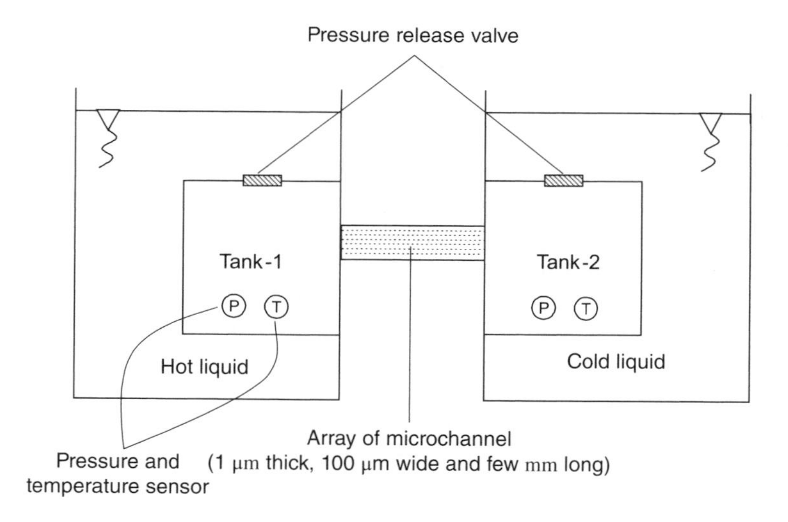

Figure 3.10 A schematic of the thermal creep experiment

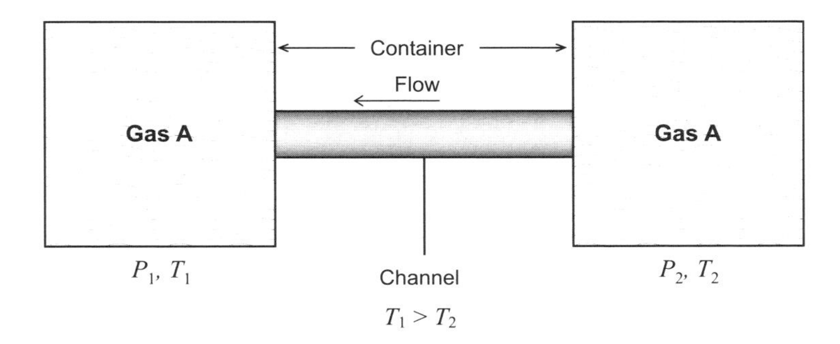

Figure 3.11 Schematic showing the thermal creep flow

experiment at *low pressures* than at atmospheric condition. The thermal creep effects can be studied by changing the *temperature* of the *temperature baths*.

Thermal creep is the phenomenon in which we are able to start *rarefied gas flows* because of tangential temperature gradients along the channel walls, where the fluid starts creeping in the direction from *cold toward hot* (see Figure 3.11). Equilibrium condition requires no flow in the channel for thick channel ($\lambda \ll h$). If channel thickness $h \sim \lambda$ (mean free path), rarefied gas effects have to be taken into account. Here, the local equilibrium mechanism is very complex, and interaction of gas molecules with the walls must be considered.

For thin channel, that is, $\lambda \geq h$, free molecular flow condition can be assumed. Here, the intermolecular collisions are negligible in comparison to the interaction of molecules with the surface. Let us assume the molecule wall interaction to be *specular* ($\sigma_v = 0$) and explain the thermal creep phenomena observed experimentally.

We can assume the density (ρ) to be directly proportional to the number of molecules per unit volume (n) as

$$\rho \propto n \tag{3.24}$$

Similarly, the average molecular speed can be assumed to be proportional to the temperature as

$$\bar{c}^2 \propto T \tag{3.25}$$

where T is the temperature and $\bar{c}$ is the average molecular speed.

The mass flux at the hot and cold sides of the channel are, respectively, equal to $m\, n_1\, \bar{c}_1$ and $m\, n_2\, \bar{c}_2$, where m is the mass of gas molecules.

Using $P = \rho RT$ and $\frac{P_1}{P_2} = 1$, the ratio of mass flux between the hot and cold sides is given by

$$\frac{\dot{m}_{\text{hot}}}{\dot{m}_{\text{cold}}} = \frac{mn_1\bar{c}_1}{mn_2\bar{c}_2} \simeq \frac{\rho_1}{\rho_2}\left(\frac{T_1}{T_2}\right)^{0.5} = \frac{P_1}{P_2}\left(\frac{T_2}{T_1}\right)^{0.5} = \left(\frac{T_2}{T_1}\right)^{0.5} \leq 1 \tag{3.26}$$

Hence, the above expression indicates a creeping flow from cold side to hot side.

The thermal creep effects, that is, flow from cold side toward the hot side in nonisothermal surfaces, also contribute toward the slip velocity. Thus, the first-order slip flow boundary condition is modified for an ideal gas flow as (Smoluchowski von Smolan, 1898)

$$u_s - u_w = \frac{2-\sigma_v}{\sigma_v}\left[\lambda\left(\frac{\partial u}{\partial y}\right)_s + \frac{3}{4}\frac{\mu}{\rho T_s}\left(\frac{\partial T}{\partial x}\right)_s\right] \tag{3.27}$$

where x and y are the streamwise and normal coordinates, ρ is the fluid density, μ is the fluid viscosity, and T_s is the temperature of the fluid adjacent to the wall. Thus, the high viscous dissipation in microflows may lead to temperature gradient of the surface, which may modify the nature of slip flow at the wall. The slip flow velocity will be higher for flow inside a channel with positive streamwise temperature gradient of the channel wall compared to the negative streamwise temperature gradient. The slip flow magnitude will also be higher for low momentum accommodation coefficient.

3.7.1 Knudsen Compressor

In the early 1900s, Knudsen built a molecular compressor by connecting a series of tubes with constrictions arranged in between each tube. By heating one side of the constriction to a very high temperature, Knudsen was able to maintain considerable pressure gradient. Large-scale Knudsen compressors have low volumetric flow rate and inefficient energy usage. However, their microscale counterparts eliminate these disadvantages and result in *low-power gas pumping systems* with *nonmoving components*.

These compressors are useful in various *microscale gas pumping applications*, that is, pumping gas samples *through gas spectrometers* for detecting *pollutants* and various *chemical or biological agents*.

3.8 Couette Flow with Slip Flow Boundary Condition

Couette flow can be used as a prototype flow to model shear-driven flows such as micromotor, microbearing, and so on. As the flow is shear driven, the pressure does not change in the streamwise direction. The compressibility effects may be important for large temperature fluctuations or at high speed. Let us consider an incompressible Couette flow with slip (Figure 3.12).

Here, the two parallel plates extend between $y = 0$ and $y = h$, and the top surface moves with a velocity U_∞. The continuity equation for this flow using the fully developed flow assumption $\left(\frac{\partial u}{\partial x} = 0\right)$ is

$$\frac{\partial v}{\partial y} = 0 \tag{3.28}$$

Using the boundary condition, $v|_{\text{wall}} = 0$, the continuity equation gives $v = 0$.

The simplified N–S equations in x and y coordinates, respectively, are

$$\rho v\frac{\partial u}{\partial y} = \mu\frac{\partial^2 u}{\partial y^2} \tag{3.29}$$

$$\rho v\frac{\partial v}{\partial y} = -\rho g - \frac{\partial p}{\partial y} + \mu\frac{\partial^2 v}{\partial y^2} \tag{3.30}$$

Equation (3.30) gives the pressure gradient in the y-direction. Integration of equation 3.29 and the use of no-slip boundary condition lead to the linear velocity profile as

$$\frac{u}{U_\infty} = \frac{y}{h} \tag{3.31}$$

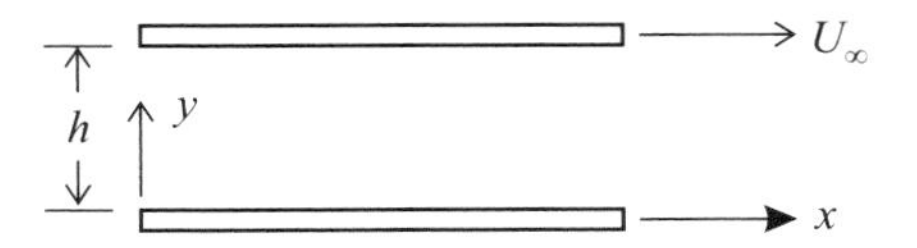

Figure 3.12 Schematic of the Couette flow

It may be noted that the simplified governing equation $\left(\frac{\partial^2 u}{\partial y^2} = 0\right)$ can also be derived using the nondimensional approach discussed in Chapter 2. The governing equation for Couette flow is the same as that of Poiseuille flow between two parallel plates with zero pressure gradient, that is, $\frac{\partial P}{\partial x} = 0$. Let us derive the Couette flow velocity profile with slip flow boundary condition.

Here, we have to use first-order slip flow boundary condition for the slip flow as the linear velocity profile of the Couette flow makes it impossible to incorporate higher order slip effects since $\frac{\partial^2 U}{\partial y^2} = 0$ for Couette flow.

We use the first-order slip flow boundary condition as

$$u_s - u_w = \frac{2 - \sigma_v}{\sigma_v} \lambda \left.\frac{\partial u}{\partial y}\right|_{y=0 \text{ and } h} \tag{3.32}$$

The boundary conditions at the lower and upper walls can be written using the wall velocity as follows:

On the lower wall, $(y = 0)$: $u_w = 0$.
On the upper wall, $(y = h)$: $u_w = U_\infty$.

The integration of the momentum equation (3.29) gives

$$u = C_1 y + C_2 \tag{3.33}$$

The application of boundary condition at the lower wall gives

$$u = \frac{2 - \sigma_v}{\sigma_v} \lambda \left.\frac{\partial u}{\partial y}\right|_{y=0} = C_2 \tag{3.34}$$

The boundary condition at the upper wall gives

$$u - U_\infty = \frac{2 - \sigma_v}{\sigma_v} \lambda \left.\frac{\partial u}{\partial y}\right|_{y=h} \tag{3.35}$$

Using equations (3.33) and (3.35), we have

$$u = C_1 h + C_2 = \frac{2 - \sigma_v}{\sigma_v} \lambda \left.\frac{\partial u}{\partial y}\right|_{y=h} + U_\infty \tag{3.36}$$

Substituting C_2 from equation (3.34) in this equation, we get

$$C_1 h + \frac{2 - \sigma_v}{\sigma_v} \lambda \left.\frac{\partial u}{\partial y}\right|_{y=0} = \frac{2 - \sigma_v}{\sigma_v} \lambda \left.\frac{\partial u}{\partial y}\right|_{y=h} + U_\infty \tag{3.37}$$

We observe from equation (3.33) that

$$\frac{\partial u}{\partial y} = C_1$$

It may be noted that due to different direction of the wall normal to the bottom surface and the top surface, we have

$$\left.\frac{\partial u}{\partial y}\right|_{y=0} = -\left.\frac{\partial u}{\partial y}\right|_{y=h}$$

Thus, equation (3.37) can be written as

$$C_1 h + \frac{2-\sigma_v}{\sigma_v}\lambda C_1 + \frac{2-\sigma_v}{\sigma_v}\lambda C_1 = U_\infty$$

$$C_1\left(h + 2\left(\frac{2-\sigma_v}{\sigma_v}\lambda\right)\right) = U_\infty$$

$$C_1 = \frac{U_\infty}{h + 2\lambda\left(\frac{2-\sigma_v}{\sigma_v}\right)} \tag{3.38}$$

Hence, $C_2 = \frac{2-\sigma_v}{\sigma_v}\lambda\left.\frac{\partial u}{\partial y}\right|_{y=0} = \frac{2-\sigma_v}{\sigma_v}\lambda C_1$

Thus,

$$C_2 = \frac{\frac{2-\sigma_v}{\sigma_v}\lambda U_\infty}{h + 2\lambda\left(\frac{2-\sigma_v}{\sigma_v}\right)} \tag{3.39}$$

Now, substituting C_1 and C_2 in equation (3.33), we have

$$u = \frac{U_\infty y}{h + 2\lambda\left(\frac{2-\sigma_v}{\sigma_v}\right)} + \frac{\frac{2-\sigma_v}{\sigma_v}\lambda U_\infty}{h + 2\lambda\left(\frac{2-\sigma_v}{\sigma_v}\right)}$$

We can also write this equation in the dimensionless form as

$$\frac{u}{U_\infty} = \frac{\frac{y}{h}}{1 + 2Kn\left(\frac{2-\sigma_v}{\sigma_v}\right)} + \frac{\frac{2-\sigma_v}{\sigma_v}Kn}{1 + 2Kn\frac{(2-\sigma_v)}{\sigma_v}}$$

Thus,

$$\frac{u}{U_\infty} = \frac{\frac{y}{h} + \frac{2-\sigma_v}{\sigma_v}Kn}{1 + 2Kn\left(\frac{2-\sigma_v}{\sigma_v}\right)} \tag{3.40}$$

The above-mentioned velocity solution indicates not only a velocity slip on the wall but also a correction to the slope of the profile (see Figure 3.13) for the Couette flow with slip flow boundary condition. The velocity profile as a function of Knudsen number and momentum accommodation coefficient has been shown in Figure 3.14. There is an increase in slip velocity

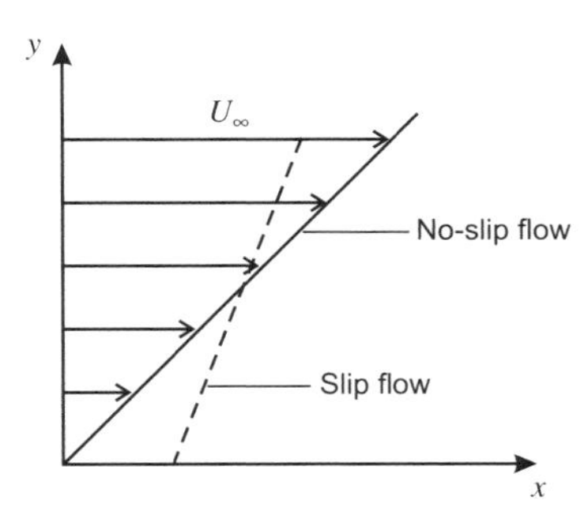

Figure 3.13 Comparison of Couette flow velocity profile using the slip flow boundary condition with that of the no-slip flow boundary condition

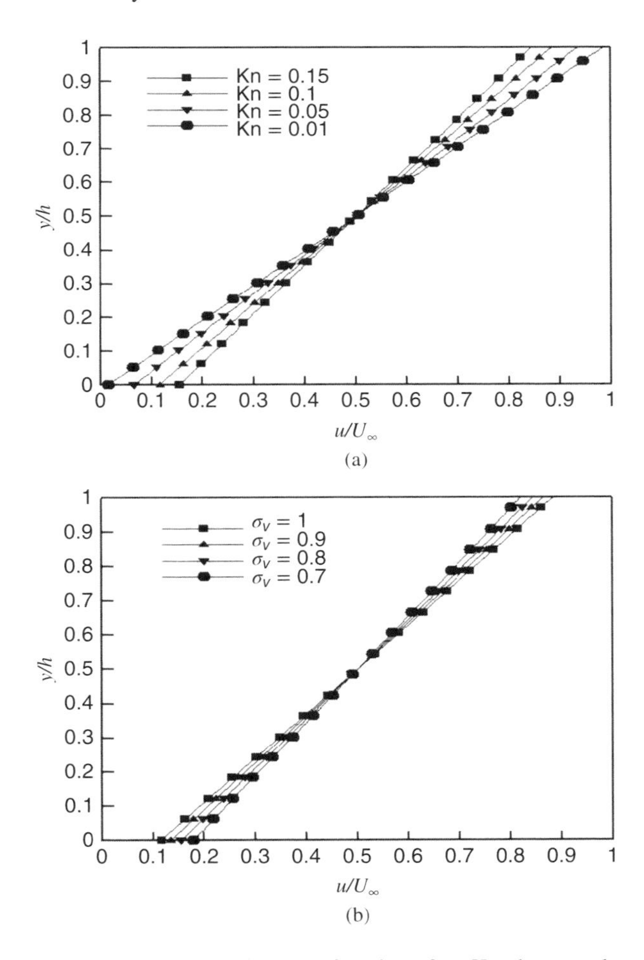

Figure 3.14 Velocity profile in a Couette flow as a function of (a) Knudsen number with fixed $\sigma_v = 0.8$ and (b) tangential accommodation coefficient with fixed $Kn = 0.15$

with an increase in Knudsen number for a fixed momentum accommodation coefficient case (see Figure 3.14(a)). Similarly, the slip velocity increases with the decrease in accommodation coefficient. There is also a change in the slope of the linear velocity profile with the change in accommodation coefficient and Knudsen number.

The volume flow rate can be calculated as

$$\dot{Q} = \int_0^h u dy \tag{3.41}$$

$$\frac{\dot{Q}}{U_\infty h} = \frac{1}{2} \tag{3.42}$$

The slip flow solution gives the same volume flow rate as that of the no-slip case, that is, the volume flow rate is independent of Knudsen number. Let us also compare the skin friction coefficient of slip flow with that of no-slip flow.

$$c_f = \frac{\tau_w}{\frac{1}{2}\rho U_\infty^2} \tag{3.43}$$

Here, τ_w is the wall shear stress.

For the no-slip boundary condition case, equation (3.43) can be simplified as

$$(c_f)_{\text{no-slip}} = \frac{\mu U_\infty}{h} \times \frac{1}{\frac{1}{2}\rho U_\infty^2} = \frac{2}{\frac{\rho U_\infty h}{\mu}} = \frac{2}{Re} \tag{3.44}$$

where $Re = \dfrac{\rho U_\infty h}{\mu}$

For the slip flow boundary condition case:

$$(c_f)_{\text{slip}} = \frac{2}{Re}\left(\frac{1}{1 + 2\left(\frac{2-\sigma_v}{\sigma_v}\right) Kn}\right) \tag{3.45}$$

Hence,

$$\frac{(c_f)_{\text{slip}}}{(c_f)_{\text{no-slip}}} = \frac{1}{1 + 2\left(\frac{2-\sigma_v}{\sigma_v}\right) Kn} \tag{3.46}$$

In general, there is a general decrease in skin friction coefficient compared to the no-slip flow. As an example for $\sigma_v = 0.8$ and $Kn = 0.05$, there will be 13% decrease in skin friction coefficient compared to the no-slip flow. This is also an indication that the work done in overcoming friction for microbearing is expected to be less compared to that in conventional lubrication.

3.9 Compressibility Effect in Microscale Flows

Compressibility is a measure of the relative change in volume of a fluid as a response to pressure change, which is defined as

$$\beta = -\frac{1}{\forall}\frac{\partial \forall}{\partial P} \tag{3.47}$$

where $\forall$ is the volume and P is the pressure. Bulk modulus (K) is defined as the inverse of compressibility. The bulk modulus (K) of water is equal to 2.2×10^9 Pa and that of air is equal to 1.42×10^5 Pa. Higher value of bulk modulus indicates less compressibility of water in comparison to the air.

According to conventional practice, the flow of a compressible fluid such as air is treated as incompressible when Mach number is less than 0.3. However, there are some situations when Mach number is exceedingly small and the flow is compressible. It may be remembered from heat transfer studies that for strong wall heating or cooling, the density changes sufficiently and the incompressible approximation breaks down even at low speeds. In some microdevices, the pressure may change significantly due to viscous effects even though the speed is less than Mach number of 0.3. There will be a strong density change corresponding to the high pressure changes. This fact must be taken into account while writing the continuity equation of motion. The full continuity equation is written as

$$\frac{D\rho}{Dt} + \rho\vec{\nabla} \cdot \vec{V} = 0 \tag{3.48}$$

where $\frac{D}{Dt}$ is the substantial derivative $\left(\frac{\partial}{\partial t} + \vec{V} \cdot \vec{\nabla}\right)$. For incompressible approximation, $\left(\frac{1}{\rho}\frac{D\rho}{Dt}\right)$ is very small, that is, the density changes following a fluid particle is very small. Note that density may change from one particle to another without violating the incompressible approximation, that is, variable density/temperature/salinity flow is often treated as incompressible.

Let us consider the state principle of thermodynamics, where one can express the density changes of a simple system in terms of the change in pressure and temperature as

$$\rho = \rho(P, T)$$

The chain rule of calculus gives

$$\frac{D\rho}{Dt} = \frac{D\rho}{DP} \cdot \frac{DP}{Dt} + \frac{D\rho}{DT}\frac{DT}{Dt} \tag{3.49}$$

We can rewrite this equation as

$$\frac{1}{\rho}\frac{D\rho}{Dt} = \alpha\frac{DP}{Dt} - \beta\frac{DT}{Dt} \tag{3.50}$$

where

$$\alpha(P, T) = \left.\frac{1}{\rho}\frac{\partial\rho}{\partial P}\right|_T \tag{3.51a}$$

$$\beta(P, T) = \left.-\frac{1}{\rho}\frac{\partial\rho}{\partial T}\right|_P \tag{3.51b}$$

where α and β are known, respectively, as isothermal compressibility coefficient and bulk expansion coefficient. For ideal gas ($P = \rho RT$), $\alpha = \frac{1}{P}$, and $\beta = \frac{1}{T}$.

Thus, we have

$$\frac{1}{\rho}\frac{D\rho}{Dt} = \frac{1}{P}\frac{DP}{Dt} - \frac{1}{T}\frac{DT}{Dt} \tag{3.52}$$

This expression indicates that the flow should be treated as compressible if pressure and temperature changes are sufficiently strong. Note the first term in the right-hand side can be significant even at low Mach number, which has been shown in the following section by nondimensionalization of flow equations through the parallel plate. Similarly, the second term in right-hand side can also be strong even at low Mach number because of strong wall heating or cooling. Experiments in gaseous microducts have confirmed that the pressure gradients in long microchannels are not constant, which is consistent with the compressible flow behavior.

3.9.1 Compressibility Effects of Flow between Parallel Plates

The effect of compressibility on flow in microsystems has been discussed in this section. The term $\frac{DP}{Dt}$ in equation (3.52) is composed of two parts; that is, $\frac{\partial P}{\partial t} + \vec{v} \cdot \vec{\nabla} P$, where the first part is a temporal contribution to the change in pressure and the second part is a convective contribution to the change in pressure. Let us discuss the second contribution from a typical example of flow between parallel plates (Figure 3.15) using the control volume approach.

The conservation of mass between the inlet and outlet can be written as

$$\rho_i u_i = \rho_0 u_0 \tag{3.53}$$

Here, ρ and u are the channel average density and the velocity, respectively.

The momentum equation in streamwise direction can be written as

$$(P_i - P_0)h - 2L\tau = \dot{m}(u_0 - u_i) \tag{3.54}$$

where the left-hand side of this equation denotes the force and the right-hand side denotes the change in momentum. Here, τ is the shear stress and $\dot{m}$ is the mass flow rate.

Dividing both sides of this equation by (hP_0), we get

$$\frac{(P_i - P_0)}{P_0} = \frac{\Delta P}{P_0} = 2\frac{L}{h}\frac{\tau}{P_0} + \frac{\dot{m}(u_0 - u_i)}{hP_0} \tag{3.55}$$

Using $\dot{m} = \rho_0 u_0 h$, the last term on the right-hand side of this equation is

$$\begin{aligned} \frac{\rho_0 u_0 h(u_0 - u_i)}{hP_0} &= \frac{\rho_0 u_0 u_i}{P_0}\left(\frac{u_0}{u_i} - 1\right) \\ &= \frac{\rho_0 u_0 u_i}{P_0}\left(\frac{P_i}{P_0} - 1\right) = \frac{\rho_0 u_0 u_i}{\rho_0 RT}\left(\frac{P_i - P_0}{P_0}\right) \end{aligned} \tag{3.56}$$

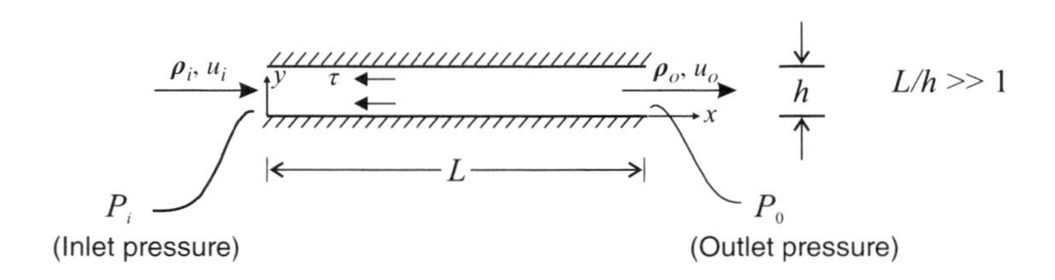

Figure 3.15 A schematic of flow between two parallel plates

It may be noted that we have used the equation of state and conservation of mass for the simplification of equation (3.56) given as

$$\frac{u_0}{u_i} = \frac{\rho_i}{\rho_0} = \frac{P_i}{P_0} \tag{3.57}$$

Now we rewrite equation (3.56) as

$$\frac{\rho_0 u_0 u_i}{\rho_0 RT}\left(\frac{P_i - P_0}{P_0}\right) = \frac{\gamma u_0 u_i}{\gamma RT}\left(\frac{\Delta P}{P_0}\right) = \gamma M_0 M_i \frac{\Delta P}{P_0} \tag{3.58}$$

where γ is the specific heat ratio (C_P/C_V).

It may be noted that for the simplification of equation (3.58), we have used the speed of sound C_s as

$$C_s^2 = \gamma RT \tag{3.59}$$

By substitution of equation (3.59) in equation (3.55), we have

$$\frac{\Delta P}{P_0} = 2\frac{L}{h}\frac{\tau}{P_0} + \gamma M_0 M_i \frac{\Delta P}{P_0}$$

$$\Rightarrow \frac{\Delta P}{P_0}(1 - \gamma M_0 M_i) = 2\frac{L}{h}\frac{\tau}{P_0} \tag{3.60}$$

Let us approximate the shear stress as

$$\tau \sim \frac{\mu u}{h} \tag{3.61}$$

We can use the following expressions from the kinetic theory of gases:

$$\frac{\mu}{\rho} = \frac{\lambda}{2}\overline{\upsilon}_m \tag{3.62}$$

where $\overline{\upsilon}_m$ is the mean thermal speed, λ is the mean free path, and μ is the dynamic viscosity.

$$\overline{\upsilon}_m = \sqrt{\frac{8}{\pi\gamma}}C_s \tag{3.63}$$

$$P_0 = \rho_0 RT \tag{3.64}$$

We have

$$\frac{\mu}{P_0} \sim \lambda / C_s \tag{3.65}$$

Substituting in equation (3.60), we get

$$\frac{\Delta P}{P_0}(1 - \gamma M_0 M_i) \simeq 2\frac{L}{h}M_0 Kn_0 \tag{3.66}$$

Note: The typical aspect ratio for microsystem is $L/h \sim 10^3 - 10^4$. The Knudsen number range for slip flow regime is $10^{-3} < Kn < 10^{-1}$. The change in pressure in the microchannel is significant, and there will be significant change in density in microscale flow. Thus, the equation (3.66) indicates the relative importance of *compressibility effects* in the slip flow regime.

3.10 Slip Flow between Two Parallel Plates

In the previous chapter, we have discussed about the simplification of N–S equation on the basis of nondimensionalization of the governing equation. Let us relook at the simplification of N–S equation on the basis of the order of magnitude approach. For the momentum equation of the flow between two parallel plates, it may be approximated as

$$\frac{\text{Inertia term}}{\text{Diffusion term}} = \frac{\rho u \frac{\partial u}{\partial x}}{\mu \frac{\partial^2 u}{\partial y^2}} \sim \frac{\rho u^2/L}{\mu u/h^2} = \frac{\rho u h}{\mu}\left(\frac{h}{L}\right) = Re\left(\frac{h}{L}\right) \tag{3.67}$$

In microchannel flow, Re is small and $h/l \approx 10^{-3}$ to 10^{-4}. Thus, the inertial effects are small in microflows.

We can also use the estimate shown in equation (3.66) considering the compressibility effect for simplification as follows.

$$\begin{aligned}\frac{\text{Inertia term}}{\text{Diffusion term}} &= \frac{\frac{\Delta P}{P_0}(\gamma M_0 M_i)}{\frac{L}{h}M_0 Kn_0} \simeq \frac{M_i}{Kn_0}\cdot\frac{h}{L}\cdot\frac{\Delta P}{P_0} \\ &\simeq \frac{M_i}{Kn_i}\cdot\frac{h}{L}\cdot\frac{\Delta P}{P_i} \\ &\text{(Using } P_0\cdot Kn_0 = P_i\cdot Kn_i \text{ from} \\ &\text{definition of } \lambda)\end{aligned} \tag{3.68}$$

Using the kinetic theory of gas approximation for μ and v_m (equations (3.62) and (3.63)), we have

$$\frac{\mu}{\rho} \sim \lambda C_s \Rightarrow \lambda \sim \frac{\mu}{\rho C_s} \tag{3.69}$$

Hence,

$$Kn = \frac{\lambda}{h} \sim \frac{\mu}{\rho C_s h} = \frac{\mu U_\infty}{\rho V_\infty h Cs} = \frac{M}{Re} \tag{3.70}$$

Thus, substituting Kn expression into equation (3.68), we have

$$\frac{\text{Inertia term}}{\text{Diffusion term}} \simeq \frac{h}{L}\cdot Re\cdot\frac{\Delta P}{P_i} \tag{3.71}$$

The above-mentioned two estimates, that is, equations (3.67) and (3.71), are similar except the presence of $\frac{\Delta P}{P_i} = \frac{P_i - P_0}{P_i}$, which is usually smaller than unity.

Therefore, for low Re flows ($Re \leq 0(1)$) in large aspect ratio channels ($L/h \gg 1$), the inertial effects in the momentum equation can be neglected.

Hence, under such condition, the x-component of momentum equation reduces to

$$\frac{dp}{dx} = \mu \frac{d^2u}{dy^2} \tag{3.72}$$

$$\frac{d^2u}{dy^2} = \frac{1}{\mu}\frac{dp}{dx} \tag{3.73}$$

Integrating this equation, we have

$$\frac{du}{dy} = \frac{1}{\mu}\frac{dp}{dx}y + C_1 \tag{3.74}$$

$$u = \frac{1}{\mu}\frac{dp}{dx}\frac{y^2}{2} + C_1 y + C_2 \tag{3.75}$$

With reference to Figure 3.15, the boundary conditions can be written as

$$B.C.1 : \text{at } y = h/2, \frac{du}{dy} = 0$$

$$\therefore \quad C_1 = -\frac{1}{\mu}\frac{dp}{dx}\frac{h}{2}$$

$$B.C.2 : \text{at } y = 0, u = \frac{2-\sigma_v}{\sigma_v}\left[\lambda\left(\frac{\partial u}{\partial y}\right)_{y=0} + \frac{\lambda^2}{2}\left(\frac{\partial^2 u}{\partial y^2}\right)_{y=0}\right] \text{ (Using } u_w = 0 \text{ in equation (3.12))}$$

Using B.C.2, we get

$$C_2 = \frac{2-\sigma_v}{\sigma_v}\left[\lambda\left(\frac{\partial u}{\partial y}\right)_{y=0} + \frac{\lambda^2}{2}\left(\frac{\partial^2 u}{\partial y^2}\right)_{y=0}\right] \tag{3.76}$$

Substituting C_1 and C_2 in equation (3.75) and using equations (3.73) and (3.74), the velocity profile can be written as

$$u = \frac{1}{\mu}\frac{dp}{dx}\frac{y^2}{2} - \frac{1}{\mu}\frac{dp}{dx}\frac{h}{2}y + \frac{2-\sigma_v}{\sigma_v}\left[-\lambda\frac{1}{\mu}\frac{dp}{dx}\frac{h}{2} + \frac{\lambda^2}{2}\frac{1}{\mu}\frac{dp}{dx}\right]$$

$$= \frac{h^2}{2\mu}\frac{dp}{dx}\left[\frac{y^2}{h^2} - \frac{y}{h} + \frac{2-\sigma_v}{\sigma_v}\left(\frac{\lambda^2}{h^2} - \frac{\lambda}{h}\right)\right]$$

$$u = \frac{h^2}{2\mu}\frac{dp}{dx}\left[\frac{y^2}{h^2} - \frac{y}{h} + \frac{2-\sigma_v}{\sigma_v}(Kn^2 - Kn)\right] \tag{3.77}$$

Figure 3.16(a,b) shows the velocity profile as a function of momentum accommodation coefficient and Knudsen number for the flow between two parallel plates. Both increase in Knudsen number for constant accommodation coefficient and increase in accommodation coefficient

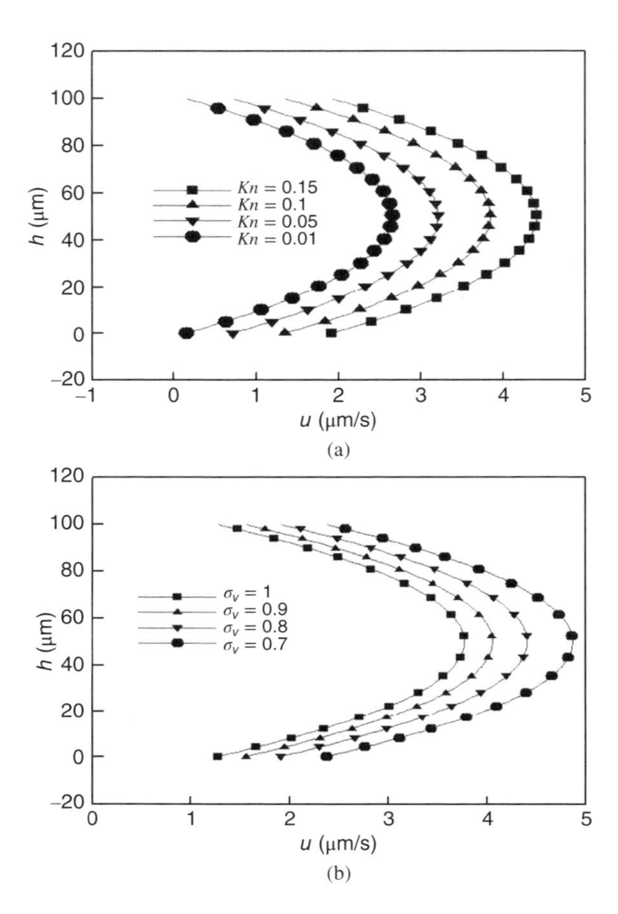

Figure 3.16 Poiseuille flow velocity profile inside two parallel plates of length $L = 0.05$ m, height $h = 100$ μm, $\mu = 10^{-3}$ N–S/m^2 by imposed pressure gradient $\Delta P = 0.01$ N/m^2 for (a) different Knudsen numbers with fixed $\sigma_v = 0.8$ and (b) different tangential accommodation coefficients σ_v with fixed $Kn = 0.15$

for constant Knudsen number lead to increase in the slip velocity. There is also simultaneous increase in maximum velocity and average velocity with an increase in Knudsen number and accommodation coefficient. Overall, Figure 3.16(a,b) shows an increase in the flow rate for the same pressure gradient because of the presence of slip at the wall. We can calculate the mass flow rate with the slip flow boundary condition as

$$\dot{m}_{\text{slip}} = \rho \int_0^h u(y)dy$$

$$\dot{m}_{\text{slip}} = \frac{h^3 P_0^2}{24\mu RTL}\left[(\pi^2 - 1) + 12\frac{2-\sigma_v}{\sigma_v}(Kn_0(\pi - 1) - Kn_0^2 \ln \pi)\right] \tag{3.78}$$

where $\pi = \frac{P_i}{P_0}$ and we have assumed the gas law with $\rho = \frac{P}{RT}$ and $P \cdot Kn = Kn_0 \cdot P_0$

Hence, the corresponding flow rate without slip effects for $Kn_0 \simeq 0$ is given by

$$\dot{m}_{\text{no-slip}} = \frac{h^3 P_0^2}{24\mu RTL}(\pi^2 - 1) \tag{3.79}$$

The ratio of mass flow rate with slip effect and without slip effect can be written as

$$\frac{\dot{m}_{\text{slip}}}{\dot{m}_{\text{no-slip}}} = 1 + 12\frac{2-\sigma_v}{\sigma_v}\frac{Kn_0}{\pi + 1} - 12\frac{2-\sigma_v}{\sigma_v}Kn_0^2\frac{\ln\ \pi}{\pi^2 - 1} \tag{3.80}$$

Note: (1) Figure 3.17 compares the ratio of mass flow rate with slip boundary condition with that of no-slip boundary condition as a function of the pressure ratio and Knudsen number. At $Kn \to 0$, the ratio is equal to 1.0, indicating the validity of the no-slip boundary condition. The mass flow rate ratio increases with an increase in the Knudsen number value.

(2) The slip flow boundary condition trend follows the experimental data. Equation (3.77) shows that the contribution of second-order term in the slip flow boundary condition is contrary to the first-order term. The effect of second-order slip correction is to reduce the increase in mass flow rate due to first-order slip leading to a closer comparison with the experiment.

(3) The mass flow rate ratio also depends on the magnitude of pressure gradient. The increase in mass flow rate due to slip reduces with increase in the pressure gradient value.

(4) Figure 3.18 shows the slip velocity variation along the channel surface for different channel dimensions. It shows the increase in slip velocity toward the downstream direction. It also increases with the reduction in the transverse dimension. Owing to the *pressure drop along* the channel, the *local mean free path increases* resulting in an *increase in the local Kn*. Also the *density* along the channel *decreases*, and thus the average velocity in the channel increases in downstream direction. These two effects are responsible for an increase in the slip velocity along the channel walls. In addition to that, the increase in Knudsen number may contribute toward the increase in the influence of compressibility effect.

3.11 Fluid Flow Modeling

Depending on the Knudsen number value, different flow regime classifications for gas flow have been proposed as follows:

$Kn \to 0$	:	Euler equation (negligible molecular diffusion)
$Kn \leq 10^{-3}$	:	N–S equation with no-slip boundary condition
$10^{-3} \leq Kn \leq 10^{-1}$	:	N–S equation with slip flow boundary condition
$10^{-1} < Kn \leq 10$	:	Transition regime
$Kn > 10$	:	Free molecular flow

Figure 3.19 shows the fluid flow regimes prevalent in different MEMS devices. It clearly demonstrates the limitation of the N–S equation for microscale flow modeling. It can be seen that most of the microsystems with gaseous flow works in the slip flow regime. Some flows inside microchannel, micropump, micronozzle, and microvalve operate in transitional regime. Free molecular flow is observed for Couette flow between hard disk and read-write head with a gap of about 100 nm.

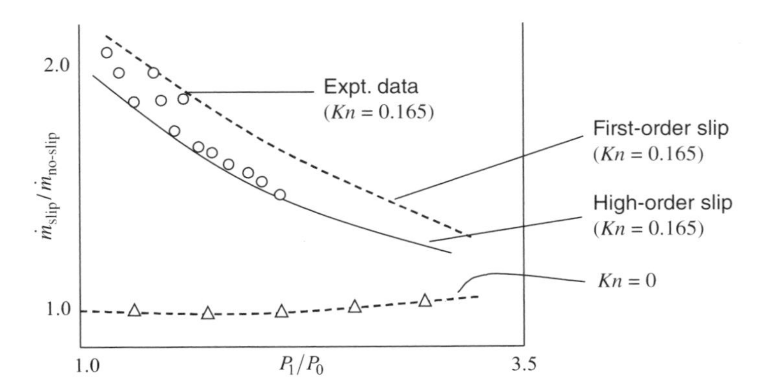

Figure 3.17 Variation of mass flow rate normalized with no-slip flow rate as a function of pressure ratio

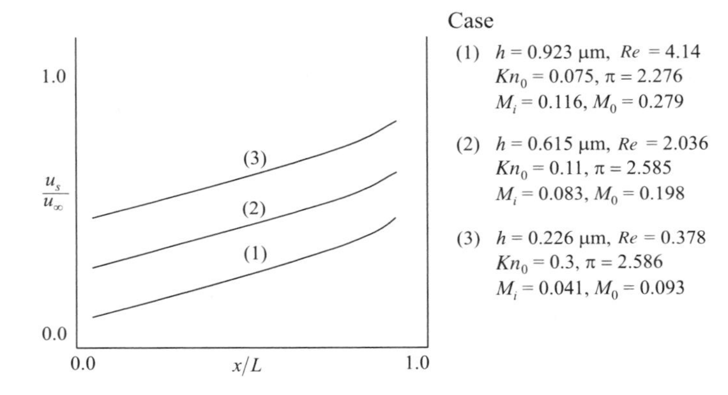

Figure 3.18 Velocity slip variation on channel surface for different transverse dimensions (h) of the channel

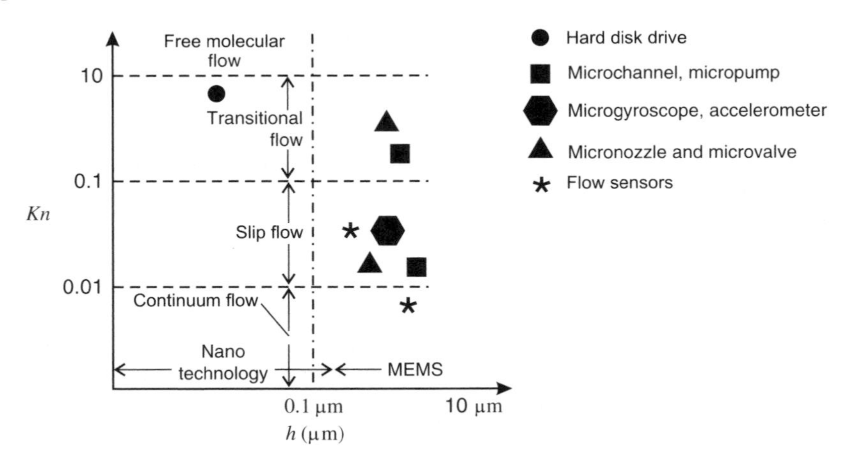

Figure 3.19 Operational range of various MEMS devices as a function of Knudsen number (Kn) and length scale (h)

Flow inside a microchannel may operate in multiple flow regimes. Let us consider a long microchannel with the entrance pressure to be atmospheric and exit conditions to be near vacuum. As air flows down the channel, pressure drops to overcome viscous forces in the channel. For prevailing isothermal conditions, density also drops. The conservation of mass requires the flow to accelerate down the constant area tube. The fluid *acceleration affects* the pressure gradient, resulting in *nonlinear pressure drop* in the channel. The Mach number increases down the tube, limited only by the choked flow condition. Mean free path increases with the corresponding increase in Knudsen number. The simple duct flow manifests all possible complexity in flows. Note that similar situation may take place when the entrance pressure is high, that is, at 7 atm and exit pressure is atmospheric.

Let us consider a microdevice of characteristic length equal to 1 μm at standard temperature $T = 288$ K. The mean free path can be calculated using $\lambda = \frac{KT}{\sqrt{2}\pi P\sigma^2}$, where K is the Boltzmann constant, σ is the molecular dia, T is the temperature, and P is the pressure. We get $\lambda = 0.065$ μm at $P = 1.01 \times 10^5$ N/m^2, which gives $Kn = 0.065$ and the flow is in slip flow regime. Note that for flow of light gas like helium, the Kn value is about three times larger than that of airflow. The modeling of fluid flow is carried out using either molecular modeling approach or continuum approximation.

Figure 3.20 presents different classifications of fluid flow modeling. We will briefly discuss these modeling schemes in the following section. Figure 3.21 shows the validity range of different gas modeling approaches. Here, ρ_0 and n_0 are the gas density and the number density, respectively, at standard conditions. These findings are based on rarefied gas flows, which are extended to microgas flows. In the dilute gas region, the Boltzmann equation is valid. The continuum approach requires that the sampling volume be in thermodynamic equilibrium. The frequency of intermolecular collisions inside the sampling volume must be high enough for the existence of thermodynamic equilibrium. This implies that the mean free path must be small in comparison to the characteristic length of the sampling volume. As a consequence, the

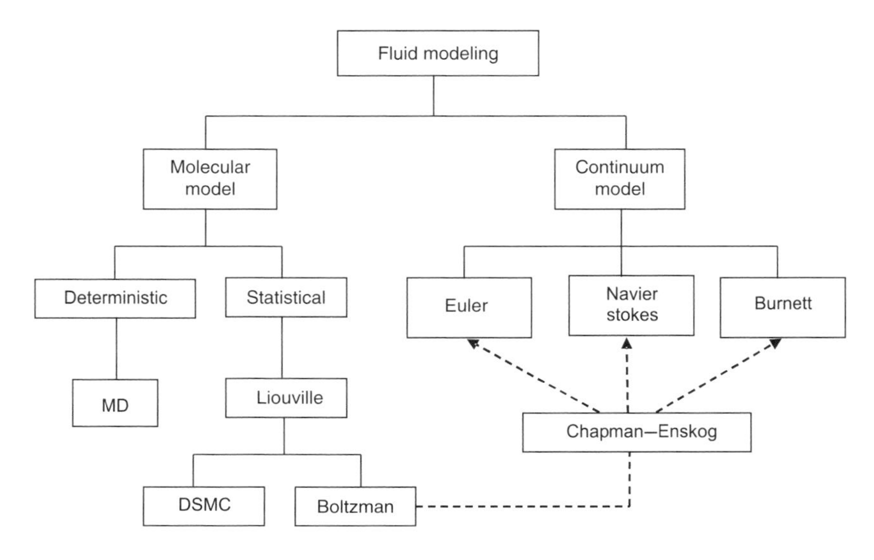

Figure 3.20 Different schemes for modeling of fluid flow

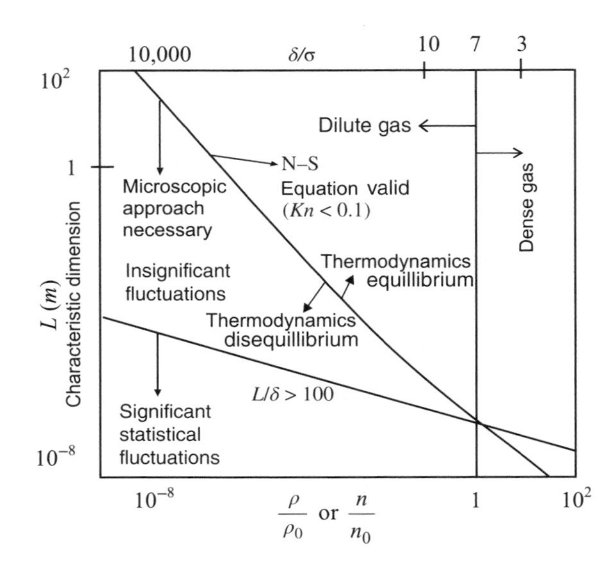

Figure 3.21 Validity range of different gas flow modeling approaches. Here, ρ_0 and n_0 are the mass density and the number density at standard condition of 1-atm pressure, δ is the mean molecular spacing, and σ is the mean molecular diameter

thermodynamic equilibrium requires $\frac{\lambda}{L} \ll 1$, that is, $Kn \ll 1$. The continuum approach hold until it breaks down at $Kn = 0.1$. For gas with $Kn > 0.1$, a microscopic approach is necessary, which recognizes the molecular nature of gases.

Note: (1) The N–S equations break down before the level of statistical fluctuations becomes significant.

(2) In a dense gas, significant fluctuations may be present even when the N–S equation is still valid.

The following sections discuss different molecular modeling approaches considering the microscale flow applications.

3.11.1 Continuum-Based Model

For mathematical modeling using continuum concept, it has been assumed that the flow quantities such as pressure, velocity, and fluid properties, that is, density, vary continuously from one point to another. However, continuum assumption may be questionable due to particulate nature of matter. Let us define density variation as the function of control volume size, $\Delta\forall$. Density at a point is defined as

$$\rho = \lim_{\Delta\forall \to 0} \frac{\delta m}{\delta\forall}$$

where δm is the mass of the element constructed around the point of interest.

Figure 3.22 shows the variation of ρ as a function of $\Delta\forall$. If $\Delta\forall$ is very large, ρ is affected by the inhomogeneities in the fluid arising from varying composition and temperature. If $\Delta\forall$ is very small, random fluctuations in position of atoms/molecules change their number inside

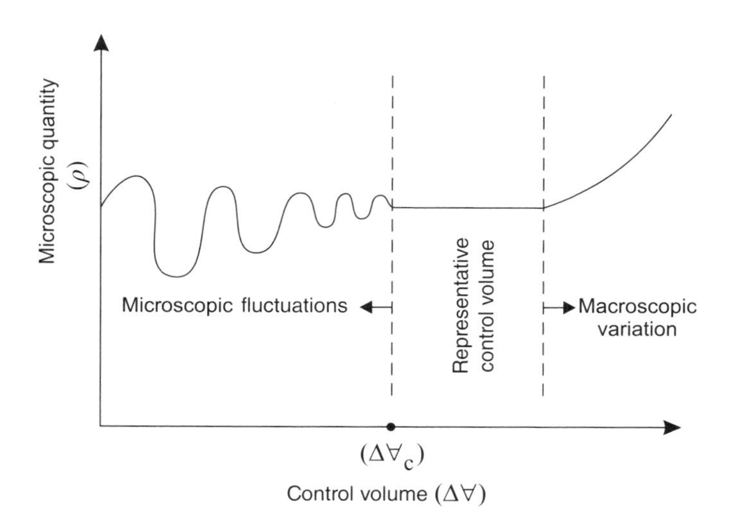

Figure 3.22 Schematic showing the representative control volume for the validity of continuum assumption

the control volume from one instant to another. Thus, density value fluctuates. The continuum limit $\Delta\forall_c$ is defined as the smallest magnitude of $\Delta\forall$ below which the statistical fluctuation becomes significant. Continuum-based fluid modeling has been used in various macroscale flow applications. In the following sections, we discuss other flow modeling approaches for microscale flow applications.

3.11.2 Deterministic Molecular Models

Molecular dynamics is used in various applications: (a) determination of properties of liquids, (b) plasma physics, (c) defects in solids, (d) biomolecules, and (e) friction. Depending on the local flow conditions, the nature of the molecular models recognize the fluid as a myriad of discrete particles: molecules, atoms, ions, and electrons (see Figure 3.23). The goal

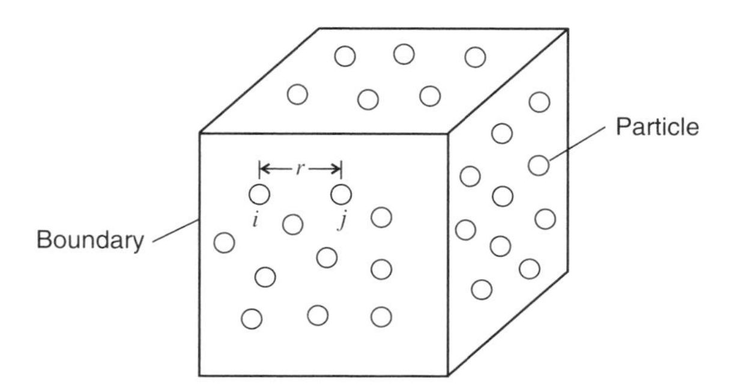

Figure 3.23 Schematic showing the domain for molecular dynamics simulation

is to determine the position, velocity, and state of all particles at all times. The macroscopic properties at any location in the flow can be computed from the discrete-particle information by a suitable averaging or weighted averaging process. The most fundamental of the molecular models is the deterministic one.

There are three main steps in a typical deterministic MD limitation, which are discussed in the following sections.

3.11.2.1 Setup Initial Conditions and Geometry

In this step, a *set of molecules* (N) are introduced in a two- or three-dimensional *regular lattice*, and each molecule is assigned a *random velocity*. To achieve faster equilibration, atoms can be assigned velocities with the equilibrium velocity, that is, *Maxwell distribution* at a *specified temperature*.

$$f(c) = \left(\frac{m}{2\pi K_B T}\right)^{3/2} \exp\left[\frac{-m}{2K_B T}c^2\right] \tag{3.81}$$

Here, $f(c)$ is the probability of a molecule having speed in the range of c and $c \pm dc$, m is the mass of the molecule, K_B is the Boltzmann's constant, and T is the temperature.

3.11.2.2 Specification of Intermolecular Potential

In this step, a *potential energy function* is specified in order to describe the interaction between molecules (particles). A commonly used model potential is the Lennard-Jones potential of the form (L-J 6–12) (Figure 3.24) given as

$$V_{ij}(r) = 4\varepsilon\left[c_{ij}\left(\frac{r}{\sigma}\right)^{-12} - d_{ij}\left(\frac{r}{\sigma}\right)^{-6}\right] \tag{3.82}$$

Here, V_{ij} is the potential energy between two particles i and j, r is the distance between the two molecules, ε is the energy scale, σ is the characteristic length scale, c_{ij} and d_{ij} are parameters dependent on the particular fluid–solid combination. The first term in the right-hand side of this equation represents strong repulsive force when two molecules are at extremely close range compared to the molecular length scale. This short-range repulsive force prevents overlap of the molecules in physical space. The second part in right-hand side represents the weak van der Waals attractive force, which commences when molecules are sufficiently close, that is, several times σ. This negative part of the potential represents the attractive polarization interaction of neutral, spherical symmetric particles. The power of σ in this equation is derivable from the consideration of quantum mechanics. The power for the repulsive part is found empirically. The L-J potential is zero at very large distance, it has a weak negative peak at r slightly larger than σ, it is equal to zero at $r = \sigma$ and is infinite as $r \to 0$.

For ease in computational effort, the L-J potential equation is also used in a modified form as

$$V_{ij}(r) = 4\varepsilon\left[c_{ij}\left(\frac{r}{\sigma}\right)^{-12} - d_{ij}\left(\frac{r}{\sigma}\right)^{-6} - c_{ij}\left(\frac{r_c}{\sigma}\right)^{-12} + d_{ij}\left(\frac{r_c}{\sigma}\right)^{-6}\right] \tag{3.83}$$

Here, r_c *is the cutoff* radius with typical values in the range 2.2σ–2.5σ. In this equation, the first term (with c_{ij}) represents a *short-range repulsive force* and the second term (d_{ij}) represents

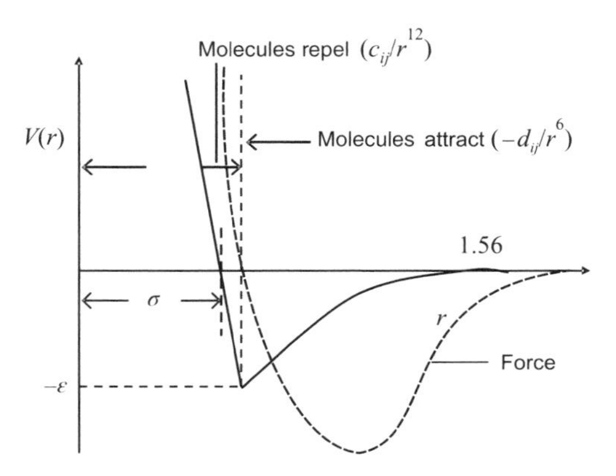

Figure 3.24 Potential energy (V) with respect to the intermolecular distance r based on the Lennard-Jones potential

an *attractive polarization interaction*. The rest of two terms in this equation guarantee that the *potential* is zero outside the cutoff radius r_c. Thus, there is no need to include particle–particle interaction for $r > r_c$. This reduces the computational effort during simulation.

The dynamics of the *wall boundary* can be described using the same potential with different values of the energy, length, and cutoff radius. In flows with active boundaries, for example, microchannels with elastomeric walls, appropriate refinements should be employed.

The force field resulting from the L-J 6–12 potential (equation (3.83)) can be calculated as

$$F_{ij}(r) = -\frac{\partial V_{ij}}{\partial r} = \frac{48\varepsilon}{\sigma}\left[c_{ij}\left(\frac{r}{\sigma}\right)^{-13} - \frac{d_{ij}}{2}\left(\frac{r}{\sigma}\right)^{-7} - c_{ij}\left(\frac{r_c}{\sigma}\right)^{-13} + \frac{d_{ij}}{2}\left(\frac{r_c}{\sigma}\right)^{-7}\right] \tag{3.84}$$

There are other commonly employed intermolecular potentials such as square-well, Buckingham, and Coulomb similar to that of the L-J potential.

3.11.2.3 Integration of Newton's Equation for Motion

The third step in the MD method involves the time integration of the equations of motion

$$m\frac{d^2 r_i}{dt^2} = -\sum_{j\neq i}\frac{\partial V_{ij}}{\partial r_i} \tag{3.85}$$

The location of the particle (r_i) is determined using

$$\vec{r}_i(t+\Delta t) = \vec{r}_i(t) + \vec{v}_i(t)\Delta t + \frac{1}{2}\vec{a}(t)\Delta t^2 \tag{3.86}$$

Here, $\vec{a} = \vec{F}/m$ (force/mass).

The Eulerian velocity is computed as a time average of N_i molecules contained in the bin (i) as follows

$$v(x) = \frac{1}{N_i}\left\langle \sum_j \frac{dx_j}{dt}\right\rangle \tag{3.87}$$

Similar expressions are used for density, viscosity, diffusion coefficient, and stress tensor.

An approximate formula is available for stress tensor computation by Koplik and Banavar (1995):

$$\tau(x) = \frac{1}{V_i}\left\langle \sum_j m\left[\frac{dx_j}{dt} - \upsilon(x)\right]^2 + \sum_{j<i} r_{ij}f_{ij} \right\rangle \tag{3.88}$$

where f is the instantaneous intermolecular force. The temperature is calculated using

$$T = \frac{1}{3N_i K_B}\left\langle \sum_j^{N_i} mV_j^2 \right\rangle \tag{3.89}$$

The selection of the bin and the corresponding spatiotemporal averaging is very important for the accuracy and efficiency of the MD simulation.

3.11.2.4 MD-Continuum Coupling

MD simulations can only be employed for short time and very small length scales because of their large computational requirements. The coupling of MD to N–S equation would extend the range of applicability of both approaches. The molecular modeling is computationally intensive. There are different flow regimes in a microchannel flow. Therefore, hybrid scheme is used for the modeling of microflow.

In the two-domain coupling, we have a region in which MD simulation is performed, the region where the incompressible N–S equation is solved, and overlap region where both descriptions are valid (Figure 3.25).

In order to terminate the MD region, constraints should be imposed of the form

$$\sum_{n=1}^{N_i} P_n - M_i \upsilon_i = 0 \tag{3.90}$$

N_i : Total no of particles in the i-bin.
P_n : Momentum of nth particle in the υ-direction.
M_i : Mass of the ith bin.
υ_i : Velocity of the ith bin.

MD simulation are highly *inefficient for dilute* gases where the molecular interactions are infrequent. The simulations are *more suited for dense gases and liquids*. They are reserved for

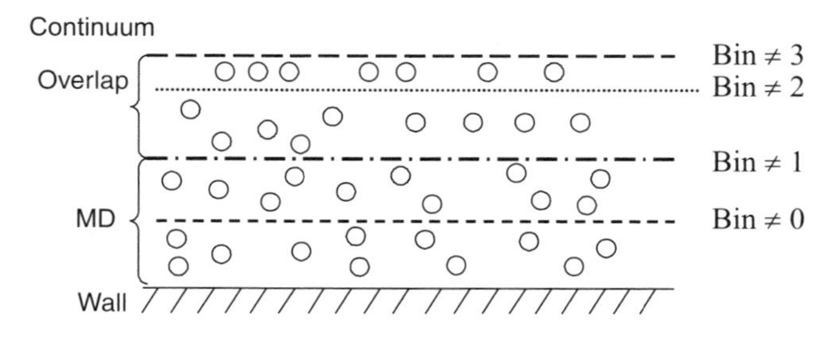

Figure 3.25 Domain for MD-continuum coupling

situations where the continuum approach or statistical methods are inadequate. Figure 3.25 shows the scheme for coupling between molecular modeling and continuum modeling for reduction in the computational time. Here, the molecular modeling is carried out in the near-wall region, and continuum modeling is carried out at away from the wall region. There is an overlap region where both continuum and molecular modelings are carried out. The results of the molecular model are used as a boundary condition for the continuum model. More details about continuum and molecular hybrid methods can be found in Nie *et al.* (2004) and Li *et al.* (2010).

3.11.2.5 Computational Complexity

The limitation for MD simulation is the number of molecules N that can realistically be modeled in a computer. Since the computation for any particular molecule requires the consideration of all other molecules as potential collision partners, the amount of computation required is proportional to N^2. Some saving in computer time is achieved by cutting off the weak tail of the potential at say $r_c = 2.5\sigma$ and shifting the potential by a linear term in r so that force goes smoothly to zero at the cut off. As a result, only nearby molecules are treated as potential collision partners and the computation time for N molecules no longer scale with N^2.

MD simulation can also be performed by a *linked-cell* list that follows a data structure similar to that of the domain decomposition techniques. In particular, the domain is subdivided into smaller subdomain (cells) so that molecules within a subdomain interact only with molecules in the same subdomain and its nearest neighbors. More details on MD simulation can be found in Allen and Tildesley (1987) and Rapaport (2004).

3.11.3 Statistical Molecular Model

The aim of statistical approach is to explain the macroscopic behavior of matter in terms of the behavior of the constituent molecules, that is, in terms of motions and interactions of a large number of particles.

The motion of molecules can be expressed using Newton's second law as

$$\ddot{x}_i = F_i \tag{3.91}$$

x_i : position vector of the ith particle ($i = 1, 2, \ldots, N$).
F_i : force on the ith particle divided by the mass of the particle.

We can also define

$$\zeta_i = x_i \tag{3.92}$$

where ζ_i is the velocity vector.

To obtain the actual dynamics of the problem, we have to solve $6N$ number of unknowns (x_i, ζ_i) using first-order differential equations and $6N$ initial conditions, that is,

$$x_i(0) = x_i^0$$
$$\dot{x}_i(0) = \zeta_i(0) = \zeta_i^0$$

The one difficulty is to supply the *initial data* x_i^0 and ζ_i^0. The other issue is that the detailed information may be unnecessary as we are interested in the *average information*, that is, the pressure exerted on a wall by a gas at a given density and temperature.

In statistical approach, instead of having a *definite position and velocity* of a particle, we compute the *probability of finding* a molecule at a particular *position* and *state*. Once we solve the conservation equation for *probability distribution*, we can compute important statistical averaged quantities, that is, *momentum* and energy of the molecules.

Probability density $P(x)$ is defined such that $P(x)dx^n$ is the probability that x lies between x and $x + dx^n$, where dx^n denotes the volume of an infinitesimal cell equal to the product $dx \cdots dx_n$.

The probability density $P(x)$ is used to compute average. For example, the average of quantity $\phi(x)$ can be written as

$$\langle \phi(x) \rangle = \overline{\phi(x)} = \int_x P(x)\phi(x) d^n x \tag{3.93}$$

3.11.3.1 Joint Probability Density ($P(x_K, \zeta_K, t)$)

The probability that the Kth molecule has position x_K and velocity ζ_K, that is, it is located between x and $x_K + dx_K$ with velocity between ζ_K and $\zeta_K + d\zeta_K$ is known as joint probability density.

3.11.4 Liouville Equation

The starting point of the statistical mechanics is the *Liouville* equation, named after French mathematician Joseph Liouville. It describes the time evolution of the phase space function as

$$\frac{\partial P}{\partial t} + \sum_{j=1}^{N} \zeta_j \cdot \frac{\partial P}{\partial x_j} + \sum_{j=1}^{N} F_j \cdot \frac{\partial P}{\partial \zeta_j} = 0 \tag{3.94}$$

Here, $P(x, \zeta, t)$ is the probability of finding a molecule at the point x at time t with speed ζ, F is the external force, and N is the number of molecules. It is understandable that such an equation is not tractable for realistic number of particles. Simpler distribution function is obtained by the integration of the Liouville equation.

By integrating the above-mentioned equation, we get the Boltzmann equation, that is, the fundamental relation of kinetic theory of gases: *monoatomic gas* molecules are assumed. For monoatomic gas, there is no internal degree of freedom (rotation), that is, the state of each molecule is completely described by three space coordinates and three velocity coordinates. The only mode of a monoatomic gas is translation, while for diatomic gases rotation also contributes. The fluid is also restricted to *dilute gases* and *molecular chaos* is assumed.

Dilute gas approximation requires the average distance between the molecules to be an order of magnitude larger than their diameter, σ, that is, $\frac{\delta}{\sigma} \gg 1$. This almost guarantees that all collisions between molecules are *binary collisions*, avoiding the complexity of modeling multiple encounters. Dissociation and ionization phenomena involve triple collisions.

Molecular chaos restriction improves the accuracy of computing the macroscopic quantities from the microscopic information. The volume over which averages are computed has to have a sufficient number of molecules to reduce statistical errors. Molecular chaos limit requires the *length scale-L for the averaging process to be at least 100 times the average distance* between the molecules (i.e., typical averaging over at least 1 million molecules). Note that the statistical fluctuation with standard deviation of 0.1% will be observed for 1 million molecules and 3% if 1000 molecules are used.

3.11.5 Boltzmann Equation

Boltzmann equation is obtained by integrating Liouville equation. Boltzmann equation is based on the following assumptions:

1. The density is low so that only binary collisions are considered.
2. Molecular chaos exist.
3. Collision is instantaneous.

Consider a gas in which an external force F acts and assume that no collisions take place between the gas molecules. In time dt, the velocity ζ of any molecule will change to $\zeta + Fdt$, and its position will change from x to $x + \zeta dt$. Thus, the number of molecules $P(x, \zeta, t)dx\,d\zeta$ is equal to the number of molecules $P(x + \zeta t, \zeta + Fdt, t + dt)dx\,d\zeta$, that is,

$$P(x + \zeta dt, \zeta + Fdt, t + dt)dx\,d\zeta - P(x, \zeta, t)dx\,d\zeta = 0 \tag{3.95}$$

If, however, collision do occur between the molecules, there will be a net difference between the number of molecules, which can be written as $J(P)dxd\zeta dt$, where $J(P)$ is the collision operator. Thus, the evolution of the distribution function can be written as

$$P(x + \zeta dt, \zeta + Fdt, t + dt)dx\,d\zeta - P(x, \zeta, t)dx\,d\zeta = J(P)dx\,d\zeta\,dt \tag{3.96}$$

Dividing the equation by $dxd\zeta dt$ and letting $dt \to 0$, we have

$$\frac{P(x + \zeta dt, \zeta + Fdt, t + dt) - P(x, \zeta, t)}{dt} = J(P) \tag{3.97}$$

Using $\lim_{dt \to 0}$, we have

$$\frac{\partial P}{\partial t} + \zeta\frac{\partial P}{\partial x} + F\frac{\partial P}{\partial \zeta} = J(P) \tag{3.98}$$

For monoatomic gas molecules undergoing binary collisions, the Boltzmann equation can be expressed as

$$\frac{\partial(nf)}{\partial t} + \zeta_j\frac{\partial(nf)}{\partial x_j} + F_j\frac{\partial(nf)}{\partial \zeta_j} = J(f, f^*) \tag{3.99}$$

Here, n is the number density and f is the normalized velocity distribution function, that is, $dn = nfd\zeta$.

The first term in the left-hand side of above-specified equation is the rate of change of molecules. The second term in the left-hand side is the convection of molecules across the surface by molecular velocity $\left(\zeta \cdot \frac{\partial(nf)}{\partial x}\right)$. The third term in the left-hand side is the convection of molecules across the surface as a result of the external force, F $\left(F \cdot \frac{\delta(nf)}{\delta\zeta}\right)$. The term in the right-hand side represents the collisions of two molecules, that is, the scattering of molecules into and out as a result of intermolecular collisions.

$$J(f,f^*) = \int_{-\infty}^{\infty}\int_{0}^{4\pi} n^2(f^*f_1^* - ff_1)\zeta_r\sigma d\Omega(d\zeta)_1 \tag{3.100}$$

where * indicates post-collision values; ζ_r, relative speed between two molecules; Ω, solid angle; and f and f_1 represent two different molecules.

Here, the integration is carried out over the three-dimensional velocity space R^3 and the hemisphere, which includes the particle moving away from each other after collision:

$$d\zeta = d\zeta_1\, d\zeta_2\, d\zeta_3$$

Once a solution for f is obtained, macroscopic quantities, that is, ρ, u, and T, can be computed using

$$\rho = mn = m\int (nf)d\zeta$$

$$u_i = \int \zeta_i f d\zeta$$

$$T = \frac{1}{3K_B}\int \zeta_i\zeta_i f d\zeta$$

For nondimensionalization, we can use

Characteristic length: L
Characteristic speed: $\left[2\left(\frac{K}{m}\right)T\right]^{1/2}$
K_B : Boltzmann constant
m : molecular mass
T : temperature
Characteristic number density: n_0.

The nondimensionalized Boltzmann equation is

$$\frac{\partial\hat{f}}{\partial\hat{t}} + \hat{\zeta}_j\frac{\partial\hat{f}}{\partial\hat{x}_j} + \hat{F}_j\frac{\partial\hat{f}}{\partial\hat{\zeta}_j} = \frac{1}{Kn}\hat{J}(\hat{f},\hat{f}^*) \quad j = 1, 2, 3 \tag{3.101}$$

where variables denoted with ^ are nondimensional.

The conservation of mass, momentum, and energy can be derived by *multiplying* the above-specified Boltzmann equation with molecular mass, momentum, and energy, respectively, and then integrating over all possible molecular velocities. Subject to the restriction of *dilute gas*, the Boltzmann equation is valid for all Kn from 0 to ∞.

As $Kn \to \infty$, molecular collisions become unimportant. This is the *free molecular* regime depicted as $Kn > 10$, where the only important collision is between gas molecules and the solid surface of an obstacle or conduit. Analytical solutions are then possible for simple geometries, and numerical solutions for *complicated geometries* are straightforward once the fluid of specific characteristic is modeled.

One example is when we neglect molecular diffusion, that is, $Re \to \infty$ corresponding to Euler equation where

$Kn \to 0$: (Hydrodynamic limit)

Here, the number of molecular collisions is so large that the flow reaches the equilibrium state in a time short compared to the macroscopic timescale.

The velocity distribution function is everywhere of the local equilibrium or Maxwellian form as

$$\hat{f}(0) = \frac{n}{n_0}\pi^{-3/2}\exp\left[-\left(\hat{\zeta} - \hat{u}\right)^2\right] \tag{3.102}$$

where $\hat{\zeta}$ is the dimensionless speed of the molecules; $\hat{u}$ is the dimensionless speed of the flow.

At this Kn number, the flow is isentropic and heat conduction, viscous diffusion, and dissipation vanish from the continuum conservation relation viewpoint.

Analytical solution of the Boltzmann equation is difficult mostly because of the *nonlinearity of the collision integral*. Simple model of this integral have been proposed to facilitate analytical solutions.

3.11.5.1 Solution of Boltzmann's Equation

One of the solution approaches of the Boltzmann equation is *Chapman–Enskog method*. Here, the velocity distribution function is expanded into a perturbation series with the Knudsen number being the small parameter.

$$\hat{f} = \hat{f}^0 + Kn\,\hat{f}^{(1)} + Kn^2\hat{f}^{(2)} + \cdots \tag{3.103}$$

The functions $\hat{f}^{(1)}$ and $\hat{f}^{(2)}$ depend on the collision function model using gas density and temperature, and should satisfy the moment equation. The above-mentioned equation is *substituted for the Boltzmann's equation*, and a set of inhomogeneous linear equations is obtained by *equating terms of equal order*. The use of distribution functions $\hat{f}^{(1)}, \hat{f}^{(2)}$, and so on leads to the determination of transport terms needed to close the continuum equations appropriate to the particular level of approximation. The *continuum stress tensor* and *heat flux* vector can be written in terms of the distribution function ($f^{(1)}$). This can be further simplified in terms of macroscopic velocity and temperature derivatives.

In addition, assuming a model for the *molecular interaction* we can obtain explicit expressions for the transport coefficients of the momentum and energy equation. For example, for the *hard sphere molecules model*, we have

$$\mu = \frac{5}{16}\frac{\sqrt{\pi m K_B T}}{\pi d^2} \tag{3.104}$$

where d is the molecular diameter. Other models such as variable hard sphere (VHS) and variable soft sphere (VFS) are also available in the literature.

For the first-order solution of the Boltzmann equation using Chapman–Enskog method, we get the constitutive laws of the N–S equation as

$$\sigma_{ij}^{\text{N–S}} = -\mu\left(\frac{\partial u_j}{\partial x_i} + \frac{\partial u_i}{\partial x_j}\right) + \mu\frac{2}{3}\frac{\partial u_m}{\partial x_m}\partial_{ij} - \zeta\frac{\partial u_m}{\partial x_m}\partial ij \tag{3.105}$$

where μ is the first coefficient of viscosity and ζ is the second coefficient of viscosity.

For the second-order solution or expansion, we get the Burnett level stress tensor as

$$\sigma_{ij}^{B} = -2\mu\overline{\frac{\partial u_i}{\partial x_j}} + \frac{\mu^2}{p}\left[w_1\frac{\partial u_K}{\partial x_K}\overline{\frac{\partial u_i}{\partial x_i}} + w_2\left(\frac{D}{Dt}\overline{\frac{\partial u_i}{\partial x_j}} - 2\overline{\overline{\frac{\partial u_i}{\partial x_K}}\frac{\partial u_K}{\partial x_j}}\right)\right.$$

$$\left. + w_3R\overline{\frac{\partial^2 T}{\partial x_i \partial x_j}} + w_4\frac{1}{\rho T}\overline{\frac{\partial p}{\partial x_i}} + w_5\frac{R}{T}\overline{\frac{\partial T}{\partial x_i}\frac{\partial T}{\partial x_j}} + w_6\overline{\overline{\frac{\partial u_i}{\partial x_K}}\overline{\frac{\partial u_K}{\partial x_j}}}\right] \tag{3.106}$$

where a bar over a tensor designates

$$\overline{f}_{ij} = (f_{ij} + f_{ij})/2 - \delta_{ij}/3f_{mm} \tag{3.107}$$

The coefficients w_i depend on the gas model and R is the specified gas constant. Since the Burnett equations are obtained by a second-order Chapman–Enskog expansion in Kn, they require *second-order slip boundary condition*. However, it may be noted that it has been observed that the second-order slip b.c. are inaccurate for $Kn > 0.2$. The Burnett equation can be used to obtain analytical/numerical solutions for at least a portion of the transition regime for a monoatomic gas.

3.11.5.2 Burnett Equation in Microchannel

We start with the basic conservation equation for mass, momentum, and energy as

$$\frac{\partial \rho}{\partial t} + \vec{\nabla}\cdot(\rho\vec{v}) = 0 \tag{3.108}$$

$$\frac{\partial}{\partial t}(\rho\vec{v}) + \vec{\nabla}\cdot[\rho\vec{v}\vec{v} - \overline{\overline{\sigma}}] = \vec{f} \tag{3.109}$$

$$\frac{\partial E}{\partial t} + \vec{\nabla}\cdot[E\vec{v} - \sigma\vec{v} + \vec{q}] = \vec{f}\cdot\vec{v} \tag{3.110}$$

where f is the external force; $E = \rho(e + \frac{1}{2}\vec{v}\cdot\vec{v})$ (total specific energy); e is the internal specific energy.

Note that using the constitutive equations for stress tension $\overline{\overline{\sigma}}$

$$\overline{\overline{\sigma}} = -P\overline{\overline{I}} + \overline{\overline{\tau}} \tag{3.111}$$

$$\overline{\overline{\tau}} = \mu[\vec{\nabla}\vec{v} + (\vec{\nabla}\vec{v}^T)] + \zeta(\vec{\nabla}\cdot\vec{v})\overline{\overline{I}} \tag{3.112}$$

where $\overline{\overline{I}}$ is the unit tensor, μ and ζ are the first and second coefficients of viscosity.

We can obtain

$$\rho\frac{D\vec{v}}{Dt} = -\vec{\nabla}P + \vec{\nabla}\cdot\overline{\overline{\tau}} + \vec{f} \tag{3.113}$$

$$\rho\frac{De}{Dt} = -P\vec{\nabla}\cdot\vec{v} - \vec{\nabla}\cdot\vec{q} + \phi \tag{3.114}$$

where $\phi = \overline{\overline{\tau}}\cdot\vec{\nabla}\vec{v}$ is the viscous dissipation function.

For *2D compressible* flow, the above-specified set of equations are written as

$$\frac{\partial}{\partial t}\begin{pmatrix}\rho\\ \rho u\\ \rho v\\ E\end{pmatrix}+\frac{\partial}{\partial x}\begin{pmatrix}\rho u\\ \rho u^2+P+\sigma_{11}\\ \rho uv+\sigma_{12}\\ (E+P+\sigma_{11})u+\sigma_{12}\cdot v+q_1\end{pmatrix}$$

$$+\frac{\partial}{\partial y}\begin{pmatrix}\rho v\\ \rho uv+\sigma_{21}\\ \rho v^2+P+\sigma_{22}\\ (E+P+\sigma_{22})\cdot v+\sigma_{21}\cdot u+q_2\end{pmatrix}=0 \tag{3.115}$$

For the microchannel of height h and length L, we have

$$\varepsilon = h/L \ll 1$$

Neglecting any *temperature gradients* in the gas and also neglecting any term greater than $0(\varepsilon)$, we can write the Burnett momentum equation as

$$P_x\left[1-\left(\frac{w_2}{3}+\frac{w_6}{12}\right)\frac{\gamma\pi}{2}\quad Kn_0^2M_0^2\left(\frac{P_0}{P}\right)^2(u_y)^2\right]=u_{yy}+\varepsilon(\cdots) \tag{3.116}$$

$$P_y\left[1+\left(\frac{w_6}{12}-\frac{2w_2}{3}\right)\frac{\gamma\pi}{2}\quad Kn_0^2M_0^2\left(\frac{P_0}{P}\right)^2(u_y)^2\right]$$
$$=\left(\frac{w_6}{6}-\frac{4w_2}{3}\right)\sqrt{\frac{\gamma\pi}{2}}M_0Kn_0\left(\frac{P_0}{P}\right)u_yu_{yy}+\varepsilon(\cdots) \tag{3.117}$$

where the nondimensionalization has been carried out with exit condition (P_0, u_0), and Kn_0 and M_0 are Knudsen number and Mach number at the outlet, respectively. Note that P_x and P_y indicate pressure gradient in the x- and y-direction, respectively.

For Maxwell's gas model:

$$(w_1, w_2, w_6) = (10/3, 2, 8) \quad \text{[Schamberg (1947)]}$$

We have the x- and y-component of momentum equation as

$$P_x\left[1-\frac{2}{3}\gamma\pi\, Kn_0^2M_0^2\left(\frac{P_0}{P}\right)^2u_y^2\right]=u_{yy}+0(\varepsilon) \tag{3.118}$$

$$P_y\left[1+\frac{\gamma\pi}{3}\,Kn_0^2M_0^2\left(\frac{P_0}{P}\right)^2u_y^2\right]=\frac{4}{3}\sqrt{\gamma\pi/2}M_0Kn_0\frac{P_0}{P}u_yu_{yy}+0(\varepsilon) \tag{3.119}$$

The term $Kn_0^2M_0^2(P_0/P)^2$ is relatively small for *low Mach number* flows in the early transition *regime* (i.e., $Kn < 1$). Hence, Burnett equation reduces to

$$P_x = u_{yy} \tag{3.120}$$

$$P_y=\frac{4}{3}\sqrt{\frac{\gamma\pi}{2}}Kn_0M_0\left(\frac{P_0}{P}\right)u_yu_{yy} \tag{3.121}$$

Hence, the *streamwise Burnett equation* is reduced to the form obtained in the N–S limit. The cross flow momentum equation shows that the pressure gradient in that direction is balanced by the Burnett normal stress, which in the case of *continuum is* identically *zero* on the flat surface.

The Burnett equation can be used to obtain numerical/analytical solution for at least a portion of *the transition regime* ($0.1 < Kn < 10$) for a *monatomic gas*. But their *complexities* have precluded much results for the realistic geometries.

In the transition regime, the molecularly based Boltzmann equation cannot easily be solved unless the *nonlinear collision integral is simplified.* MD solution as mentioned earlier is not suitable *for dilute gases.* The best approach for the transition regime is the direct simulation Monte Carlo (DSMC). This approach is introduced in a later section.

3.11.5.3 Approximate Forms of Boltzmann Equation

BGK model (Bhatnagar, Gross, and Krook) proposed an approximate formulation of the Boltzmann equation by simplifying the collision integral.

In this method, the collision integral is approximated as

$$J(f, f^*) = \gamma_*(f^{(0)} - f) \tag{3.122}$$

Hence, BGK model without external forcing is

$$\frac{\partial f}{\partial t} + \zeta_j \frac{\partial f}{\partial \hat{x}_j} = \gamma_*(f^{(0)} - f) \tag{3.123}$$

With nondimensionalization, Kn appears as denominator of right-hand side and hence at $Kn \to \infty$ it goes to zero.

Here, $\hat{f}^{(0)}$ is the Maxwell distribution and γ_* is the collision frequency, which is assumed to be independent of the molecular velocity v but is a function of spatial coordinates and time. Collision frequency is zero for free molecular flow and hence the right-hand side is zero.

$$\gamma_* = \frac{\overline{v}}{\lambda} = \frac{\text{Thermal velocity}}{\text{Mean free path}} = \frac{\sqrt{8K_BT/(\pi m)}}{\sqrt{2}/(2n\pi d^2)} \tag{3.124}$$

In general, numerical simulation using full Boltzmann equation suggests that the BGK model is accurate for *isothermal flows*. However, for *nonisothermal flows*, corrections for the Prandtl number (for collision frequency) need to be introduced.

3.11.5.4 Linearized Boltzmann Equation

In linearized Boltzmann approach, the distribution function is written as

$$f(\vec{x}, \vec{v}, t) = f_0(n_0, T_0)[1 + h(\vec{x}, \vec{v}, t)] \tag{3.125}$$

where f_0 is the absolute Maxwell's distribution corresponding to equilibrium state (n_0, T_0) and h is the perturbation distribution function.

The linearized Boltzmann equation is written as

$$\frac{\partial h}{\partial t} + \vec{v}.\frac{\partial h}{\partial t} - \tilde{Q}h = 0 \tag{3.126}$$

where $\tilde{Q}h$ is the linearized collision term (an integral involving $f_0 h$).

3.11.6 Direct Simulation Monte Carlo (DSMC) Method

It is not easy to use molecular based Boltzmann equation in the transition regime because of the complexity of the collision integral. MD simulation is also not suitable for dilute gases. DSMC is a statistical computational approach for solving rarefied gas dynamics problems unlike MD simulations. It treats gas as discrete particles subject to dilute gas and molecular chaos assumption. It is valid for all ranges of Knudsen number. It is quite expensive for $Kn < 0$, where luckily N–S equation is valid. DSMC is an ideal method for transition regime. In DSMC, a large number of random particles are tracked due to their interactions resulting in modification of states and positions. Here, the molecular motions and the intermolecular collisions are uncoupled over small time intervals. This leads to a significant improvement in computation time, which is now proportional to N instead of N^2 for MD simulation. The particle motions are modeled deterministically, and the collisions are treated probabilistically. Each simulated molecule represents a large number of actual molecules. For example, a volume of $10\ \mu m \times 10\ \mu m \times 10\ \mu m$ contains about 10^{10} molecules. In DSMC method, using a million of simulated molecules is sufficient, which improves the computational resource requirement significantly. However, it should be remembered that if the ratio of actual to simulated molecules is high, statistical scatter of the solution is increased.

The basic steps during DSMC simulation can be divided into four steps. The total space is divided into cells similar to the finite volume method used in computational fluid dynamics simulation. The size of the DSMC cell is chosen proportional to the mean free path, that is, $\Delta x_c \simeq \lambda/3$. The four steps in DSMC are

1. Particle motion
2. Particle indexing
3. Particle collision simulation
4. Calculation of macroscopic properties

In the first step, the motion of the simulated molecules is carried out for time Δt less than mean collision time, Δt_c. Value of time steps larger than Δt will result in traveling of molecules through several cells prior to cell-based collision calculation. During this stage, the molecules advance in space and may go through wall collisions or leave the computational domain through outflow boundaries. Therefore, the boundary conditions are enforced at this level. The wall surface–molecule interaction is modeled by applying conservation laws on individual molecules contrary to velocity distribution function used in Boltzmann algorithm. A prior knowledge of accommodation coefficients is required. Many complicated process, that is, chemical reactions, radiation effects, and ionized flow effects can be included.

In the second step, the tracking of particles with proper indexing is carried out. The new cell locations of the molecules are indexed properly. In the third step, the collision process is simulated probabilistically. The collision pairs are selected within the subcell. The collision of molecules near to each other is ascertained and thus accuracy of the simulation is maintained. In the final step, macroscopic flow properties are calculated at the cell centers. Note that macroscopic fluid velocity is obtained by averaging the molecular velocities for a long time. The 2–5 orders of magnitude difference between the molecular and average speeds results in large statistical noise. Thus, it requires very long time averaging for microflow simulation.

In DSMC, there is a long-time requirement for achieving steady state. Let us consider a gas flow in a channel of length 1 mm and height 1 μm with an average speed of 1 mm/s. The convection timescale for the macroscopic disturbance to travel from the inflow to outflow is equal to 1 s. The viscous timescale for the problem is $t_{\text{visc}} \simeq \sqrt{h^2/\gamma}$. For air ($\gamma = 10^{-6}\text{m}^2/s$), $t_{\text{visc}} \simeq 10^{-3}$ s, which is three orders of magnitude less than convective timescale. The mean collision time of air at standard condition is the order of the 10^{-10} s. Hence, the DSMC step must run about 10^{10} time steps for settling to a steady-state condition.

The number of simulated molecules is also quite high for DSMC calculation. Let us look at a physical domain of size 1 mm × 1 μm × 1 μm at $Kn = 0.1$. One cell size should be about $\lambda/3$. Here, $Kn = \frac{\lambda}{L} = 0.1, \therefore L = 10\lambda$. Hence, number of cell/micron $= \frac{10\lambda}{\lambda/3} = 30$. We need about 20 simulated molecules per cell. Hence, we need to simulate about 5.4×10^{10} molecules. Now we can see the number of time steps combined with a large number of molecules demands high computational requirement.

Problems

3.1 Water ($\mu = 1 \times 10^{-3}$ kg/m-s) flows in a capillary of 8 μm diameter and 5 cm long under pressure difference, $\Delta P = 0.005$ Pa. The capillary is hydrophobic having slip length equal to 100 nm. Calculate the flow rate through the capillary in zeptoliters per second. *Note*: 1 l = 0.001 m³.

3.2 The implementation of second-order slip boundary condition requires obtaining the second derivative of the tangential velocity in the normal direction to the surface ($\partial^2 u/\partial n^2$), which may lead to computational difficulties. To circumvent this problem, the following general velocity slip boundary condition in nondimensional form is used.

$$u_s - u_w = \frac{2-\sigma_v}{\sigma_v}\frac{Kn}{1-b\times Kn}\left(\frac{\partial u}{\partial n}\right)_s$$

where b is a general slip coefficient. The value of b can be determined so that for $|bKn| < 1$, the above boundary condition matches exactly the slip boundary condition given by

$$u_s - u_w = \frac{2-\sigma_v}{\sigma_v}\left[Kn\left(\frac{\partial u}{\partial n}\right)_s + \frac{Kn^2}{z}\left(\frac{\partial^2 u}{\partial n^2}\right)_s + \cdots\right]$$

For a two-dimensional incompressible isothermal flow between parallel plates separated by the distance h in the slip flow regime, the velocity distribution is:

$$u(y) = \frac{h^2}{2\mu}\frac{dp}{dx}\left[\frac{y^2}{h^2} - \frac{y}{h} - \frac{2-\sigma_v}{\sigma_v}\frac{Kn}{1-b\times Kn}\right]$$

Find the value of b.

3.3 (a) Consider the micro-Couette flow (fluid density, ρ and viscosity, μ) between two plates with top plate moving at velocity U_0 and bottom plate stationary. The separation distance between the two plates is equal to h. Assume the slip flow boundary condition to be valid at the wall. Derive the expression for skin friction coefficient C_f as a function of Reynolds number ($Re = \rho U_0 h/\mu$), momentum accommodation coefficient σ_v, and Knudsen number (Kn).

(b) Based on the results derived earlier, comment on the performance of microbearing in comparison to the conventional lubrication.

3.4 (a) Derive equation (3.16), that is, the relationship between slip flow rate (Q_{slip}) and no-slip flow rate (Q_{th}) for slip length (L_S) of a pipe flow with radius R.

(b) Air flows between two parallel plates with a separation distance of 50 μm. The Knudsen number and the momentum accommodation coefficient of the flow system are equal to 0.1 and 0.85, respectively. Calculate the slip velocity and slip flow rate.

$$\rho = 1.23\ \text{kg/m}^3,$$

$$\frac{dp}{dx} = -1.0 \times 10^3\ \text{pa/m},$$

$$\mu = 20.0 \times 10^{-6}\ \text{Pa-s}$$

3.5 (a) Calculate the flow rate of air per unit width between the two parallel plates with $h = 50$ μm, pressure difference $\Delta P = 50$ Pa, length of the parallel plate channel $L = 5$ cm, momentum accommodation coefficient $\sigma_v = 0.8$, mean free path of air $\lambda = 5$ μm, viscosity of air $\mu = 20 \times 10^{-6}$ Pa-s, and density of air $\rho = 1.2$ kg/m^3 under the first-order slip flow boundary condition of problem 1(b)

(b) Calculate the ratio of flow rate with the slip flow boundary condition to that with the no-slip boundary condition.

(c) Calculate the slip velocity and slip length.

3.6 (a) In a microbearing configuration, a shaft of 1000 μm diameter and 1 cm long rotates at 120 rpm inside a bearing housing of 1050 μm diameter. Air occupies the gap between the shaft and bearing having mean free path, $\lambda = 2.5$ μm. Calculate the power consumed in overcoming friction for the above arrangement with both no-slip and slip flow boundary conditions of the Couette flow between the gap. The viscosity of air $\mu = 20 \times 10^{-6}$ Pa-s density, $\rho = 1.23$ kg/m^3 and the momentum accommodation coefficient $\sigma_v = 0.8$.

(b) Calculate the ratio of viscous dissipation for the slip flow boundary condition and the no-slip boundary condition and comment on the results.

3.7 Consider flow of helium gas in a microchannel of height $h = 100$ μm, width $w = 1$ mm, and length $L = 1$ cm under the upstream pressure of 1.0×10^5 pascal and downstream pressure of 0.1×10^5 pascal at constant temperature $T = 298$ K. The diameter of helium molecule is 62×10^{-12} m. What is the nature of flow regime at streamwise location, $x = 0$ mm, 2 mm, 5 mm, and 10 mm of the microchannel?

$$\rho = 0.1615\ \text{kg/m}^3,$$

$$\mu = 19.86 \times 10^{-6}\ \text{Pa-s}$$

References

Allen MP and Tildesley DJ 1987 *Computer Simulation of Liquids*. Clarendon Press, Oxford.

Barrat J and Bocquet L 1999 Large slip effect at a nonwetting fluid–solid interface. *Phys. Rev. Lett.*, **82**, pp. 4671–4674.

Boehnke UC, Remmler T, Motschmann H, Wurlitze S, Hauwede H, and Fischer TM 1999 Partial air wetting on solvophobic surfaces in polar liquids. *J. Colloid Interface Sci.*, **211**, pp. 243–251.

Cheng JT and Giordano N 2002 Fluid flow through nanometer-scale channels. *Phys. Rev. E*, **65**, pp. 1–5.

Choi C-H, Westin JA, and Breuer KS 2003 Apparent slip flows in hydrophilic and hydrophobic micro channels. *Phys. Fluids*, **15**, pp. 2897–2902.

Churaev NV, Sobolev VD, and Somov AN 2002 Electrokinetic properties of methylated quartz capillaries. *Adv. Colloid Interface Sci.*, **96**, pp. 265–278.

Koplik J and Banavar JR 1995 Continuum deductions from molecular hydrodynamics. *Annu. Rev. Fluid Mech.*, **27**, pp. 257–292.

Li YU, Xu J, and Li D 2010 Molecular dynamics simulation of nanoscale liquid flows. *Microfluid. Nanofluid.*, **9**, pp. 1011–1031.

Neto C, Craig VSJ, and Williams DRM 2003 Evidence of shear-dependent boundary slip in Newtonian liquids. *Eur. Phys. J. E*, **12**, pp. S71–S74.

Nie XB, Chen SY, Weinan E, and Robbins MO 2004 A continuum and molecular dynamics hybrid method for micro-and nano-fluid flow. *J. Fluid Mech.*, **500**, pp. 55–64.

Ou J and Rothstein JP 2005 Direct velocity measurements of the flow past drag-reducing ultrahydrophobic surfaces. *Phys. Fluids*, **17**, pp. 103606–103610.

Rapaport DC 2004 *The Art of Molecular Dynamics*. Cambridge University Press.

Smoluchowski von Smolan M 1898 Ueber Warmeleitung in Verdunnten gasen. *Ann. Phys. Chem.*, **64**, pp. 101–130.

Schamberg R 1947 The fundamental differential equations and the boundary conditions for high speed slip flow and their application to several specific problems. PhD thesis, California Institute of Technology.

Thompson PA and Troian SM 1997 A general boundary condition for liquid flow at solid surfaces. *Nature*, **389**, pp. 360–362.

Tretheway D and Meinhart C 2002 Apparent fluid slip at hydrophobic microchannel walls. *Phys. Fluids*, **14**, pp. L9–L12.

Supplemental Reading

Beskok A and Karniadakis G 1999 A model for flows in channels, pipes and ducts at micro and nano scales. *Microscale Thermophys. Eng.*, **3**, pp. 43–77.

Blake TD 1990 Slip between a liquid and a solid. DM Tolstoi's theory reconsidered. *Colloids Surf.*, **47**, pp. 135–145.

Gad-el-Hak M 1999 The fluid mechanics of microdevices: the freeman scholar lecture. *J. Fluids Eng.*, **121**(5), pp. 5–33.

Karniadakis G, Beskok A, and Aluru N 2005 *Microflows and Nanoflows, Fundamentals and Simulation*. Springer-Verlag.

Tabeling P 2005 *Introduction to Microfluidics*. Oxford University Press.

Yang J and Kwok DY 2003 Effect of liquid slip in electrokinetic parallel-plate microchannel flow. *J. Colloid Interface Sci.*, **260**, pp. 225–233.

4

Diffusion, Dispersion, and Mixing

4.1 Introduction

The microreactors require mixing and dispersion of various reactants. In a lab-on-a-chip system, it is often required to mix reactants. One example application is fragmentation of proteins by mixing them with enzymes. This procedure is used for the identification of proteins by mass spectrometry. The role played by hydrodynamic instabilities and turbulence on mixing between two streams is limited or nonexistent due to the low Reynolds number in microsystem. The detailed understanding of diffusion and dispersion is necessary to investigate the mixing issues in microsystems. These issues have been discussed in this chapter.

4.2 Random Walk Model of Diffusion

We can explain the diffusion process by means of a simple example of random walk. Let us consider a solution consisting of solute (particles) dissolved in a solvent (liquid). During diffusion, there is a motion of solute from its high to low region of concentration. Brownian motion and thermally induced random motion of the particles are responsible for this motion of the solute. Figure 4.1 shows the 1-D random walk of particles from a fixed location with time. The four particles shown in the figure show random upward and downward motions. The locations of the particles at a given instant of time are different from each other. Let us look at the random walk in 1-D, that is, the solute particle moves randomly in the x-direction with step length, $\Delta x_i = \pm l$. There is equal probability to select the +ve and −ve sign. If each step is performed with the same time step τ, the particle after N-steps starting from $x = 0$ will reach position $x(N)$ given by

$$x(N) = \sum_{i=1}^{N} l_i, \quad \text{where } l_i = \pm l \tag{4.1}$$

As there is equal probability of $+l$ and $-l$, the mean value of $x(N)$ is equal to zero. Hence, $x(N)$ does not reveal the diffusion kinematics. Standard deviation of a quantity is the measure of data spread out defined as

$$\sigma = \sqrt{\frac{1}{N} \sum_{i=1}^{N} (x_i - \bar{x})^2} \tag{4.2}$$

Transport Phenomena in Microfluidic Systems, First Edition. Pradipta Kumar Panigrahi.

Companion Website: www.wiley.com/go/panigrahi/microfluidic

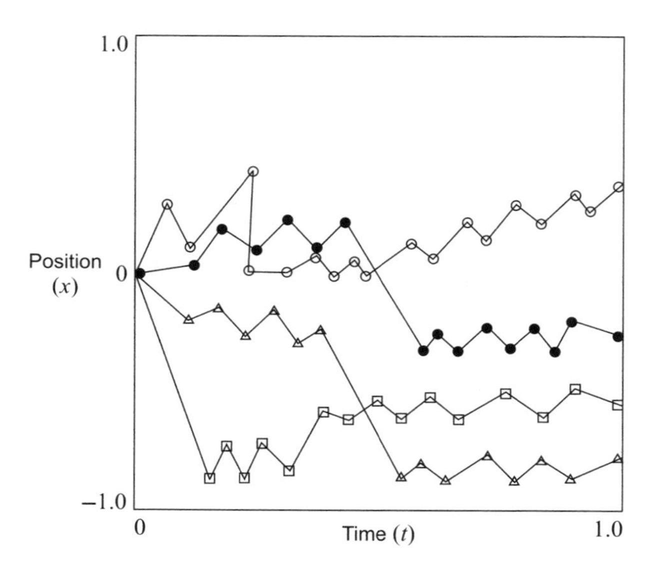

Figure 4.1 Random one-dimensional location of four particles with time starting from the same fixed location

The square of standard deviation of the final particle position can also be written as

$$\sigma^2 = \sum_{i=1}^{N} l_i^2 = Nl^2 \tag{4.3}$$

The diffusion length for N-steps in 1-D can be defined as square root of the standard deviation given as

$$l_{\text{diff}}(N) = \sqrt{N}l \tag{4.4}$$

The total time t elapsed after N equal time steps of τ is given by

$$t = N\tau \tag{4.5}$$

Substituting equation (4.4) in equation (4.5), we have

$$l_{\text{diff}}(N) = \sqrt{\frac{t}{\tau}}l = \sqrt{\frac{l^2}{\tau}t} = \sqrt{Dt} \tag{4.6}$$

Here, we have introduced the parameter diffusion constant (D) given as l^2/τ.
The diffusion time constant can also be written as

$$t_{\text{diff}}(l) = \frac{l^2}{D} \tag{4.7}$$

Diffusion is an extremely slow process for mixing over macroscopic distances. Even in small microfluidic system, diffusion is still a very slow process. For example, let us consider a microfluidic distance of 100 μm. For diffusion of small ions in water ($D \approx 2 \times 10^{-9}$ m^2/s), the diffusion time is equal to 5 s. For diffusion of sugar molecules in water ($D \simeq 5 \times 10^{-10}$ m^2/s), the diffusion time is equal to 20 s. For 5000-base-pair DNA molecules in water ($D \simeq 1 \times 10^{-12}$ m^2/s), the diffusion time is 10^4 s $\simeq$ 3 h.

4.3 Stokes–Einstein Law

The random walk model discussed in the previous section assumes the diffusion of molecules to evolve in a free environment. It assumes the collisions between molecules and helps in understanding the diffusion of one gas molecule to another. However, when the small particles or molecules are immersed in a dense phase, that is, liquid of viscosity μ, the role of dense phase must be explicitly brought out. The Stokes–Einstein law describes the diffusion coefficient D as

$$D = \frac{KT}{6\pi R\mu} \tag{4.8}$$

where R is the radius of diffusing particle, μ is the viscosity of the fluid, K is the Boltzmann constant, and T is the temperature. For oxygen molecules ($R = 1.73$Å) in water, this law gives $D = 1.3 \times 10^{-5}$ cm^2/s at 25°C. This is 30% lower than the measured value and thus estimates proper order of magnitude. The Stokes–Einstein formula becomes imprecise in highly concentrated solutions and in non-Newtonian polymeric fluids.

4.4 Fick's Law of Diffusion

In the previous section, we described the diffusion process using the motion of individual particles/molecules. Let us consider now a group of diffusers or particles instead of isolated walkers. The mass concentration (ρ) is defined as

$$\rho = \frac{\delta m}{\delta \forall} \tag{4.9}$$

Here, δm is the mass of collection of diffusing particles contained in volume $\delta\forall$. The mass flux through surface area δA is defined as

$$J = \frac{\delta m}{\delta t \delta A} \tag{4.10}$$

Fick's law describes the relation between the material flux and concentration as

$$J = -D\nabla\rho \tag{4.11}$$

where D is the coefficient of diffusion.
For gases, the diffusion coefficient D, is estimated as

$$D \approx \frac{1}{3}\lambda\bar{c} \tag{4.12}$$

where $\bar{c}$ is the thermal speed and λ is the mean free path. For gases at normal conditions, $D = 0.15$ cm^2/s which agrees with measurements. For liquids, the diffusivity is lower as the viscosity of the surrounding fluid slows down the displacement of molecules. For example, the diffusivity of fluorescein and sucrose in water at 20 °C is equal to 2×10^{-6} cm²/s and 4.6×10^{-10} cm²/s, respectively. Small diffusion coefficient of sucrose is due to large size of the molecules as large molecules have difficulty in displacing themselves. The Stokes–Einstein equation predicts the above diffusion coefficient values. The correlation of diffusion coefficient to size of molecules is also seen from this equation.

4.5 Diffusivity and Mass Transport Nomenclature

Transport of mass or diffusion of mass takes place in a fluid mixture of two or more species whenever there is a spatial gradient in the proportion of the mixture, that is, a concentration gradient. Consider a two-component or binary system, for example, a glass of water into which a drop of colored dye is injected. As is known from experience, the dye will diffuse outward from the point of injection where the concentration is highest compared to the other portions of the water where there is no dye. The transport of dye molecules is equal and opposite to the transport of the water molecules, and after a sufficient time, an equilibrium state is achieved with a uniform mixture of dye and water. In a solution or mixture, there are a variety of ways of defining concentration. The two basic concentration units used are mass concentration and molar concentration. Let a solution be made up of i number of species. The concentration parameters can be expressed in a variety of ways as follows:

$$\text{Mass concentration } (\rho_i) = \frac{m_i}{\forall} = \frac{\text{Mass of species, } i}{\text{Volume of solution}} (\text{kg/m}^3)$$

$$\text{Molar concentration } (c_i) = \frac{n_i}{\forall} = \frac{\text{No. of moles of species, } i}{\text{Volume of solution}} (\text{mol/m}^3)$$

The related dimensionless concentrations are

$$\text{Mass fraction } (w_i) = \frac{\rho_i}{\rho} = \frac{\text{Mass concentration of species, } i}{\text{Mass density of solution}}$$

$$\text{Molar (mole) fraction } (x_i) = \frac{c_i}{c} = \frac{\text{Molar concentration of species, } i}{\text{Molar density of solution}}$$

The mean molar mass of the mixture is $\overline{M} = \frac{\rho}{c}$

Quantities of interest for diffusive transport are mass flux and molar flux:

$$j_i = \text{Mass flux} = \rho_i u_i (\text{kg/m}^2\text{-s}) \tag{4.13}$$

$$j_i^\star = \text{Molar flux} = c_i u_i (\text{mol/m}^2\text{-s}) \tag{4.14}$$

There are a number of definitions of the average velocity characterizing the bulk motion of a multicomponent system, where each species is moving at a different speed. Often, the solvent velocity is used as a reference velocity. Another reference velocity is the mass average velocity defined as

$$u = \frac{1}{\rho} \Sigma \rho_i u_i \tag{4.15}$$

A molar average velocity may be correspondingly defined as

$$u^* = \frac{1}{c}\Sigma c_i u_i \tag{4.16}$$

In equations (4.15) and (4.16), the terms inside the summation represent the individual mass and molar fluxes of each species "i" with respect to fixed coordinates, respectively.

In flow systems, the mass and molar fluxes with each species motion in reference to the mass average and molar average velocities are written as

$$J_i = \rho_i(u_i - u) \tag{4.17}$$

$$J_i^\star = c_i(u_i - u^\star) \tag{4.18}$$

Fick's first law of diffusion states that there is a linear relation between the species flux and the concentration gradient. For a binary system, using equations (4.13)–(4.18) and definition of mass/mole fraction, we have

$$J_1 = j_1 - \rho_1 u = -D_{12}\nabla\rho_1 = -\rho D_{12}\nabla w_1 \tag{4.19}$$

$$J_1^\star = j_1^\star - c_1 u^\star = -D_{12}\nabla c_1 = -cD_{12}\nabla x_1 \tag{4.20}$$

where $D_{12} = D_{21}$ is the mass diffusivity or mass diffusion coefficient (m^2/s). In a binary mixture, $J_1 = J_2$ and $J_1^\star = J_2^\star$.

4.6 Governing Equation for Multicomponent System

The conservation equations (mass, momentum, energy) are primarily considered for fluids of uniform and homogeneous composition. Here, we examine how these conservation equations change when two or more species are present and when chemical reactions may also take place. In a multicomponent mixture, transfer of mass takes place whenever there is a spatial gradient in the mixture properties, even in the absence of body forces that act differently upon different species. In fluid flows, the mass transfer will generally be accompanied by a transport of momentum and may further be combined with a transport of heat. For a multicomponent fluid, conservation relations can be written for individual species. Let u_i be the species velocities and ρ_i the species density where the index i, is used to represent the ith species.

For single-component system, the *mass conservation equation* is

$$\frac{\partial\rho}{\partial t} + \vec{\nabla}\cdot\rho\vec{u} = 0 \tag{4.21}$$

For multicomponent system, it modifies to

$$\frac{\partial\rho_i}{\partial t} + \vec{\nabla}\cdot\rho_i\vec{u}_i = 0 \tag{4.22}$$

Suppose if species are produced by chemical reaction, say, at a mass rate r_i kilogram per cubic meter second, then we have

$$\frac{\partial\rho_i}{\partial t} + \vec{\nabla}\cdot\rho_i\vec{u}_i = r_i \tag{4.23}$$

Using mass average velocity ($\rho u = \Sigma \rho_i u_i$) and flux with respect to mass average velocity ($J_i = \rho_i(u_i - u)$) in the above equation, we write

$$\frac{\partial \rho_i}{\partial t} + \vec{\nabla} \cdot \rho_i \vec{u} = -\vec{\nabla} \cdot \vec{J}_i + r_i \tag{4.24}$$

The summation of continuity equation (4.23) for all i species gives

$$\frac{\partial \Sigma \rho_i}{\partial t} + \vec{\nabla} \cdot \Sigma \rho_i \vec{u}_i = \Sigma r_i \tag{4.25}$$

Now, using $\rho = \Sigma \rho_i$ and $\rho \vec{u} = \Sigma \rho_i \vec{u}_i, \Sigma r_i = 0$ (mass is conserved in chemical reaction), we get

$$\frac{\partial \rho}{\partial t} + \vec{\nabla} \cdot \rho \vec{u} = 0 \tag{4.26}$$

Note that equation (4.26) is identical to the mass conservation equation (4.21) for pure fluid, that is, single-component system.

If we are considering a binary system than using equation (4.9) for Fick's law, equation (4.24) can be written as

$$\frac{\partial \rho_1}{\partial t} + \vec{\nabla} \cdot \rho_1 \vec{u} = \vec{\nabla} \cdot \left(\rho D_{12} \vec{\nabla} w_1 \right) + r_1 \tag{4.27}$$

Here, r_1 is the mass rate of production of species-1 per unit volume (kilogram per cubic meter second).

Equivalently, in terms of molar fluxes using equation (4.20), we can write

$$\frac{\partial c_1}{\partial t} + \vec{\nabla} \cdot c_1 \vec{u} = \vec{\nabla} \cdot \left(C D_{12} \vec{\nabla} x_1 \right) + R_1 \tag{4.28}$$

Here, R_1 is the molar rate of production of species-1 per unit volume (molar per cubic meter second).

When $\vec{\nabla} \cdot \vec{u} = 0$ for incompressible flow, equation (4.27) can be simplified as

$$\frac{\partial \rho_1}{\partial t} + \vec{u} \cdot \vec{\nabla} \rho_1 = D_{12} \nabla^2 \rho_1 + r_1 \tag{4.29}$$

The corresponding relation for molar concentration can be obtained by dividing both the sides of equation (4.29) by molecular weight, M_1, to give

$$\frac{\partial c_1}{\partial t} + \vec{u} \cdot \vec{\nabla} c_1 = D_{12} \nabla^2 c_1 + R_1 \tag{4.30}$$

When r_1 or $R_1 = 0$, equation (4.29) or (4.30) is termed as *convective diffusion* equation.

In addition, if velocity $u = 0$, the equation reduces to the ordinary diffusion equation, which is also referred to as *Fick's second law of diffusion*.

The *momentum equation* as represented by the Navier–Stokes equation (N–S equation) is not restricted to a single-component fluid but is valid for a multicomponent solution or mixture as long as the external body force is such that each species is acted upon by the same external force as in the case with gravity. The reason for there being no distinction between various contributions to stress tensor associated with diffusive transport is that the phenomenological relation for the stress is unaltered by the presence of concentration gradients. The *energy relation* is also unchanged for a multicomponent mixture as long as the external force on each species is the same. However, when concentration gradients are present, the phenomenological relation for heat flux vector $\vec{q}$ does change from that for a pure fluid. For fluid mixtures, the energy flux must be modified to incorporate an added flux arising from the interdiffusion of the species, that is,

$$q = \Sigma h_i J_i - K\nabla T \tag{4.31}$$

here, h_i = partial specific enthalpy of ith species and J_i = mass flux relative to the mass average velocity. The first term in the right-hand side of the above equation represents the transport of energy caused by interdiffusion and includes the transport of chemical potential energy.

4.7 Characteristic Parameters

We analyze the dimensionless form of the governing equations to obtain important characteristic similarity parameters. This can facilitate subsequent analysis over restricted ranges of parameters.

Mass conservation equation

$$\frac{\partial \rho}{\partial t} + \vec{\nabla} \cdot \rho \vec{u} = 0 \tag{4.32}$$

For steady, incompressible, and constant density fluid flow case, we write the mass conservation equation as

$$\vec{\nabla} \cdot \vec{u} = 0 \tag{4.33}$$

Navier–Stokes equation

For incompressible flow and constant viscosity (μ), we have

$$\frac{D\vec{u}}{Dt} = -\frac{\vec{\nabla} P}{\rho} + \frac{\mu \nabla^2 \vec{u}}{\rho} + \vec{g} \tag{4.34}$$

Convective diffusion equation

For binary solution:

$$\frac{DC}{Dt} = D_{12} \nabla^2 C \tag{4.35}$$

Energy equation

$$\frac{\rho Dh}{Dt} = \frac{DP}{Dt} + K\nabla^2 T + \phi \tag{4.36}$$

For incompressible case, $\frac{DP}{Dt} = 0$. If we assume the viscous dissipation ϕ, to be negligible compared to the average energy level, the above equation simplifies to

$$\frac{DT}{Dt} = \alpha \nabla^2 T \tag{4.37}$$

Here, we also have assumed the heat flux due to interdiffusion to be negligible. Let us assume a forced convection flow with L, V, and τ, respectively, as the length, characteristic speed, and characteristic time of the problem. The reduced dimensionless variables of the problem are $x = Lx^*$, $u = Vu^*$, $t = \tau t^*$, $P - P_0 = \rho V^2 p^*$, $T - T_0 = (T_0 - T_w)T^*$, and $c - c_0 = (c_0 - c_w)c^*$.

The subscripts 0 represent the free stream condition and w represent the value at a wall or surface. Introducing the above dimensionless variables, in equations (4.33)–(4.35) and (4.37), we have

$$\nabla^* \cdot u^* = 0 \tag{4.38}$$

$$St\frac{\partial u^*}{\partial t^*} + u^* \cdot \nabla^* u^* = \frac{1}{Re}\nabla^{*2}u^* - \nabla^* P^* + \frac{1}{Fr} \tag{4.39}$$

$$St\frac{\partial c^*}{\partial t^*} + u^* \cdot \nabla^* c^* = \frac{1}{Pe_D}\nabla^{*2}c \tag{4.40}$$

$$St\frac{\partial T^*}{\partial t^*} + u^* \cdot \nabla^* T^* = \frac{1}{Pe_T}\nabla^{*2}T \tag{4.41}$$

Here, dimensionless parameters St and Fr are expressed as

$$St = \text{Strouhal number} = \frac{L}{\tau V} = \frac{L/V}{\tau} = \frac{\text{Flow time scale}}{\text{Unsteady time scale}}$$

$$Fr = \text{Froude number} = \frac{V^2}{gL} = \frac{\rho V^2 L^2}{\rho g L^3} = \frac{\text{Inertia force}}{\text{Gravitational force}}$$

The Strouhal number is a measure of unsteadiness of flow motion. The Froude number is important for free surface flow:

$$Re = \text{Reynolds number} = \frac{\rho VL}{\mu} = \frac{\rho L^2 V^2}{\mu LV} = \frac{\text{Inertia force}}{\text{Viscous force}}$$

$$Pe_T = \text{Peclet number (thermal)} = \frac{VL}{\alpha}$$

$$= \frac{\rho C_P V(T_0 - T_w)}{K(T_0 - T_w)/L} = \frac{\text{Heat transported by convection}}{\text{Heat transported by conduction}}$$

$$Pe_D = \text{Peclet number (diffusion)} = \frac{VL}{D_{12}} = \frac{V(c_0 - c_w)}{D_{12}(c_0 - c_w)/L}$$

$$= \frac{\text{Mass transported by convection}}{\text{Mass transported by diffusion}}$$

The Peclet number also can be written as ratio of time scale. For a microchannel of transverse dimension "h" and axial dimension "L" it is customary to use h as the length scale. Thus, we have

$$Pe_D = \frac{Vh}{D_{12}} = \frac{\frac{h^2}{D_{12}}}{\frac{h}{V}} = \frac{\text{diffusion time } (\tau_{\text{diff}})}{\text{convection time } (\tau_{\text{conv}})}$$

For high Peclet number, we have $\tau_{\text{conv}} \ll \tau_{\text{diff}}$. This indicates that convection is faster than diffusion and we are in the *convection-dominated regime*. For low Peclet number, diffusion happens faster than convection and we are in the *diffusion-dominated regime*.

Peclet numbers can also be written as

$$Pe_T = \frac{VL}{\nu}\frac{\nu}{\alpha} = Re \cdot Pr \quad (Pr = \text{Prandtl no.})$$

In highly viscous fluids, where the Prandtl number is large, heat transfer by convection dominates over conduction provided the Reynolds number is not small. In liquid metal, the Prandtl number is small, so conduction heat transfer is dominant:

$$Pe_D = \frac{VL}{\nu}\frac{\nu}{D} = Re \cdot Sc \quad (Sc = \text{Schmidt number})$$

For large Schmidt number (common in liquids), convection dominates over diffusion at moderate and even at low Reynolds number. The importance of expressing the Peclet number this way is that Prandtl and Schmidt numbers are *properties of the fluid*, whereas the Reynolds number is a *property of the flow*.

In a microsystem, there is no characteristic order of the Peclet number magnitude. For liquid, $D = 10^{-5}$ cm^2/s with velocity of the order of 1 mm/s and channel of 100 μm height, the Peclet number is of the order of 100. For the same liquid, flowing at 10 μm/s in a channel of 1 μm height, the Peclet number is equal to 10^{-2}. Both these situations are common, that is, not exceptional in microfluidic systems.

4.8 Diffusion Equation

In this section, we develop the concentration distribution for two simple cases, that is, point source diffusion and planar source diffusion. If the velocity field $u = 0$, convection is absent, we have the diffusion equation:

$$\frac{\partial c}{\partial t} = D\nabla^2 c \tag{4.42}$$

4.8.1 Fixed Planar Source Diffusion

Let us first consider a case with a constant source, diffusing inside a fluid domain. The initial condition states that the concentration is zero, that is, at $t = 0$; $c = 0$. Subsequently, a source of n_0 number of molecules is introduced, which can be represented as delta function, that is,

$$\text{at} \quad t = 0^+, \quad c(x) = n_0\delta(x)$$

Note that the Dirac delta function is defined by $\delta(x) = 0$ for $x \neq 0$ and $\int_{-\infty}^{\infty} \delta(x)\delta x = 1$:

$$c(x, t = 0^+) = n_0\delta(x) \tag{4.43}$$

This condition is analogous to injection of fixed n_0 number of molecules at position $x = 0$ at time $t = 0$ in the middle of *infinitely thin* and *infinitely long water* filled tube along the x-axis. Note that this situation may be similar to injection of any particular source material inside a microchannel (see Figure 4.2). The initial point like concentration acts as the source of diffusion and written as delta function.

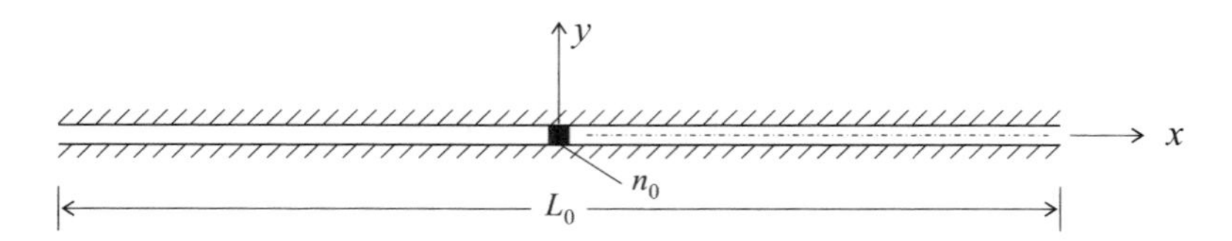

Figure 4.2 An infinitely thin long microchannel filled with water and a fixed amount of source material (n_0) introduced at time, $t = 0^+$

At any time, t, the total concentration of the source material is constant, which is given as

$$\int_{-\infty}^{\infty} c(x)dx = n_0 \tag{4.44}$$

The solution of the diffusion equation (4.42) with the above boundary condition will provide the concentration distribution as a function of space (x) and time (t). Let us define a new variable (ζ) as

$$\zeta = \frac{x^2}{t} \tag{4.45}$$

Therefore, $c(x, t) = v(\zeta)$.

We can write the individual terms of the diffusion equation by applying chain rule as

$$\frac{\partial c}{\partial t} = \frac{\partial v}{\partial t} = \frac{\partial v}{\partial \zeta} \cdot \frac{\partial \zeta}{\partial t} = v' \left(\frac{-x^2}{t^2} \right) \tag{4.46}$$

$$\frac{\partial c}{\partial x} = \frac{\partial v}{\partial x} = \frac{\partial v}{\partial \zeta} \cdot \frac{\partial \zeta}{\partial x} = v' \left(\frac{2x}{t} \right) \tag{4.47}$$

$$\frac{\partial^2 c}{\partial x^2} = \frac{\partial}{\partial x} \left(v' \left(\frac{2x}{t} \right) \right) = \frac{\partial v'}{\partial x} \cdot \frac{2x}{t} + v' \cdot \frac{2}{t}$$

$$= \frac{\partial v'}{\partial \zeta} \cdot \frac{\partial \zeta}{\partial x} \cdot \frac{2x}{t} + v' \cdot \frac{2}{t} = v'' \frac{4x^2}{t^2} + v' \frac{2}{t} \tag{4.48}$$

Here, $'$ indicates the derivative with respect to ζ. Substituting in the governing equation,

$$\frac{\partial c}{\partial t} = D \frac{\partial^2 c}{\partial x^2} \tag{4.49}$$

Here, we have neglected the transverse diffusion as it can reach uniform concentration in transverse y-direction due to small diffusion time. We have

$$v' \left(\frac{-x^2}{t^2} \right) = D v'' \frac{4x^2}{t^2} + \frac{2D}{t} v'$$

$$v'' 4\zeta D + v'(2D + \zeta) = 0$$

$$v'' + v' \frac{2D + \zeta}{4\zeta D} = 0 \tag{4.50}$$

Integrating the above equation, we have

$$v' = C_1 \zeta^{-(1/2)} e^{-\zeta/4D}$$

$$v(\zeta) = C_1 \int_0^{\zeta} \zeta^{-(1/2)} e^{-\zeta/4D} d\zeta \tag{4.51}$$

Note: If $c(t, x)$ is a solution of the diffusion equation, its partial derivative is also a solution as the equation is linear. Hence, $\frac{dv}{dx}$ is also a solution:

$$\frac{\partial v}{\partial x} = \frac{\partial v}{\partial \zeta} \cdot \frac{\partial \zeta}{\partial x} = C_1 \zeta^{-(1/2)} e^{-\zeta/4D} \frac{2x}{t}$$

$$= C_1 \frac{\sqrt{t}}{x} e^{-x^2/4Dt} \frac{2x}{t} = \frac{C_1}{\sqrt{t}} e^{-x^2/4Dt} \tag{4.52}$$

It has to satisfy equation (4.44), that is,

$$\int_{-\infty}^{\infty} c(x)dx = n_0$$

Substituting equation (4.52) in the above equation gives

$$C_1 = \frac{n_0}{\sqrt{4\pi D}} \tag{4.53}$$

Therefore, the solution can be written by substituting C_1 in equation (4.52) as

$$c(x, t) = \frac{n_0}{\sqrt{4\pi Dt}} e^{-x^2/4Dt} \tag{4.54}$$

Figure 4.3 shows the concentration distribution after injecting ink molecules at position $x = 0$ and time $t = 0$ in the middle of an infinitely thin and infinitely long water-filled tube

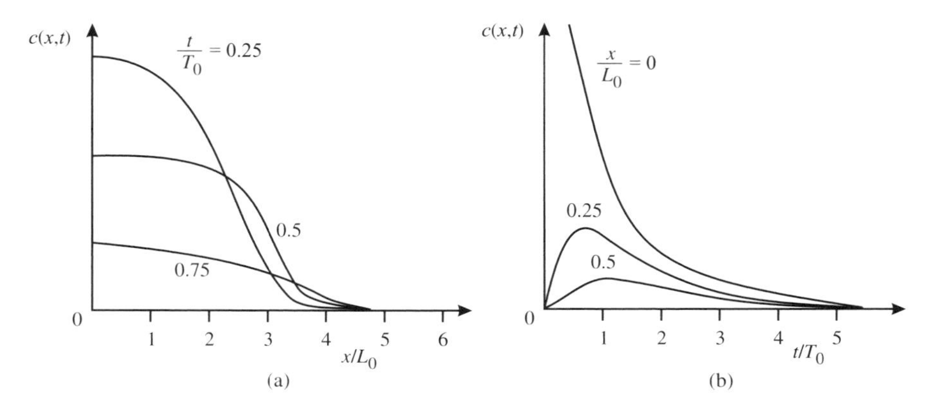

Figure 4.3 The concentration distribution for the point source 1-D diffusion, where L_0 is the length scale in the x-direction and time scale, $T_0 = L_0^2/D$: (a) lengthwise dependence of concentration distribution at three time instants and (b) the time dependence of concentration at three streamwise positions

along the x-axis. Figure 4.3(a) shows that the ink concentration at the point of injection reduces with an increase in time. The streamwise extent of the ink also increases with time. Figure 4.3(b) shows that the maxima in the concentration of the ink at any streamwise location is observed at a later time, that is, sometimes after injection of the ink.

4.8.2 Constant Planar Source Diffusion

The other variation of diffusion problem is when a constant planar source is diffused. The initial concentration is equal to zero. The geometry is the same as the previous problem, but the boundary condition is different, that is,

$$c(x = 0, t > 0) = c_0 \tag{4.55}$$

Here, the concentration at $x = 0$ always remains constant contrary to the previous example, where a fixed concentration is introduced once. This problem is analogous to transient conduction in semi-infinite solid with constant surface temperature boundary condition. The detailed solution procedure can be found in regular heat transfer book. The solution for the above problem can be obtained by using $\zeta = \frac{x}{\sqrt{4Dt}}$. The governing equation in partial differential form will be transformed to ordinary differential equation. The solution of the governing equation is

$$c(x, t) = c_0 \ erfc\left(\frac{x}{\sqrt{4Dt}}\right) \tag{4.56}$$

Figure 4.4 shows the diffusion of a constant planar source, where at time $t = 0$ a source fills the half space such that the density or concentration at boundary plane $x = 0$ remains constant at all later times. Figure 4.4(a) shows that with an increase in time, the concentration font diffuses to a larger x-location. Figure 4.4(b) shows the variation of concentration at a particular x-location as a function of elapsed time. For large time, the concentration reaches the initial concentration, c_0, and the time to reach the equilibrium concentration, c_0, increases with distance, x, from the source location.

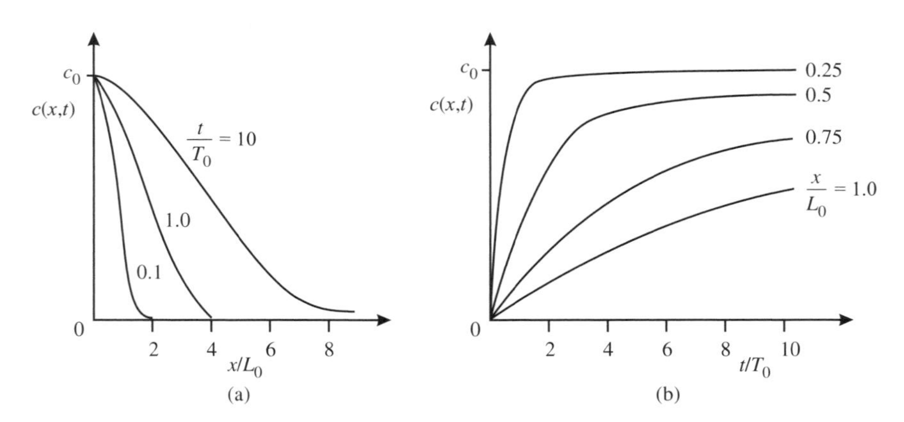

Figure 4.4 The concentration distribution in case of constant planar source diffusion, where concentration at $x = 0$ remains constant at all later times: (a) the x-dependence of concentration at three time instants and (b) the time dependence of concentration at four streamwise locations. Here, L_0 is the length scale and $T_0 = L_0^2/D$ is the time scale

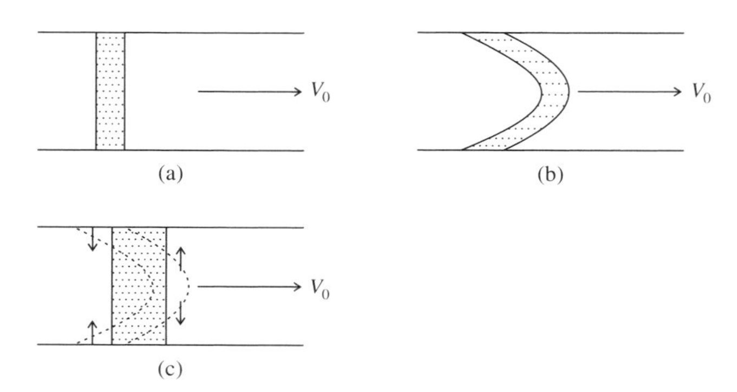

Figure 4.5 Example of Taylor dispersion in a microchannel with a steady Poiseuille flow: (a) initial flat concentration of the solute, (b) stretching of the solute to a paraboloid-shaped plug neglecting diffusion, and (c) solute plug with diffusion indicated by vertical arrows

4.9 Taylor Dispersion

Combined convection and diffusion occurs when some solute is placed in a solution moving with nonzero velocity. This is known as the Taylor dispersion problem. The schematic of Taylor dispersion problem is shown in Figure 4.5. In Figure 4.5, a homogeneous band of solute is placed in a microchannel at time, $t = 0$. The concentration profile disperses due to combined influence of convection from the Poiseuille flow and diffusion because of the concentration gradients. For negligible diffusion, the solute becomes stretched into paraboloid-shaped band due to the Poiseuille flow (Figure 4.5(b)). However, diffusion brings solute particles from the high-concentration region near the center to the low-concentration region near the wall at the front end of the plug, whereas, the back end brings the high concentration from the wall region to the low-concentration region near the center. This leads to a uniform plug section of the channel (Figure 4.5(c)). The plug moves downstream at a speed equal to the average Poiseuille flow velocity, V_0.

4.9.1 Taylor Dispersion in Microchannels

Taylor dispersion phenomenon in microchannel consists of dispersion of a passive tracer in the flow. The schematic of a Taylor dispersion problem is shown in Figure 4.6.

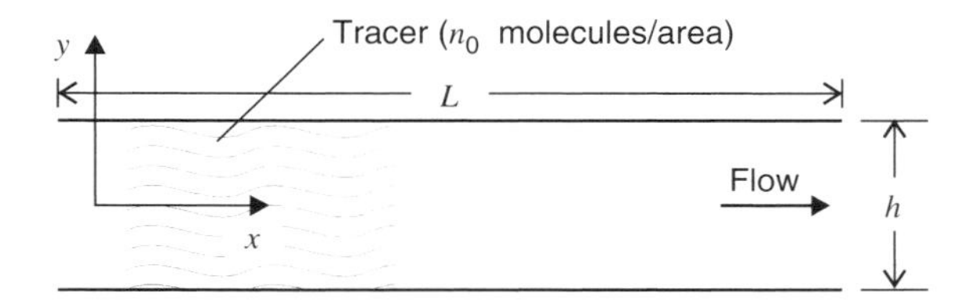

Figure 4.6 The schematic for Taylor dispersion flow in a microchannel separated by distance, h, between two planes

The convection–diffusion equation for the tracer, that is, species-1, can be written as

$$\frac{\partial c_1}{\partial t} + \vec{V} \cdot \vec{\nabla} c_1 = D_{12} \nabla^2 c_1 \tag{4.57}$$

Expanding the above equation in Cartesian coordinate, we have

$$\frac{\partial c_1}{\partial t} + u\frac{\partial c_1}{\partial x} + v\frac{\partial c_1}{\partial y} + w\frac{\partial c_1}{\partial z} = D_{12}\left(\frac{\partial^2 c_1}{\partial x^2} + \frac{\partial^2 c_1}{\partial y^2} + \frac{\partial^2 c_1}{\partial z^2}\right) \tag{4.58}$$

Using $u = u(y)$, $v = w = 0$, and $\frac{\partial}{\partial z} = 0$ for the microchannel flow, equation (4.58) simplifies to

$$\frac{\partial c_1}{\partial t} + u(y)\frac{\partial c_1}{\partial x} = D_{12}\left(\frac{\partial^2 c_1}{\partial x^2} + \frac{\partial^2 c_1}{\partial y^2}\right) \tag{4.59}$$

As $L \gg h$, in the large-time limit ($t \gg \tau_{\text{diff}}^{\text{rad}}$), the transverse diffusion term can be neglected compared to the axial diffusion term. Here, $\tau_{\text{diff}}^{\text{rad}} = \frac{h^2}{D_{12}}$ is the time to move distance, h, by transverse diffusion. This means that diffusion process has sufficient time to act over the short transverse distance, h, yielding a uniform concentration profile in y-direction, that is, $c(x, y, t)$ can be replaced by $c(x, t)$. Thus, the governing equation becomes

$$\frac{\partial c_1}{\partial t} + u(y)\frac{\partial c_1}{\partial x} = D_{12}\frac{\partial^2 c_1}{\partial x^2} \tag{4.60}$$

This equation can be integrated as c does not depend on y. We do not perform the integration here, which is rather long. The final solution of the problem is

$$c(x, t) = \frac{n_0}{(\pi D_{\text{eff}} t)^{1/2}} \exp\left[-\frac{(x - V_0 t)^2}{4D_{\text{eff}} t}\right] \tag{4.61}$$

Here, V_0 is the mean velocity of the Poiseuille flow; and n_0 is the molecules per area at time $t = 0$:

$$D_{\text{eff}} = D_{12}\left(1 + \beta\left(\frac{V_0 h}{D_{12}}\right)^2\right) \tag{4.62}$$

where β is a coefficient that depends on the shape of the channel. For a channel between two parallel plates, we have

$$\beta = \frac{1}{210}$$

For a capillary with circular cross-section, we have

$$\beta = \frac{1}{48}$$

Note that equation (4.61) is similar to equation (4.54), where the plug region is moving at speed V_0 and x is replaced by $x - V_0 t$. The Taylor dispersion flow is analogous to the diffusion of fixed planar source with change in frame of reference, that is, the solvent flows at average velocity, V_o.

Based on the effective diffusion coefficient, D_{eff}, one can introduce a time scale τ_{Taylor} for broadening a sample to width of w in a channel of radius h (for microcapillary):

$$\tau_{\text{Taylor}} = \frac{w^2}{D_{\text{eff}}} = \frac{w^2}{D_{12}\left(1 + \frac{1}{48}\left(\frac{V_0 a}{D_{12}}\right)^2\right)} \tag{4.63}$$

This expression indicates that for a given velocity, the Taylor time scale increases with reduction in the capillary radius. Hence, Taylor dispersion acts slower as the channel radius is decreased.

Note: (1) For Taylor dispersion, there is an effective diffusion coefficient which increases proportionally with square of the velocity of the flow. The spot spreads much faster in the direction of the flow than if molecular diffusion is the only mechanism responsible.

(2) The effective diffusion can be approximated as

$$D_{\text{eff}} \sim Pe^2 D_{12}$$

where the Peclet number $Pe = Uh/D$. In microsystems, the Peclet number can attain values between 0.1 and 100. For high Pe case, diffusion along the flow can be increased several orders of magnitude with respect to the molecular diffusion.

(3) Effective diffusion can also be understood from heuristic point of view. Let us imagine a narrow capillary of radius a to be divided into three concentric cylinders of size $a/3$. The radial diffusion time, t, for size of $a/3$ is equal to $t \equiv a^2/9D_{12}$. The radial jump from one shell to another shell transforms the radial motion to the axial motion or displacement. If the central shell moves at velocity V_0, the liquid shells may be assumed to move with respect to each other at velocity $V_0/2$. The axial displacement of the species during that time is equal to $\left(\frac{V_0}{2} \times t = \frac{V_0 a^2}{18D_{12}}\right)$. The sum of the above contribution can be written as

$$\begin{aligned} l_{\text{eff}}^2 &= D_{\text{eff}} t \simeq (l_{\text{diff}})^2 + (l_{\text{conv}})^2 = (D_{12} t)^2 + \left(\frac{V_0}{2} t\right)^2 \\ &= D_{12} t + \left(\frac{V_0}{2}\right)^2 \left(\frac{a^2}{9D_{12}}\right) t = \left(D_{12} + \frac{V_0^2 a^2}{36D_{12}}\right) t \end{aligned} \tag{4.64}$$

Thus, we have effective diffusion coefficient as

$$D_{\text{eff}} \simeq D_{12} + \frac{V_0^2 a^2}{36D_{12}} \tag{4.65}$$

(4) In microfluidics, the radial length scale, h, is much smaller than the axial length scale, L. For a characteristic velocity, V, there are four possible time scales:

$\tau_{\text{diff}}^{\text{rad}} = \dfrac{h^2}{D_{12}}$ = time to move radial distance, h, by radial diffusion.

$\tau_{\text{diff}}^{\text{ax}} = \dfrac{L^2}{D_{12}}$ = time to move axial distance, L, by axial diffusion.

$\tau_{\text{conv}}^{\text{rad}} = \dfrac{h}{V}$ = time to move radial distance, h, by axial convection.

$\tau_{\text{conv}}^{\text{ax}} = \dfrac{L}{V}$ = time to move axial distance, L, by axial convection.

For the Taylor dispersion to be valid, we should satisfy the condition

$$\tau_{\text{diff}}^{\text{rad}} \ll \tau_{\text{conv}}^{\text{ax}}$$

For the convection to play an active role, we should satisfy the condition

$$\tau_{\text{conv}}^{ax} \ll \tau_{\text{diff}}^{\text{rad}}$$

4.9.2 H-Filter

A microfluidic device known as H-filter, which was developed in the mid-1990, works using magnitude of different time scales. This device allows continuous extraction of molecular analytes from fluids containing interfacing particles, that is, blood cells, bacteria, microorganisms, dusts, viruses, and so on. It can also be used to separate small-sized analytes from the large-sized analytes. Its main advantage is that it does not need any membrane filter or similar component. The Reynolds number in most of the microfluidic channels are usually less than 1. Therefore, no convective mixing of fluid occurs. Hence, the solvents, solutes, and suspended particles only move in a direction transverse to the flow direction by diffusion. We know that diffusion coefficient roughly scales with the inverse of the size of the molecules and depends to some extent on the shape of the molecule. Small molecules have large diffusion coefficient compared to larger molecules. Therefore, they move longer average distance per time than large molecules. This difference in diffusion coefficient can be used to separate molecules or large particles over time. The other factor is the time spent inside the channel, which is proportional to the length of the channel. Proper design of a microfluidic channel allows controlled extraction of analytes with high diffusion coefficients from components of the sample with lower diffusion coefficient.

Figure 4.7 shows the schematic of an H-filter. The sample solution has two different solutes. The small ions are shown as black dots and the white spots represent DNA molecules. Both the channels have the same length (L), width (w), and height (h). Two time scales are important

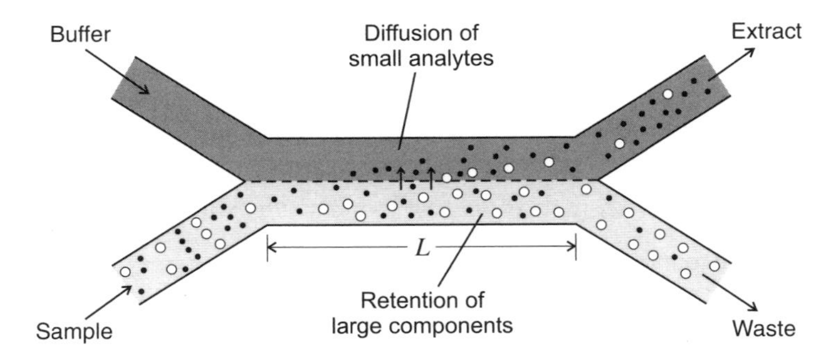

Figure 4.7 Schematic of an H-filter for separation of small-sized analytes from large-sized analytes

for the operation of this device. The convection time scale ($\tau_{\text{convection}}$) refers to the time taken for convection from the inlet to the outlet of the channel. The diffusion time scale refers to the time it takes to diffuse across the half-width of the channel:

$$\tau_{\text{convection}} = \frac{L}{V} \tag{4.66}$$

$$\tau_{\text{diffusion}} = \frac{\left(\frac{1}{2}w\right)^2}{D_{12}} = \frac{w^2}{4D_{12}} \tag{4.67}$$

Here, V is the average velocity and D_{12} is the diffusion coefficient. If a solute has $\tau_{\text{convection}} \ll \tau_{\text{diffusion}}$, it has enough time to act and will remain in the original stream. If $\tau_{\text{convection}} \geq \tau_{\text{diffusion}}$, the solute has time to diffuse across the central channel. For the H-filter to separate two different solutes from each other, the following conditions should be satisfied:

$$\tau_{\text{convection},1} \ll \tau_{\text{diffusion},1}$$

$$\tau_{\text{convection},2} \geq \tau_{\text{diffusion},2}$$

Let us consider an example having $L = 1$ mm, $V = 0.5$ mm/s, $D_{12,1} = 2 \times 10^{-9}$ m^2/s, $D_{12,2} = 4 \times 10^{-9}$ m^2/s, $h = 1$ μm, and $w = 100$ μm. The different time scales for two species are

$$\tau_{\text{convection},1} = 2\ \text{s},\ \tau_{\text{diffusion},1} = 1.25\ \text{s}$$

$$\tau_{\text{convection},2} = 2\ \text{s},\ \tau_{\text{diffusion},2} = 0.625\ \text{s}$$

The optimal solution for the successful operation of the H-filter is

$$\tau_{\text{diffusion},2} \ll \tau_{\text{convection}} \ll \tau_{\text{diffusion},1}$$

The optimal condition is not satisfied for the above example. Therefore, we need to change either the length or the average speed of the system. If we change the length L to 1 cm and reduce the velocity to 0.1 mm/s, we have

$$\tau_{\text{convection},1} = \tau_{\text{convection},2} = \frac{10\ \text{mm}}{0.1\ \text{mm/s}} = 100\ \text{s}$$

This satisfies the required condition of H-filter for separating the small-sized solute, that is, ion from the sample solution as the convection time scales lies between diffusion time scale of species-1 and species-2.

4.10 Micromixer

The Reynolds number in a microdevice is traditionally very low, because of small flow rate and channel dimension. Therefore, the flow instability leading to transition and turbulence does not take place. Other mechanism for initiation of mixing between two streams is necessary. The following sections present different schemes for initiating and enhancing micromixing.

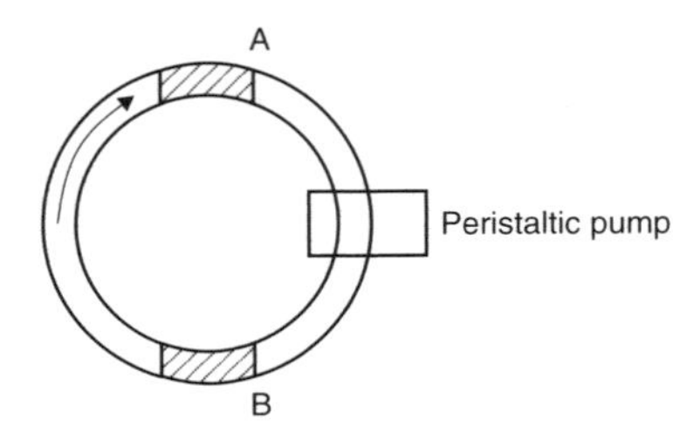

Figure 4.8 A circular micromixer operating on principle of Taylor dispersion

4.10.1 Ring-Shaped Micromixer

For miniaturized chemical analysis, there are often samples of given size but not fixed permanent flow rate, where two samples A and B are required to be mixed with each other. A ring-shaped micromixer as shown in Figure 4.8 can be used for this situation, where a peristaltic pump drives the flow around a ring. The Taylor dispersion operates along the ring, and after some rotations, the two spots "A" and "B" mix completely. The speed of rotation of the fluids is U, and the transverse distance, that is, the height of the channel, is b. The effective diffusivity is

$$D_{\text{eff}} \sim Pe^2 D = \left(\frac{Ub}{D}\right)^2 D \tag{4.68}$$

The increase in effective diffusivity, D_{eff}, due to rotation will lead to reduction in Taylor time scale, τ_{Taylor} (equation (4.63)).

4.10.2 Micromixer Based on Size Reduction

Two fluid streams can be mixed by diffusion using the principle of size reduction. As diffusion mixing is given by $\tau_{\text{diff}} \sim \frac{h^2}{D}$, where h is the height of the channel. One possible scheme is to pass the mixing stream through large number of microchannels in parallel as shown in Figure 4.9.

4.10.3 Hydrodynamics Focusing

Another route to reduce the size of the mixing stream is *hydrodynamic focusing*. Here, the size of the liquid stream is reduced by a control flow from the side as shown in Figure 4.10. At high flow ratio between the two streams, the control fluid layer can be reduced to the order of a few nanometers. The diffusion mixing time as low as a few microseconds can be obtained by this route. If the fluid streams A and B are miscible with each other, very fast mixing between the two streams is achieved by this route.

4.10.4 Chaotic Mixing

The position of the species particle develops depending on the local velocity field, which can be obtained by solving (neglecting molecular diffusion) the following equation:

$$\frac{d\vec{r}}{dt} = \vec{V}(\vec{r}, t) \tag{4.69}$$

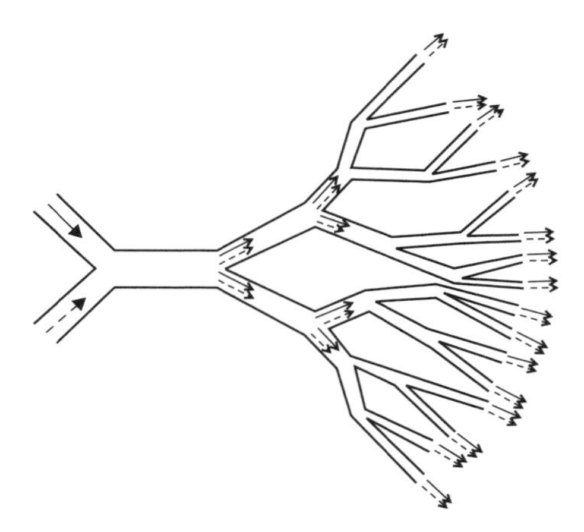

Figure 4.9 A micromixer with the principal channel divided into large number of smaller cross-sectional channels

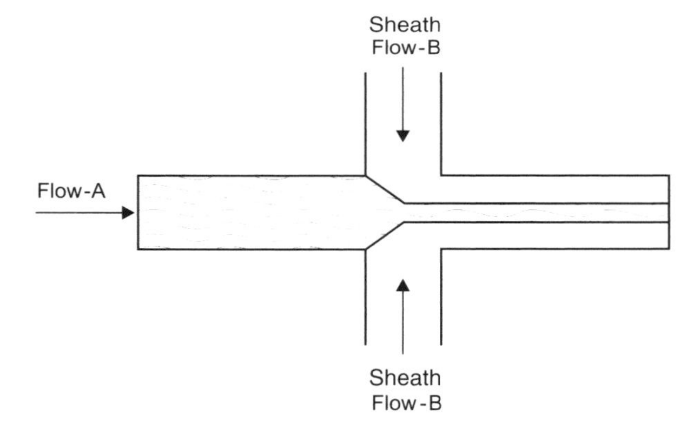

Figure 4.10 Hydrodynamic focusing by a control flow

This is a dynamical system defined by a system of equations. The solution of this equation provides trajectories of the species $r(t)$ in physical space (x, y, z). If the solution of this equation is sensitive to initial conditions, we get chaotic system, where two particles situated in neighboring trajectories diverge from each other as shown in Figure 4.11. The Lyapunov exponent of a dynamical system quantifies the average of the maximum expansion rate for a pair of particles advected in the flow. If there exists a direction in which the exponent is positive, the system is then considered to be chaotic.

A system is chaotic, if the trajectories of particle separate from each other exponentially. The intensive movement of the species in the surrounding fluid leads to strong diffusive exchanges between the species resulting in effective mixing. In regular advection, transport of species takes place parallel to main flow direction and there is no transverse transport of species. Chaotic advection leads to transport in other directions and thus could improve mixing significantly. Chaotic advection can be induced either passively by suitable modification of the geometry to provide three-dimensionality or actively by external excitation

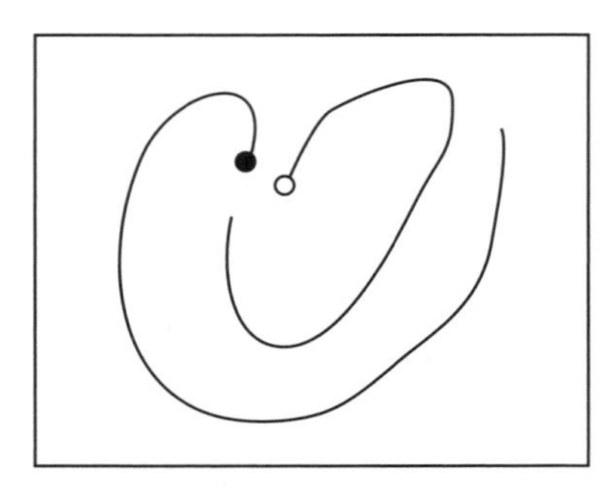

Figure 4.11 Evolution of two particles initially very close to each other in a chaotic system, which subsequently separate from each other

to provide time variation in flow field. Depending on the design, chaotic advection sets in at different Reynolds numbers. Slanted ribs, cylindrical posts, grooves, serpentine passage, and herringbone grooves are used to twist the flow and generate chaotic advection (see Figure 4.12) in *passive micromixer*.

In *active micromixer*, pressure disturbance, electrohydrodynamic disturbance, dielectrophoretic disturbance, electrokinetic disturbance, and magnetohydrodynamic disturbances are used to disturb the laminar streams (see Figure 4.13). The pressure pulse divides the

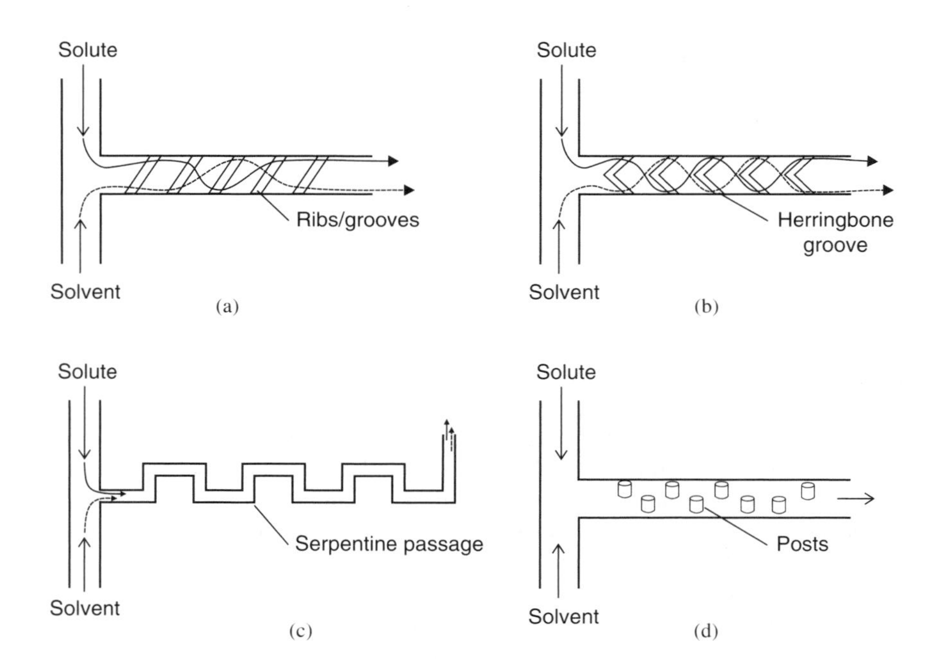

Figure 4.12 Different schemes for initiating chaotic advection in microscale flow: (a) slanted ribs/grooves, (b) herringbone grooves, (c) serpentine passage, and (d) posts or cylindrical obstacles on the channel

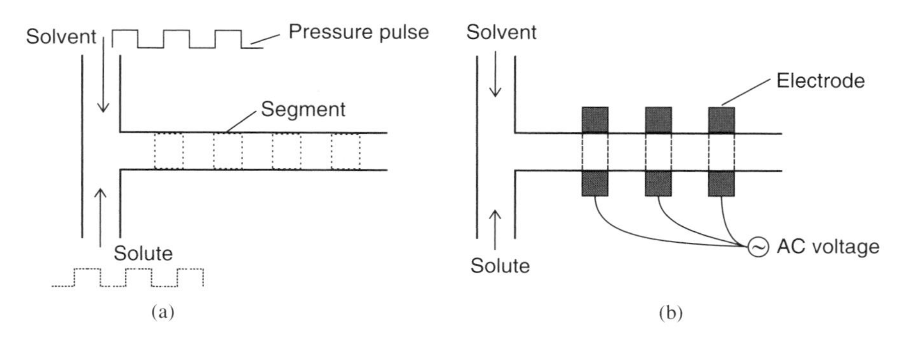

Figure 4.13 Schematic showing active mixing strategy: (a) pressure disturbance leading to segmentation and (b) dielectrophoretic, electrohydrodynamic, electrokinetic, and magnetohydrodynamic disturbance generated by the electrode

solvent and solute stream into segments. Subsequently, mixing takes place through diffusion in flow direction. The effective diffusion coefficient is defined on the basis of Taylor dispersion. The segment length depends on the pulsing frequency. The concentration distribution in a microchannel can be obtained by solving

$$\frac{\partial C}{\partial t} + u\frac{\partial C}{\partial x} = D_{\text{eff}}\frac{\partial^2 C}{\partial x^2} \tag{4.70}$$

where u is the velocity profile in streamwise x-direction, which can be assumed similar to the Poiseuille flow, and D_{eff} is the Taylor dispersion coefficient, which is dependent on the channel geometry.

In electrohydrodynamic disturbance, electrodes are placed on the mixing channel. By changing the voltage or frequency of the electrodes, good mixing is achieved. In dielectrophoresis, embedded particles in the flow field are polarized inside a nonuniform electric field. Particles move to and from the electrode depending on the nature of electrical actuation and chaotic advection is generated. In external magnetic field, the species transport can be influenced by magnetohydrodynamic effect. Applied voltage on the electrode generates the Lorentz force, which in turn induces transverse movement in the mixing channel.

4.10.5 Droplet Formation and Chaotic Advection

Basic configuration similar to hydrodynamic focusing can be used to form droplets of solvent and solute (see Figure 4.14(a)). The middle inlet is used to flow carrier fluid, which is immiscible to both solute and solvent. The solute and solvent enter from opposite sides. The droplet formation process depends on the capillary number $Ca = u\mu/\sigma$ and sample fraction (x):

$$x = \frac{\dot{Q}_{\text{solvent}} + \dot{Q}_{\text{solute}}}{\dot{Q}_{\text{solvent}} + \dot{Q}_{\text{solute}} + \dot{Q}_{\text{carrier}}} \tag{4.71}$$

where μ is the viscosity and $\dot{Q}$ is the flow rate.

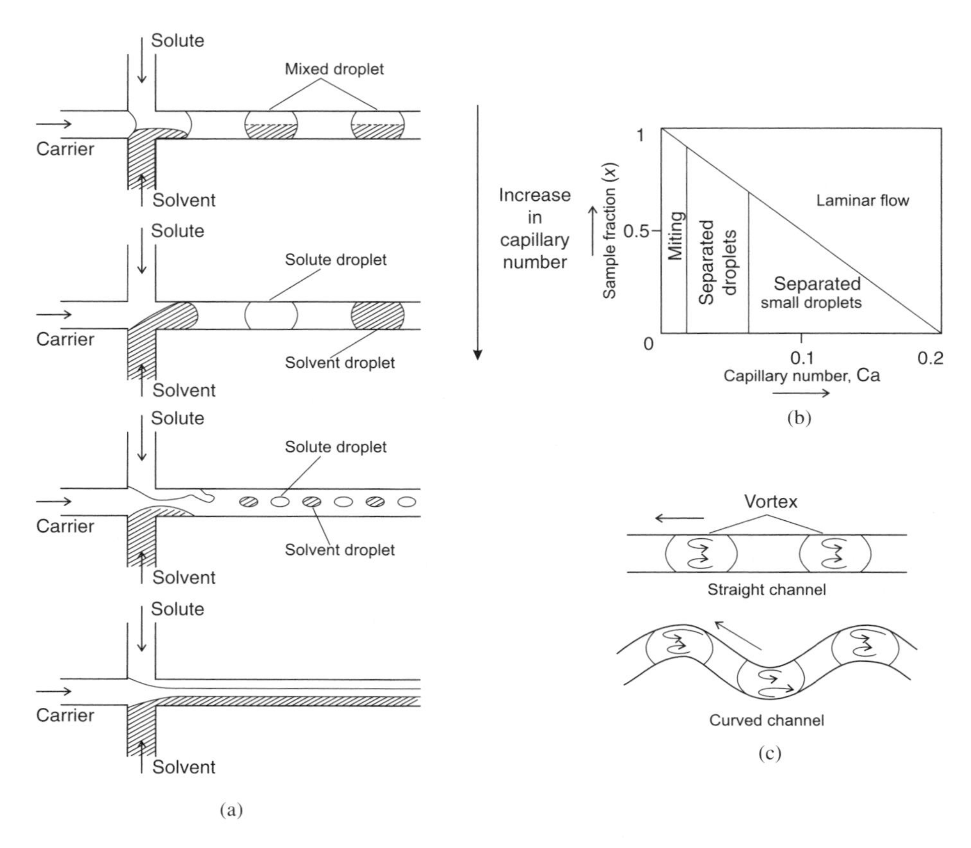

Figure 4.14 (a) Different flow patterns of microdroplet with an increase in capillary number, (b) droplet formation regime as function of capillary number, and (c) vortex structures inside droplet

At low capillary number, the solute and solvent merge into a sample droplet (Figure 4.14(a), top). The droplets form separately without any merging and mixing with an increase in capillary number at the same sample fraction, x (Figure 4.14(a), second from the top). The alternate droplets become smaller and unstable with a further increase in the capillary number (Figure 4.14(a), third from the top). The three streams flow side by side with a further increase in the capillary number, that is, at high capillary number and high sample fraction (Figure 4.14(a), fourth from the top). Figure 4.14(b) shows the different flow regimes of droplet formation as a function of capillary number and sample fraction.

The solute and solvent inside the mixed droplet mix with each other and the nature of mixing is dependent on the flow conditions, that is, channel geometry. In case of straight channel, the two vortices containing the solvent and solute exist in each half of the droplet without any significant mixing between the two. For microchannel having turns, the flow pattern is periodical, and chaotic advection may set in inside the droplet leading to faster mixing (see Figure 4.14(c)).

4.11 Convective Diffusion

This section considers the transport of mass by convective diffusion in an isothermal solution containing one or more uncharged molecular species. The system may involve chemical, physical–chemical, or biological reaction or it may be nonreacting. If it is reacting and the chemical, physical, and biological changes take place in the bulk of the fluid, the reaction is termed as *homogeneous*. If it takes place only in a restricted region, such as at bounding surfaces or phase interfaces, it is termed as *heterogeneous*. In homogeneous reactions, species are produced and its production rates enter into the conservation of mass equation for a multicomponent flow. On the other hand, for a heterogeneous reaction, the species production enters only in the boundary conditions at the reaction surface. A wide range of phenomena are incorporated under the umbrella of heterogeneous reactions. Examples would be chemically catalyzed reactions at a solid surface, for example, catalytic converter used in locomotives and diesel generators, adsorption and desorption at solid and liquid surfaces, dissolution and precipitation of materials from solutions and melts, enzyme–substrate reaction at surfaces, and electrode–electrolyte electrochemical reactions. Many cases of microfabrication during deposition and etching processes are also examples of heterogeneous reaction.

Heterogeneous reactions involve several steps. The first is the transfer of reacting species to the reaction surface. The second step is the heterogeneous reaction itself. This step is often composed of a series of substeps that may include diffusion of the reactants through the material, adsorption on the surface, chemical reaction, desorption of products, and diffusion of products out of the material. The third step is the transfer of the products away from the reaction surface into the bulk phase. The overall rate is controlled by the rate of the slowest step. If the rate limiting step is either step 1 or step 3, which involves the introduction or removal of reactants, then the reaction is said to be *diffusion controlled*. On the other hand, if step 2 involving the chemical, physical, or biological transformations is the slow step, then the rate is determined by the kinetics of the given process and is known as *kinetics controlled*. Those cases where the rates of the diffusion and reaction steps are comparable are sometimes termed *mixed heterogeneous* reactions. The following section discusses the species transport phenomena due to heterogeneous reaction.

4.11.1 Convective Diffusion Layer

The region in which convective diffusion plays a dominant role is known as the convective diffusion layer. We describe the relevant governing equations and important nondimensional number in this section.

The general convective diffusion equation is

$$\frac{\partial c}{\partial t} + \vec{u} \cdot \vec{\nabla} c = D\nabla^2 c \tag{4.72}$$

For 2-D case, we have

$$\frac{\partial c}{\partial t} + u\frac{\partial c}{\partial x} + v\frac{\partial c}{\partial y} = D\frac{\partial^2 c}{\partial y^2} + D\frac{\partial^2 c}{\partial x^2} \tag{4.73}$$

In the following section, we discuss the convective diffusion layer from order of magnitude estimate. In the subsequent section, we develop analytical expressions for soluble and reacting wall flow.

4.11.2 Order of Magnitude Estimate

The order of magnitude estimation can facilitate the understanding of convective-diffusion phenomena. Let us look at a special case for high Schmidt number, which is true for most of the liquids.

4.11.2.1 External Flow Situation

It may be noted that for dilute solutions, $Sc = \nu/D = 10^3$. Hence, $Pe_D = Re \times Sc$ is generally large. The nondimensional convective diffusion equation is

$$St\frac{\partial c^*}{\partial t^*} + u^* \cdot \nabla^* c^* = \frac{1}{Pe_D}\nabla^{*2} c^* \tag{4.74}$$

For $Pe_D \to \infty$, the right-hand side of the above equation is approximated equal to zero and we have

$$\frac{Dc}{Dt} = 0 \tag{4.75}$$

The above equation indicates that c is a constant following a fluid particle. Suppose we have $c = c_0$ in the free stream and the wall boundary condition for a reacting surface is $c_w = 0$. The solution of equation (4.75) cannot satisfy the boundary conditions at the reaction surface. Evidently, near the surface, there must be a thin diffusion boundary layer of thickness δ_D within which the concentration changes rapidly (see Figure 4.15). This reasoning parallels the Prandtl boundary layer argument for viscous flow past a solid boundary at high Reynolds number. If δ_v is the Prandtl viscous boundary layer thickness for steady unbounded laminar flow, we know that

$$\frac{\delta_v}{L} \sim (Re)^{-(1/2)} \sim \left(\frac{\nu}{VL}\right)^{(1/2)} \tag{4.76}$$

For large Schmidt number ($\nu \gg D$), viscous boundary layer thickness should be considerably larger than diffusion boundary layer thickness ($\delta_v \gg \delta_D$). Let us obtain a relationship between the viscous boundary layer thickness and diffusion boundary layer thickness.

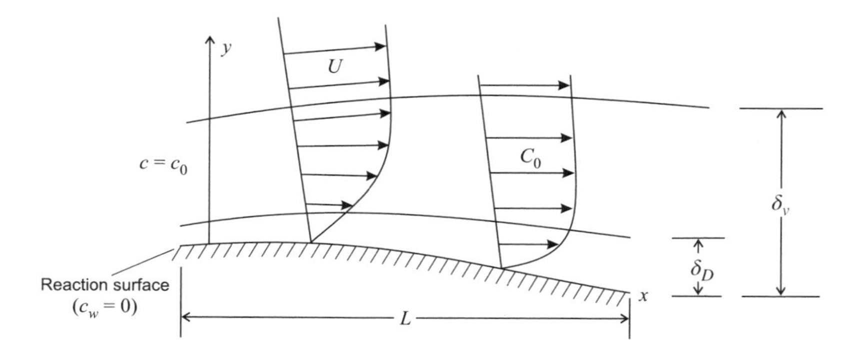

Figure 4.15 The schematic showing the development of viscous and diffusion boundary layers over a reactive wall surface

The convective diffusion equation in the diffusion boundary layer is

$$\frac{\partial c}{\partial t} + u\frac{\partial c}{\partial x} + v\frac{\partial c}{\partial y} = D\frac{\partial^2 c}{\partial x^2} + D\frac{\partial^2 c}{\partial y^2} \tag{4.77}$$

From order of magnitude analogy, we can write

$$\frac{\partial^2 c}{\partial y^2} \Big/ \frac{\partial^2 c}{\partial x^2} = \left(\frac{L}{\delta_D}\right)^2 \gg 1 \tag{4.78}$$

So $\frac{\partial^2 c}{\partial x^2}$ contribution can be neglected and thus the above equation reduces to

$$\frac{\partial c}{\partial t} + u\frac{\partial c}{\partial x} + v\frac{\partial c}{\partial y} = D\frac{\partial^2 c}{\partial y^2} \tag{4.79}$$

As $\delta_D \ll \delta_v$, the velocity profile in the diffusion layer must be that corresponding to the viscous layer profile near the wall. At small distances, the velocity profile is assumed linear in y as

$$\frac{u}{U} \simeq \frac{y}{\delta_v} \tag{4.80}$$

Hence, using the above estimate, the order of magnitude estimate can be written as

$$u\frac{\partial c}{\partial x} \sim \left(\frac{Uy}{\delta_v}\right)\frac{c_0 - c_w}{x} \sim \frac{U\delta_D}{\delta_v}\frac{c_0 - c_w}{x}$$ (setting, $y = \delta_D$ i.e., the edge of diffusion boundary layer)

Using $\frac{\delta_v^2}{x} \sim \frac{\nu}{U}$ from viscous boundary layer theory in the above estimate, we have

$$u\frac{\partial c}{\partial x} \sim \frac{\nu\delta_D(c_0 - c_w)}{\delta_v^3} \tag{4.81}$$

The mass conservation equation in 2-D is

$$\frac{\partial u}{\partial x} + \frac{\partial v}{\partial y} = 0 \tag{4.82}$$

We can assume that $u\frac{\partial c}{\partial x}$ and $v\frac{\partial c}{\partial y}$ are of the same order of magnitude on the basis of equation (4.82). Hence, from equation (4.79) for steady case $\left(\frac{\partial c}{\partial t} = 0\right)$, we have

$$u\frac{\partial c}{\partial x} \sim v\frac{\partial c}{\partial y} \sim D\frac{\partial^2 c}{\partial y^2}$$

Equating the order of magnitude of diffusion term with convection term (equation (4.81)), we have

$$D\frac{(c_0 - c_w)}{\delta_D^2} \sim \frac{\nu\delta_D(c_0 - c_w)}{\delta_v^3}$$

$$\delta_D \sim \left(\frac{D}{\nu}\right)^{1/3}\delta_v \equiv \frac{\delta_v}{(S_c)^{1/3}} \tag{4.83}$$

Hence, from equation (4.83), it may be noted that the diffusion layer thickness is of the order of one tenth that of the viscous layer for $Sc = 10^3$. Equation 4.83 also shows that the diffusion layer growth is parabolic in the streamwise distance, x, which is similar to that of the viscous boundary layer. It may be noted that similar expressions are derived in case of thermal boundary layer thickness (δ_T), that is,

$$\delta_T = \frac{\delta_v}{(Pr)^{1/3}} \tag{4.84}$$

4.11.2.2 Internal Flow Situation

From the velocity boundary layer equation, we can compare the order of magnitude of individual terms for internal flow situation (see Figure 4.16) as

$$\frac{\partial u}{\partial t} \sim u\frac{\partial u}{\partial x} \sim \nu\frac{\partial^2 u}{\partial y^2}$$

Characteristic time for viscosity to diffuse to the center of the channel = $\frac{h^2}{\nu}$. The characteristic time of convection of the particle = L_U/U. At the fully developed velocity profile location, equating the two time scales, we have

$$\frac{h^2}{\nu} \simeq \frac{L_U}{U}$$

$$\frac{L_U}{h} \simeq \frac{Uh}{\nu}$$

For a straight channel, the development length, when the two boundary layers meet at the center of the channel, can be derived using the flat plate boundary layer estimate as

$$\frac{L_U}{h} \sim 0.16\frac{Uh}{\nu}$$

We can write similar behavior for the concentration boundary layer.

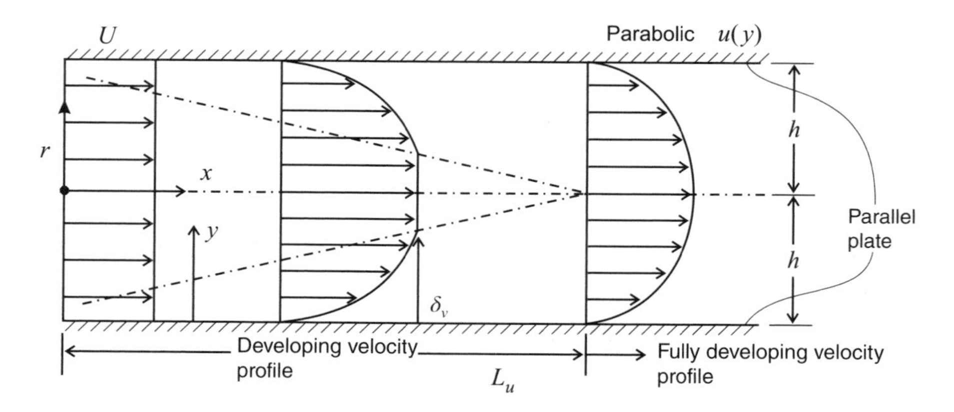

Figure 4.16 Schematic of a developing velocity boundary layer in the internal flow situation

As $\nu \gg D$ for diffusion in dilute solution, the development length is very much longer for the concentration profile than that for the velocity profile. For fully developed channel flow ($v = 0$, $u = u(y)$), using the coordinate transformation, we have

$$u = u_{\max}\left(1 - \frac{r^2}{h^2}\right) = u_{\max}\left(1 - \frac{(y-h)^2}{h^2}\right)$$

$$u = u_{\max}\left(\frac{h^2 - y^2 - h^2 + 2yh}{h^2}\right) = u_{\max}\left(\frac{2yh - y^2}{h^2}\right) \tag{4.85}$$

and

$$u_{\max} = \frac{3}{2}U = -\frac{h^2}{2\mu}\frac{dp}{dx} \tag{4.86}$$

At close to the wall, we can approximate from equation (4.85) as

$$u \simeq u_{\max}\left(\frac{2y}{h}\right) \tag{4.87}$$

We can assume that velocity profile is fully developed and developing diffusion boundary layer thickness δ_D is small in comparison to the channel half-height. Let us estimate δ_D by substituting the above linear profile to

$$D\frac{\partial^2 c}{\partial y^2} \sim u\frac{\partial c}{\partial x} \tag{4.88}$$

$$D\frac{c_0 - c_w}{\delta_D^2} \sim u_{\max}\frac{2\delta_D}{h}\frac{c_0 - c_w}{x} \tag{4.89}$$

which gives

$$\frac{\delta_D}{x} \sim \left(\frac{h}{x}\right)^{2/3}\left(\frac{D}{u_{\max}h}\right)^{1/3} \tag{4.90}$$

Note: The diffusion layer thickness grows as the cubic root of the streamwise distance rather than the square root as in unbounded flow past a surface. Moreover, the growth is independent of the kinematic viscosity and therefore of the Schmidt number, although it does depend on viscosity through $u_{\max}$. Setting $\delta_D = h$, $x = L_D$ and assuming $u_{\max} \sim U$, the development length is

$$\frac{L_D}{h} \sim \frac{Uh}{D} \equiv Pe_D$$

The development length of a diffusion boundary layer is of the same order of magnitude as the Peclet number (diffusion).

4.12 Detailed Analysis

In this section, the detailed analysis of reacting or soluble flat plate in external and internal flow situation is presented. The analytical expression for diffusion boundary layer thickness is compared with that from the order of magnitude estimate derived in the previous section.

4.12.1 *Flow Past a Reacting Flat Plate*

Here, we concentrate on general similarity criteria and behavior that can be deduced from the well-known Blasius solution, for flow past a nonreacting flat plate. For high Reynolds number boundary layer flow past a flat plate with no gravitational force in the streamwise direction, for constant viscosity, the boundary layer equation is given by

$$\frac{Du}{Dt} = \nu \frac{\partial u^2}{\partial y^2} \tag{4.91}$$

The corresponding convection diffusion equation is

$$\frac{Dc}{Dt} = D \frac{\partial^2 c}{\partial y^2} \tag{4.92}$$

Let the nondimensional velocity and concentration be defined as

$$u^* = \frac{u}{U} \qquad \text{and} \qquad c^* = \frac{c - c_w}{c_0 - c_w}$$

where U is the velocity at the edge of the viscous boundary layer, c_0 is the solute concentration at the edge of diffusion boundary layer, and c_w is the concentration at the wall.

The boundary conditions in nondimensional form are

$$\begin{aligned} u^* = 1,\ c^* = 1 \quad &\text{as} \quad y \to \infty \\ u^* = 0,\ c^* = 0 \quad &\text{as} \quad y = 0 \end{aligned}$$

Case-I

$Sc = 1 (\nu = D$, Dilute gases)

In this case, the governing equations and boundary conditions are identical for velocity and concentration. Thus, we can write

$$c^* = u^*$$

or

$$c = c_w + (c_0 - c_w)\frac{u}{U}$$

Let us use this result for an analogy between skin friction and mass transfer coefficient. Skin friction coefficient is defined as

$$C_f = \frac{(\tau_{yx})_w}{\frac{1}{2}pU^2} = \frac{\mu\left(\frac{du}{dy}\right)_w}{\frac{1}{2}\rho U^2} = \frac{2\nu}{U}\left(\frac{\partial u^*}{\partial y}\right)_w \tag{4.93}$$

Mass transfer coefficient may be defined as

$$C_{\text{diff}} = \frac{J_w}{(c_0 - c_w)_U} = \frac{D\frac{\partial c}{\partial y}}{(c_0 - c_w)U} = \frac{D}{U}\left(\frac{\partial c^*}{\partial y}\right)_w \tag{4.94}$$

Using $u^* = c^*$ and $\nu = D$, we get

$$C_{\text{diff}} = \frac{C_f}{2} \quad \text{(Reynolds analogy)} \tag{4.95}$$

Note: The above analogy is frequently used to denote a correspondence between heat and momentum transfer. The same agreement as the above can be applied to heat transfer for a constant wall temperature with $\nu = \alpha$, where we get similar equation as the above by replacing *dimensionless heat transfer coefficient in place of mass transfer coefficient*.

Case-II

$Sc = \nu/D \gg 1$ (Dilute solutions)

In this case, the diffusion boundary layer is embedded in the viscous boundary layer and the velocity it sees is that close to the wall. The Blasius solution for flat plate boundary layer in the series form is

$$u = U\left[0.332\eta - \frac{0.028}{12}\eta^4 + 0\left(\eta^7\right)\right] \tag{4.96}$$

$$v = \left(\frac{\nu U}{x}\right)^{1/2}\left[0.083\eta^2 + 0\left(\eta^5\right)\right] \tag{4.97}$$

where the similarity coordinate is

$$\eta = y\left(\frac{U}{\nu x}\right)^{1/2} \tag{4.98}$$

If the mass transfer rate is small, then the velocity profile is unaltered by diffusion and the above series expression for velocity can be applied in the convective-diffusion equation. Here, Sc being large, we can assume that convection dominates over diffusion. We assume that at close to the wall, η is small and higher-order terms are neglected. We rewrite equations (4.96) and (4.97)) as

$$u = U[0.332\eta]$$

$$u = 0.332yU\left(\frac{U}{\nu x}\right)^{1/2} = a(x)y \tag{4.99}$$

$$v = \left(\frac{\nu U}{x}\right)^{1/2} 0.083\eta^2$$

$$v = \left(\frac{\nu U}{x}\right)^{1/2} 0.083y^2\frac{U}{\nu x} = b(x)y^2 \tag{4.100}$$

Substituting the above two equations into the boundary layer equation, we have

$$u\frac{\partial u}{\partial x} + v\frac{\partial u}{\partial y} = \nu\frac{\partial^2 u}{\partial y^2} \tag{4.101}$$

$$a(x)y\frac{\partial u}{\partial x} + b(x)y^2\frac{\partial u}{\partial y} = \nu\frac{\partial^2 u}{\partial y^2} \tag{4.102}$$

Dividing both the sides of the above equation by U, we get

$$\Rightarrow a(x)y\frac{\partial u^*}{\partial x} + b(x)y^2\frac{\partial u^*}{\partial y} = \nu\frac{\partial^2 u^*}{\partial y^2} \tag{4.103}$$

Similarly, for concentration, we have

$$a(x)y\frac{\partial c^*}{\partial x} + b(x)y^2\frac{\partial c^*}{\partial y} = D\frac{\partial^2 c^*}{\partial y^2} \tag{4.104}$$

Let us transform the coordinate by using the dimensionless variable, ξ, as

$$\xi = y\left(\frac{\nu}{D}\right)^{1/3}$$

Rewriting the above concentration boundary layer equation (equation (4.104)) in terms of ξ, we have

$$a(x)\xi\left(\frac{D}{\nu}\right)^{1/3}\frac{\partial c^*}{\partial x} + b(x)\xi^2\left(\frac{D}{\nu}\right)^{2/3}\left(\frac{\partial c^*}{\partial \xi}\right)\left(\frac{\nu}{D}\right)^{1/3} = D\frac{\partial^2 c^*}{\partial \xi^2}\left(\frac{\nu}{D}\right)^{2/3}$$

$$\left(a(x)\xi\frac{\partial c^*}{\partial x} + b(x)\xi^2\frac{\partial c^*}{\partial \xi}\right)\left(\frac{D}{\nu}\right)^{1/3} = \nu\frac{\partial^2 c^*}{\partial \xi^2}\left(\frac{D}{\nu}\right)^{1/3}$$

$$a(x)\xi\frac{\partial c^*}{\partial x} + b(x)\xi^2\frac{\partial c^*}{\partial \xi} = \nu\frac{\partial^2 c^*}{\partial \xi^2} \tag{4.105}$$

As the above nondimensional concentration equation (equation (4.105)) is similar to the nondimensional velocity governing equation (equation (4.104)), we can write

$$c^*(x, \xi) = u^*(x, y)$$

We write the expression for skin friction coefficient and mass transfer coefficient as

$$C_f = \frac{2\nu}{U}\left(\frac{\partial u^*}{\partial y}\right)_w \tag{4.106}$$

$$C_{\text{diff}} = \frac{D}{U}\left(\frac{\partial c^*}{\partial y}\right)_w = \frac{D}{U}\left(\frac{\nu}{D}\right)^{1/3}\left(\frac{\partial c^*}{\partial \xi}\right)_w \tag{4.107}$$

Using

$$\frac{\partial c *}{\partial \xi} = \frac{\partial u^*}{\partial y}$$

we have

$$C_{\text{diff}} = \frac{C_f}{2}\frac{1}{(Sc)^{2/3}} \tag{4.108}$$

The above equation is known as the Chilton–Colburn analogy.

Note: (1) The above analogy is derived for large Schmidt numbers. But it is found to be quite accurate down to the Schmidt number of 0.5.

(2) The above arrangement could have been applied to heat and momentum transfer with C and D replaced by T and α, respectively. The resulting formula would be the same as that above, but the Schmidt number is replaced by Prandtl number and the dimensionless diffusion coefficient is replaced by dimensionless heat transfer coefficient. Let us derive the expression for diffusion boundary layer thickness.

The Nernst relation is written as

$$\delta_D = \frac{D(c_0 - c_w)}{j_w^*} \tag{4.109}$$

Blasius flat plate solution is

$$C_f = \frac{0.664}{(Re_x)^{1/2}} \tag{4.110}$$

Combining the above three equations for C_{diff}, δ_D, and C_f, we have

$$\frac{\delta_D}{x} = \frac{3}{Sc^{1/3}} \frac{1}{Re_x 1/2} \tag{4.111}$$

We recall that for the flat plate boundary layer, the viscous boundary layer thickness (δ_v) is expressed as

$$\frac{\delta_V}{x} = \frac{5.0}{(Re_x)^{1/2}} \tag{4.112}$$

Therefore, from the above two equations, we have

$$\delta_D = \frac{0.6}{(Sc)^{1/3}} \delta_V \tag{4.113}$$

Note: From order of magnitude analysis, we have observed earlier

$$\delta_D \sim \frac{\delta_V}{Sc^{1/3}}$$

Hence, the detailed analysis provides similar conclusion as that of the order of magnitude analysis.

4.12.2 Channel Flow with Soluble or Rapidly Reacting Walls

For a soluble surface, where the dissolution process is much more rapid than the removal of dissolved particles, we have $c_w = c_{\text{sat}}$, where c_{sat} is the equilibrium concentration of the saturated solution in the liquid layer at the surface. The schematic of a channel wall with soluble walls is shown in Figure 4.17. The complementary of this problem is the one for reacting wall in the fast reaction limit where $c_w = 0$ and concentration at the channel center is the bulk or the free stream value c_0. The reaction is rapid indicating that the problem is diffusion controlled and therefore we investigate the convective-diffusion phenomena for this problem.

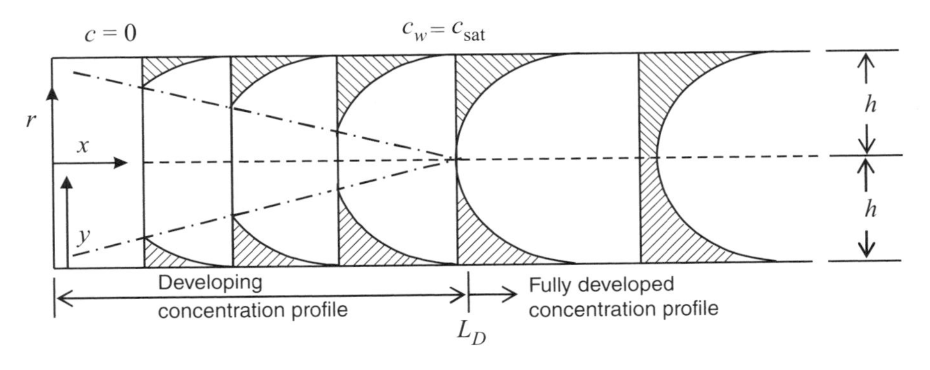

Figure 4.17 The schematic of a channel with soluble walls

Similarity variable:
We have seen in the last example that near the wall the velocity distribution can be approximated as

$$u \simeq u_{\max}\left(\frac{2y}{h}\right) \tag{4.114}$$

Using this approximation with

$$D\frac{\partial^2 c}{\partial y^2} \sim u\frac{\partial c}{\partial x}$$

We can write

$$D\frac{\partial^2 c}{\partial y^2} \sim \frac{2y}{h}u_{\max}\frac{\partial c}{\partial x} \tag{4.115}$$

Based on the above equation, we can write

$$\frac{c}{c_{\text{sat}}} \equiv c^* = f\left(x, y, \frac{u_{\max}}{Dh}\right) \tag{4.116}$$

Using the nondimensional combination of the variables in the right-hand side, we have

$$c^* = f(\eta)$$

where

$$\eta = \text{similarity variable} = y\left(\frac{2u_{\max}}{hDx}\right)^{1/3}$$

where the factor of "2" has been introduced for convenience. The general governing equation for convective diffusion is

$$u\frac{\partial c}{\partial x} + v\frac{\partial c}{\partial y} = D\frac{\partial^2 c}{\partial y^2} \tag{4.117}$$

The boundary conditions are

$$c = c_{\text{sat}} \text{ at } y = 0 \text{ and } c = 0, \text{ as } y \to \infty$$

The boundary condition of $c = 0$ at $y = h$ is replaced by the asymptotic one of $c = 0$ as $y \to \infty$.

Dividing the governing equation (equation (4.117)) by c_{sat} in both the sides, we have

$$u\frac{\partial c^*}{\partial x} + v\frac{\partial c^*}{\partial y} = D\frac{\partial^2 c^*}{\partial y^2} \tag{4.118}$$

We can now write the individual terms of the governing equation in terms of the similarity variables as

$$\frac{\partial c^*}{\partial x} = \frac{\partial c^*}{\partial \eta}\frac{\partial \eta}{\partial x} = -\frac{\eta}{3x}\frac{dc^*}{d\eta} \tag{4.119}$$

$$\frac{\partial c^*}{\partial y} = \frac{\partial c^*}{\partial \eta}\frac{\partial \eta}{\partial y} = \left(\frac{2u_{\max}}{hDx}\right)^{1/3}\frac{dc}{d\eta} \tag{4.120}$$

$$\frac{\partial^2 c^*}{\partial y^2} = \left(\frac{2u_{\max}}{hDx}\right)^{2/3}\frac{d^2 c^*}{dn^2} \tag{4.121}$$

With $v = 0$ for fully developed flow case, equation (4.118) simplifies to

$$u\frac{\partial c^*}{\partial x} = D\frac{\partial^2 c^*}{\partial y^2} \tag{4.122}$$

We assume the conditions at close to the entrance with $y/h \ll 1$ and the velocity profile to be approximated by the profile near the wall.
Substituting equations (4.119) and (4.121) in equation (4.122), we have

$$\left(\frac{2y}{h}u_{\max}\right)\left(-\frac{\eta}{3x}\frac{dc^*}{d\eta}\right) = D\left(\frac{2u_{\max}}{hDx}\right)^{2/3}\frac{d^2c^*}{d\eta^2}$$

$$\frac{\eta^2}{3}\frac{dc^*}{d\eta} + \frac{d^2c^*}{d\eta^2} = 0 \tag{4.123}$$

The boundary conditions in similarity coordinates are

$$c^* = 1 \text{ at } \eta = 0 \text{ and } c^* = 0 \text{ as } \eta \to \infty$$

Equation (4.123) can be written as

$$\frac{d}{d\eta}\left(\ln\frac{dc^*}{d\eta}\right) = -\frac{n^2}{3} \tag{4.124}$$

Integrating the above equation twice gives

$$c^* = A\int_0^{\eta} e^{-n^3/9}d\eta + B \tag{4.125}$$

Using the boundary conditions and evaluating constants A and B, we get

$$c^* = 1 - \frac{\int_0^{\eta} e^{-n^3/9}d\eta}{\int_0^{\infty} e^{-n^3/9}d\eta} \tag{4.126}$$

The denominator in the above equation can be written as

$$\int_0^{\infty} e^{-n^3/9}d\eta = \int_0^{\infty} e^{-\left(n/\sqrt[3]{9}\right)^3}d\eta$$

Let

$$u = \frac{\eta}{\sqrt[3]{9}} \qquad \therefore \qquad du = \frac{d\eta}{\sqrt[3]{9}}$$

Substituting in the above integral, we get

$$\int_0^{\infty} e^{-u^3}\sqrt[3]{9}du = (9)^{1/3}\int_0^{\infty} e^{-u^3}du$$

Let

$$u^3 = z \qquad \therefore \qquad du = \frac{dz}{3u^2}$$

Thus, we get

$$(9)^{1/3}\int_0^\infty e^{-z}\frac{dz}{3u^2} = (9)^{1/3}\int_0^\infty e^{-z}\frac{dz}{3(z)^{2/3}}$$

$$= \frac{(9)^{1/3}}{3}\int_0^\infty e^{-z}z^{-\frac{2}{3}}dz = \frac{(9)^{1/3}}{3}\int_0^\infty z^{1/3-1}e^{-z}dz$$

$$= \frac{(9)^{1/3}}{3}\Gamma\left(\frac{1}{3}\right) = 1.850$$

where the gamma function is defined as

$$\Gamma(\eta) = \int_0^\infty z^{n-1}e^{-z}dz$$

Thus, it follows that

$$c^* = \frac{c}{c_{sat}} = 1 - 0.538\int_0^\eta e^{-\eta^3/9}d\eta \tag{4.127}$$

Note: The solution for rapidly reacting wall ($c_w = 0$ and $c = c_0$ at channel center) is simply the complement of the above solution, that is,

$$\frac{c}{c_0} = 0.538\int_0^\eta e^{\frac{-\eta^3}{9}}d\eta \tag{4.128}$$

We know that

$$\frac{\partial c^*}{\partial y} = \frac{dc^*}{d\eta}\frac{\partial n}{\partial y} = \left(\frac{2u_{max}}{hDx}\right)^{1/3}\frac{dc^*}{d\eta}$$

Therefore, the diffusion flux is written as

$$j^* = -D\left(\frac{dc}{dy}\right)_w = -c_{sat}D\left(\frac{2u_{max}}{hDx}\right)^{1/3}\left(\frac{dc^*}{d\eta}\right)_w \tag{4.129}$$

From the solution for

$$\left(\frac{c}{c_{sat}} = 1 - 0.538\int_0^\eta e^{\frac{-\eta^3}{9}d\eta}\right) \tag{4.130}$$

we have

$$j^* = 0.678c_{sat}D\left(\frac{u_{max}}{hDx}\right)^{1/3} \tag{4.131}$$

The mass flux can also be written in terms of mass transfer coefficient (h_m) as

$$j^* \simeq h_m(c_o - c_w) \tag{4.132}$$

Using linear approximation known as the Nernst layer approximation, that is,

$$j^* \simeq \frac{D(c_0 - c_w)}{\delta_D} \tag{4.133}$$

we have

$$j^* \simeq \frac{Dc_{sat}}{\delta_D}(\because \quad c_0 = 0, c_w = c_{sat}) \tag{4.134}$$

Equating the two equations for j^* (equations (4.131) and (4.134)), we have

$$\frac{\delta_D}{x} = 1.475\left(\frac{h}{x}\right)^{2/3}\left(\frac{D}{u_{\max}h}\right)^{1/3} \tag{4.135}$$

Remember from order of magnitude analysis, we have derived in the earlier section

$$\frac{\delta_D}{x} \sim \left(\frac{h}{x}\right)^{2/3}\left(\frac{D}{u_{\max}h}\right)^{1/3}$$

Hence, the detailed analytical solution and order of magnitude calculation match with each other.

4.13 Reverse Osmosis

An example of reverse osmosis can be used as an illustration of mixed heterogeneous reaction. Reverse osmosis is a *pressure-driven membrane process* used to separate relatively pure solvents, most often water from solutions containing salts and dissolved organic molecules. The solvent passes through the membrane under the action of hydrostatic pressure leaving the dissolved materials behind.

The membrane can be *semipermeable asymmetric membrane* made from *cellulose acetate* or *composite membranes* made with dense thin *polymer coating* on a *polystyrene support* (see Figure 4.18). *Asymmetric cellulose acetate membrane* consists of a thin *rejecting "skin"* about 0.1–0.5 μm thick integral with a much thicker porous substrate of 50–100 μm thick. The rejecting screen is permeable to water and relatively impermeable to various dissolved impurities, that is, salt ions and other small molecules that cannot be filtered. The permeability of the membrane depends on the construction of the membrane and the solute size. The skin offers hydraulic resistance to the flow. The porous substrate gives the membrane strength but offers

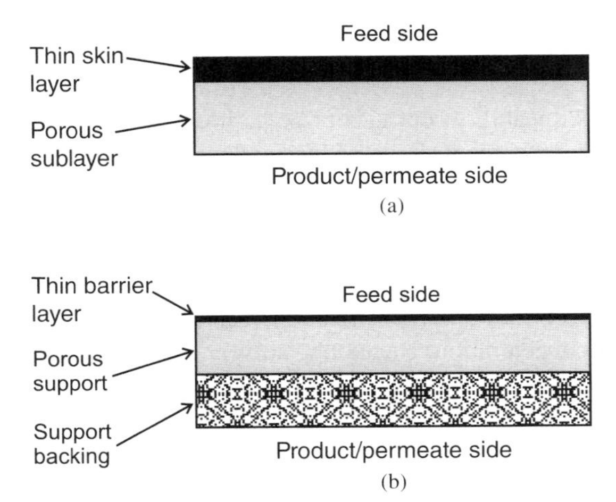

Figure 4.18 Schematic of (a) asymmetric membrane and (b) thin-film composite membrane

Table 4.1 The osmotic pressure of NaCl at different concentrations

Dissolved species	Concentration (kg/m^3)	Osmotic pressure (MPa)
NaCl ($M = 58.5$)	50	4.609
NaCl ($M = 58.5$)	10	0.844
NaCl ($M = 58.5$)	05	0.421

almost no hydraulic resistance. The dense rejecting skin of the composite membrane can be up to 10 times thinner than the cellulose acetate membranes.

When an ideal semipermeable membrane separates an aqueous organic or inorganic solution from pure water, the tendency to equalize concentrations would result in the flow of pure water through the membrane to the solution. The pressure needed to stop the flow is called the *osmotic pressure*. If the pressure on the solution is increased beyond the osmotic pressure, then the flow would be reversed and the fresh water would pass from the solution through the membrane, therefore the name *reverse osmosis*. Osmotic pressure is a *property of the solution* and does not in any way depend on the properties of the membrane.

Osmosis pressure data for some aqueous solutions at standard temperature are shown in Table 4.1. It shows that with an increase in NaCl concentration, the osmotic pressure increases.

The solvent (water) flux through the membrane may be written as

$$j_A^* = A(\Delta P - \Delta\pi) \tag{4.136}$$

where A is the solvent permeability coefficient; ΔP, hydrostatic pressure difference across the membranes, and $\Delta\pi$, osmotic pressure difference corresponding to the solute concentrations immediately adjacent to the membrane surface on both the sides. The driving force for the solute flux, j_s^*, can be written as

$$j_s^* = B\Delta c_w = c_w(1 - R_s) \tag{4.137}$$

where B is the solute permeability coefficient; Δc_w, difference in solute concentration immediately adjacent to the membrane wall on the feed and product sides; c_w, concentration on the feed side; and R_s, solute reject coefficient.

As a consequence of the passage of solvent, say, water through the membrane, the solute is carried out to the membrane surface. Hence, the concentration at the membrane surface tends to be higher than in the bulk of the liquid. This phenomenon is called *concentration polarization*. Due to the concentration polarization, the *osmotic pressure increases* because of the increase in solute concentration. Hence, the solvent flux is decreased because the *effective driving pressure* is reduced based on equation (4.136). Another effect is the increase in the solute concentration in the product side for *leaky membranes* as the flux of the solute across the membrane is proportional to the difference in solute concentration of both sides. Therefore, the concentration distribution of the solute inside the reverse osmosis channel, that is, concentration polarization, influences its performance and has been discussed in the following section.

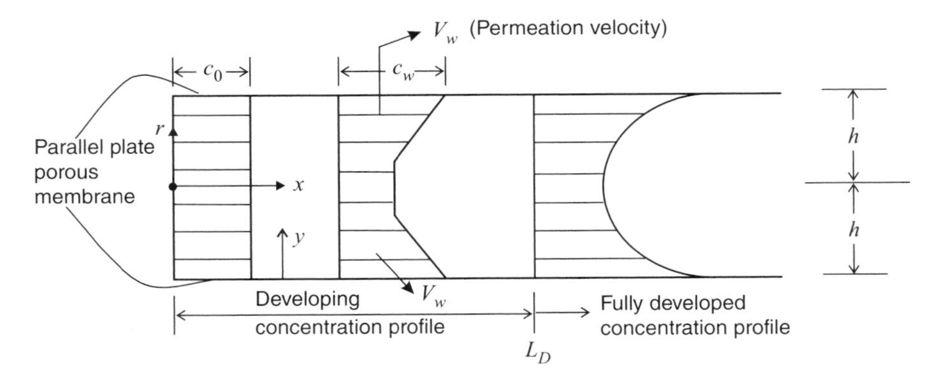

Figure 4.19 A channel flow undergoing reverse osmosis process

4.13.1 Reverse Osmosis Channel Flow

Here, we discuss the analytical solution of a reverse osmosis process in a channel flow situation. Let us assume that the velocity profile is fully developed at channel inlet and the channel wall membrane is totally impermeable to the solute ($R_s = 1$). The permeation velocity of the solvent brings the solute toward the wall resulting in an equilibrium concentration, c_w, at close to the wall (see Figure 4.19).

In contrast to convection diffusion equation treated in the channel flow problem with soluble or rapidly reacting walls, it is necessary to include here the lateral convection term to account for the product removal through the membrane walls (see Figure 4.19):

$$u\frac{\partial c}{\partial x} + v\frac{\partial c}{\partial y} = D\frac{\partial^2 c}{\partial y^2} \tag{4.138}$$

Due to constant permeation velocity (v_w), the velocity in the x-direction is altered causing streamwise variation of bulk velocity. The Reynolds number based on v_w is small $\left(\frac{hv_w}{\nu} \ll 1\right)$, and therefore, the u-velocity distribution has the same form as impermeable one and the v-velocity is proportional to the v_w. The N–S equation in the x-direction is written as

$$\rho\left(u\frac{\partial u}{\partial x} + v\frac{\partial u}{\partial y}\right) = -\frac{dp}{dx} + \mu\left(\frac{\partial^2 u}{\partial y^2} + \frac{\partial^2 u}{\partial x^2}\right) \tag{4.139}$$

For *impermeable wall*, because of the fully developed condition, $\frac{\partial u}{\partial x} \sim 0$ and $v \sim 0$, $\frac{\partial^2 u}{\partial x^2} \sim 0$. Thus, we get the simplified N–S equation in the x-direction as

$$\frac{\partial P}{\partial x} = \mu\frac{d^2 u}{dy^2} \tag{4.140}$$

The governing equation in $x - r$ coordinate can be written as

$$\frac{\partial P}{\partial x} = \mu\frac{d^2 u}{dr^2} \tag{4.141}$$

$$u = \frac{1}{\mu}\frac{dp}{dx}\frac{r^2}{2} + c_1 r + c_2 \tag{4.142}$$

Boundary condition:
At centerline, that is, at $r = 0$, $\frac{du}{dy} = 0$. Thus, we get $c_1 = 0$.
The no-slip boundary condition is at $r = h, u = 0 \Rightarrow c_2 = -\frac{1}{\mu}\frac{dp}{dx}\frac{h^2}{2}$

$$u = \frac{1}{\mu}\frac{dp}{dx}\frac{h^2}{2}\left[\frac{r^2}{h^2} - 1\right] \tag{4.143}$$

We can derive the expression for average velocity (u_{rmav}) by integration, that is,

$$u_{rmav} \times 2h = 2\int_0^h \frac{1}{\mu}\frac{dp}{dx}\frac{h^2}{2}\left[\frac{r^2}{h^2} - 1\right] dr$$

$$= \frac{2}{3\mu}\frac{dp}{dx}h^3$$

$$\therefore \quad u_{rmav} = -\frac{1}{3\mu}\frac{dp}{dx}h^2$$

$$u = -\frac{3}{2}u_{rmav}\frac{r^2}{h^2} - 1 = \frac{3}{2}u_{rmav}\left[1 - \frac{r^2}{h^2}\right] \tag{4.144}$$

The above derivation holds good for impermeable wall. For *permeable wall*, the average velocity will vary in the x-direction, that is,

$$u_{rmav} = f(x)$$

If we consider a differential element of size Δx of the channel, we can write the average velocity at the end of the differential element equal to $u_{rmav} + \frac{\partial}{\partial x}(u_{rmav})\Delta x$, where u_{rmav} is the velocity at the inlet of differential element.
Mass conservation gives

$$\left(\frac{\partial}{\partial x}u_{rmav}\Delta x\right) 2h = 2v_w \Delta x \tag{4.145}$$

The left-hand side of the above equation is the mass output in the streamwise direction. The right-hand side is the mass input due to the permeation through the wall in the y-direction. The above equation simplifies to

$$\frac{\partial u_{rmav}}{\partial x} = \frac{v_w}{h} \tag{4.146}$$

The two-dimensional continuity equation in differential form is

$$\frac{\partial u}{\partial x} + \frac{\partial v}{\partial r} = 0$$

Here, r" refers to the coordinate system at the center of the channel.
Using equation (4.144), we can write the above continuity equation as

$$\frac{3}{2}\left(1 - \frac{r^2}{h^2}\right)\frac{du_{rmav}}{dx} + \frac{\partial v}{\partial r} = 0 \tag{4.147}$$

Using equation (4.146) in the above equation, we can write

$$dv = \frac{3}{2}\left(1 - \frac{r^2}{h^2}\right)\frac{v_w}{h}dr$$

$$v = \frac{3}{2}\frac{v_w}{h}\left(r - \frac{r^3}{3h^2}\right) = \frac{v_w}{2h}\left(3r - \frac{r^3}{h^2}\right)$$

$$v = \frac{v_w r}{2h}\left(3 - \frac{r^2}{h^2}\right) \tag{4.148}$$

Thus, we have

$$u = \frac{3}{2}u_{rmav}\left(1 - \frac{r^2}{h^2}\right) \tag{4.149}$$

$$v = v_w\frac{r}{2h}\left(3 - \frac{r^2}{h^2}\right) \tag{4.150}$$

For developing region, $y \ll h$, that is, $\frac{y}{h} \ll 1$, substituting in equations (4.149) and (4.150) with $r = y - h$ and assuming small y, we have

$$u \approx U_{\max}\left(\frac{2y}{h}\right) = 3u_{rmav}\frac{y}{h} \tag{4.151}$$

$$v = -v_w \tag{4.152}$$

Initial condition: $c = c_0$ *at* $x = 0$
Boundary condition: From the conservation of solute flux applied across the membrane:

$$-v_w c_w - D\left(\frac{\partial c}{\partial y}\right)_w = -(1 - R_s)v_w c_w \tag{4.153}$$

where $-v_w c_w$ is the bulk flow of solute toward the membrane; $-D\left(\frac{dc}{dy}\right)_w$, diffusion flux of solute away from the membrane toward the bulk fluid; and $(1 - R_s)v_w c_w$, solute permeation through the membrane. The above boundary condition indicates that the bulk flow of solute toward the membrane minus the diffusional flux of the solute away from the membrane is equal to the solute permeation through the membrane. Thus, we get

$$R_s v_w c_w = -D\left.\frac{\partial c}{\partial y}\right|_w \tag{4.154}$$

Assuming $R_s \simeq 1$, that is, total impermeable wall for the solute, we have

$$v_w c_w = -D\left.\frac{\partial c}{\partial y}\right|_w \tag{4.155}$$

Nondimensionalizing the above equation, we have

$$\frac{v_w}{D/h}c_w^* = -\left(\frac{\partial c^*}{\partial y^*}\right)_w \tag{4.156}$$

From the above nondimensionalized boundary condition, we can define the dimensionless number, Damkohler number, as

$$\text{Damkohler number} = (Da) = \frac{v_w}{D/h} = \frac{\text{Permeation velocity}}{\text{Diffusion velocity}}$$

The above number can also be interpreted as the Peclet number for mass transfer:

$$Pe = \frac{v_w h}{D} = \frac{\text{Mass transfer by permeation}}{\text{Mass transfer by diffusion}}$$

Note: The Damkohler number can also be derived from the order of reaction concept:

$$R_i = Kc_i^\gamma$$

Here, R_i is the production of species; K, rate constant; c_i, concentration; and γ, order of reaction. The species reaction rate is often expressed empirically by the power law relation. When surface is impermeable, $v_w = 0$:

$$(j_i^*)_w = D_i\left(\frac{\partial c_i}{\partial y}\right)_w$$

$$\therefore \quad D\left(\frac{\partial c}{\partial y}\right)_w = Kc_w^\gamma \tag{4.157}$$

Let the nondimensionalization parameters are defined as: $c^* = \dfrac{c}{c_0}$, $y^* = \dfrac{y}{h}$.

We can rewrite the above governing equation as

$$\frac{1}{Da}\left(\frac{\partial c^*}{\partial y^*}\right)_w = c_w^{*\gamma}$$

where

$$Da = \frac{Kc_0^{\gamma-1}}{D/h} = \frac{\text{Reaction velocity}}{\text{Diffusion velocity}}$$

When $Da \gg 1$:

$$c_w \approx 0$$

This implies that the rate of species production by reaction is large compared to the rate of mass transfer by diffusion. The meaning of this condition is that all particles approaching the surface react instantaneously.

When $Da \ll 1$:

$$\left(\frac{dc}{dy}\right)_w = 0$$

Here, the rate of species production by reaction is small compared with the diffusion flux. In this limit, the concentration is everywhere constant and equal to c_0. We should recognize that the mass transfer through a semipermeable membrane is directly analogous to that of a mixed heterogeneous reaction.

Using the simplified boundary condition based on the solute flux derived above (equation (4.155)), that is,

$$v_w c_w = -D\frac{dc}{dy}\bigg|_w$$

We can have an order of magnitude estimate as

$$\frac{c_w - c_0}{c_0} \sim \frac{v_w \delta_D}{D} \tag{4.158}$$

Using the estimate for δ_D of the developing concentration boundary layer in a channel, that is,

$$\frac{\delta_D}{x} \sim \left(\frac{h}{x}\right)^{2/3}\left(\frac{D}{u_{\max}h}\right)^{1/3}$$

We can write equation (4.158) as

$$\begin{aligned}\frac{c_w - c_0}{c_0} &\sim \frac{v_w}{D}x\left(\frac{h}{x}\right)^{2/3}\left(\frac{D}{u_{rmav}h}\right)^{1/3} \\ &\sim \left(\frac{v_w h}{D}\right)\frac{x^{1/3}}{h^{1/3}}\left(\frac{D}{v_{rmav}h}\right)^{1/3} \sim \zeta^{1/3}\end{aligned} \tag{4.159}$$

where

$$\zeta = \left(\frac{v_w h}{D}\right)^3\left(\frac{x}{h}\right)\left(\frac{D}{3v_{rmav}h}\right)$$

Note: 3 has been added for later convenience.

Looking at the above boundary condition (equation (4.159)), we propose the solution of the convective diffusion equation as

$$\frac{c - c_0}{c_0} \equiv c^* = \zeta^{1/3} f(\eta), \tag{4.160}$$

where the similarly variable, η, is defined similar to that for reacting wall case as

$$\eta = y\left(\frac{3u_{rmav}}{hDx}\right)^{1/3} = \frac{y}{\zeta^{1/3}}\left(\frac{v_w}{D}\right)$$

Using chain rule, the partial derivative of $c^*(\zeta, \eta)$ can be written as

$$\frac{\partial c^*}{\partial x} = \frac{\partial c^*}{\partial \zeta}\frac{d\zeta}{dx} + \frac{\partial c^*}{\partial \eta}\frac{d\eta}{dx} \tag{4.161}$$

We have

$$\frac{\partial c^*}{\partial x} = \frac{\zeta^{1/3}}{3x}\left(f - \eta\frac{df}{d\eta}\right) \tag{4.162}$$

$$\frac{\partial c^*}{\partial y} = \frac{\partial c^*}{\partial n}\cdot\frac{d\eta}{dy} = \left(\frac{3u_{rmav}}{hDx}\right)^{1/3}\zeta^{1/3}\frac{df}{d\eta} \tag{4.163}$$

$$\frac{\partial^2 c^*}{\partial y^2} = \left(\frac{3u_{rmav}}{hDx}\right)^{2/3}\zeta^{1/3}\frac{d^2 f}{d\eta^2} \tag{4.164}$$

The convective diffusion equation in dimensionless form is

$$u\frac{\partial c^*}{\partial x} + v\frac{\partial c^*}{\partial y} = D\frac{\partial^2 c^*}{\partial y^2} \tag{4.165}$$

Using the approximation for developing boundary layer $u = 3u_{rmav}\frac{y}{h}$, $v = v_w$, and the expressions for derivative terms from above, we get

$$3u_{rmav}\frac{y}{h}\frac{\zeta^{1/3}}{3x}\left(f - \eta\frac{df}{d\eta}\right) - v_w\left(\frac{3u_{rmav}}{hDx}\right)^{1/3}\zeta^{1/3}\frac{df}{d\eta} = D\left(\frac{3u_{rmav}}{hDx}\right)^{2/3}\zeta^{1/3}\frac{d^2f}{d\eta^2}$$

$$\left(\frac{3u_{rmav}}{hDx}\right)\frac{y}{3}\left(f - \eta\frac{df}{d\eta}\right) - \frac{v_w}{D}\left(\frac{3u_{rmav}}{hDx}\right)^{1/3}\frac{df}{d\eta} = \left(\frac{3u_{rmav}}{hDx}\right)^{2/3}\frac{d^2f}{d\eta^2}$$

$$\frac{\eta}{3}\left(f - \eta\frac{df}{d\eta}\right) - \frac{v_w}{D}\left(\frac{3u_{rmav}}{hDx}\right)^{-(1/3)}\frac{df}{d\eta} = \frac{d^2f}{d\eta^2}$$

$$\frac{n}{3}\left(f - \eta\frac{df}{dn}\right) - \frac{v_w}{D}\zeta^{1/3}\frac{D}{v_w}\frac{df}{d\eta} = \frac{d^2f}{d\eta^2}$$

$$\frac{\eta}{3}\left(f - \eta\frac{df}{d\eta}\right) - \zeta^{1/3}\frac{df}{d\eta} = \frac{d^2f}{d\eta^2} \tag{4.166}$$

Let us assume $\zeta^{1/3} \to 0$ near the inlet. This assumption is reasonable from the expression of ζ as x is small at the inlet:

$$\zeta = \left(\frac{v_w h}{D}\right)^3\left(\frac{D}{3u_{rmav}h}\right)\left(\frac{x}{h}\right) \tag{4.167}$$

This assumption also indicates that normal convection term is neglected. We get the differential equation:

$$\frac{d^2f}{d\eta^2} + \frac{\eta^2}{3}\frac{df}{d\eta} - \frac{\eta}{3}f = 0 \tag{4.168}$$

Boundary Conditions

Using the earlier derived nondimensional solution

$$\frac{c - c_0}{c_0} = \zeta^{1/3}f(n)$$

we have

$$f \to 0 \quad \text{as} \quad n \to \infty \tag{B.C.1}$$

Using

$$\frac{\partial c^*}{\partial y} = \left(\frac{3u_{rmav}}{hDx}\right)^{1/3}\zeta^{1/3}\frac{df}{dn} \quad \text{and} \quad -D\frac{\partial c}{\partial y} = v_w c_w$$

we have

$$\frac{df}{d\eta} = -1 \quad \text{as} \quad \eta = 0 \tag{B.C.2}$$

This also requires that $\zeta^{1/3} \to 0$, that is, small value of ζ which is appropriate for developing layer near the inlet. Observing that $f = \eta$ is a solution of the above equation (equation (4.168)), we can propose $f = \eta g$ as a solution. Substituting this in equation (4.168), we get

$$\eta \frac{d^2 g}{d\eta^2} + \left(\frac{\eta^3}{3} + 2\right) \frac{dg}{d\eta} = 0 \tag{4.169}$$

Integrating twice and using the boundary condition at $\eta \to \infty$ (B.C.1)

$$g = -A \left(\frac{e^{-n^3/9}}{n}\right) - \frac{1}{3} \int_{\infty}^{n} n e^{-n^3/9} d\eta \tag{4.170}$$

Using boundary condition at $\eta \to 0$ (B.C.2), we have

$$A = -\left(\frac{1}{3} \int_{\infty}^{n} n e^{-n^3/9} d\eta\right)^{-1} = \frac{9^{1/3}}{-\Gamma\left(\frac{2}{3}\right)} = -1.536 \tag{4.171}$$

Hence,

$$f(\eta) = 1.536 \left(e^{-\eta^3/9} - \frac{\eta}{3} \int_{\infty}^{\eta} \eta e^{-\eta^{3/9}} d\eta\right) \tag{4.172}$$

where $f(0) = 1.536$. The variation of $f(\eta)$ has been presented in Figure 4.20. It should be noted that $\frac{c-c_0}{c_0} = \zeta^{1/3} f(\eta)$ and Figure 4.20 shows the nature of concentration distribution near the wall. The concentration distribution changes in the streamwise direction with an increase in ζ.

4.13.1.1 Mixing Cup Concentration

Mixing cup concentration varies along the channel and is defined as the concentration that would be measured at a streamwise location if the channel is chopped off. For a parallel

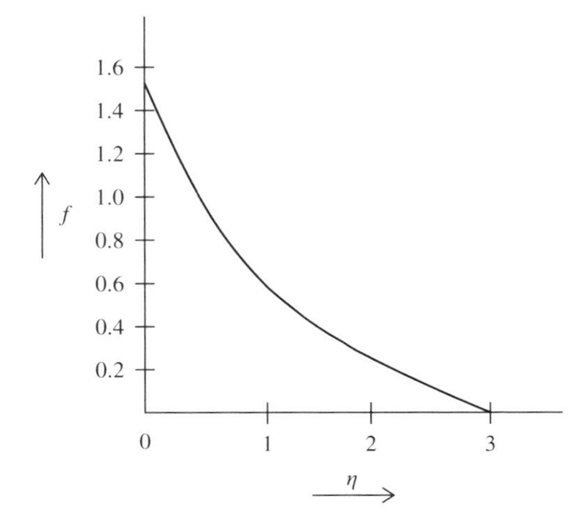

Figure 4.20 Similarity solution for concentration defect in developing boundary layer at inlet region of a reverse osmosis channel flow

channel at distance x, the fraction of solvent removed is $\frac{v_w x}{u_{rmav} h}$. The fraction of solvent left is $1 - \frac{v_w x}{u_{rmav} h}$. The solute concentration at inlet is c_0. If the membrane is completely rejecting, the concentration must be increased in proportion to the amount of water removed. Hence, the mixing cup concentration can be written as

$$c_m = \frac{1}{hu_{rmav}} \int_0^h ucdy \tag{4.173}$$

It also can be approximated as

$$c_m \simeq c_0 \left(1 - \frac{v_w x}{u_{rmav} h}\right)^{-1} \tag{4.174}$$

Let us assume that for a channel undergoing reverse osmosis, $v_w/u_{rmav} = 0.1$.

At $x/h = 1.0$, the above equation gives $\frac{c_m}{c_0} = 1.1$

At $x/h = 5.0$, $\frac{c_m}{c_0} = 2.0$

At $x/h = 8.0$, $c_m = 5c_0$.

At downstream locations, the increase in mixing cup concentration will also lead to corresponding increase in wall concentration, c_w.

4.13.1.2 Dimensionless Concentration Polarization Parameter

Dimensionless concentration polarization parameter measures the excess solute concentration at the membrane relative to the so-called mixing cup concentration, c_m:

$$\Gamma = \frac{c_w}{c_m} - 1 \tag{4.175}$$

Figure 4.21 shows the concentration polarization of a reverse osmosis channel with complete rejection membrane. It shows how the concentration polarization increases in the streamwise direction of the reverse osmosis channel flow. Let us investigate an example to understand the effect of concentration polarization on reverse osmosis channel flow.

Suppose NaCl solution of concentration 10 kg/m^3 flow through a reverse osmosis channel having concentration in the permeate side equal to 5 kg/m^3. The water permeability coefficient of water through the semipermeable membrane is equal to 0.5 kg/m^2-MPa and $\Delta P = 10$ MPa is applied from the feed side. The concentration of salt increases to 50 kg/m^3 at a downstream location.

The water flux at the inlet is $m_A = A(\Delta P - \Delta\pi)$

Using data from Table 4.1, we have water flux at the inlet region as

$$m_A(\text{inlet}) = (0.5\ \text{kg/(m}^2\text{-MPa)})(10 - 0.844 - 0.421)\ \text{MPa} = 4.3675\ \text{kg/m}^2$$

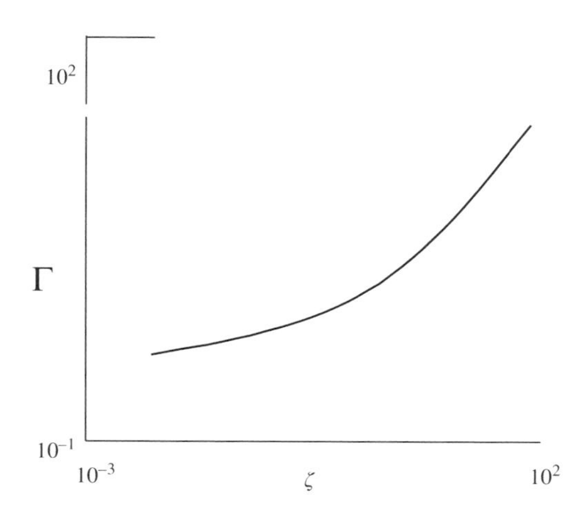

Figure 4.21 Concentration polarization development in downstream direction in a parallel membrane reverse osmosis channel with laminar flow and complete rejection

Similarly, the water flux at the downstream location is

$$m_A(\text{downstream}) = 0.5 \times (10 - 4.609 - 0.421) = 2.485\ \text{kg/m}^2$$

The above sample calculation indicates the influence of concentration polarization in reducing the solvent flux and thus reduction in the performance of the device.

Problems

4.1 An infinitely thin dye source is ejected in a long water-filled tube of radius, $a = 10\ \mu$m. The dye stream occupies the total width of the channel. The mean velocity of water inside the tube is equal to 1 cm/s. The molecular diffusivity of dye in water is equal to 5×10^{-6} cm^2/s. Calculate the time required for the dye stream to reach thickness, $w = 1$ mm.

4.2 A dilute solution having the Schmidt number equal to 1000 flows on the top of a reacting flat plate with length, $L = 1$ m, that is, in external flow configuration. The viscous boundary layer thickness is equal to 1 cm at streamwise location, $x = 0.5$ m, from the leading edge of the plate. Estimate the value of diffusion layer thickness at the same location, that is, at $x = 0.5$ m.

4.3 A reverse osmosis channel is used for purification of dissolved salt from water. The salt (NaCl) concentration of water in the feed side is equal to 10 kg/m^3 and in the product side is equal to 5 kg/m^3. The permeability coefficient of water through the semipermeable is equal to 0.5 kg/(m^2-MPa). A pressure, $\Delta P = 8$ MPa, is applied from the feed side during operation. The concentration polarization of the membrane leads to an increase in the concentration of salt at the outlet of channel equal to 50 kg/m^3. Calculate the percentage decrease in water flux at the outlet of the channel in comparison to that at the inlet of the channel. The osmotic pressure is equal to 4.609, 0.844, and 0.421 MPa at NaCl concentration of 50, 10, and 5 kg/m^3, respectively.

4.4 In the class, we have derived the mass transfer from a channel with soluble wall at constant concentration into a flowing solvent with a fully developed velocity profile that is initially free of solute. In deriving the solution, we have assumed the solute concentration zero outside the concentration boundary layer near the wall in the entrance region. Sufficiently far downstream of the channel entrance, this will no longer be the case. By considering a control volume of axial length L along a long circular tube of radius a into which a solute-free liquid flows and out of which flows liquid plus solute, derive that the following criteria need to be satisfied to ensure the validity of the solution, where U is the mean fluid velocity:

$$\frac{Ua^2}{DL} \gg 1$$

4.5 In a so-called unstirred batch-operated reverse osmosis system, a long cylinder holding a salt solution is closed by a semipermeable membrane at one end and a piston at the other. The pressure applied by the piston initially is that corresponding to osmotic equilibrium; then at time $t = 0$, the pressure is suddenly increased to a predetermined value at which it is maintained. The result is that there will be a flow through the membrane, which in general will be time dependent, and neglecting any wall effects, there will be a concentration variation in the solution that will depend on time and distance into the solution measured from the membrane.

With $v_w(t)$, the permeation velocity through the membrane, C_0, the initial salt concentration in the solution, and C_p, the salt concentration on the product side of the membrane, write down the equation governing the salt diffusion, together with the boundary and initial conditions. Assume that the rejection coefficient $R_s = 1 - \dfrac{C_p}{C_w}$ is constant, but it may be different from unity.

4.6 Water is contained between two infinite parallel plates separated by a small distance $h = 10^{-3}$ m. The bottom plate is held stationary and the top plate is moved at a constant velocity $U = 10^{-3}$ m/s so that a simple shear flow is generated between the plates. A thin band of a dye of thickness $\Delta = 10^{-4}$ m is injected between and perpendicular to the plates extending fully across the gap. The band depth is very deep and may be supposed to be infinite. The dye concentration is $C_0 = 10$ mol/m^3 and its molecular diffusion coefficient in water is $D = 10^{-9}$ m^2/s.

From the convective diffusion equation for the dye concentration, make an order of magnitude estimate of time for which molecular diffusion in the direction of motion will no longer be important and the times for which lateral molecular diffusion is negligible.

4.7 A chemical (1 μg) is introduced into an open channel (10 micron width) filled with water. There is no flow current and the chemical is transported through diffusion only. Assume that the chemical mixes rapidly across the width and stays on the surface of the channel. A detector is located at 10 mm from the delivery location of the chemical for measuring the concentration of the chemical. The diffusion constant $D = 0.004\ \mathrm{m^2/s}$:

1. How long does it take for the chemical to reach the detection point?
2. What is the concentration of the chemical at the detection point?
3. Calculate the time required by the chemical to reach the detection point if the flow current toward the detection point $V = 0.01\ \mu\mathrm{m/s}$.
4. If the flow current is in opposite direction, that is, from the detection point to the delivery point, it is possible that the chemical will never reach the detection point. Determine the velocity, V, at which this will happen.

Note: Here, "reach" means that the detection point is inside $\pm 2\sigma$ of the center of the chemical patch, where the variance $\sigma = (2Dt)^{0.5}$.

4.8 1 μg of sucrose is introduced in a microchannel of cross-section (50 μm × 500 μm) filled with water. The diffusivity (D) of sucrose is equal to $4.6 \times 10^{-10}\ \mathrm{cm^2/s}$. The length of the microchannel is 1 cm. (a) Calculate the time when the concentration of sucrose is at 2000 s and 200,000 s inside the microchannel.

4.9 Schematic of an H-filter to separate protein from a sample solution containing ribosome and protein is shown in the figure below. The diffusivity of protein is equal to 4×10^{-9} $\mathrm{m^2/s}$ and diffusivity of ribosome is equal to $1 \times 10^{-9}\ \mathrm{m^2/s}$. The channel size, w, is equal to 100 μm and the average velocity of the flow $V = 1$ mm/s. Calculate the range of L, that is, minimum and maximum limit length, L, required for the successful operation of the H-filter.

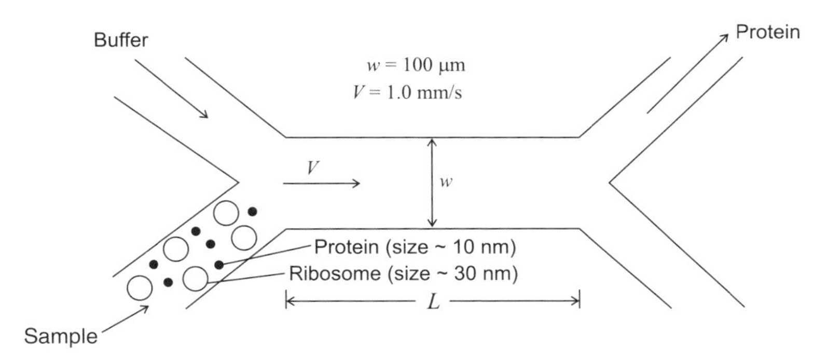

4.10 Consider a circular micromixer with mean diameter, D, equal to 2 cm. One microgram of dye molecule is injected at one end of the micromixer as shown below. (1) Calculate the diffusion time for the dye to diffuse to opposite end of the micromixer, that is, move by 180°. (2) If a peristaltic pump mounted on the micromixer imposes an average velocity, $V = 1$ cm/s, estimate the diffusion time of the chemical for diffusing to opposite end of the micromixer. Assume the molecular diffusivity of dye, D_{12}, in water to be equal to 2×10^{-6} cm^2/s. The diameter, a, of the microcapillary is equal to 1 mm.

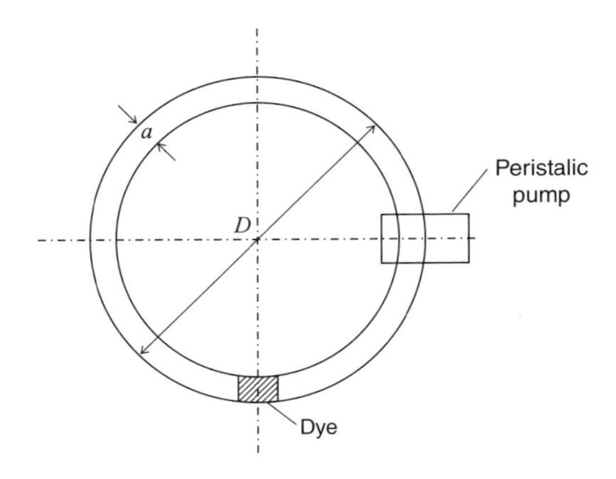

Supplemental Reading

Berthier J and Silberzan P 2006 *Microfluidics for Biotechnology*. Artech House, Inc.
Bruus H 2007 *Theoretical Microfluidics*. Oxford University Press.
Probstein RF 1994 *Physicochemical Hydrodynamics: An Introduction*. John Wiley & Sons, Inc..
Tabeling P 2005 *Introduction to Microfluidics*. Oxford University Press.

5

Surface Tension-Dominated Flows

In microfluidics, surface effects play a dominant role. Surface effects are also known as capillary effects. The capillary has been named after the Latin word *capillus* for hair. This chapter presents some of the flow physics related to surface tension-dominated flows and their application to microdevices.

5.1 Surface Tension

The effects of surface and interfacial tensions give rise to many phenomena in the liquid behavior. But all complex physical–chemical interactions involved are not understood even today. The familiar examples of surface tension are:

1. The *thin capillary tube* in which liquid rises to a height greater than that of the pool in which it is placed.
2. Break up into drops of a stream of water flowing out of a *faucet*. This physics is the basis of the *ink-jet printer* or *gel encapsulation* process to *encase perfume* or crystal or monoclonal antibodies.
3. *Liquid drop* remaining stationary when placed on a solid surface or spreading of water drop when placed on a clean glass surface.

Figure 5.1 explains the physical origin of surface tension. At the liquid–gas interface, the gas has little attraction between the molecules. The liquid molecules at the interface are attracted inward and in-sideward direction. There is no outward attraction to balance the pull, because by comparison there are not many liquid molecules inside the gas region. The attraction between the liquid molecules prevents a small fraction of them from escaping (vaporizing) into the gas. As a result, the liquid molecules at the surface are attracted inward and normal to the liquid–gas interface, which is equivalent to the tendency of the surface to contract or shrink. The *quantity*, σ, is thus called as the surface tension, that is, *force per unit length tending* to *contract the surface*. The *surface tension decreases* with increasing water temperature, that is, $\frac{\delta\sigma}{\delta T} < 0$. In practice, σ decreases very nearly linearly with increasing temperature.

Surface-active materials that are adsorbed at an interface in the form of an oriented monomolecular layer (monolayer) are termed as surfactants (see Figure 5.2). The *surfactants*

Transport Phenomena in Microfluidic Systems, First Edition. Pradipta Kumar Panigrahi.

Companion Website: www.wiley.com/go/panigrahi/microfluidic

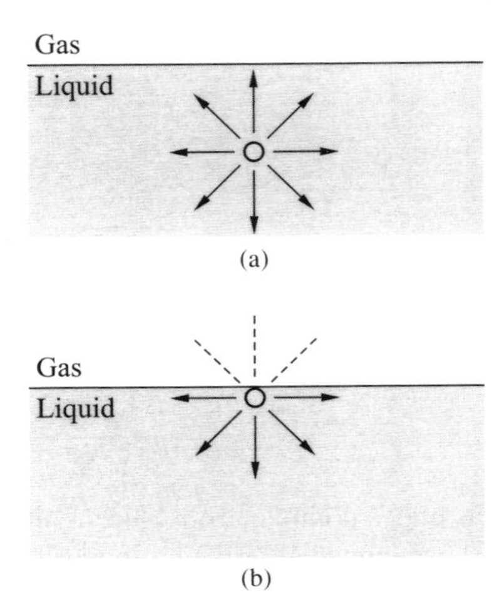

Figure 5.1 Schematic showing the origin of surface tension for a liquid–gas interface. (a) A molecule in the bulk of the liquid forms chemical bonds (arrows) with the neighboring molecules surrounding it. (b) A molecule at the surface of the liquid misses the chemical bonds in the direction of the surface (dashed lines). Consequently, the energy of surface molecules is higher than that of bulk molecules, and the formation of such an interface costs energy

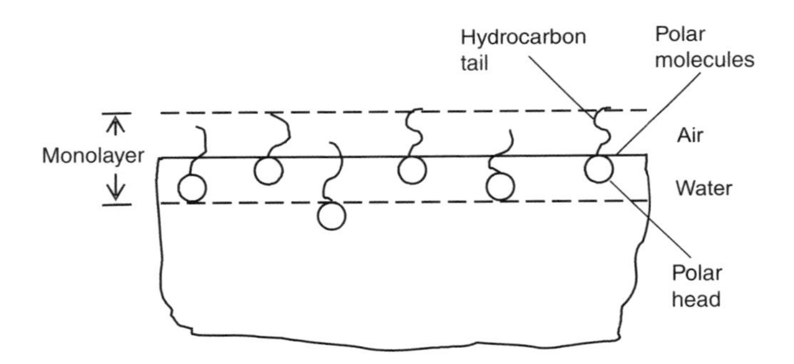

Figure 5.2 Schematic of surfactant monolayer at the water–air interface

are capable of *reducing the surface tension of water or aqueous solutions* by up to five orders of magnitude. Typical aqueous surfactants such as detergents are *organic compounds* with *long-chain hydrocarbon tail* and *polar head*. It is well known that hydrocarbons are relatively insoluble, water being highly polar. In Figure 5.2, the surfactant behavior is shown, where the polar molecules adsorb at the surface as a monolayer with the hydrocarbon tails up and polar heads down. The number of unbalanced water molecule at the surface decreases. This leads to the reduction in total energy of the surface and therefore decrease in surface tension. The adsorbed molecules *lower the surface tension*.

The above discussion for a liquid–gas interface is also applicable to a *liquid–liquid* interface between two immiscible liquids with an interfacial tension acting at the interface. As before, there is an imbalance of *intermolecular forces*, although smaller. The magnitude of the interfacial tension usually lies between the surface tensions of each liquid with gas.

5.2 Gibbs Free Energy and Surface Tension

Gibbs free energy (G) is the measure of work produced from thermodynamic system at constant temperature and pressure. It helps in understanding the capillary effects. Gibbs free energy is the energy of a system, where the thermodynamic control parameters are pressure, P, and temperature, T. It is defined as

$$G(P,T) = U + PV - TS$$

where U is the internal energy, P is the pressure, V is the volume, T is the temperature, and S is the entropy.

In equilibrium or quasiequilibrium situation, the Gibbs free energy attains a minimum value. If the system under consideration consists of two subsystems divided by a free surface, the total Gibbs energy is the sum of energy contribution G_i for each subsystems and the free energy of the surface. The free surface of the system can be described in terms of several variables, ε_i, that is, position, volume, and geometrical shape. The change in Gibbs free energy due to the change in the interface variable should be equal to zero at the equilibrium condition as

$$\partial G = \left(\sum_i \frac{\partial_\varepsilon G_i}{\partial \varepsilon_i} \right) \partial \varepsilon_i = 0 \tag{5.1}$$

where subscript i indicates the number of variables.

5.2.1 *Definition*

The surface tension (σ) is defined as the Gibbs free energy per unit area for fixed pressure and temperature:

$$\sigma = \left(\frac{\partial G}{\partial A} \right)_{P,T} \left(\frac{\mathrm{J}}{\mathrm{m}^2} = \frac{\mathrm{N}}{\mathrm{m}} = \mathrm{Pa\text{-}m} \right) \tag{5.2}$$

The interpretation of surface tension as force per unit length (N/m) can be visualized by considering a flat rectangular surface of length L and width w. Suppose we keep the width constant and stretch the surface by an amount ΔL, that is, from L to $L + \Delta L$. The work done by the external force F is $F\Delta L$ for creating the new surface area $w\Delta L$. The change in Gibbs free energy from equation (5.2) can be written as $\Delta G = \sigma w \Delta L$

Equating the change in Gibbs free energy with the work done by external forces, we have

$$\Delta G = \sigma w \Delta L = F \Delta L$$

$$\sigma = \frac{F}{w} \tag{5.3}$$

5.3 Microscopic Model of Surface Tension

Surface tension depends on the two materials on each side of the surface, that is, solids, liquid, or gases. A molecule in the bulk of the fluid forms chemical bond with the neighboring molecules gaining certain amount of binding energy. A liquid molecule at the gas–liquid interface cannot form as many bonds due to the absence of some neighboring molecules in the gas. The lack of chemical bonds leads to higher energy for the surface molecules. This is known as

Table 5.1 Measured values of the surface tension, σ, at liquid–vapor interfaces and the contact angle, θ, at liquid–solid–air contact lines

Liquid	σ[mJ/m^2]	Liquid	Solid	θ
Water	72.9	Water	SiO_2	52.3°
Mercury	486.5	Water	Au	0.0°
Benzene	28.9	Water	Pt	40.0°
Methanol	22.5	Water	PMMA	73.7°
Blood	60.0	Mercury	Glass	140.0°

All values are at 20 °C.

surface tension. It costs energy to form surface. This physical model can be used to estimate an order of magnitude of surface tension for a liquid–gas interface.

Let us consider a water–air interface. If we assume cubic geometry of water molecules, six neighboring molecules are present inside bulk of the fluid for each molecule. A surface molecule has five nearest liquid molecules with one missing above in the gas. The area covered by a single molecule is roughly $A \approx (0.3\ \text{nm})^2$. The typical intermolecular bond energy, ΔE, in a liquid is of the order of couple of thermal energies, $\Delta E \approx 2K_BT \approx 50$ meV.

Hence, from equation (5.2), we have

$$\sigma = \frac{2K_BT}{A} = \frac{50\ \text{meV}}{(0.3\ \text{nm})^2} = 90\ \text{mJ/m}^2 \tag{5.4}$$

Note that for water–air interface at 20 °C, $\sigma = 72.9$ mJ/m^2, which is close the microscopic estimation of equation (5.4).

Table 5.1 presents the surface tension values of different liquids.

5.4 Young–Laplace Equation

We have restricted our discussion to *plane interfaces* till now. However, because of the existence of surface tension, there will be a tendency to curve the interface. As a consequence of this, there must be a *pressure difference across the surface*. This pressure difference is given by Young–Laplace equation as

$$\Delta P = \sigma\left(\frac{1}{R_1} + \frac{1}{R_2}\right) \tag{5.5}$$

Here, R_1 and R_2 are the radii of curvature of the surface along any two orthogonal tangents (principal radius of curvature). ΔP is the difference in fluid pressure across the curved surface. The above equation is applicable to arbitrarily shaped surfaces where the radius of curvature may change spatially. Young (1805) and Laplace (1806) independently proposed the above equation.

In the special case of spherical bubble or drop of either a liquid in gas, gas in liquid, or immiscible liquids, we can write

$$\Delta P = P_i - P_e = \frac{2\sigma}{a} \tag{5.6}$$

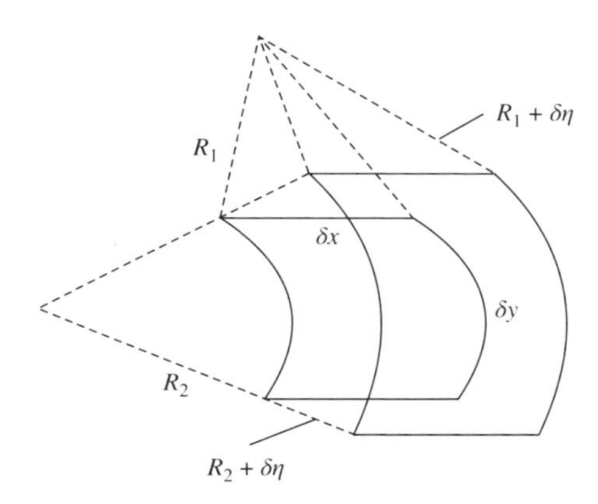

Figure 5.3 Schematic describing the pressure drop across a curved interface

Here, P_i is the internal pressure, P_e is the external pressure, and a is the bubble or drop radius.

The Young–Laplace equation can also be obtained using Gibbs free energy concept. There is a Young–Laplace pressure drop ΔP_{surf} across a curved interface due to the surface tension effect. The minimum energy condition can be used to derive the expression for ΔP_{surf}. A small piece of curved surface with area $A = \delta x \times \delta y$ in equilibrium is shown in Figure 5.3. Let the area be expanded through a small displacement. The local radius of curvature for the x-tangent and y-tangent direction changes from R_1 to $R_1 + \delta\eta$ and R_2 to $R_2 + \delta\eta$, respectively. The side length δx and δy are changed similarly to $\left(\delta x + \frac{\delta x}{R_1}\delta n\right)$ and $\left(\delta y + \frac{\delta y}{R_2}\delta\eta\right)$, respectively. Hence, the modified area can be written as

$$
\begin{aligned}
A_{\text{mod}} &= \left(1 + \frac{1}{R_1}\delta\eta\right)\delta x\left(1 + \frac{1}{R_2}\delta\eta\right)\delta y \\
&= \left(1 + \frac{\delta\eta}{R_1}\right)\left(1 + \frac{\delta\eta}{R_2}\right)\delta x\delta y \\
&= \left(1 + \frac{\delta n}{R_1}\right)\left(1 + \frac{\delta n}{R_2}\right)A
\end{aligned}
$$

Neglecting the higher-order terms $\frac{\delta\eta}{R_1}\frac{\delta\eta}{R_2}$, we can write

$$
\text{Change in area} = \partial A \approx \left(\frac{\delta\eta}{R_1} + \frac{\delta\eta}{R_2}\right)A
$$

The change in surface area due to stretching will increase the surface energy based on equation (5.2) as

$$
\delta G_{\text{surf}} = \sigma\partial A
$$

There will be an increase in volume due to stretching of the interface as $\partial\forall = A\delta n$. The increase in volume will also lead to decrease in pressure volume energy as

$$
\partial G_{\text{PV}} = P\delta\forall = \Delta P_{\text{surf}}A\delta\eta
$$

Hence, the minimum total surface energy condition at equilibrium can be written as

$$\partial G = \sigma \partial A - \Delta P_{\text{surf}} \ A \delta \eta = 0$$

$$\therefore \quad \sigma \left(\frac{\delta n}{R_1} + \frac{\delta n}{R_2} \right) A - \Delta P_{\text{surf}} \ A \delta \eta = 0$$

$$\therefore \quad \Delta P_{\text{surf}} = \sigma \left(\frac{1}{R_1} + \frac{1}{R_2} \right) \tag{5.7}$$

It is important to note the sign convention used here. The pressure is highest in the convex medium, that is, the medium where the centers of the curvature circles are placed.

5.5 Contact Angle

Another fundamental concept in the theory of surface effects in microfluidics is the contact angle that appears at the contact line between three different phases, typically the solid wall of a channel and two immiscible fluids inside that channel. The two concepts, that is, contact angle and surface tension, allow understanding of the capillary forces that act on two-fluid flows inside microchannels of lab-on-a-chip (LOC) systems.

5.5.1 *Definition of Contact Angle*

The contact angle, θ, is defined as the angle between the solid–liquid and the liquid–gas interface at the contact line where three immiscible phases meet, as illustrated in Figure 5.4(a). At equilibrium, θ is determined by the three surface tensions $\sigma_{\text{sl}}, \sigma_{\text{lg}}$, and σ_{sg} for the solid–liquid, liquid–gas, and solid–gas interfaces. Some typical values of contact angles, θ, between different solid–liquid combinations are listed in Table 5.1. Systems with contact angles $\theta < 90°$ are called *hydrophilic* (water loving), while those with $\theta > 90°$ are called *hydrophobic* (water fearing). The contact angle is well defined in equilibrium state. However, it turns out to depend in a complicated way for the dynamical state of a moving contact line. One can, for example, observe that the contact angle at the advancing edge of a moving liquid drop on a substrate is different from that at the receding edge (see Figure 5.4(b)). The contact angle for this situation is known as dynamic contact angle.

5.5.2 *Young's Equation for Contact Angle*

The expression for the contact angle in equilibrium can be derived using the free energy minimum condition. Let us consider the system sketched in Figure 5.5, where in equilibrium, a flat interface between a liquid and a gas forms the angle θ with the surface of a solid substrate. Let the liquid–gas interface be tilted by an infinitesimal angle around an axis parallel to the contact line. As a result, the contact line moves by the distance δl while maintaining same contact angle θ (see Figure 5.5). The change in free energy takes place due to the changes in interface areas near the contact line. It can be seen from Figure 5.5 that the change of the interface areas is proportional to $+\delta l$, $+\delta l \cos\theta$, and $-\delta l$ for the solid–liquid, liquid–gas, and

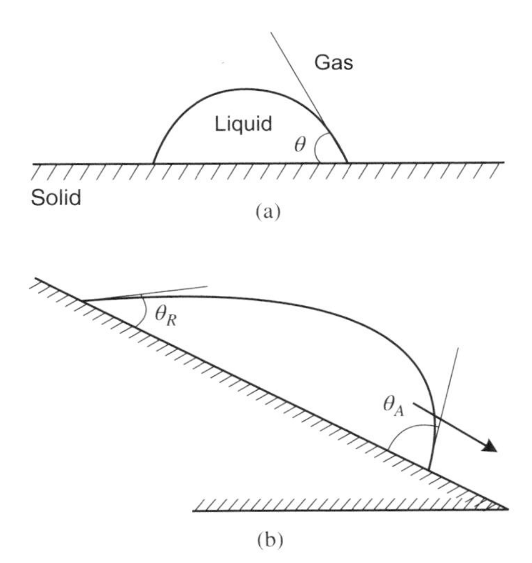

Figure 5.4 (a) Definition of the contact angle and (b) the schematic of a liquid drop moving on an inclined plane

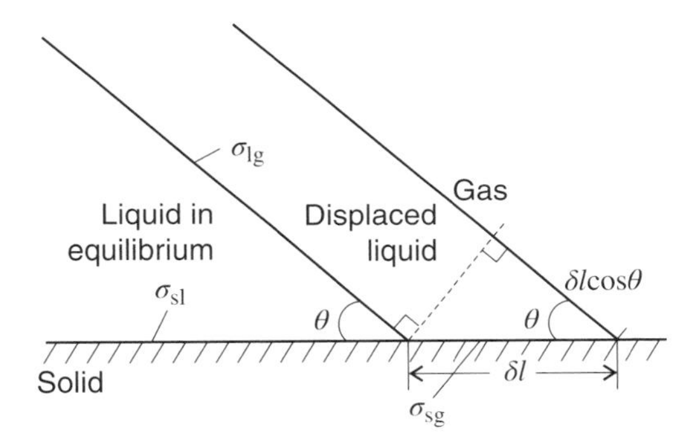

Figure 5.5 A sketch of the small displacement δl of the contact line away from the equilibrium position. The change of the interface areas is proportional to $+\delta l$, $\delta l \cos\theta$, and $-\delta l$ for the solid–liquid, liquid–gas, and solid–gas interface, respectively

solid–gas interface, respectively. The energy balance at equilibrium for the Gibbs energy per unit width, w, along the contact line becomes

$$\frac{1}{w}\delta G = \sigma_{\rm sl}\delta l + \sigma_{\rm lg}\delta l\cos\theta - \sigma_{\rm sg}\delta l = 0 \tag{5.8}$$

It may be noted that the Gibbs free energy contribution from all interfaces is considered in the calculation of total Gibbs energy in equation (5.8). After simplification, one gets Young's equation for the contact angle θ as

$$\cos\theta = \frac{\sigma_{\rm sg} - \sigma_{\rm sl}}{\sigma_{\rm lg}} \tag{5.9}$$

5.6 Dynamic Contact Angle

We have discussed contact angle from static point of view, that is, the interface between solid, liquid, and gas does not change their locations with time. However, in most of the dynamic systems, the interface has a finite velocity. This section presents the influence of the interface motion on the contact angle variation. Figure 5.6 shows the schematic of a liquid plug moving in a capillary tube. The contact angle between a flowing fluid front and the solid surface is different at the advancing and receding front. The different advancing and receding contact angle of Figure 5.6 reflects the balance between the capillary and viscous forces. From the order of magnitude analysis, the drag force of the flow on a plug can be deduced as

$$F_{\text{drag}} = \tau dA = \mu \frac{\partial u}{\partial n} dA \simeq \mu U l \tag{5.10}$$

where μ is the dynamic viscosity, l is the axial dimension of the plug, U is the velocity of the plug, and $\frac{\partial u}{\partial n}$ is the velocity gradient in wall-normal direction.

The capillary/wetting force is

$$F_{\text{cap}} \simeq \sigma l \tag{5.11}$$

Hence, combining equations (5.10) and (5.11), we have

$$\frac{F_{\text{drag}}}{F_{\text{cap}}} \simeq \frac{\mu U}{\sigma} \tag{5.12}$$

The above ratio is known as *capillary number* (Ca), which is a nondimensional quantity expressing the relative importance of the viscous and capillary forces. Figure 5.7 shows the experimental data of difference between the dynamic (θ_d) and static (θ_s) contact angles of an interface of silicon oil in 2 mm diameter capillary as a function of capillary number. The figure shows that the difference between the dynamic and static contact angles increases with increase in capillary number.

The relationship between the static and dynamic contact angles has been proposed based on the correlation from experiments as

$$\theta_d^3 - \theta_s^3 = A\, Ca \tag{5.13}$$

where θ_d and θ_s are the dynamic and static *contact angles*, respectively; A is a constant; and θ is in radians. The above law is known as Tanner's law. Figure 5.7 shows the curve fitting plots of the experimental data according to Tanner's law for different values of constant, A.

For microflows with the velocity in the range of 1 μm/s to 0.1 m/s, viscosity $\mu \simeq 10^{-3}$ kg-m/s and $\sigma \sim 50 \times 10^{-3}$ N/m, the typical values of $Ca = 2 \times 10^{-8} - -2 \times 10^{-3}$,

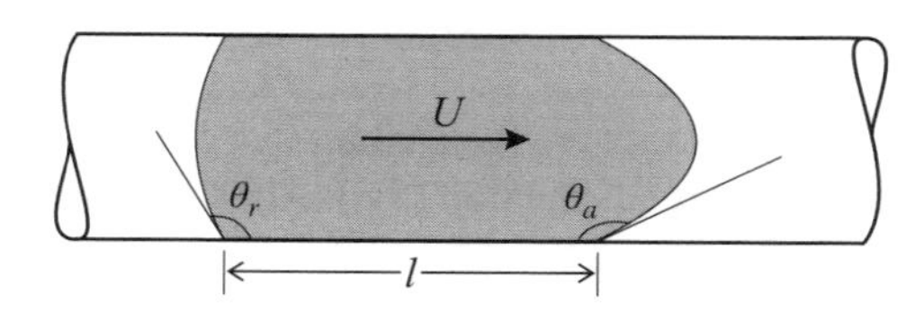

Figure 5.6 A liquid plug in a capillary tube with its advancing θa and receding θ_r contact angles

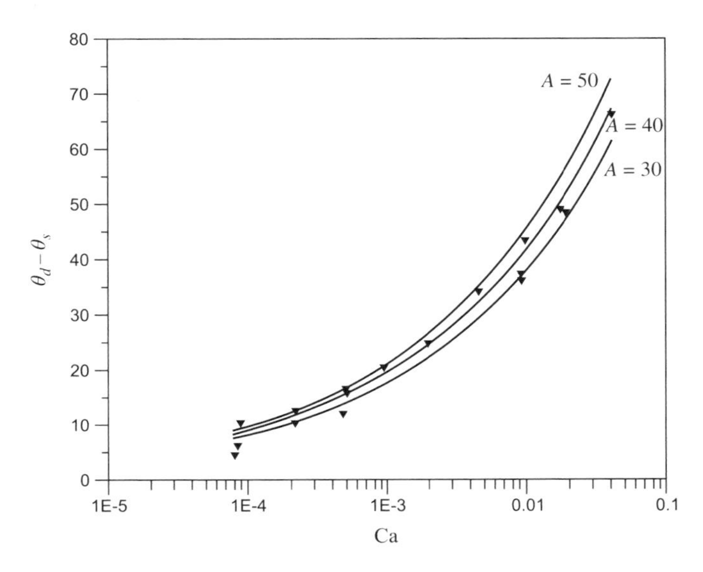

Figure 5.7 The dynamic contact angle data of silicon oil inside a 2 mm diameter capillary as a function of capillary number. The solid line represents the Tanner's law profile for different constants

which is quite small. For small capillary number, we can simplify the dynamic contact angle expression as

$$\theta_d = (\theta_s^3 + A\ Ca)^{1/3} = \theta_s\left(1 + \frac{1}{3}\frac{A\ Ca}{\theta_s^3}\right) \tag{5.14}$$

$$\therefore\ \theta_d - \theta_s = \frac{1}{3}\frac{A\ Ca}{\theta_s^2} \tag{5.15}$$

Hence, the difference between the dynamic and static contact angles is linearly related to the capillary number for microflows at small capillary number.

Note that capillary number is signed and depends on the direction of the velocity. Hence, the advancing contact angle is larger than the static contact angle, and the receding contact angle is smaller than the static contact angle. Figure 5.8 shows the schematic view of different contact angles. The interface velocity has same direction as the contact angle for advancing liquid front. However, contact angle and receding front velocity are in opposite direction.

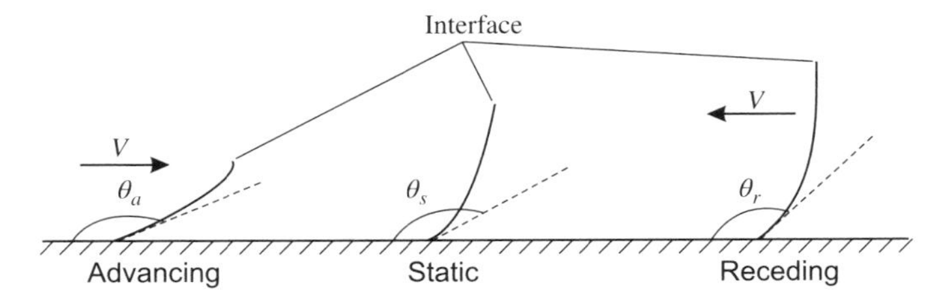

Figure 5.8 Schematic of the advancing (θ_a), static (θ_s), and receding (θ_r) contact angles

5.7 Superhydrophobicity and Superhydrophilicity

In several microsystems, it is important to obtain hydrophobic contact surface in some situations. Hydrophobic contact can reduce hydrodynamic drag at the wall. Hydrophobic contact also prevents cross contamination of one drop by another. In nature, we see raindrops roll along the surface of the leaves without any adhesion preventing rotting of the leaves. This superhydrophobicity is attributed to the surface roughness. Microscopic view shows these leaves to have high roughness, trapping air bubbles in between the roughness elements (see Figure 5.9). A sample SEM image shown in Figure 5.10 demonstrates the nature of roughness of a lotus leaf surface. Let us study the effect of surface roughness and inhomogeneities on the contact angle of a liquid drop on a rough surface in the following section.

5.7.1 Effect of Roughness

The lotus leaf example discussed in the previous section shows that the roughness may amplify the character of hydrophilic or hydrophobic contact. Figure 5.11 shows the schematic of a

Figure 5.9 Schematic showing the microscopic view of droplet on the top of a lotus leaf

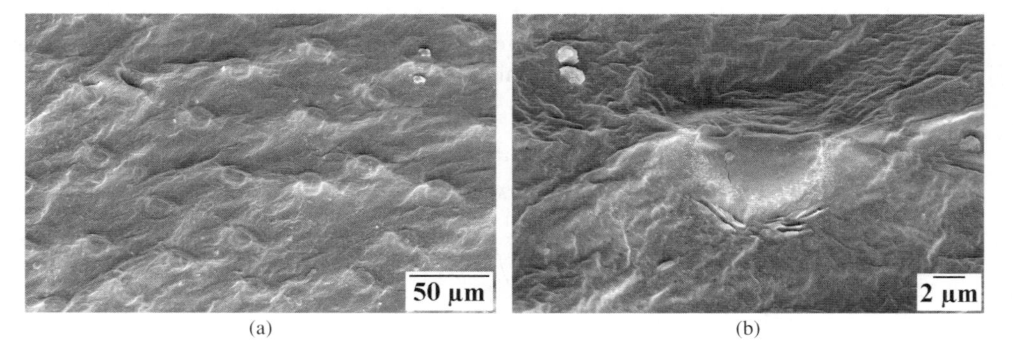

Figure 5.10 Sample SEM images of a lotus leaf surface with low resolution (a) and high resolution (b) demonstrating the nature of roughness

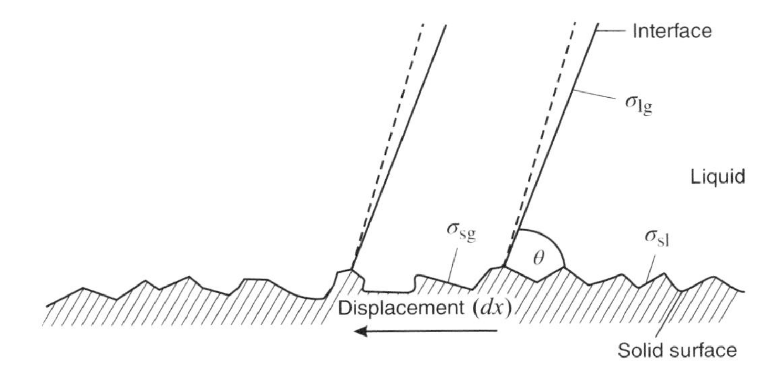

Figure 5.11 Microscopic view of a liquid drop on a rough solid surface

liquid drop in contact with a solid surface. The roughness of the solid wall modifies the contact between the liquid and the solid. We assume that the size of the roughness is very small. As a general rule, the roughness size should be smaller than the mean interaction distance between the liquid molecules and the solid wall. The molecules of the liquid interact macroscopically with a plane surface and microscopically with a rough surface. Let θ be the contact angle with a rough surface and θ^* be the contact angle with a smooth surface.

In Figure 5.11, let the contact line be moved or displaced for a distance dx on the solid surface. Let us define "r" as a roughness parameter and $r > 1$ for a rough surface as the interface interacts with the solid surface more than that of a plane surface. The actual distance traveled along the solid surface is equal to rdx.

The work done by different forces acting on the contact line due to the displacement can be written similar to equation (5.8) as

$$dW = dE = \sum F_x d_x = (\sigma_{sl} - \sigma_{sg})rdx + \sigma_{lg}\cos\theta dx \tag{5.16}$$

Note that r is not used in the second term of right-hand side as it does not involve any roughness contribution. For equilibrium, the energy should be minimum. Hence, we have

$$\frac{dE}{dx} = 0 \tag{5.17}$$

Thus, we obtain from equations (5.16) and (5.17):

$$\sigma_{lg}\cos\theta = (\sigma_{sg} - \sigma_{sl})r \tag{5.18}$$

For a smooth surface ($r = 1$), we can write

$$\sigma_{lg}\cos\theta^* = \sigma_{sg} - \sigma_{sl} \tag{5.19}$$

It may be noted that equation (5.19) is same as Young's equation for contact angle equation (5.8). Substituting equation (5.19) in equation (5.18), we have

$$\cos\theta = r\cos\theta^* \tag{5.20}$$

This is known as *Wenzel's law*.

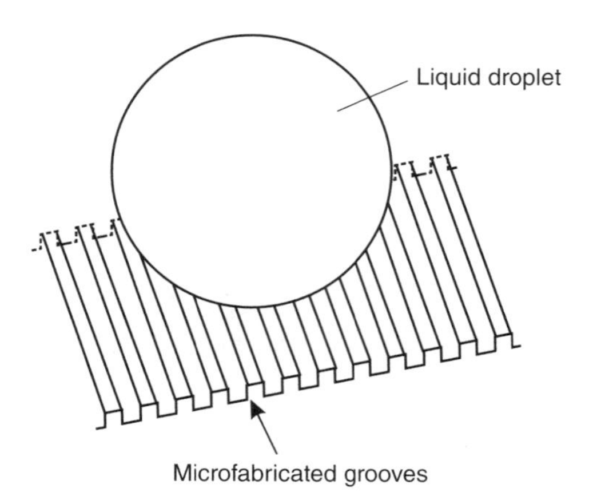

Figure 5.12 A schematic showing the superhydrophobic surface fabricated by etching microgrooves on a hydrophobic substrate

As $r > 1$ for rough surface, we have

$$|\cos\theta^*| < |\cos\theta| \tag{5.21}$$

If the contact is hydrophobic, $\theta^* > 90°$ for a smooth surface. Thus, $\theta > \theta^*$, that is, the contact is more hydrophobic due to the surface roughness. If the contact is hydrophilic, $\theta^* < 90°$ for smooth surface, then the contact is more hydrophilic ($\theta < \theta^*$) due to surface roughness. Note that in microfluidics, superhydrophobic surface is commonly microfabricated by etching microgrooves or micropillars on the substrate (see Figure 5.12).

5.7.2 Effect of Surface Inhomogeneities

When sufficient care is not taken during microfabrication, chemical inhomogeneities may develop on the microfabricated surface. Suppose Teflon is deposited on a rough substrate, the surface should be hydrophobic. However, due to chemical inhomogeneity, the hydrophobicity may not be as significant as expected. The other situation may be the chemical cleaning of a microdevice during use. The chemical cleaning also may lead to chemical inhomogeneity. Let us explain the influence of chemical inhomogeneity on the wetting property. Figure 5.13 shows the contact line of a drop on an inhomogeneous surface. The solid wall constitutes microscopic inclusion of two different materials. Let us assume the heterogeneities to be small compared to the interaction distance between liquid molecules and solid walls. Let θ_1 and θ_2 be the contact angles for each material at microscopic size. The surface fractions of the two materials are denoted as f_1 and f_2, respectively.

The energy to move the interface for a distance, dx, on the solid surface can be written similar to equation (5.16) as

$$dE = dW = \left(\sigma_{\mathrm{sl}} - \sigma_{\mathrm{sg}}\right)_1 f_1 dx + (\sigma_{\mathrm{sl}} - \sigma_{\mathrm{sg}})_2 f_2 dx + \sigma_{\mathrm{lg}} \cos\theta dx \tag{5.22}$$

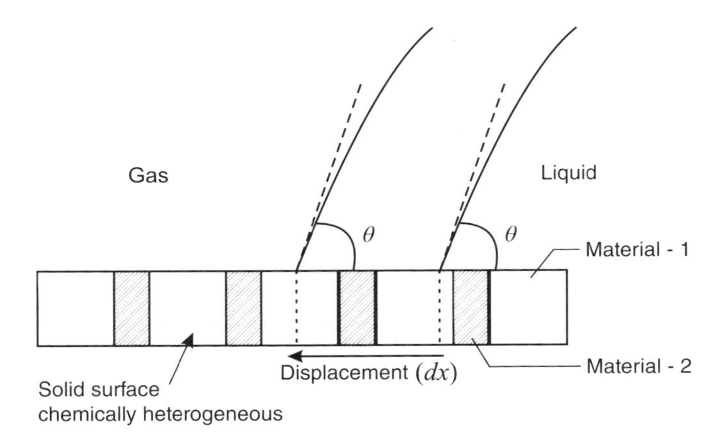

Figure 5.13 The displacement of the contact line of a drop on an inhomogeneous solid surface

It may be noted that both material 1 and material 2 participate in work done due to the smaller size of heterogeneities compared to the interaction distance between liquid molecules. The equilibrium is obtained for minimum value of E. Thus, on simplification, we have

$$\sigma_{lg} \cos \theta = (\sigma_{sg} - \sigma_{sl})_1 f_1 + (\sigma_{sg} - \sigma_{sl})_2 f_2 \tag{5.23}$$

From the Young's law, we have

$$\sigma_{lg} \cos \theta = \sigma_{sg} - \sigma_{sl} \tag{5.24}$$

Substituting equation (5.24) in equation (5.23), we can write

$$\sigma_{lg} \cos \theta = (\sigma_{lg} \cos \theta)_1 f_1 + (\sigma_{lg} \cos \theta)_2 f_2 \tag{5.25}$$

As σ_{lg} for surface 1 and surface 2 is same, we can write

$$\cos \theta = f_1 \cos \theta_1 + f_2 \cos \theta_2 \tag{5.26}$$

The above expression can be generalized as

$$\cos \theta = \sum_i f_i \cos \theta_i \tag{5.27}$$

This is known as *Cassie–Baxter relation*. The Teflon layer when deposited on a rough substrate may be porous and inhomogeneous. Hence, the wetting parameter gets modified according to Cassie–Baxter relation, and the gain in hydrophobicity is not as significant as expected.

5.7.3 Effect of Surfactant

Surfactants are very often added to biological samples to prevent the formation of aggregates. Surfactants also prevent target molecules to stick to the solid walls of the microsystem.

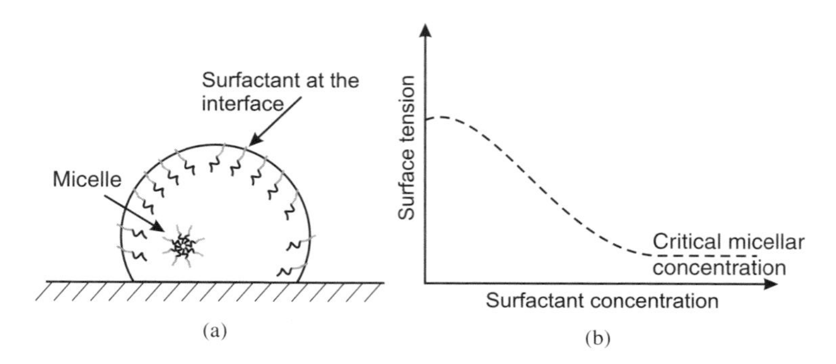

Figure 5.14 (a) The schematic of the surfactant distribution in a liquid drop and (b) the influence of surfactant concentration on surface tension

They are long molecules with hydrophilic head and a hydrophobic tail. Figure 5.14(a) shows the distribution of surfactants in a liquid drop. The surfactants gather on the interface between the liquid and the surrounding gas due to its amphiphilic nature. The presence of surfactant lowers the surface tension of the liquid. When the interface gets saturated with surfactants above a critical concentration, the surfactant molecules form micelles in the bulk of the fluid. Note that pure water has surface tension equal to 72 mN/m, which reduces to 30 mN/m at the critical concentration limit. Figure 5.14(b) shows the variation of surface tension as a function of the surfactant concentration. The surface tension value remains constant after a critical surfactant concentration.

5.7.4 Motion of Drops at Boundary of Hydrophilic–Hydrophobic Surface

The power of capillary forces occurring in a drop can be demonstrated from the motion of a drop up a step. Figure 5.15 shows the sequence of motion of a microdrop over a step separated by hydrophilic and hydrophobic surface. The drop initially deposited at the boundary region

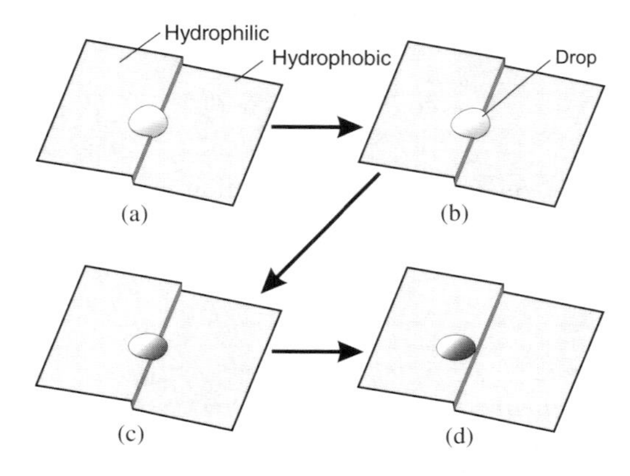

Figure 5.15 The sequence of motion of a drop at the boundary of a step between hydrophilic (top) and hydrophobic (bottom) surface

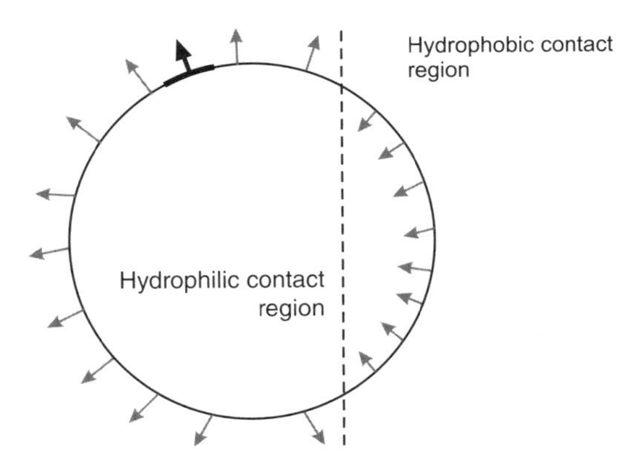

Figure 5.16 The force distribution on a water droplet at the boundary between a hydrophilic/ hydrophobic contact

of the two surfaces is not in equilibrium due to the imbalance of the surface tension forces between the hydrophilic and hydrophobic contact region. The drop moves up by the pull due to the hydrophilic forces of the upper plate and push by the hydrophobic forces of the lower plate. The motion of the drop continues till the drop is entirely on the upper plate.

Figure 5.16 shows the force distribution of a water droplet above a hydrophilic and hydrophobic contact. The surface tension force on the hydrophilic surface is toward the left side as $\theta < 90°$. The surface tension on the right hydrophobic surface is also to the left side as $\theta > 90°$. Hence, the resulting force is toward the left hydrophilic side, and the drop moves to the left. The motion stops when the resulting force at the contact surface is zero and the drop is entirely on the hydrophilic region. Continuous movement of droplet is possible by creating a wettability gradient on the surface by chemical modification.

5.8 Microdrops

Microdrops are a common feature of various biotechnology applications. For example, thousands of microwells of DNA microarrays contain drop of biological fluid (Figure 5.17).

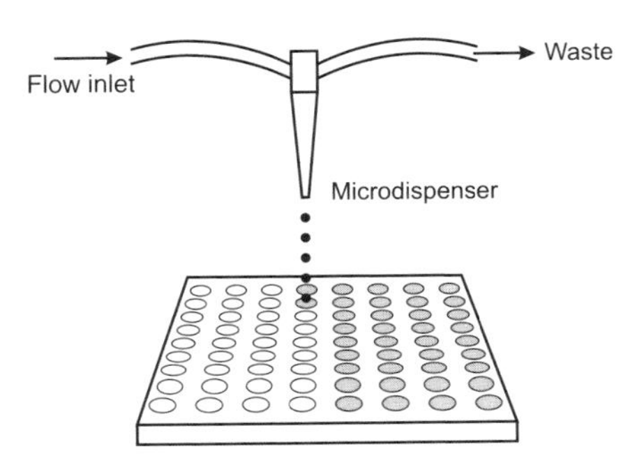

Figure 5.17 Dispensing of microdrops to wells of a microplate of DNA microarray

A drop dispenser is used to deposit the liquid in each well. Microdrops are advantageous for minimizing the surface between the liquid and the solid walls. This helps in preventing nonspecific adsorptions and adherence of target particles from solid walls on the microsystem. These particles can perturb the light source and hamper in the detection of the real signal. Other advantage of microdrops is the use of small amounts of liquid in comparison to classical microflows. It is important to know the behavior of microdrops in various situations as various surface tension-related behaviors play important role in microdrop applications. Microdrops should easily detach from the tip of the pipette. Microdrops should remain inside the microcusps without flowing, which can contaminate the neighboring cusps. Microdrops should follow the electrode line of the electrowetting device. The behavior of liquid droplets on a surface is discussed in the following section.

5.8.1 Wetting

Wetting characterizes the contact of a liquid with a solid surface. It can be of two types: (a) total wetting and (b) partial wetting (see Figure 5.18). In total wetting, the liquid film spreads out on the solid surface. Liquid stays as drops for partial wetting case. *Spreading coefficient*, S, provides the criterion for wetting. It is defined as

$$S = \sigma_{sg} - (\sigma_{sl} + \sigma_{lg}) \tag{5.28}$$

When $\sigma_{sg} < \sigma_{sl} + \sigma_{lg}$, a droplet with a finite contact angle minimizes the free energy of the system, and this state is known as partial wetting. If $\sigma_{sg} = \sigma_{sl} + \sigma_{lg}$, the contact angle is zero. Here, the system will be in equilibrium when a microscopic uniform liquid layer covers the whole surface, and this state is called complete wetting.

Using Young's equation, we can write

$$\begin{aligned} S &= \sigma_{sg} - \sigma_{sl} - \sigma_{lg} \\ &= \sigma_{lg}\cos\theta - \sigma_{lg} \\ &= \sigma_{lg}(\cos\theta - 1) \end{aligned} \tag{5.29}$$

S is zero for total wetting, and S is negative for partial wetting.

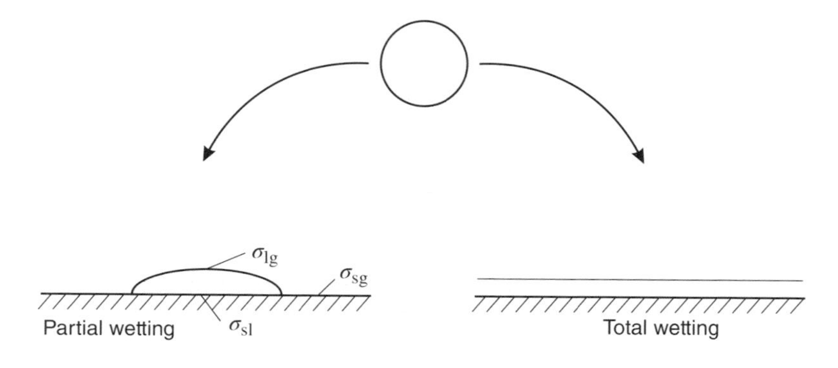

Figure 5.18 Schematic showing partial and total wetting

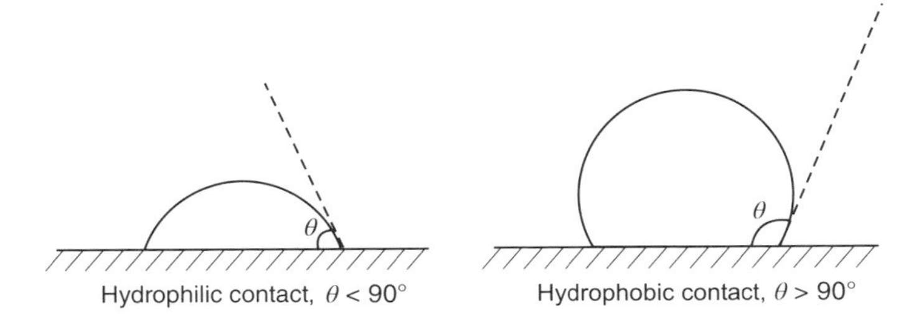

Figure 5.19 Schematic showing the hydrophilic and hydrophobic contact of a droplet

In the case of partial wetting, there is a line on which all three phases come together. This line is called the contact line. The contact of a water droplet on a solid is classified as hydrophilic or hydrophobic depending on the value of the contact angle. Figure 5.19 shows the schematic of hydrophilic contact ($\theta < 90°$) and hydrophobic contact ($\theta > 90°$) of a water droplet. The spreading coefficient S is more negative for a hydrophobic surface compared to the hydrophilic surface. In general, we should say a liquid to be "wetting" or "not wetting" based on the contact angle of a drop on the surface.

We have already observed the influence of surface tension force on the movement of droplet over the hydrophilic–hydrophobic surface boundary. Let us relate the criterion of wetting to the surface tension force balance consideration. Figure 5.20 shows the three interfaces of a microdrop at hydrophilic and hydrophobic contact. The x-axis is tangential to the solid surface at the contact line, and the y-axis is in the direction perpendicular to it. Taking projection of the forces in the x-direction, we have

$$\sigma_{lg} \cos\theta = \sigma_{sg} - \sigma_{sl} \tag{5.30}$$

or

$$\theta = \cos^{-1} \frac{\sigma_{sg} - \sigma_{sl}}{\sigma_{lg}} \tag{5.31}$$

It may be noted that equation (5.31) is same as the Young's equation of contact angle (equation (5.9)).

Sometimes, one observes unexpected changes in contact angle with time while conducting experiment with biological liquids. This is due to the fact that biological liquids are

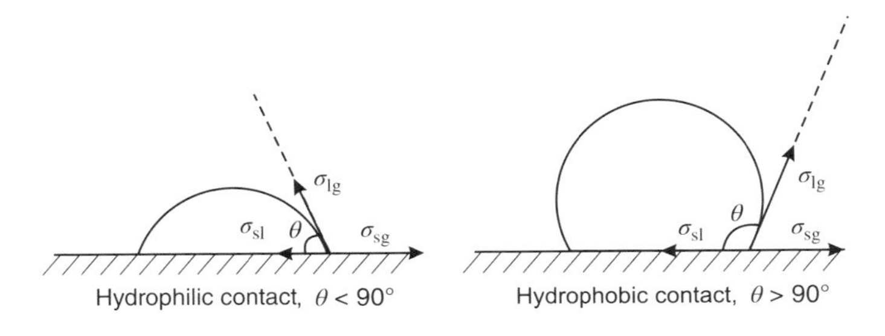

Figure 5.20 Schematic view of forces at the hydrophilic and hydrophobic contact line

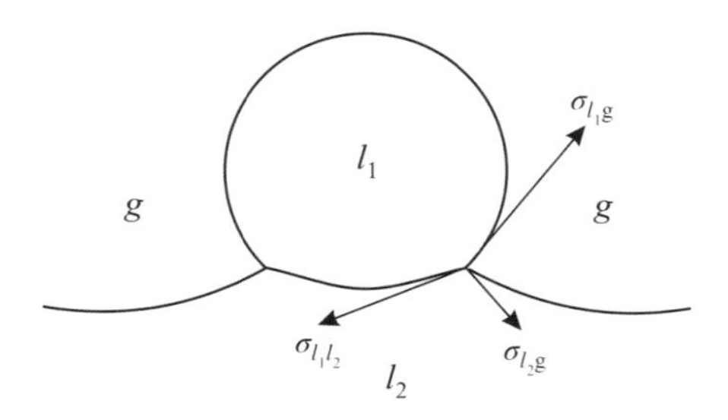

Figure 5.21 Schematic of a microdrop of oil on top of water layer

nonhomogeneous and can deposit layers of chemical molecules on the solid wall, thus progressively changing the value of surface tension, σ_{sl} and θ.

Note that we have obtained the contact angle using the projection of forces on the x-axis. The question may arise about the projection of forces on the y-axis. In fact, for a liquid surface, the vertical force is balanced by the reaction of the solid. If the solid is replaced by liquid (say, oil), then the surface is distorted to balance the net forces in y-direction (see Figure 5.21). Hence, the resultant forces are zero in both the x- and y-direction. Figure 5.21 shows all the interfaces between adjacent phases to be curved for liquid–liquid–gas interface in contrast to the solid–liquid–gas interface.

5.9 Capillary Rise and Dimensionless Numbers

Capillarity can be observed in a fine tube open at both ends placed vertically in a pool of liquid exposed to the atmosphere. For some cases, the liquid attains a level in the tube above the level of the pool (see Figure 5.22). The actual rise velocity of the free surface of the liquid in the

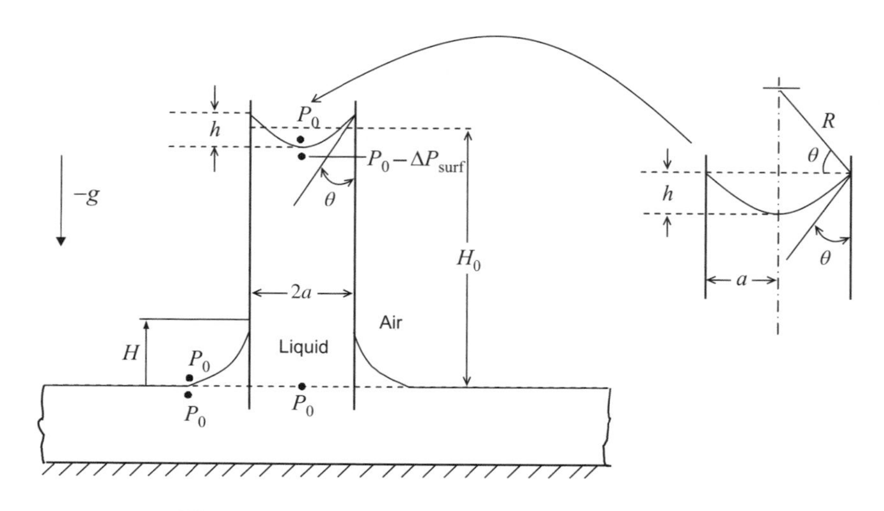

Figure 5.22 The schematic showing the capillary rise

tube from the level of the pool is one simple example of capillary motion. Generally, capillary motion is the flow that is governed by the forces associated with surface tension.

If the capillary is circular in cross-section with radius a, the meniscus will be approximately hemispherical with a constant radius of curvature, $a/\cos\theta$, where θ is the static contact angle. Departure from the hemisphericity is associated with the variation in liquid pressure over the surface due to the difference in gravitational force over the meniscus height, h. A measure of the hydrostatic gravitational force to the surface tension force is expressed by the nondimensional Bond number as

$$Bo = \frac{\text{Gravitational force}}{\text{Surface tension force}} = \frac{\rho g L^3}{\sigma L} = \frac{\rho g L^2}{\sigma}$$

If for the meniscus the characteristic length is h (Figure 5.22), the criterion for hemisphericity is that the Bo is very small:

$$\frac{\rho g h^2}{\sigma} \ll 1 \quad \text{or} \quad h \ll \left(\frac{\sigma}{\rho g}\right)^{1/2} \tag{5.32}$$

The Young–Laplace equation can be applied to determine the equilibrium height, H_0, that the column of liquid will attain, recognizing that the pressure under the meniscus controls the height. Assuming hemispherical meniscus, we have $R_1 = R_2 = a/\cos\theta$. Young–Laplace equation gives the pressure difference across the interface, ΔP_{surf} as $\frac{2\sigma\cos\theta}{a}$.

For hydrostatic equilibrium, the balance between the pressure at the surface of the reservoir and the interface gives

$$\rho g H_0 = \frac{2\sigma\cos\theta}{a}$$

$$H_0 = 2\left(\frac{\sigma}{\rho g}\right)\left(\frac{\cos\theta}{a}\right) \tag{5.33}$$

Note: For mercury ($\theta = 140°$), the capillary height will fall, not rise.

For small capillaries, H_0 is relatively large due to smaller value of capillary radius. For water, $\sigma = 73$ mN/m. If $a = 0.1$ mm, we have $H_0 = 0.15$ m, which is a large value for accurate measurement of height. Thus, the capillary rise method is one of the most accurate means for the *measurement of surface tension*. Quite significant rise heights can be obtained in microchannels. Using surface tension data from Table 5.1, we find $H_0 = 4.2$ cm for water in a 100 μm radius PMMA polymer channel and $H_0 = 42$ cm for radius, $a = 10$ μm.

Dividing "a" in both sides of equation (5.33), the above equation indicates $\rho g \ll \sigma/a^2$ for $a \ll H_0$. Thus, Bond number is very small, indicating that surface tension dominates over gravitation. For $Bo = 1$, the length scale of the capillary equals the capillary rise.

Because it is relatively easy to measure accurately the geometrical quantities a, H, and $\cos\theta$, this equation is one of the most accurate ways to measure surface tension:

$$\sigma = \frac{\rho g}{2}\frac{aH_0}{\cos\theta} \tag{5.34}$$

Two other nondimensional numbers used in the context of capillary flows are

$$Ca = \text{Capillary Number} = \frac{\text{Viscous Force}}{\text{Surface Tension Force}} = \frac{\mu V L}{\sigma L} = \frac{\mu V}{\sigma}$$

when, $Ca = 1$, the imposed velocity equals the intrinsic viscosity–surface velocity (σ/μ):

$$St = \text{Stokes number} = \frac{\text{Viscous force}}{\text{Gravitational force}} = \frac{\mu VL}{\rho g L^3} = \frac{\mu V}{\rho g L^2}$$

The Stokes number is the ratio of Ca and Bo. It is introduced when both length scale and velocity scale data are available.

5.9.1 Capillary Rise Time

Capillary rise time is the time required to reach equilibrium capillary rise height, H_0. The performance of many microsystems is dependent on the transient behavior of the capillary flow. Therefore, we intend to calculate the rate at which the capillary will rise to the equilibrium height. Let us assume the velocity profile at any instant of time to be fully developed Poiseuille profile. This assumption is justified from the fact that the developing length is expected to be very small compared to the length of the capillary due to the small diameter of the capillary tube. The instantaneous average velocity of the interface can be expressed from Poiseuille flow relation and the movement of interface as

$$v = \frac{dH}{dt} = \frac{a^2 \Delta P}{8\mu H} \tag{5.35}$$

where a is the capillary radius and H is the instantaneous capillary rise height.

Let us consider the force balance equation between a point immediately below the interface and a point at the same level as the reservoir surface. The pressure at point below the interface, $(P_0 - \Delta P)$, is expressed by Young–Laplace equation, and the water surface of the reservoir is at atmospheric pressure, P_0. The corresponding unbalance pressure difference driving the flow is

$$\Delta P = \frac{2\sigma \cos\theta}{a} - \rho g H; \text{ when } H < H_0 \tag{5.36}$$

where H_0 is the equilibrium capillary rise height.

Substituting ΔP from equation (5.36) in equation (5.35), we get

$$\frac{\mu}{\sigma}\frac{dH}{dt} = \frac{1}{8}\left[2\left(\frac{a}{H}\right)\cos\theta - \frac{\rho g a^2}{\sigma}\right] \tag{5.37}$$

Here, the left-hand side of the above equation is the capillary number $\left(\frac{\mu v}{\sigma}\right)$, and the second term in the right-hand side is the Bond number $\left(\frac{\rho g a^2}{\sigma}\right)$.

This equation can be integrated for calculating the capillary advancement time of the meniscus. Assuming low Bond number, the above differential equation can easily be integrated by separation of HdH and dt resulting in

$$H(t) = \sqrt{\frac{a\sigma\cos\theta}{2\mu}t} = a\sqrt{\frac{t}{\tau_{\text{adv}}}} \tag{5.38}$$

where

$$\tau_{\text{adv}} = \frac{2\mu a}{\sigma\cos\theta} = \frac{4\mu}{\Delta P}$$

The above equation indicates that the capillary rise rate increases with decrease in capillary diameter and increase (τ_{adv} is small) in surface tension.

5.10 Coating Flows

We discuss an example where surface tension force plays an active role in the overall dynamics of the problem. A coating flow is a fluid flow in which a large surface area is covered with one or more thin uniform liquid layers. Examples of technical importance include manufacture of *synthetic membranes*, *photographic film*, and *painting*. The hydrodynamic problem of coating flows is to obtain the relation between the *film-forming geometry*, the *liquid properties*, the speed *of application of the film*, and the *desired coating* thickness.

The dip coating can be analyzed by considering the simplified example of the vertically upward withdrawal of an infinite flat plate from a liquid pool. The analysis is carried out in the limit of *small values* of the ratio of viscous forces to surface tension forces (*capillary number*), and gravity drainage down the plate is neglected. When the capillary number is small, the final coating thickness (δ_f) is small compared to the overall length scale of the flow field (R) (see Figure 5.23). In this case, viscous effects are important over the length scale, δ_f, of the *lubrication film region* but are negligible in the *static meniscus region* where the shape of the meniscus is controlled by *surface tension* and the hydrostatic *pressure field*.

Let us assume that in the lubrication film region, $Re = \rho UR/\mu$ is small, that is, $Re \ll 1$. Thus, inertia terms are negligible. It is also assumed that $Bo = \rho gR^2/\sigma$ is small, that is, $Bo \ll 1$. Hence, gravity drainage ρg may be dropped from the governing equation.

Therefore, the Navier–Stokes equation (N–S equation) reduces to balance between the pressure gradient and the gradient in shear as in lubrication theory:

$$\rho\frac{Du}{Dt} = -\nabla P + \mu\nabla^2 u + \rho g \tag{5.39}$$

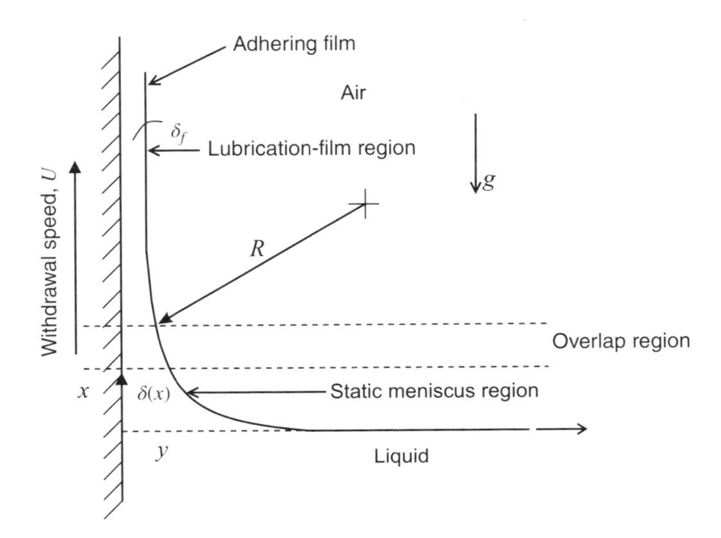

Figure 5.23 Schematic of dip coating process

Assuming one of the principal curvatures of the interface to be equal to zero because of the infinite plate geometry, using Young–Laplace equation in static meniscus region, we can write

$$-\Delta P = \frac{\sigma}{R} \tag{5.40}$$

For a plane curve, $y = f(x)$, the curvature (K) is expressed as

$$\kappa = \frac{1}{R} = \frac{|y''|}{(1 + y'^2)^{3/2}} \tag{5.41}$$

where $'$ indicates derivative with respect to "x."
Therefore, for the meniscus, we can write

$$\frac{1}{R} = \frac{\delta''}{(1 + \delta'^2)^{3/2}} \simeq \delta'' \quad (\text{Assuming} \quad \delta'^2 \ll 1) \tag{5.42}$$

Here, $'$ denotes differential with respect to "x" coordinate measured vertically upward from the pool level.

Thus, the pressure difference across the interface can be written from equation (5.40) as

$$\Delta P \simeq -\sigma\delta'' \tag{5.43}$$

Now, the N–S equation (5.39) can be written in simplified form after neglecting the inertia force (small Re) and gravity force (small Bo) as

$$\begin{aligned} -\nabla P + \mu\nabla^2 u = 0 \\ \sigma\frac{d^3\delta}{dx^3} + \mu\frac{\partial^2 u}{\partial y^2} = 0 \end{aligned} \tag{5.44}$$

where u is the vertical velocity of the fluid in the film.

The first term in the above equation is the surface tension force, and the second term is the viscous force.

Using boundary conditions, $u = U$, at $y = 0$ (no-slip condition),

$$\mu\frac{\partial u}{\partial y} = 0 \quad \text{at} \quad y = \delta(x) \quad (\text{free surface stress is small})$$

Integrating the N–S equation, we have

$$u = U - \frac{\sigma}{\mu}\frac{d^3\delta}{dx^3}\left(\frac{y^2}{2} - \delta y\right) \tag{5.45}$$

The volume flow rate per unit width of the plate is

$$\tilde{Q} = \int_0^{\delta(x)} u dy = U\delta + \frac{\sigma}{\mu}\frac{d^3\delta}{dx^3}\frac{\delta^3}{3} \tag{5.46}$$

The film is of constant thickness and parallel to the plate at far from the static meniscus region. In this region, the flow rate of liquid is equal to the product of film thickness and withdrawal velocity as

$$\tilde{Q} = \delta_f U \tag{5.47}$$

Eliminating $\tilde{Q}$ from equations (5.46) and (5.47), we have

$$\delta^3 \frac{d^3\delta}{dx^3} + \left(\frac{3\mu U}{\sigma}\right)\delta = \left(\frac{3\mu U}{\sigma}\right)\delta_f \tag{5.48}$$

Here, $\mu U/\sigma$ is the capillary number, Ca. Using reduced variables,

$$\eta = \frac{\delta}{\delta_f} \quad \text{and}$$

$$\xi = \frac{x}{\delta_f}\left(\frac{3\mu U}{\sigma}\right)^{1/3}$$

Equation (5.48) becomes

$$\eta^3 \frac{d^3\eta}{d\xi^3} = 1 - \eta \tag{5.49}$$

The equation (5.49) after numerical integration gives

$$\delta_f = 0.946(Ca)^{2/3}\left(\frac{\sigma}{\rho g}\right)^{1/2} \tag{5.50}$$

Hence, the equation (5.50) indicates that the coating thickness increases with increase in capillary number during the coating process. For a fixed capillary number, the coating thickness reduces with increase in the density of the fluid.

5.11 Enhanced Oil Recovery

One important *enhanced oil recovery* procedure is the displacement of oil in porous strata by foam flooding. Foam is a dispersion of liquid containing surfactant in a gas such as CO_2, air, N_2, or steam. Injection of gas and surfactant solution generates foam in situ. Such flows are complicated due to the factors like bubble size and the tendency for the bubbles to block a large percentage of the flow patterns in the complex porous geometry. To better understand this phenomenon, a situation where air bubbles are used to blow a viscous liquid out of a circular capillary leaving a thin film deposited on the inside wall is considered. The influence of surfactant on the flooding problem will be discussed in a later section.

The analogy with dip coating problem discussed in the previous section can be seen in a reference frame in which the bubble moving at velocity U into the stationary tube is considered analogous to the stationary wall moving with a velocity $-U$ (see Figure 5.24). The bubble air–liquid interface forms itself into a round meniscus at its front end that travels down the tube until it reaches the end. After the meniscus has passed any point, it leaves behind a liquid film that is essentially at rest with a constant pressure along its length. In the case of *small enough Reynolds number* and a *very small capillary radius*, gravitational and inertia forces may be neglected, and the flow depends only on a balance *between viscous* and *surface tension forces*. Moreover, for any given length of liquids, the *rate of outflow* depends essentially on the *applied pressure*, whereas the *amount* of *fluid left* in the tube after the air has reached the end depends essentially on the *surface tension*. Note that $\frac{1}{2}$ of the meniscus of the bubble

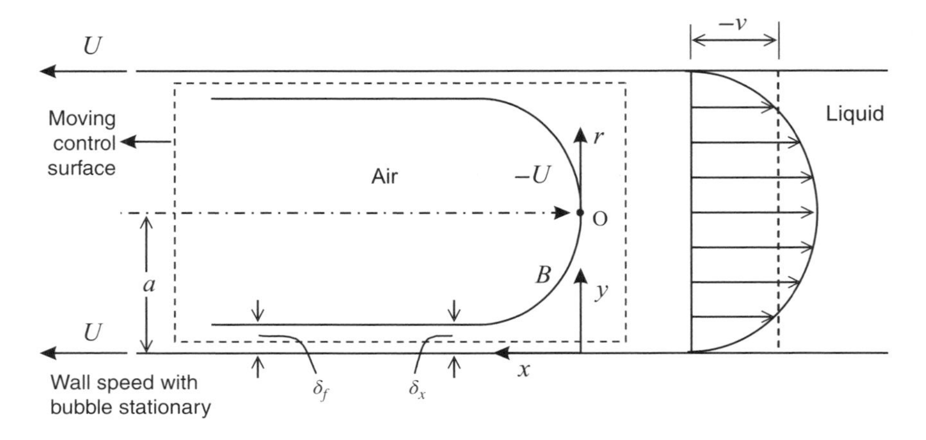

Figure 5.24 Formation of a liquid film in a circular capillary by blowing an air bubble

is similar to that of the dip coating process (see Figure 5.23). Thus, the expression for δ_f is same as the dip coating process (equation (5.50)). The expression for δ_f was obtained by independent researchers for separate situations, that is, dip coating (Levich, 1962) and bubble motion (Bretherton, 1961). The dip coating process is analogous to the air bubble motion inside the capillary after transformation of frame of reference. Reduction of surface tension by foam flooding leads to decrease in oil film thickness and thus enhanced oil recovery.

5.12 Classification of Surface Tension Gradient-Driven Flow

Similar to the role played by pressure gradient in most of the traditional macrodevices, surface tension gradient plays an important role in many microdevices. This section presents the theoretical background for the flows driven by surface tension gradient. In establishing the matching condition for N–S equation at the interface between two immiscible fluids, we mostly assume that the surface tension is a constant. However, there are many cases where the surface tension varies in space. Especially, gradients in the concentration of surfactants (such as soap) at the interface and temperature gradients imply presence of gradient in the surface tension, σ.

Just as gradients in the pressure field imply a gradient force per volume, $-\nabla p$, so does a gradient in the surface tension imply a gradient force per area, $+\nabla\sigma$. The difference in sign between the two gradient forces is due to the fact that pressure forces tend to maximize volume, whereas surface tension forces tend to minimize area. The surface tension gradient force is known as the Marangoni force:

$$f_{\mathrm{Maran}} = \nabla\sigma \tag{5.51}$$

One can get an idea of the strength of temperature-induced Marangoni forces by noting that heating up a water–air interface by 5 °C from 20 °C to 25 °C will lower the surface tension gradient by 0.8 mJ/m^2 from 72.9 mJ/m^2 to 72.1 mJ/m^2. The shorter a distance over which one can maintain this temperature gradient, the stronger is the Marangoni force. Therefore, in microsystem, one can hope for a sufficiently large effect of surface tension gradient compared to other forces.

The Marangoni force can be used as a micropropulsion system, as some bacteria actually do in nature. The principle is simple. The bacteria emit some surfactant that lowers the surface tension behind the little body. The body is pushed forward as the interface tries to minimize the region of high surface tension (without surfactant) while maximizing the region of low surface tension (with surfactant). One can build a little boat illustrating this principle by attaching a piece of soap at the end of a stick. As the soap dissolves, the stick moves forward.

Spatial variations in surface tension at a liquid–gas interface result in *added tangential stresses* at the interface and hence a *surface tractive force* that acts on the adjoining fluid, giving rise to fluid motions in the underlying basic fluid. The motion induced by tangential gradients of surface tension is usually termed as *Marangoni effect*. The flow due to surface tension gradient can be classified based on the factor responsible for generation of surface tension gradient as follows:

1. Spatial variations in temperature at the interface. The resulting flow is termed as *thermocapillary* flow.
2. Spatial variation in surface concentration of an impurity or additive. The resulting flow is termed as *diffusocapillary* flow.
3. Spatial variation in electric charge or surface potential. The resulting flow is termed as *electrocapillary* flow.

The surface tension-driven flows play an active role in many technically important fields, that is, *metal processing*, *crystal growth*, and displacement of *oil from porous* strata by technique of foam flooding.

5.13 Boundary Conditions

We must know the *concentration*, *temperature*, and *charge* distributions at the interface in order to define the surface tension variation required to solve the hydrodynamic problem. However, these distributions are themselves coupled to the equations of conservation of mass, energy, and charge through an appropriate *interfacial boundary* conditions. The boundary conditions are obtained from the requirement that the *tangential shear stress must be continuous* across the interface and the net *normal force component* must balance the interfacial *pressure difference due to surface tension*.

Let us assume that an interface is established between two fluids denoted by α and β, respectively. Denoting f^α and f^β as the *force per unit area* exerted on the interface from both *viscous stresses* and *pressure* associated with boundary fluids α and β, respectively, the forces for each fluid may be written as

$$\vec{f}^\alpha = \vec{f}_s^\alpha + \vec{f}_n^\alpha = \vec{n} \cdot \bar{\bar{\tau}}^\alpha + \vec{n}p^\alpha \tag{5.52}$$

$$\vec{f}^\beta = \vec{f}_s^\beta + \vec{f}_n^\beta = -\vec{n} \cdot \bar{\bar{\tau}}^\beta - \vec{n}p^\beta \tag{5.53}$$

Here, $\vec{n}$ is the unit normal vector, $\bar{\bar{\tau}}$ is the shear stress tensor, and p is the pressure. The first term in the right-hand side is the tangential component, and the second term is the normal component of the force.

If the surface tension varies along the interface, a tangential force per unit area exists on the interface due to Marangoni effect, given by

$$f_s = \nabla_s \sigma \tag{5.54}$$

where ∇_s is the surface gradient. The positive sign of $\nabla_s \sigma$ indicates that the liquid tends to move in a direction from *lower to higher* surface tension.

The tangential force balance on the interface will give

$$f_s^\alpha + f_s^\beta + \nabla_s \sigma = 0 \tag{5.55}$$

The normal force balance on the interface using Young–Laplace equation gives

$$f_n^\alpha + f_n^\beta = \sigma \left(\frac{1}{R_1} + \frac{1}{R_2} \right) \tag{5.56}$$

5.14 Thermocapillary Motion

This section presents an example, which demonstrates the flow driven by surface tension gradient. Figure 5.25 shows the schematic of an example demonstrating the thermocapillary motion. The two side walls of the pan are kept at different temperatures. The temperature difference between the two side walls influences the surface tension distribution. A schematic of the possible surface tension distribution is shown in Figure 5.25. The surface tension reduces at higher temperature (T_1) compared to that at lower temperature (T_2). This section reports the influence of this surface tension gradient on the curvature of the fluid interface, that is, the height distribution and the resulting fluid flow beneath it. The pan is much longer than the width, and therefore, the flow nonuniformities due to the side wall effects are small, and the flow is assumed essentially two-dimensional (2-D).

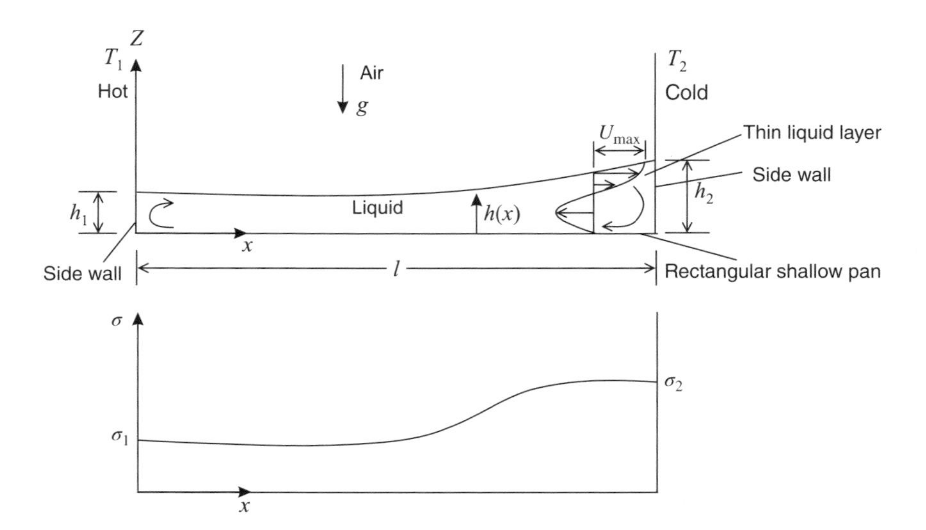

Figure 5.25 Schematic showing the thermocapillary motion

The N–S equation for incompressible flow with constant viscosity is

$$\frac{D\vec{V}}{Dt} = -\frac{\vec{\nabla}P}{\rho} + \frac{\mu\nabla^2\vec{V}}{\rho} + \vec{g}$$

Assuming steady, 2-D, negligible inertial terms and with lateral velocity gradients small compared to vertical gradients, the momentum equation in the x-direction becomes

$$\frac{dp}{dx} = \mu\frac{d^2u}{dz^2} \tag{5.57}$$

The z-momentum equation assuming vanishing z-component velocity due to absence of any driving force in this direction becomes

$$\frac{dp}{dz} = -\rho g \tag{5.58}$$

It should be recognized that the liquid surface layer is set in motion by the surface tension force which is balanced by the motion of the fluid in the opposite direction below the surface. Thus, there is no net flux across any cross-section, and the continuity equation gives

$$\int_0^{h(x)} u(z)dz = 0 \tag{5.59}$$

Boundary conditions

1. At the wall,

$$u = 0 \quad \text{at} \quad z = 0 \text{ (no-slip condition)}$$

2. At the interface,

$$f_s^\alpha + f_s^\beta + \nabla_s\sigma = 0$$

$$\Rightarrow \mu\frac{du}{dz} = \frac{d\sigma}{dx} \quad \text{at} \quad z = h(x)$$

Note that the air viscosity being negligibly small compared to that of the liquid, the viscous force from air is neglected in the above equation.

3. The other boundary condition is the continuity of pressure at the surface:

$$p = p_a \quad \text{at} \quad z = h(x)$$

Integrating equation (5.57) and using boundary conditions 1 and 2, we get

$$\mu u = \left(\frac{d\sigma}{dx} - h\frac{\partial p}{\partial x}\right)z + \frac{1}{2}\frac{\partial p}{\partial x}z^2 \tag{5.60}$$

Integrating equation (5.58) and using boundary conditions (3), we get

$$p = p_a + \rho g(h - z) \tag{5.61}$$

From equation (5.61), we can write

$$\frac{\partial p}{\partial x} = \rho g\frac{dh}{dx} \tag{5.62}$$

Substituting equation (5.60) in continuity equation (5.59) and integrating, we have

$$\left.\frac{d\sigma}{dx}\frac{z^2}{2} - \frac{hz^2}{2}\frac{\partial p}{\partial x} + \frac{1}{2}\frac{\partial p}{\partial x}\frac{z^3}{3}\right|_0^{h(x)} = 0$$

$$\frac{d\sigma}{dx}\frac{h^2}{2} - \frac{h^3}{2}\frac{\partial p}{\partial x} + \frac{h^3}{6}\frac{\partial p}{\partial x} = 0$$

$$\frac{d\sigma}{dx} - \frac{2}{3}h\frac{\partial p}{\partial x} = 0 \tag{5.63}$$

Using equation (5.62) in equation (5.63), we write

$$\frac{d\sigma}{dx} = \frac{2}{3}\rho g h\frac{dh}{dx} \tag{5.64}$$

Integrating the above equation and using $\sigma = \sigma_1$ and $h = h_1$ at $x = 0$, we have

$$\sigma - \sigma_1 = \frac{\rho g}{3}(h^2 - h_1^2) \tag{5.65}$$

Equation (5.65) indicates that the variation in σ automatically requires a corresponding variation in h.

Using equation (5.64) in equation (5.60), we have

$$u = \frac{z}{2\mu}\left(\frac{3}{2}\frac{z}{h} - 1\right)\frac{d\sigma}{dx} \tag{5.66}$$

Equation (5.66) indicates that the liquid velocity has a maximum value at the interface:

$$u_{\text{max}} = \frac{h}{4\mu}\frac{d\sigma}{dx} \tag{5.67}$$

The velocity reverses direction, that is, the velocity is equal to zero at $z = \frac{2h}{3}$. Similarly, the maximum negative value can be found using $\frac{\partial u}{\partial z} = 0$ and $\frac{\partial^2 u}{\partial z^2} = +\text{ve}$, which gives $z = \frac{h}{3}$. The velocity profile because of this thermocapillary convection has been shown in Figure 5.25.

Note: We have assumed the Reynolds number to be small for inertial effect to be negligible while simplifying the N–S equation. Let us say, small means “small compared to unity,” that is, $Re \leq 0.1$:

$$Re_h = \frac{h_1 u_{\text{max}}}{\gamma} \leq 0.1$$

$$\frac{h_1^2}{4\mu\gamma}\frac{d\sigma}{dx} \leq 0.1$$

$$h_1^2 \ll \frac{0.4\mu\gamma}{\frac{d\sigma}{dx}} \tag{5.68}$$

For water, $\frac{\partial\sigma}{\partial T} \approx -0.15$ mN/m-K

If $\frac{\partial T}{\partial x} = -100$ K/m, we have $\frac{\partial\sigma}{\partial x} = 15$ mN/m^2.

At 20 °C, we can use $\mu = 1.002 \times 10^{-3}$N-s/m^2 and $\gamma = 1.004 \times 10^{-6}$ m^2/s. Thus, equation (5.68) will give $h_1 = 163$ μm. Using h_1, we can calculate the Bond number as Bond number, $\frac{\rho g h^2}{\sigma} = 3.6 \times 10^{-3}$ for $\sigma = 73$ mN/m

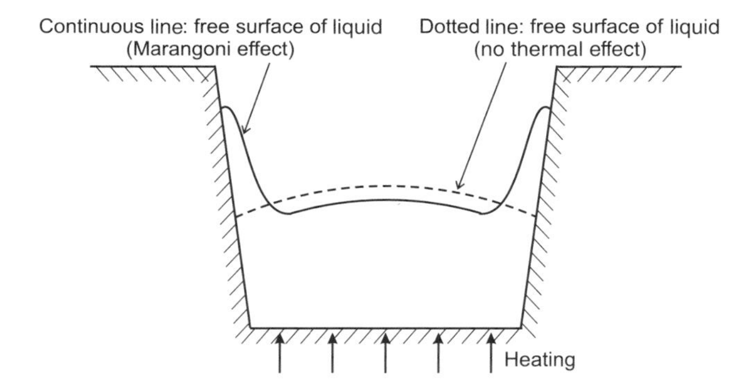

Figure 5.26 Marangoni effect in a microwell of a DNA array

Hence, the small Bond number value indicates strong effect of surface tractive force relative to the force of gravity.

The surface tension gradients can be important in very *thin liquid layers* or in *reduced* gravity *environment (microgravity)*. For example, a crystal grown from its melt under reduced gravity is governed by convection driven by thermally induced surface tension gradients rather than buoyancy forces.

5.14.1 DNA Arrays

The Marangoni effect plays a crucial role in the DNA arrays used for recognition of DNA segments. For identification of DNA of some virus, microplate with many microwells or cusps is used. Each well is grafted with a predetermined DNA sequence aiming at a specific target. During matching of the target DNA and its complementary sequence, there is a binding between the two segments due to hydrogen bonds. The target can be identified by fluorescence when it binds in a well. During operation, the liquid drops are deposited inside the microwell, followed by heating.

The heating of the drop lowers the surface tension at the walls, and the liquid rises along the walls due to increased capillary force. Therefore, cusps are designed such that the liquid cannot exit by capillary force and overflow into the neighboring cusps. Figure 5.26 shows the behavior of drop under Marangoni effect in a DNA array microwell. The interface is spherical in the case of uniform temperature. The interface is deformed by the Marangoni effect due to heating of the drop.

5.15 Diffusocapillary Flow

In the previous section, an air bubble motion inside a circular capillary was discussed assuming constant interfacial tension during oil recovery operation. But in reality, surfactants are employed in enhanced oil recovery problem, where surfactants are injected to the underground oil well. For oil recovery, gases (say, CO_2) are injected to the oil reservoir. The gas bubbles travel through the pores of the oil reservoir due to the application of pressure (see Figure 5.24).

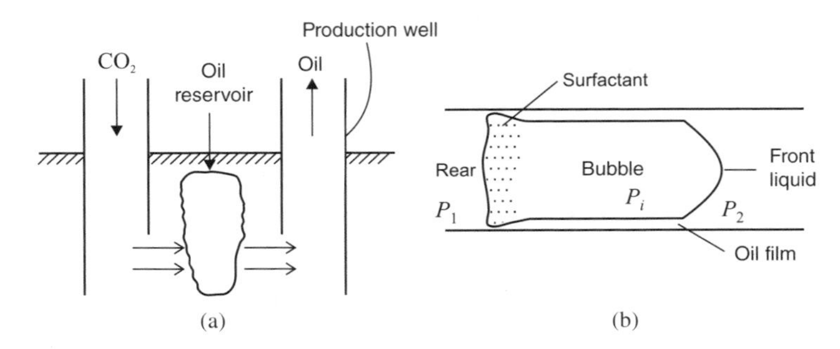

Figure 5.27 Schematic of an (a) oil recovery system and (b) bubble shape inside the capillary due to accumulation of surfactant

During the movement of gas bubbles, the oil film is collected at the production well located outside the reservoir. As the bubble travels down the capillary, the flow in the surrounding fluid causes a *nonuniform distribution of surfactant* on the bubble surface. This nonuniform surfactant distribution determines the surface forces, since the surface tension gradient depends on the local surface concentration of the adsorbed material. The *film thickness* and the total *pressure drop* to drive the bubble are dependent on the surfactant distribution. This kind of flow is known as diffusocapillary flow. The surfactant is swept toward the rear of the bubble where it accumulates.

The pressure difference across the front interface may be given by

$$P_i - P_2 = \frac{2\sigma_{\text{front}}}{a} \tag{5.69}$$

where "P_i" is the pressure inside the bubble, "a" is the capillary radius, and σ_{front} is the surface tension of the front interface.

Similarly, the pressure difference across the rear interface can be written as

$$P_i - P_1 = \frac{2\sigma_{\text{rear}}}{a} \tag{5.70}$$

Using the above two equations, the pressure difference across the bubble can be written as

$$P_1 - P_2 = \frac{2}{a}(\sigma_{\text{front}} - \sigma_{\text{rear}}) \tag{5.71}$$

Equation (5.71) indicates the presence of net forward force due to the difference in surface tension between the front and rear interface of the bubble. Therefore, the pressure force requirement for oil recovery reduces due to diffusocapillary force because of surfactant distribution.

It may be noted that equation (5.50) also indicates that the film thickness reduces due to decrease in surface tension. A schematic of oil recovery system and the role of bubble formation in enhanced oil recovery have been shown in Figure 5.27(b).

5.16 Electrowetting

It has been observed that ions and dipoles redistribute in the liquid due to the application of electric potential. This redistribution can cause a change in the wetting properties of the drop.

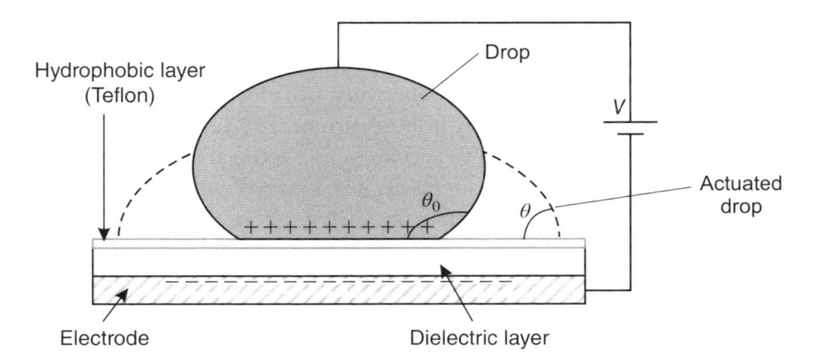

Figure 5.28 The schematic showing the principle of electrowetting of a drop. The liquid is much better conductor than the insulating layer. The interface is polarized without generating any crossing current

This phenomenon is known as electrowetting. A hydrophobic surface like Teflon can behave like a hydrophilic surface due to the application of electric potential. Electrowetting can be used for the creation, transportation, and merging of droplet for digital microfluidic systems. Electrowetting can also be used in micromixers.

Figure 5.28 shows the principle of electrowetting, where the wetting angle decreases due to the application of electric field (both dc and ac). Lippmann's law states that the surface tension σ_{sl} of an electrically conductive liquid changes when the drop is placed in an electric field. Electric charges migrate to the liquid/substrate interface, that is, toward the contact line. The surface tension changes due to the charge redistribution, and according to the Lippmann's law, the solid–liquid surface tension can be expressed using the relation

$$\sigma_{\text{sl}}(V) = \sigma_{\text{sl},0} - \frac{1}{2}CV^2 \tag{5.72}$$

where C is the capacitance (Farad (coulomb/volt)) of the dielectric layer separating the bottom electrode from the liquid, V is the electric potential (joule/coulomb), and subscript 0 refers to the unactuated state.

According to Young's law, we have

$$\cos\theta_0 = \frac{\sigma_{\text{sg}} - \sigma_{\text{sl},0}}{\sigma_{\text{lg}}} \tag{5.73}$$

$$\cos\theta = \frac{\sigma_{\text{sg}} - \sigma_{\text{sl}}(V)}{\sigma_{\text{lg}}} \tag{5.74}$$

where θ_0 and θ are the unactuated and actuated contact angles, respectively. Combining the Lippmann's equation with Young's equation, we have

$$\sigma_{\text{lg}}\cos\theta - \sigma_{\text{lg}}\cos\theta_0 = \frac{1}{2}CV^2 \tag{5.75}$$

Equation (5.72) can also be written using the expression for capacitance as

$$\sigma_{\text{sl}}(V) = \sigma_{\text{sl},0} - \frac{\varepsilon A}{2d}V^2 \tag{5.76}$$

where ε is the permittivity (Farad/meter) of the insulator dielectric, d is the thickness of the insulator, and A is the surface area. The insulator controls the capacitance between the droplet and the electrode. Here, a liquid droplet is kept between two electrodes with one dielectric layer on the surface of an electrode to prevent any flow of current. The Teflon coating on the dielectric surface makes the drop hydrophobic with contact angle more than 90°. The application of potential difference changes the contact angle to less than 90°, that is, the droplet becomes hydrophilic. In the case of alternating voltage, the voltage, V, in the above equation should be the rms value given by

$$V_{\text{rms}} = \frac{V}{\sqrt{2}} \tag{5.77}$$

5.16.1 *Electrowetting-Based Microactuator*

Direct electrical control of surface tension can be used for rapid actuation of discrete liquid droplets. The droplets can be transferred at high rates under low voltage offering advantages over continuous-flow processes. No pumps and valves are required, and these systems can be flexible, efficient, and capable of performing complex and parallel microfluidic processing.

Figure 5.29 shows the schematic of an electrowetting microactuator. Here, a polarizable and conductive liquid is sandwiched between two sets of planar electrodes. The upper plate consists of a single continuous ground electrode, and the bottom plate consists of an array of independently addressable control electrodes. A thin hydrophobic insulation covers both the electrodes. For successful operation, the system geometry and droplet volume should be controlled in such a manner that the droplet overlaps at least two control electrodes. Initially, all the electrodes are grounded, and the contact angle is equal to equilibrium contact angle of the droplet. The application of electrical potential to a control electrode underneath leads to local reduction in surface tension, σ_{sl}. Thus, there is a reduction in contact angle between the droplet and the energized electrode. The energized control electrode surface becomes hydrophilic in nature. The droplet meniscus is deformed asymmetrically, and a pressure gradient is established between the ends of the droplet, which results in bulk flow toward the energized electrode. The droplet moves away from the energized electrode. The next control electrode is energized after arrival of the droplet, and the previous energized electrode is switched off. The surface tension gradient is established again between the front and rear end

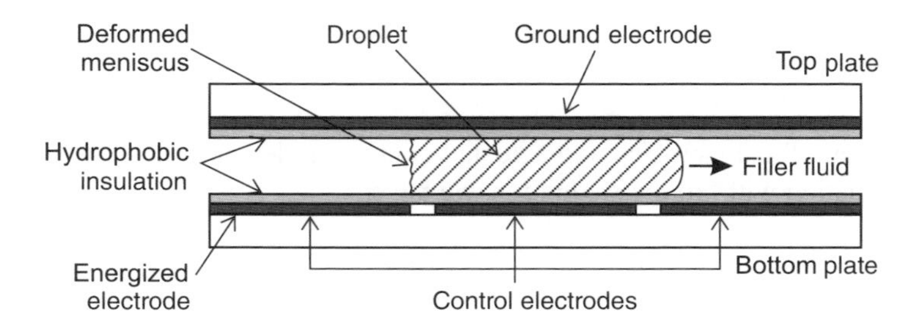

Figure 5.29 Schematic cross-section of the electrowetting microactuator

of the liquid drop which drives the droplet. Thus, the motion of the droplet can be achieved as per the process requirement. The pumping flow rate can be calculated as that discussed for the thermocapillary pumping system in a later section.

5.17 Marangoni Convection in Drops

Droplets are present in many microfluidic systems. This section presents the convective motion inside the droplet due to the variation in local properties of the liquid. Surface tension depends on temperature, concentration of chemical species at surface, and electric potential, that is,

$$\sigma = \sigma(T, C, V) \tag{5.78}$$

where T is the temperature, C is the concentration, and V is the electric potential. Thus, the change in surface tension can be written as

$$\nabla\sigma = \frac{\partial\sigma}{\partial T}dT + \frac{\partial\sigma}{\partial C}dC + \frac{\partial\sigma}{\partial V}dV \tag{5.79}$$

The classical relation between surface tension and temperature can be written as

$$\sigma = \sigma_0(1 - \beta(T - T_0)) \tag{5.80}$$

where σ_0 is the surface tension at temperature T_0 and β is a constant. If the drop is not isothermal, there is a gradient of surface tension due to the gradient of temperature. This leads to the tangential force distribution at the interface, leading to convective motion inside the drop. Viscosity diffuses this motion inside the drop. In Figure 5.30, the drop is exposed to a temperature gradient, $T_2 > T_1$. Hence, $\sigma_{\mathrm{lg}\,1} > \sigma_{\mathrm{lg}\,2}$. Therefore, the convection motion is directed from the top needle tip to the bottom solid plate.

The strength of Marangoni convection is linked to the Marangoni number defined as

$$Ma_T = \frac{\Delta\sigma R}{\rho\gamma\alpha} \tag{5.81}$$

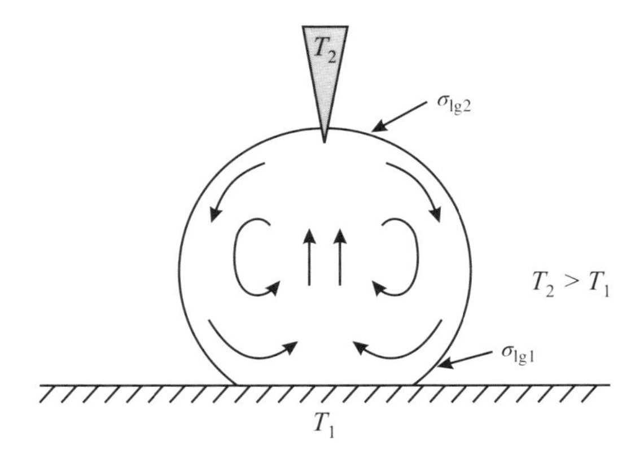

Figure 5.30 Schematic of Marangoni-driven convection inside a droplet due to temperature difference

where $\Delta\sigma$ is the variation of the surface tension, R is the radius of the drop, ρ is the density of the liquid, γ is the kinematic viscosity, and α is the thermal diffusivity. Thermal Marangoni number is defined as the ratio between the surface tension force due to temperature difference and the viscous resistive force of the fluid.

Similar to the temperature-driven Marangoni convection, concentration-driven Marangoni convection is also present in microfluidics. When surfactants are added to the fluid, they migrate to the interface, and a gradient of interface concentration may occur, leading to a convective motion at the interface. In the case of concentration-driven convection, the nondimensional Marangoni number is defined as

$$Ma_C = \frac{\Delta\sigma R}{\rho\gamma D} \tag{5.82}$$

where D is the diffusion coefficient.

Solutal Marangoni number, Ma, is the ratio between the surface tension force due to concentration difference and the viscous force of fluid.

5.18 Marangoni Instability

In the preceding section, we have examined thermocapillary and diffusocapillary flows. Not all such flows are stable, and in fact, surface tension variations at the interface can be sufficient to cause an instability. Rayleigh number is mostly used to describe the instability of buoyancy-driven flow defined as

$$Ra = \frac{g\beta\Delta T h^4}{\gamma\alpha} \tag{5.83}$$

where h is liquid layer height, ΔT is the vertical temperature gradient, β is the coefficient of thermal expansion of the fluid, γ is the kinematic viscosity, and α is the thermal diffusivity. Rayleigh number is the ratio of thermalbuoyancy force to viscous force.

Bénard in the year 1900 observed that *hexagonal convection* cells are formed within thin film of molten spermaceti of about 0.5–1 mm depth that were heated from below, with the cell spacing somewhat more than three times the liquid depth. These cells are now referred as Bénard cells (Figure 5.31). Bénard initially assumed that surface tension at the free surface of

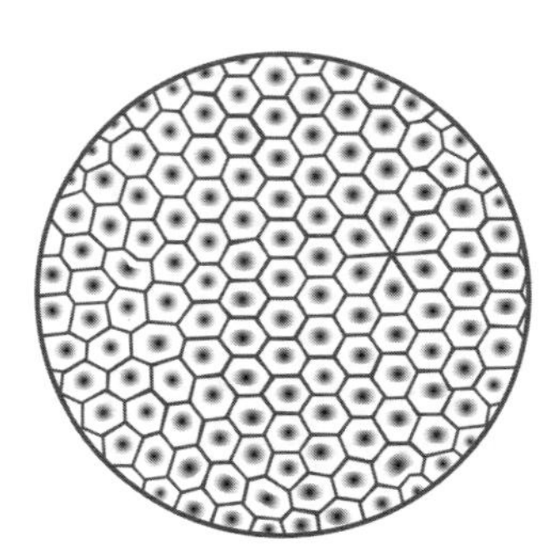

Figure 5.31 Schematic of Bénard cell seen from the top of the container, where a thin layer of liquid is heated below

the film is an important factor for the cell formation. But this idea was abandoned for some time as Rayleigh analyzed the buoyancy-driven flow of a layer of fluid heated from below and observed that if the hexagonal cells are formed, the ratio of the spacing to cell depth is almost exactly equal to that measured by Bénard. Rayleigh showed that if the cells are to form, then the vertical *adverse temperature gradient* must be sufficiently large such that a dimensionless parameter (Rayleigh number) exceeds a critical value.

Block showed that the Bénard's results were not consequence of buoyancy, but were surface tension induced. He showed that cellular convection takes place even if Ra is smaller than that required by Rayleigh theory. Block observed Bénard cells even when the *thin film is cooled from below*. Note that if cells are buoyancy induced, then in this case as gravity and density gradient are in same direction, the thin film must be stably stratified, which was not true from Block's experiment. He was also able to remove the Bénard cell by covering the surface with a *monolayer of surfactant*. Block concluded that for thin films of thickness less than 1 mm, variations in *surface tension due to temperature variations* were the cause of Bénard cell formation and not buoyancy as postulated by Rayleigh. It is now generally agreed that for films *smaller than about a few millimeters*, surface tension is the controlling force. Buoyancy is the controlling force for larger thickness, where the Rayleigh mechanism delimits the stable and *unstable regimes*.

The phenomenon of surface tension-induced Bénard cells is commonly observed in the *drying of paint* with the appearance of what is usually called an "*orange peel*" pattern. The cause of the orange peel or Bénard cells is the *surface tension gradient*-induced flow along the paint film by the rapid cooling effect at the free surface associated with the evaporation of the *volatile solvents* in the paint. Also, the orange peel effect is independent whether the paint layer is topside or underside of the plate being painted. This indicates insignificant role played by the gravity force on the instability.

The mechanism of Bénard cell formation is also termed as Marangoni instability. Let there be a *small disturbance*, which causes the film of initially uniform thickness to be heated locally at a point on the surface. This results in decreased local surface tension and development of surface tension gradient that leads to an induced motion tangential to the surface away from the point of local heating (see Figure 5.32). From mass conservation, this motion in turn induces a motion of the bulk phase toward the interface. The upwelling liquid coming from the heated region is warmer than that of the liquid–gas interface. The motion is thus reinforced creating cellular convection patterns and will be maintained if the convection overcomes shear and heat

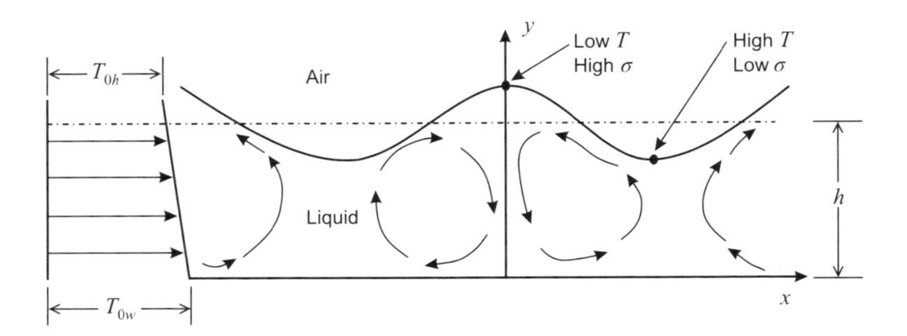

Figure 5.32 Instability mechanism for Bénard cell formation

diffusivity. Note that the curvature of the interface can be deduced from the thermocapillary convection example discussed earlier (Figure 5.25).

From linear stability analysis, two nondimensional numbers governing the instability of thin liquid film are

$$\text{Marangoni number}(Ma) = \frac{\sigma_T \beta h^2}{\mu} \tag{5.84}$$

$$\sigma_T = -\left(\frac{d\sigma}{dT}\right)_{T=T_{oh}} \tag{5.85}$$

Here, β is the vertical temperature gradient and h is the film height.

Marangoni number is the ratio of surface tension force to viscous force.

Marangoni number is a measure of the heat transfer by convection due to surface tension gradients to the bulk heat transport by conduction. Marangoni convection also depends on another nondimensional number known as Biot number:

$$\begin{aligned}\text{Biot number} &= \frac{h/k}{1/(\text{Surface heat transfer coeff})} \\ &= \frac{\text{Thermal resistance of liquid layer}}{\text{Thermal resistance of external environment}}\end{aligned}$$

Note that here h corresponds to the characteristic length scale (see Figure 5.32).

Critical Marangoni number increases with increasing values of Biot number. For a fixed temperature boundary condition at the wall surface and for $Bi = 0$, critical Marangoni number $\cong 80$. For fixed heat flux condition and $Bi = 0$, critical Marangoni number $\simeq 40$.

5.19 Micropropulsion System

Marangoni force can be used as a micropropulsion system. In fact, some bacteria in nature move based on the Marangoni force. Surface tension phenomena occur at the interface between two immiscible fluids as a consequence of altered forces on molecules in that region. It is a force exerted on the plane of the interface and of uniform strength in all directions so that an object (bacteria) suspended in the interface is subjected to equal pull in all directions, as shown in Figure 5.33.

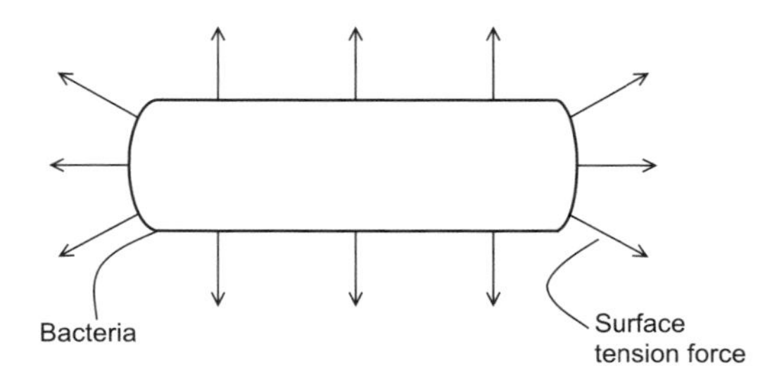

Figure 5.33 An object (bacteria) placed near an interface. Here, the paper is parallel to the interface

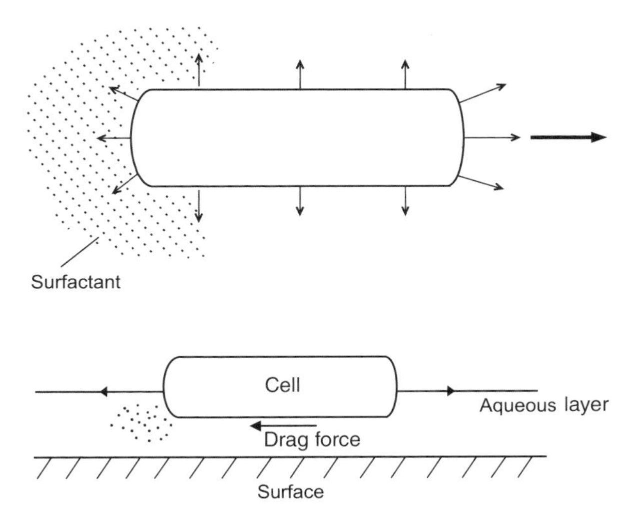

Figure 5.34 The surfactant is excreted at one end, and the net imbalance in surface tension force balances drag during motion

Figure 5.34 shows the schematic illustrating the motion of bacteria. The bacteria excrete a surfactant from sites located near the ends of the cell. The surfactant lowers the surface tension in the vicinity of the excreting site, causing an unbalanced force on the cell, which causes it to be propelled in the plane of the interface at a speed at which the drag on the cell just equals the unbalanced surface tension force. If the surfactant excretion occurs at the other end of the cell, then the movement would be in the opposite direction.

It is not necessary that the cell is located precisely at the interface. If the cell is located *below the interface*, the excreted surfactant will diffuse to the interface and set up gradients in surface tension that will give rise to convective flow patterns. The flow will carry the cell in the same direction, but at altered speed because of altered momentum balance.

Small Boat Movement
The same principle can be used for moving a little boat by adding detergent at one end or local heating (Figure 5.35). The surface tension imbalance force between the front and rear end of the boat leads to propulsion.

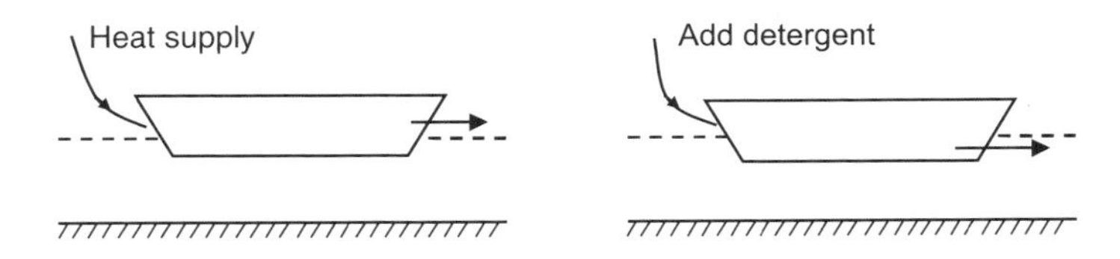

Figure 5.35 The movement of a small boat due to surface tension force

5.20 Capillary Pump

A capillary pump can be used in an LOC system for biosensing. The powerless capillary pump system is both adequate and handy for the task of delivering liquids at the specific points of the chip. A schematic of an LOC with capillary pump has been shown in Figure 5.36.

A typical LOC system shown in Figure 5.36 can be constructed by spinning a polymer layer, that is, PMMA of height $h = 20$ μm onto a glass plate. By photolithography, six circular reservoirs of radius $r = 3$ mm are etched into the PMMA layer. Three channels with rectangular cross-section of width $w = 200$ μm and height $h = 20$ μm are connected pairwise between the reservoirs. At the center of the chip, each of the three channels widens into a 1.3 mm × 1.0 mm rectangular measuring site, where the cantilever probes are dipped into the liquid. The whole chip is covered by a second glass plate to close off the microfluidic channel. Using a simple pipette, the biochemical liquid is injected into one of the large reservoirs. By capillary forces, the liquid is sucked into the microchannel leading from the reservoir to the measuring site.

The cantilever is coated with specific biomolecules. Biochemical reaction takes place at the surface when the cantilever is immersed into the biochemical solution. Mechanical surface stresses are induced due to the chemical reaction. The cantilever bends due to the stress. The bending can be detected by the piezoresistive readout built into the cantilever. The careful design of biocoating on the cantilever can selectively respond to certain biomolecules. Other reservoirs can be used for parallel sensing operation.

Here, the microchannel is placed horizontally. Therefore, the gravitational force cannot balance the capillary forces. Hence, the capillary advance will continue till there is a channel for the liquid to propagate in (Figure 5.37). The position $L = 0$ is defined as the entrance of the microchannel, that is, the input reservoir. The reservoir is very wide compared to the channel size, and therefore, no Young–Laplace pressure drop is present, that is, $p(x = 0) = P_0$. The following section presents the advancement time of the capillary front for capillary pumping.

5.20.1 Advancement Time of Capillary Pump

We assume that the gravity head of the liquid column to be negligible as the height of the microchannel is very small. Therefore, the pressure at the entrance of the microchannel has a constant value.

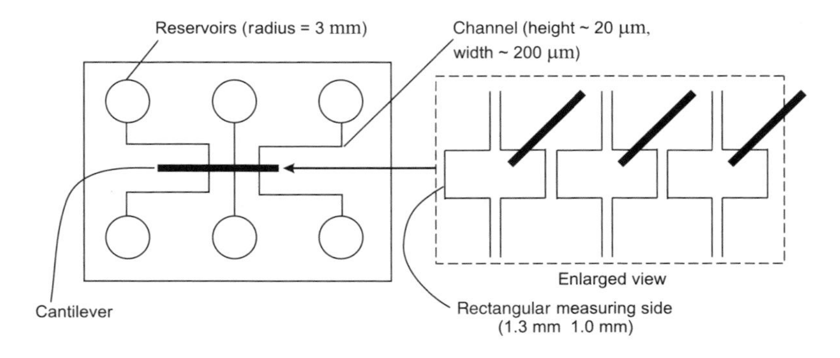

Figure 5.36 A lab-on-a-chip with capillary pump

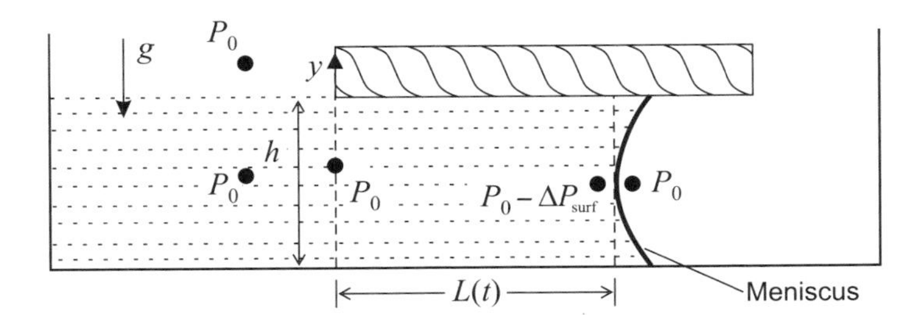

Figure 5.37 Schematic showing the principle of a capillary pump. The curved meniscus results in an uncompensated Young–Laplace under pressure, $-\Delta P_{surf}$. This pressure difference drives the liquid to the right in the microchannel

At the entrance of the microchannel, that is, at the input reservoir, the pressure drop due to Young–Laplace is zero due to the high width of the input reservoir:

$$P(x = 0) = P_0$$

The curved meniscus at position $L(t)$ leads to an Young–Laplace underpressure $-\Delta P_{surf}$ that drives the liquid to the right of the channel.

Channel cross-section is rectangular, h (height) $\times$ w (width) and $h \ll w$. Using the Hagen–Poiseuille result for flow rate,

$$Q = \frac{h^3 w \Delta P}{12\mu L} \tag{5.86}$$

The pressure drop ΔP between the entrance ($x = 0$) and the advancing meniscus ($x = L(t)$) can be obtained from the Young's–Laplace relationship as

$$\Delta P = \Delta P_{surface} = \frac{2\sigma}{h}\cos\theta \left(\begin{aligned} \text{Note: } \Delta P &= \sigma\left(\tfrac{1}{R_1} + \tfrac{1}{R_2}\right) \\ &= \sigma\left(\frac{1}{\frac{h}{2\cos\theta}} + 0\right) \\ &= \tfrac{2\sigma}{h}\cos\theta \end{aligned} \right) \tag{5.87}$$

where h is the height of the microchannel.

Note that R_2 is set equal to zero as the microchannel is of high aspect ratio rectangular cross-section, which is similar to a parallel plate geometry.

Here, θ is the contact angle. Thus, the flow rate through the capillary pump can be determined using equation (5.86) and the surface tension information from equation (5.87).

We can write the speed of the front of liquid as

$$V_0 = \frac{dL(t)}{dt} \tag{5.88}$$

From Poiseuille (equation (5.86)) flow relationship:

$$V_o = \frac{Q}{wh} = \frac{h^2 \Delta P}{12\mu L(t)} \tag{5.89}$$

Hence, equating the equations (5.88) and (5.89), we have

$$\frac{dL(t)}{dt} = \frac{h^2 \Delta P}{12\mu L(t)} \tag{5.90}$$

$$\Rightarrow LdL = \frac{h^2 \Delta P}{12\mu} dt$$

$$\Rightarrow \frac{L^2}{2} = \frac{h^2 \Delta P}{12\mu} t + \text{Constant} \tag{5.91}$$

Using initial condition at $t = 0$, $L = 0$, we get Constant = 0:

$$\therefore \quad L(t) = \sqrt{\frac{h^2 \Delta P t}{6\mu}} = h\sqrt{\frac{\Delta P t}{6\mu}} = h\sqrt{\frac{2\sigma \cos \theta t}{6\mu h}} = h\sqrt{\frac{\sigma \cos \theta t}{3\mu h}}$$

$$L(t) = h\sqrt{\frac{\sigma \cos \theta t}{3\mu h}} \tag{5.92}$$

The above equation indicates that the advancement time of the meniscus depends on the fluid properties, surface properties, and microcapillary dimension. Hence, proper tuning of microchannel properties can lead to optimization of the capillary pumping system. Specific mixing sequence of the reacting fluid from multiple reservoirs can be obtained by adjusting the channel dimension of the capillary pumping system. Thus, capillary pumping can be used for design and operation of a complex LOC system.

5.21 Thermocapillary Motion of Droplets

A thermal gradient can move drops across a horizontal surface. This method of moving drops involves creating a temperature difference along the length of the surface. This method can potentially eliminate the need for pumps to move small drops on flat surfaces. This method is also useful in a situation where the vibrations associated with a pump need to be minimized. Since the motion of the drops is not directly dependent on gravity, a thermal gradient can propel these drops in a microgravity environment. This is useful for heat transfer in a microgravity environment, where a thermal gradient can remove condensate from a surface, regardless of orientation, improving the heat transfer coefficients.

A drop of fluid resting on a surface moves from the warm side to the cold side when a thermal gradient is applied to that surface. This phenomenon is known as *thermocapillary motion*, which arises from the dependence of surface tension on temperature. Surface tension is the property of liquids that causes a liquid drop to assume the shape with the smallest surface area per unit volume. In the absence of any outside forces, a freely floating drop of liquid would assume the shape of a sphere, since this shape has the smallest surface area for a given volume. Surface tension of liquid generally decreases as temperature increases. This creates an imbalance between the forces pulling at each area element along the surface of the drop (Figure 5.38). This imbalance between forces causes a flow just beneath the surface of the drop. The liquid circulates throughout the drop, including the bottom of the drop where it is adjacent to the solid surface. Here, the liquid pushes against the surface in the direction of

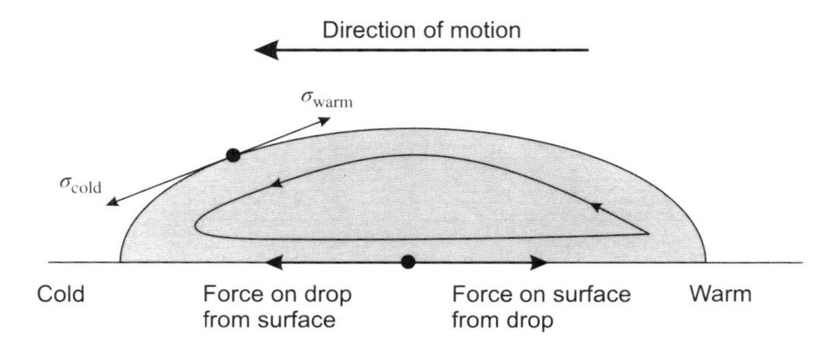

Figure 5.38 Schematic showing flow patterns inside a drop under horizontal thermal gradient

cold to warm. Because the solid surface is fixed in position and because it pushes back (due to Newton's third law), the drop moves toward the cold side of the surface. The steeper thermal gradient leads to greater velocity of the droplet.

5.22 Thermocapillary Pump

Microfluidic pump can be classified as either mechanical or nonmechanical. Mechanical pumping systems consist of thin membrane which can be deflected by electrostatic, piezoelectric, or thermopneumatic actuation. These pumping systems are usually well suited for supply of continuous streams. Nonmechanical pumps operate on electrodynamic, electroosmotic, electrowetting, and thermocapillary mechanism. Here, we discuss a nonmechanical surface tension-driven thermocapillary pump for moving discrete liquid drops in microfabricated channels. The surface tension forces for this pump are high, and therefore, controlling the surface forces can serve as a drive mechanism for a simple liquid pumping system. Let us consider an example of moving a water droplet at average velocity of 1 cm/s inside a 100 μm diameter capillary tube. The dimensionless number corresponding to this case is estimated as

$$(\text{Bond number})^{-1} = \frac{\text{Surface tension force}}{\text{Gravity force}} = \frac{\sigma}{\rho g d^2} \simeq 10^3$$

$$(\text{Capillary number})^{-1} = \frac{\text{Surface tension force}}{\text{Viscous force}} = \frac{\sigma}{\mu V} \simeq 10^4$$

$$(\text{Weber number})^{-1} = \frac{\text{Surface tension force}}{\text{Inertial force}} = \frac{\sigma}{\rho d V^2} \simeq 10^4$$

The above sample calculation demonstrates the importance of surface tension forces in typical interfacial flow condition in a microchannel.

Figure 5.39 shows the principle of thermocapillary pump of a droplet using hydrophilic ($\theta < 90°$) and hydrophobic ($\theta > 90°$) channel wall. For the hydrophilic system, heating at the receding end leads to the motion of the droplet. For the hydrophobic system, heating of the advancing end leads to the motion of the droplet. This pumping device does not require any moving parts and is applicable to discrete drops. It is compatible with any channel surface that produces a curved liquid interface, that is, silicon, glass, various polymers, and so on.

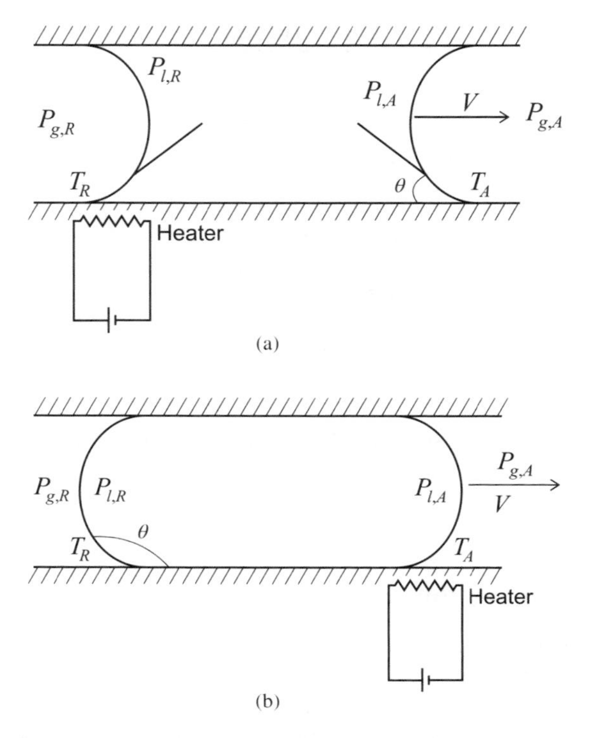

Figure 5.39 Schematic showing the thermocapillary pumping of a liquid drop on (a) hydrophilic surface and (b) hydrophobic surface

For a circular capillary, the Young–Laplace equation at the interface can be written as

$$P_g - P_l = \frac{4\sigma\cos\theta}{d} \tag{5.93}$$

where P_g is the interface pressure in the gaseous side and P_l is the interface pressure in the liquid side, θ is the contact angle, σ is the surface tension, and d is the capillary diameter. Note that the above equation is for static system. However, it is also a reasonable approximation for moving drops if the static angle is replaced by the dynamic contact angle.

For the stationary drop, there is a pressure difference between the liquid and gas phases across each interface. However, the liquid pressure between the two ends is balanced, that is, $P_{l,R} - P_{l,A} = 0$. A pressure difference is generated for liquid motion by manipulating the surface tension on one side of a liquid drop. The surface tension is modified by heating one drop interface. The decrease in surface tension because of increase in temperature can be expressed as

$$\sigma = a - bT \tag{5.94}$$

where a and b are positive empirical constants.

We can assume Poiseuille flow solution of the Navier–Stokes equation for estimating the droplet velocity. This is a reasonable assumption as the flow streamlines is only different at the very ends of the drop. Hence, for large drop length to height ratio, the validity of Poiseuille solution is assumed. The average velocity of Poiseuille flow inside capillary tube is

$$V = \frac{d^2 \Delta P}{32\,\mu \mathrm{L}} \tag{5.95}$$

The pressure difference between the two ends of the drop can be expressed as

$$\Delta P = P_{l,R} - P_{l,A} = \Delta P_c + \Delta P_h + \Delta P_e \tag{5.96}$$

where ΔP_c is the contribution due to thermally induced surface tension difference, ΔP_h is the contribution from gravity due to inclination of the capillary, and ΔP_e is the contribution due to external pressure source:

$$\Delta P_c = P_{c,A} - P_{c,R} = 4\left[\left(\frac{\sigma\cos\theta}{d}\right)_A - \left(\frac{\sigma\cos\theta}{d}\right)_R\right]$$

$$\Delta P_h = \rho g L \sin\phi$$

$$\Delta P_e = P_{g,R} - P_{g,A}$$

Here, ϕ is the inclination angle between the capillary axis and the horizontal, $P_{g,R}$ is the gas-phase pressure at the receding end, and $P_{g,A}$ is the gas-phase pressure at the advancing end. Combining the aforementioned six equations, the steady-state thermocapillary pump velocity can be written as

$$\begin{aligned} V = {} & \frac{4d}{32\mu L}[(a - bT_A)\cos\theta_A - (a - bT_R)\cos\theta_R] \\ & + \frac{\Delta P_e d^2}{32\mu L} + \frac{d^2 \rho g \sin\phi}{32\mu} \end{aligned} \tag{5.97}$$

The above equation is valid for both hydrophilic and hydrophobic systems. We can rewrite the drop velocity based on the temperature difference between the advancing and receding ends of a hydrophilic drop as

$$V = \frac{4db\cos\theta_R}{32L\mu}(\Delta T - \Delta T_{\min}) + \frac{d^2}{32\mu L}(\Delta P_e + \rho g L \sin\phi) \tag{5.98}$$

where

$$\Delta T = T_R - T_A \tag{5.99}$$

and

$$\Delta T_{\min} = (T_R - T_A)_{\min} = \left(\frac{a}{b} - T_A\right)\left(1 - \frac{\cos\theta_A}{\cos\theta_R}\right) \tag{5.100}$$

For the hydrophobic case, the above equation for velocity can be modified by exchanging A(advancing) and R(receding) subscript and adding a minus sign immediately after the equation sign. Here, ΔT_{min} is the minimum temperature difference needed to initiate the pumping in the absence of external pressure and gravity.

5.23 Taylor Flows

Microflows constitute more than single fluid in many applications. The flow of more than one fluid moving simultaneously in a microchannel is denoted as multiphase flow. It can be either one liquid and one gas phase or two different liquid phases.

Multiphase flows are very important in LOC system, where more than one liquid are brought together for further treatment or chemical analysis. Gas bubbles can be used sometimes for performing some functionality of the device due to its mechanical properties. The other importance of multiphase flow is that unwanted bubbles may appear due to the electrolysis. The molecules of interest in buffer solution are moved as liquid plugs separated by nonmiscible plugs of another liquid (see Figure 5.40). Multiphase microflows are also encountered during parallelization of biological operations, that is, screening and biodiagnostics. Figure 5.41 shows simultaneous plugs moving in parallel capillary tubes, which convey to separate biodiagnostics chambers. The plugs have two menisci, one meniscus corresponds to the advancing front and the other corresponds to the receding front.

5.23.1 Practical Applications

There are various applications of multiphase flows in microfluidics. One of the oldest and simplest is using bubble as a flow meter (Fairbrother and Stubbs, 1935). Here, the bubble extends over almost the entire cross-section of the channel, and therefore, the velocity of the bubble is nearly equal to the bubble. The velocity of the bubble can be measured visually and used for the measurement of the total flow rate. Another application is the use of gas

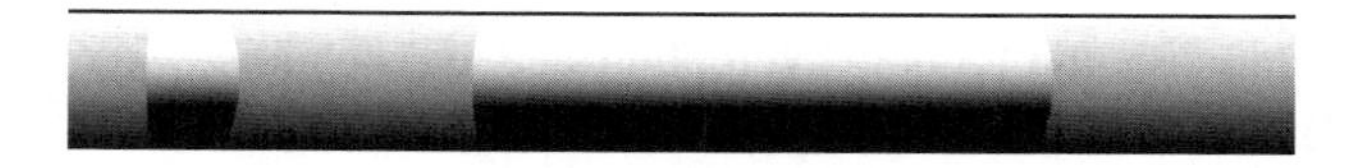

Figure 5.40 A schematic showing fluid plugs, that is, buffer fluid plugs and oil plugs

Figure 5.41 Synchronized fluid plugs moving in parallel capillaries

bubbles for improving the membrane process. Addition of gas bubbles to the capillary channel enhances the circulation leading to improving mass transfer and overall improvement of the microfabrication (Laborie *et al.*, 1998). The microfiltration efficiency is also increased by the pressure pulsing caused by the removal of filter cake.

Technician's autoanalyzer, so-called continuous-flow analyzer (Thiers *et al.*, 1971), uses multiphase flow concept for high-throughput analysis. In these machines, the samples to be analyzed are injected to a capillary separated by bubbles. The liquid slug sample is practically sealed between two bubbles. Long capillary tubes with multiple analysis section can be used with minimal mixing of consecutive samples as the bubbles prevent mixing between the samples.

Monolith multiphase chemical reactors are another example of microfluidic multiphase flow applications. The slug flow pattern enhances the mass transfer in the liquid–solid process. There is also low-pressure drop for a given specific contact area. Machado *et al.* (1999) have patented the use of monolith reactors for fast and highly exothermic nitroaromatic hydrogenation. In this process, the product is recycled through the reactors several hundreds of times, and the low-pressure-drop monolith reactor is therefore preferred.

Broekhuis *et al.* (2004) used monolith reactor for the production of sorbitol. Guenther *et al.* (2004) developed a gas–liquid separator downstream the reaction zone of microfabricated integrated system. Khan *et al.* (2004) used the plug flow characteristic for the production of colloidal silica.

5.23.2 Flow Patterns

Two-phase flows in capillaries show many types of flow patterns. Various parameters control the nature of flow patterns, that is, nature of two fluids, details of injection, use of surfactants, surface characteristic, and so on. The two-phase flow patterns can be described as liquid–liquid and liquid–gas systems based on the visual observation. Various configurations have been used for generating two-phase flows in microfluidic system, (a) T-junction and (b) cross-junction, whose schematic is shown in Figure 5.42. Two different fluids enter through the main channel and lateral channel.

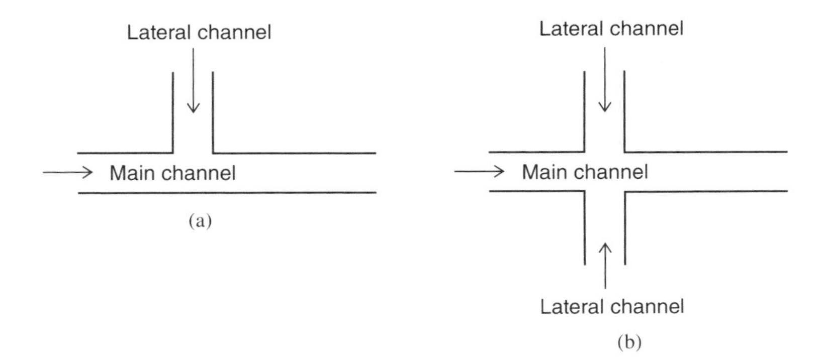

Figure 5.42 Various junction types for generation of two-phase flows in a microfluidic system: (a) T-junction and (b) cross-junction

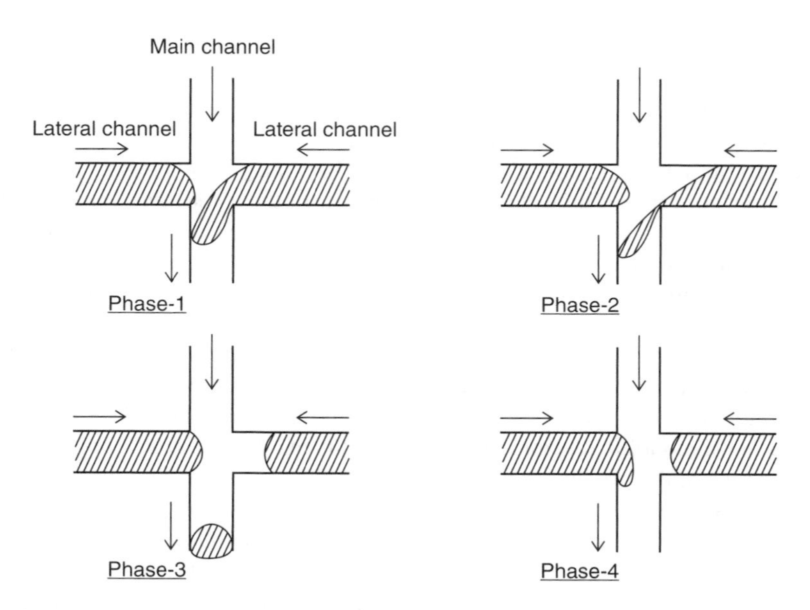

Figure 5.43 Schematic showing the formation of droplets at different phases in a cross-junction microfluidic system

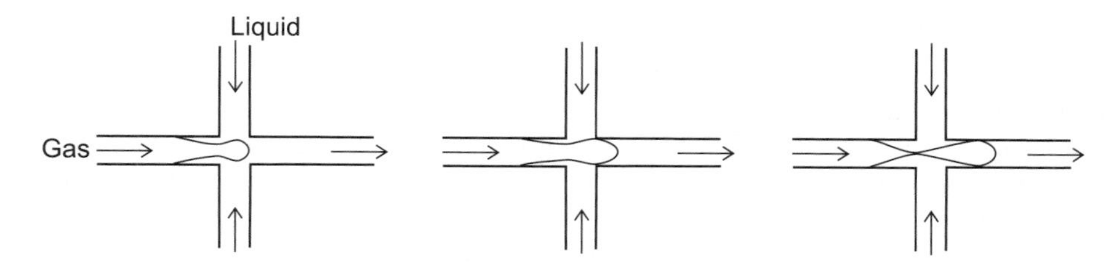

Figure 5.44 Schematic showing the formation of an air bubble

The nature of fluid and the flow conditions, that is, respective flow rate and surface characteristics, primarily influence the two-phase flow patterns inside the channel. The formation process of droplets in a cross-junction channel has been shown in Figure 5.43. Here, the drop-forming fluid is injected slowly from the lateral channel into a stronger main channel flow of the surrounding fluid. The shear from the flow forces droplets to shed from the junction, whose size depends on the relative flow rate. For very low shear, the size of the channel determines the scale of the droplets, similar to the drop falling from a pipette.

Flow focusing is another mechanism for generation of droplets and bubble. Figure 5.44 shows the formation process of air bubble. Figure 5.44 shows the formation of gas bubbles inside a surrounding liquid flow. The interface between the air and liquid deforms with time to form a small neck region, leading to breaking and formation of small bubble. Similar to formation of bubbles, liquid droplets are also formed by flow focusing mechanism. In the case of cross-junction, the flow entering through lateral channels flows much faster than the main channel fluid. The shear from the outer fluid layers causes the formation of a thin stream of inner fluid. This is named as "flow focusing." The basic mechanism that breaks a liquid jet into droplets under the action of surface tension is Rayleigh–Plateau instability. The pressure fluctuations inside the liquid due to small variation in curvature are the primary cause for generation of the instability. These small variations grow to eventually break the liquid film into drops. The interface shape approaches cylindrical geometry before the breaking point. The size of the initial cylindrical geometry jet determines the drop size. Therefore, there has been wide interest in reducing the radius of the jet before allowing it to break.

5.23.2.1 Liquid–Gas Systems

Figure 5.45 shows the representative gas–liquid flow patterns observed in capillary channels. Broadly, five different flow patterns are observed which can be classified as follows:

1. *Film flow*: In film flow (downflow only), liquid flows downward on the walls of the channel, and the gas flows through the center, either upward or downward. This takes place typically at low superficial velocities.
2. *Bubbly flow*: In bubbly flow, gas flows as small bubbles dispersed in the continuous wetting fluid. This flow pattern is observed at moderate velocities for low gas frictions, where coalescence is minimal.
3. *Taylor flow*: In Taylor flow, large long bubbles span most of the cross-section of the channel. This is also called as plug flow, slug flow, bubble-train flow, segmented flow, or intermittent flow. The relative length of the bubbles and liquid slug depends on the inlet flow conditions.
4. *Churn flow*: In churn flow, small bubbles appear at the rear of the slug. This happens at higher velocities. Further increase in velocity leads to chaotic flow patterns.
5. *Annular flow*: In annular flow, a thin wavy liquid film flows along the wall with a mist of gas and entrained liquid in the core. This flow pattern is observed at high velocities and low liquid fraction.

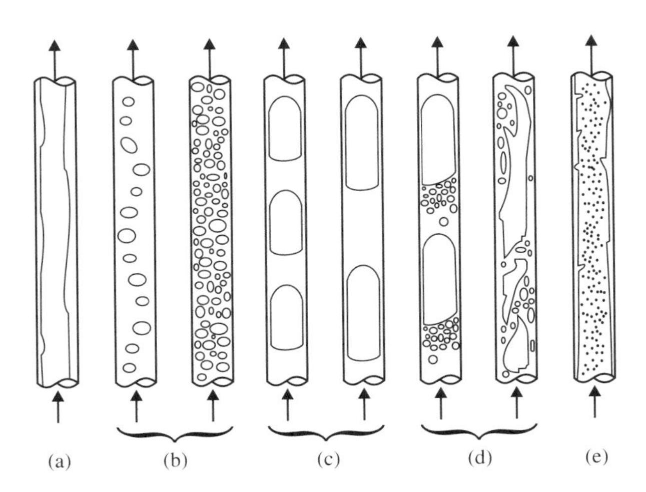

Figure 5.45 Schematic showing various gas–liquid flow patterns observed in capillary channels: (a) film flow, (b) bubbly flow, (c) segmented flow (Taylor flow), (d) churn flow, and (e) annular flow

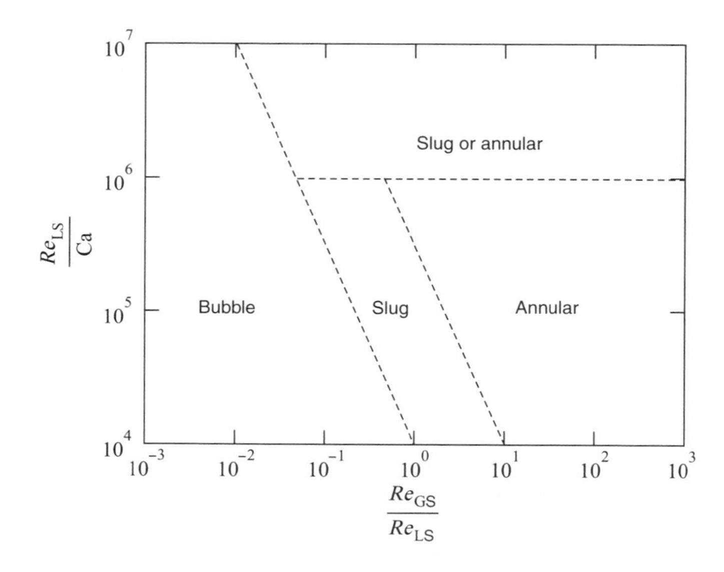

Figure 5.46 Typical flow patterns as a function of representative dimensionless number

In microcapillary configuration, different flow patterns can be generated by varying the liquid and gas flow rate. A bubbly regime is obtained for a given liquid fluid rate while increasing the gas flow rate at a low gas flow rate. Slug flow is obtained at a higher gas flow rate, followed by annular flow pattern. The transition from one regime to the other depends on the growth of disturbance and initial amplitude of disturbance. The flow pattern depends on the gas and liquid properties ($\rho_G, \mu_G, \rho_L, \mu_L, \sigma$), duct geometry ($d$), and superficial velocities of gas (u_{GS}) and liquid (u_{LS}), where subscript G corresponds to gas and subscript L corresponds to liquid. Therefore, the number of relevant dimensionless number is quite large, and the experimental flow maps are typically applicable only to specific system configuration. A typical flow map has been presented in Figure 5.46 as a function of $\left(\frac{Re_{GS}}{Re_{LS}}\right)$ and $\left(\frac{Re_{LS}}{Ca}\right)$. Here, Re_{GS} is the Reynolds number $\left(\frac{\rho_G u_{GS} d}{\mu_G}\right)$ for gas phase and Re_{LS} is the Reynolds number $\left(\frac{\rho_L u_{LS} d}{\mu_L}\right)$ for liquid phase.

Figure 5.46 shows that keeping liquid flow constant, various flow patterns are observed with increasing gas flow rate. Similarly, keeping the gas flow rate constant and varying the liquid flow rate, various two-phase flow patterns are observed.

5.23.2.2 Liquid–Liquid System

The primary difference between liquid–gas and liquid–liquid system is the wetting characteristics. Gas cannot wet the solid surface in contrast to liquid, where the wetting behavior depends on the nature of liquid and solid. Therefore, the annular regime observed in gas–liquid system observed at high flow rate is replaced by stratified regime for liquid–liquid system.

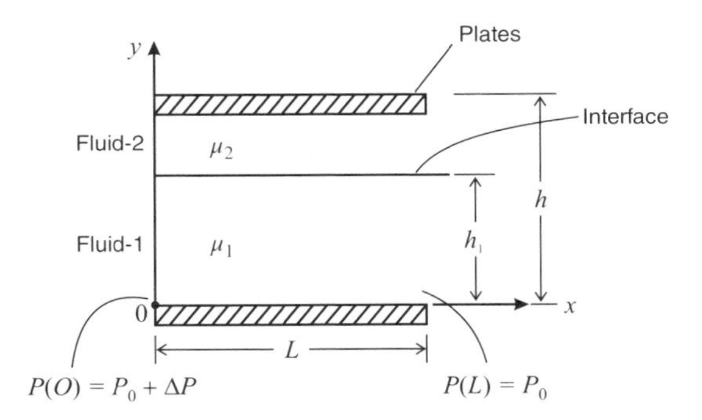

Figure 5.47 Schematic showing two-phase Poiseuille flow of fluid 1 (viscosity, μ_1) and fluid 2 (viscosity, μ_2) between two horizontally placed parallel infinite plates

Wetting characteristics play an important role for generating flow structures of one fluid inside another fluid. If one wants to generate fluid structures of a nonwetting fluid inside a second fluid, the second fluid should completely wet the walls of the microchannels. For example, it is impossible to generate oil droplets in water inside hydrophobic PMMA channel. When both fluids partially wet the walls, the flow regimes are less reproducible. The interface separating two fluids is of irregular shape. Droplets adhere to the channel walls instead of being advected by the mean flow. These observations indicate the importance of wetting behavior of fluids in generation of two-phase flow patterns. The transition zone of the flow patterns depends on whether the wetting fluid flows in main channel or lateral channel in addition to the respective flow patterns. As surface characteristics influence the flow pattern, surface treatment and use of surfactants can be adopted for controlling the two-phase flow patterns in microfluidic applications.

5.24 Two-Phase Liquid–Liquid Poiseuille Flow

In this section, we consider the analytical solution of two immiscible fluids moving between two infinite parallel plates. This example demonstrates the general solution procedure of two-phase fluid flows. Figure 5.47 shows a basic steady Poiseuille flow between two parallel infinite planar plates. The flow is driven by pressure ΔP over length L along the x-axis. The bottom layer ($0 < y < h_1$) is fluid 1 of viscosity μ_1, and the top layer ($h_1 < y < h$) is liquid 2 of viscosity μ_2. The flow field can be considered independent of the z-direction ($w = 0$) due to infinite plate assumption, that is, the plate being very large in the z-direction. The flow field can also be assumed to be fully developed, that is, independent of x-direction due to small scale of the channel (high aspect ratio). The liquid–liquid interface is assumed to be flat, that is, there is no Young–Laplace pressure distribution across the interface, and the pressure field is equal to that of a single-phase Poiseuille flow. The total velocity field can be decomposed into two parts for the respective fluids. The governing equation of the flow field is the N–S equation corresponding to two fluids.

The continuity equations for two fluids, respectively, under 2-D flow assumption are

$$\frac{\partial u_1}{\partial x} + \frac{\partial v_1}{\partial y} = 0 \quad \text{(For fluid 1)} \tag{5.101}$$

$$\frac{\partial u_2}{\partial x} + \frac{\partial v_2}{\partial y} = 0 \quad \text{(For fluid 2)} \tag{5.102}$$

Using the fully developed flow assumption $\left(\frac{\partial}{\partial x} = 0\right)$, no-slip boundary condition at the wall ($u_{\text{wall}} = 0$) and integration of the above two equations, one can derive

$$v_1 = v_2 = 0$$

Hence, the flow field can be represented as

$$u(y) = \begin{cases} u_2(y) & \text{for} \quad h_1 < y < h \\ u_1(y) & \text{for} \quad 0 < y < h_1 \end{cases}$$

The simplified N–S equations are

$$\mu_1 \frac{\partial^2 u_1}{\partial y^2} = -\frac{\Delta P}{L} \quad \text{For fluid 1} \tag{5.103}$$

$$\mu_2 \frac{\partial^2 u_2}{\partial y^2} = -\frac{\Delta P}{L} \quad \text{For fluid 2} \tag{5.104}$$

The boundary conditions are

$$\begin{aligned}
&1: u_2 = 0 \quad \text{at} \quad y = h && \text{(no-slip)}\\
&2: u_1 = 0 \quad \text{at} \quad y = 0 && \text{(no-slip)}\\
&3: u_1 = u_2 \quad \text{at} \quad y = h_1 && \text{(interface)}\\
&4: \mu_1 \left.\frac{\partial u_1}{\partial y}\right|_{y=h_1} = \mu_2 \left.\frac{\partial u_2}{\partial y}\right|_{y=h_1} && \text{(interface)}
\end{aligned}$$

The integration of the simplified N–S equations (5.103) and (5.104) with first two boundary conditions gives

$$u_2(y) = \frac{\Delta P}{2\mu_2 L}(h - y)(y - C_2) \tag{5.105}$$

$$u_1(y) = \frac{\Delta P}{2\mu_1 L}(C_1 - y)y \tag{5.106}$$

The two constants C_1 and C_2 can be determined from the last two boundary conditions as

$$C_2 = \frac{\left(\frac{\mu_1}{\mu_2} - 1\right)\left(1 - \frac{h_1}{h}\right)}{\frac{\mu_1}{\mu_2}\left(1 - \frac{h_1}{h}\right) + \frac{h_1}{h}} h_1 \tag{5.107}$$

$$C_1 = h + C_2 \tag{5.108}$$

The substitution of the above constants, C_1 and C_2, in equations (5.105) and (5.106) gives the velocity field of a two-phase Poiseuille flow of two immiscible liquids.

5.25 Hydrodynamics of Taylor Flow

In addition to bulk behavior of the two-phase flow, detailed hydrodynamics inside the liquid plug has many practical significances. The recirculating flow patterns inside the plug influences the heat transfer with the solid boundary and the mixing inside the droplet. Figure 5.48(a) shows the streamline pattern and velocity profile inside an ethanol plug. The streamline patterns show two recirculating bubbles located at either side of the centerline, with reference to the bubble velocity. The rotation direction of the recirculating bubble is opposite to each other. The u-velocity profile plot in Figure 5.48(b) shows greater deviation from the analytical profile near the ethanol–air interface compared to that away from the interface. The u-velocity profile in the central region of the plug approaches to that of the parabolic analytical profile. The v-velocity profile in Figure 5.48(c) shows opposite direction at both sides of the centerline due to the opposite rotation of the recirculation bubble. At the central region of the liquid plug, the v-velocity is close to zero, similar to that of Poiseuille flow. Most part of the plug is expected to have similar velocity distribution as fully developed Poiseuille flow inside longer plug length.

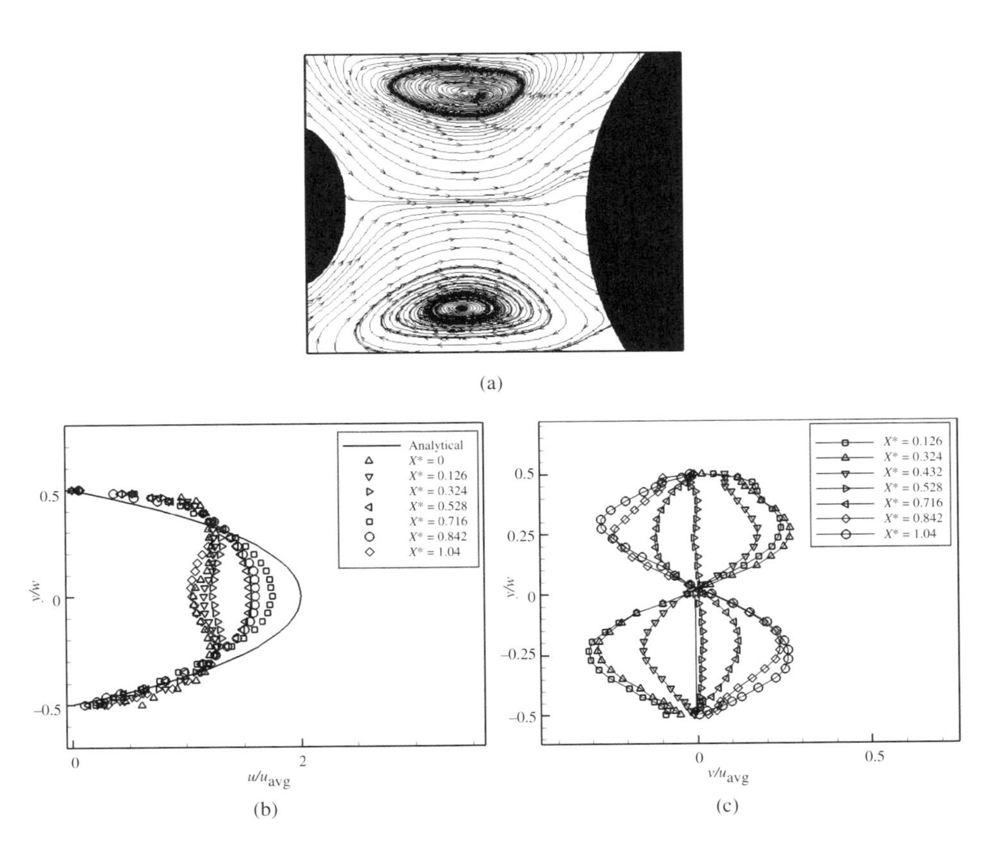

Figure 5.48 Sample results from micro-PIV measurement inside the ethanol plug for 500 μm capillary: (a) streamline patterns, (b) u-velocity profile inside the ethanol plug, and (c) the v-velocity profile inside the ethanol plug. Here, X^* indicates the normalized distance from the interface normalized by the channel width

Thus, the pressure drop is expected to be similar to that of the single-phase Poiseuille flow. However, the effectiveness for mixing and heat transfer enhancement is expected to be superior for shorter liquid plug.

5.25.1 Liquid Film Thickness

In Taylor flow, a thin film is observed near the gas bubble which prohibits the contact between gas and solid. This phenomenon is similar to the role of lubricating viscous fluid in bearing for preventing the solid–solid contact. The spherical region at the front has a constant radius r with the Laplace pressure difference across the interface $\Delta P = 2\sigma/r$ for small thickness of the film ($\delta \ll r$). The curvature in the axial direction vanishes for the flat film region, and the Laplace pressure difference is given by $\Delta P = \sigma/r$. The transitional region between the spherical cap region and the flat film region has a change in the curvature to satisfy the continuity of Laplace pressure at the interface.

Detailed analysis of N–S equation by Bretherton in the transition region results in the expression of film thickness as

$$\frac{\delta}{d} = 0.66\, Ca^{2/3} \tag{5.109}$$

Figure 5.49 shows the comparison of film thickness from Bertherton's above equation with the experimental data of Taylor (1961) and Bretherton (1961). There is a good agreement between the theory and experimental data for $10^{-4} < Ca < 10^{-2}$. The deviation between the theory and experiment at lower and higher capillary number may be attributed to (a) Marangoni effects and (b) inertial effects.

5.25.1.1 Marangoni Effects

Marangoni effects may be responsible for the difference between theory and experiment in the low capillary number regime, where surface tension effect is dominant. The presence of impurities may lead to gradient of these impurities in the gas–liquid interface. The surfactants

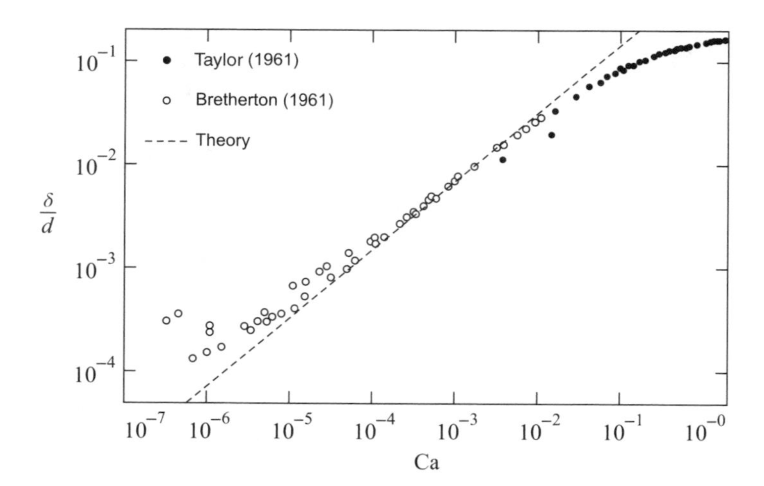

Figure 5.49 Comparison of film thickness (δ/d) between theoretical prediction of Bertherton and experimental results of Taylor and Bretherton as a function of capillary number

or contaminations have a concentration distribution along the liquid–gas interface. The Marangoni stress develops, and the assumption of no-shear boundary condition for the gas–liquid interface breaks down. This Marangoni effect may be attributed to the difference between theory and experiment at low *Ca* (see Figure 5.49).

5.25.1.2 Inertial Effects

It may be noted that the capillary number can be varied by replacing with high-viscosity fluid while keeping the velocity constant or by changing the flow velocity for the same fluid. However, most of the earlier experiments were conducted at low velocity for ease in observation. Therefore, inertial effect is negligible for these cases. Hence, Stokes flow assumption can be used. However, recent experimental studies (Aussilous and Quere, 2000) have shown increase in film thickness at higher velocities. Hence, inertial effects may be attributed to the deviation between experiment and theory at high capillary number (see Figure 5.49).

5.26 Plug Motion in Capillary

Motion of liquid plug is commonly observed in many microfluidic devices. The motion of liquid plug inside a capillary has been analyzed in this section using a lumped model. Figure 5.50 shows the decomposition of two-phase flow in lumped elements. The flow region is decomposed into two types of elements: (a) moving bulk fluid and (b) moving interfaces. The total pressure drop between A and F can be written as combination of various pressure drop contribution:

$$\begin{aligned}\Delta P_{\text{channel}} &= \Delta P_{AB} + \Delta P_{BC} + \Delta P_{CD} + \Delta P_{DE} + \Delta P_{EF} \\ &= (\Delta P_{AB} + \Delta P_{CD} + \Delta P_{EF}) + (\Delta P_{BC} + \Delta P_{DE}) \\ &= (\Delta P)_{\text{friction}} + (\Delta P)_{\text{capillary}}\end{aligned} \tag{5.110}$$

The lumped elements between A and B, C and D, and E and F have pressure drop due to friction. The pressure drop between B and C, and D and E is attributed to the interface. The advancing and receding interface contributes either positively or negatively to the capillary pressure drop, which is a function of contact angle. For a spherical interface of curvature radius R of the advancing front, the pressure difference can be written based on Young–Laplace equation as

$$\Delta P_a = \frac{2\sigma}{R} \tag{5.111}$$

Note that R is signed and the sign of pressure difference depends on the orientation of the curvature. For the capillary radius a and advancing contact angle θ_a, we have

$$\Delta P_{DE} = \Delta P_a = -\frac{2\sigma}{a}\cos\theta_a \quad \left(\text{using,} \quad \cos\theta = \frac{-a}{R}\right) \tag{5.112}$$

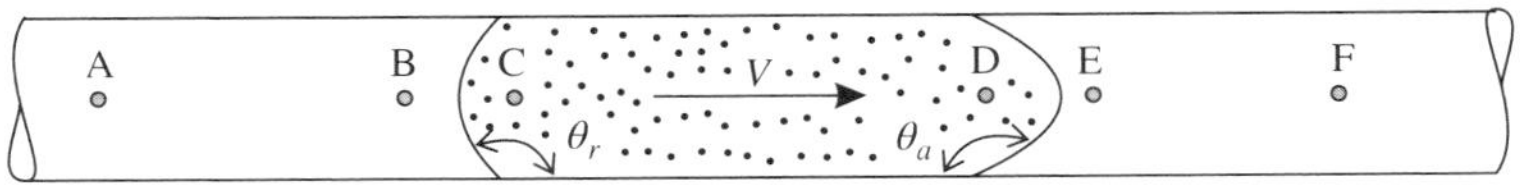

Figure 5.50 Decomposition of plug flow into different lumped elements

Similarly, for the receding front, we have

$$\Delta P_{BC} = \Delta P_r = \frac{2\sigma}{a}\cos\theta_r \tag{5.113}$$

The total capillary pressure drop is

$$\Delta P_{\text{capillary}} = \Delta P_r + \Delta P_a = \frac{2\sigma}{a}(\cos\theta_r - \cos\theta_a) \tag{5.114}$$

We can use the expression for dynamic contact angle (equation (5.14)) for the advancing and receding front as

$$\theta_a = \theta_{s,a}\left(1 + \frac{1}{3}\frac{A\,Ca}{\theta_{s,a}^3}\right) \tag{5.115}$$

and

$$\theta_r = \theta_{s,r}\left(1 - \frac{1}{3}\frac{A\,Ca}{\theta_{s,r}^3}\right) \tag{5.116}$$

where $\theta_{s,r}$ and $\theta_{s,a}$ are the static contact angles. Here, the capillary number is assumed +ve for both interface, and the sign convention is accordingly taken into consideration.

Theoretically, the contact angle should be unique for both advancing and receding front. However, two values of contact angles are observed, that is, slowing down to stop an advancing front ($\theta_{s,a}$) and slowing down to stop a receding front ($\theta_{s,r}$). In experiments, the contact angle of liquid drop deposited on the substrate does not reach a unique value, θ, independent of the way the droplet is deposited on the substrate. A small heterogeneity of the substrate leads to a significant hysteresis of the contact angle. This effect is due to the pinning of the contact line on defects of the substrate. A surface is considered relatively clean and flat if the contact angle hysteresis is less than 5°. Note that hysteresis leads to a concept of threshold force per unit length to make the contact line advance or recede given as $\sigma_{\text{lg}}(\cos\theta - \cos\theta_a)$ and $\sigma_{\text{lg}}(\cos\theta - \cos\theta_r)$, respectively. Note that the contribution from σ_{sl} and σ_{sv} is not included as their contribution to force does not include contact angle. Figure 5.51 shows this difference in contact angle known as hysteresis. Hence, the static contact is not uniquely

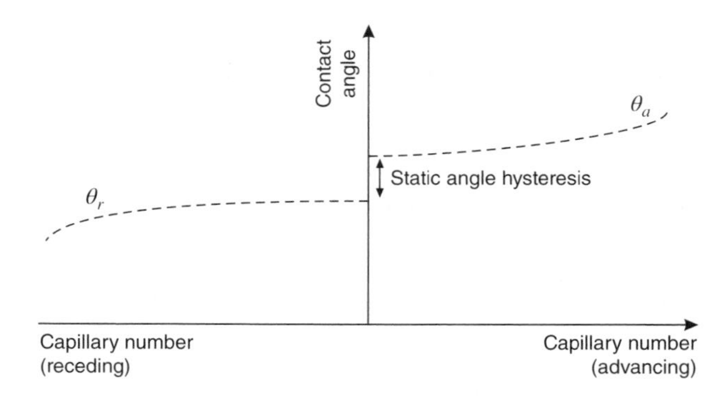

Figure 5.51 Hysteresis of static contact angle

defined. The static angle hysteresis (Figure 5.51) is zero if the surface is perfectly smooth and homogeneous. Hence, equations (5.114)–(5.116) can be used to evaluate the capillary pressure drop of equation (5.110). Let us discuss about the pressure drop due to friction.

There is pressure drop due to bulk fluid motion. According to *Washburn law*, the friction pressure drop is

$$\Delta P_{\text{friction}} = \frac{8V}{a^2}(\mu_1 L_1 + \mu_2 L_2) \tag{5.117}$$

where subscripts 1 and 2 correspond to liquid and gas plug, respectively, and V is the average liquid velocity. The total pressure drop in the capillary is

$$\Delta P_{\text{channel}} = \Delta P_{\text{capillary}} + \Delta P_{\text{friction}} \tag{5.118}$$

At low capillary number, that is, for microchannel flow, the pressure drop due to friction is much smaller than the capillary pressure. If we have N plugs, the total pressure drop due to capillary is

$$\Delta P_{\text{capillary, total}} = N\frac{2\sigma}{a}(\cos\theta_r - \cos\theta_a) \tag{5.119}$$

The capillary will block the flow if this pressure is greater than the available pressure head, that is,

$$\Delta P_{\text{capillary, total}} > P_i - P_o \quad \text{(for blockage)} \tag{5.120}$$

where P_i and P_o are the inlet and outlet pressure, respectively.

We know that the pressure distribution for single-phase, fully developed liquid flow has linear trend in streamwise direction. Let us look into the nature of pressure distribution inside a capillary during plug flow. It may be noted that from equation (5.119) that the contribution to pressure drop depends on the magnitude of the contact angle. If the advancing contact angle θ_a is larger than $\frac{\pi}{2}$, there is a positive pressure drop associated with the advancing interface. Similarly, if the receding front is smaller than $\frac{\pi}{2}$, there is a +ve contribution to the pressure drop. The dynamic contact angle also is a function of the velocity. Hence, the nature of pressure distribution in a plug flow will be dependent on the flow velocity. Figure 5.52 shows the pressure drop corresponding to the plug flow with low and high velocity. For high speed, the contact angle of the receding front changes from greater than $\frac{\pi}{2}$ to less than $\frac{\pi}{2}$. Hence, the pressure drop at low speed is negative, and the pressure drop at high speed is positive. The slope of the pressure drop in the bulk liquid and gas region is due to the friction.

5.27 Clogging Pressure

Gas bubbles may be either generated in microchannels by electrochemical process or introduced at the inlet for flow control purpose or may appear due to leakage in the system. The LOC may contain different types of contractions. These gas bubbles can be large enough to span the entire channel cross-section. They can get stuck at the channel contraction and clog the flow disturbing the measurements and functionality of the system. External pressure known as clogging pressure is applied to push the bubbles out of the system and clear the clogged channel. Let us formulate the clogging pressure relationship in a microchannel contraction.

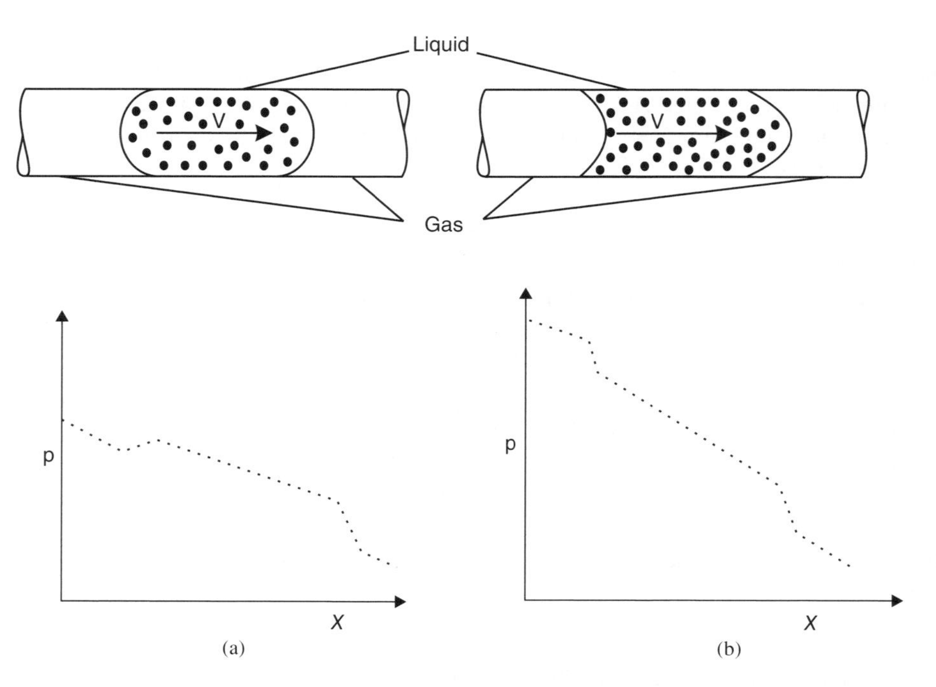

Figure 5.52 The plug moving inside a capillary and the pressure distribution (a) at low speed and (b) at high speed

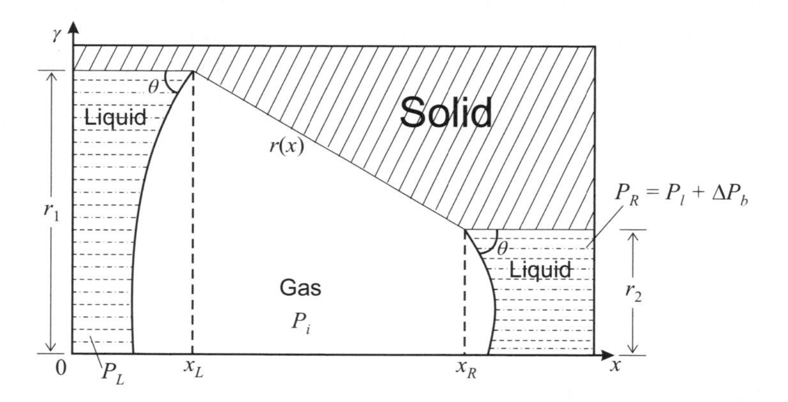

Figure 5.53 A gas bubble surrounded by a liquid inside a hydrophilic axisymmetric channel with contraction

Figure 5.53 shows a hydrophilic axisymmetric microchannel. The microchannel radius changes from r_1 to r_2 through a contraction with position-dependent channel radius $r(x)$. The local tapering $\theta_t(x)$ of the channel can be expressed as

$$\theta_t(x) = \frac{\partial}{\partial x} r(x) \tag{5.121}$$

A large bubble divides the liquid into left and right parts. The bubble forms two menisci in contact with the liquid and the bulk part in direct contact with the channel wall.

Let us assume the bubble to have quasistatic motion, that is, the velocity of the bubbles is nearly zero, and the bubble remains close to equilibrium for all bubble positions. The channel is also assumed to be smooth, that is, free from sharp corners. The pressure at the left and right of the bubble is P_L and P_R, respectively. The pressure difference across the bubble is ΔP_b. The surface tension for the liquid–gas, solid–liquid, and solid–gas interface is σ_{lg}, σ_{sl}, and σ_{sg}, respectively. The gas pressure inside the bubble is P_i. At the left meniscus ($x = x_L$), R_1 and R_2 are the main radii of curvature. The pressure difference $\Delta P = P_i - P_L$ across the interface according to the Young–Laplace equation is

$$\Delta P(x_L) = \sigma_{\mathrm{lg}}\left(\frac{1}{R_1} + \frac{1}{R_2}\right) \tag{5.122}$$

For circular tube, the interface will be spherical, and two radii of curvature are expressed as

$$R_1 = R_2 = \frac{r(x)}{\cos\theta} \tag{5.123}$$

At the taper section, the contact angle (θ) will change by the taper angle (θ_t) as

$$\theta = \theta \pm \theta_t \tag{5.124}$$

For the left interface, the contact angle increases due to the entry of the interface inside the contraction section. Similarly, for the right interface, the contact angle decreases due to the exit of the interface from the contraction section. Hence, using Young–Laplace equation for the left meniscus, it can be written as

$$\Delta P(x_L) = P_i - P_L = 2\sigma_{\mathrm{lg}}\frac{\cos(\theta + \theta_t(x_L))}{r(x_L)} \tag{5.125}$$

Similarly, for the right meniscus,

$$\Delta P(x_R) = P_i - P_R = 2\sigma_{\mathrm{lg}}\frac{\cos(\theta - \theta_t(x_R))}{r(x_R)} \tag{5.126}$$

Subtracting equation (5.125) from equation (5.126), we have

$$\begin{aligned}\Delta P_b &= P_R - P_L = \Delta P(x_L) - \Delta P(x_R)\\ &= 2\sigma_{\mathrm{lg}}\left(\frac{\cos(\theta + \theta_t(x_L))}{r(x_L)} - \frac{\cos(\theta - \theta_t(x_R))}{r(x_R)}\right)\end{aligned} \tag{5.127}$$

The above equation indicates the role played by contraction and surface tension between the liquid and gas interface on the pressure difference between the front and rear of the bubble.

The clogging pressure is defined as the minimal external pressure that must be supplied to push the bubble through the channel. Hence, clogging pressure is the maximal of the position-dependent pressure drop across the bubble:

$$P_{\mathrm{clog}} = \max(-\Delta P_b(x)) \tag{5.128}$$

5.28 Digital Microfluidics

Digital microfluidic technology attempts to operate similar to the concept of a semiconductor chip. In a semiconductor chip, the circuitry is programmed to perform different sets of instructions. LOC system also attempts to program various chemical or biological reaction protocols. The programmability is achieved by using various microvalves and micropumps. Piezoelectric components are traditionally incorporated to drive these moving components. However, fabrication and operation of these devices are not easy. Therefore, there is an attempt to use discrete liquid plugs instead of continuous liquid plugs inside the microchannel.

Figure 5.54 illustrates three different protocols for mixing of two fluids. In Figure 5.54(a), two fluids are continuously introduced, which pass through various branches before finally getting recombined before the outlet port. In Figure 5.54(b), the fluids are introduced as liquid plugs, which merge with each other before the outlet port. Figure 5.54(c) does not use any microchannel concept. Here, the droplets are manipulated on an open surface. In Figure 5.54(a), the two fluids are gradually mixed and therefore can be assumed to have no capability of programmability. The movements of liquid plugs in Figure 5.54(b) are restricted within microchannels. Therefore, this system can be assumed to have limited programming capability. The droplet movement can be easily controlled in Figure 5.54(c) and therefore has maximum programming capability. Similar to a microprocessor, various procedures, that is,

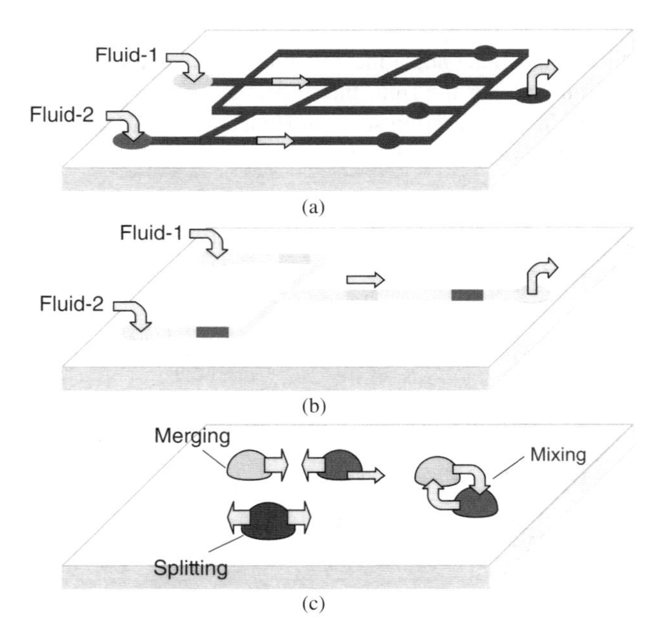

Figure 5.54 Different protocols for mixing of two fluids: (a) continuous flow, (b) plug flow, and (c) open-surface manipulation

dilution, titration, chemical synthesis, polymerase chain reaction (PCR), and so on, can be carried out in an LOC device by using the programmability concept of digital microfluidic device.

Digital microfluidics can be defined as the technology based upon micromanipulation of discrete droplets. Manipulation of droplets can be achieved by various mechanisms, that is, electrowetting, dielectrophoresis, thermocapillary transport, and so on. Digital microfluidic architecture is under software-driven electronic control unlike continuous-flow microfluidic architecture. The droplets are placed on metal electrodes. The thermal and electrical controls are achieved by passing electric current through different electrodes depending on the software logic. Figure 5.54(c) shows merging, splitting, and mixing by rotation of the droplets on the open surface over electrode arrays.

Basic liquid unit volume of a digital microfluidic architecture is fixed by the geometry of the system. The volumetric flow rate is determined by the droplet transport rate and the number of droplets transported. The use of unit volume droplet allows the microfluidic functions to be reduced to a set of basic operations. This also facilitates the accomplishment of digitization processes.

The digital microfluidic system has the following overall benefits:

1. *No moving parts*: There is no requirement of valves and pumps in the digital microfluidic systems.
2. *Controllability*: Many droplets can be independently controlled.
3. *Evaporation*: Evaporation can be controlled depending on the medium surrounding the droplets.
4. *No ohmic current*: Direct current is blocked, leading to minimization of sample heating and electrochemical reactions.
5. *Types of liquids*: This technique works well with various electrolyte solutions.
6. *Sample utilization*: No fluid is wasted in priming the channels or filling reservoirs, leading to 100% utilization of sample or reagent.
7. *Microscopy*: This technique is compatible with microscopic observation.
8. *Energy efficiency*: This technique is very energy efficient with nanowatts–microwatts of power requirement.
9. *Speed of operation*: High droplet speeds up to 25 cm/s can be achieved.
10. *Operational flexibility*: Direct computer control helps in flexible operation of the device.

Various practical applications of digital microfluidics have been proposed for chemical, biological, and medical applications. Figure 5.55 shows a binary particle separation system (Cho and Kim, 2003). The particle separation takes place in three steps. In the first step, low-level electric field is applied for isolating each type of particle based on their charge and mobility. The mother droplet is split into two daughter droplets in the second step. The daughter droplets are transported during the third step for further on-chip processing. The control/driving electrodes are energized differently during various steps of the digital microfluidic application.

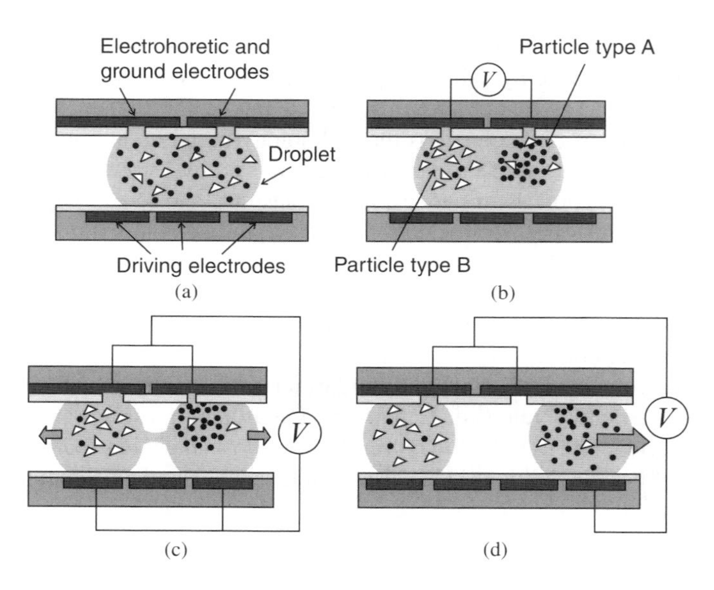

Figure 5.55 Particle separation process based on digital microfluidics: (a) initial stage, (b) charge separation within the droplet, (c) splitting of droplet, and (d) transport of split daughter droplet

Problems

5.1 A bacterium modeled as a cylinder (diameter: $A = 0.6$ μm) shown below excretes surfactant at location h, leading to the surface tension difference $\Delta\sigma = 40$ dyn/cm) between the front and rear end of the bacterium, where σ is the surface tension. Calculate the velocity U of the bacterium if the drag offered by the fluid to the bacterium motion is equal to $3\pi\mu AU$. The viscosity of the fluid is $\mu = 0.001$ Pa-s.

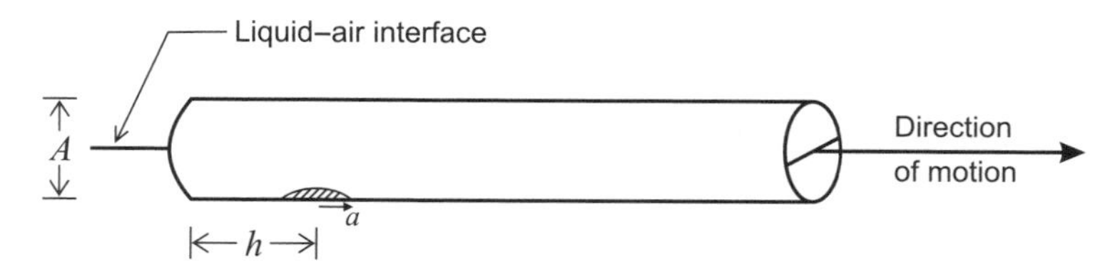

5.2 Calculate the drop velocity inside a capillary of 50 μm diameter and droplet length of 5000 μm of water and mercury at temperature difference of 10 °C, 20 °C, and 30 °C using the following data. Assume the inclination angle, $\phi = 0°$, external pressure, $\Delta Pe = 0$, cold side temperature = 25°C, static contact angle of water, $\theta = 40°$ and static contact angle of mercury = 140°. Use the following data for the calculation.

Liquid	μ (Poise)	ρ (g/cm^3)	a (dyn/cm)	b (dyn/cm- °C)	a/b (°C)	b/μ (cm/s- °C)
Water	0.009548	0.997	75.83	0.1477	513.4	15.5
Mercury	0.01552	13.54	490.6	0.2049	2394	13.2

5.3 Repeat the above problem for a microchannel having 32 μm high and 500 μm wide cross-section.

5.4 Write down the explicit expression for droplet velocity due to thermocapillary pumping action inside a microcapillary for hydrophilic and hydrophobic surface as presented inside the text.

5.5 Oil recovery is carried out by imposing compressed airflow inside the porous soil structure. Let us assume that the average pore can be assumed as circular cross-section with diameter 25 μm. The average velocity of the oil flow through the capillary is 0.3 m/s. The viscosity of oil is equal to 2.6 mPa-s, the density of oil is equal to 860 kg/m^3, and the surface tension of the oil and air interface is equal to 72.9 mN/m. (a) Calculate the thickness of the oil layer remaining on the soil surface. (b) Let the flooding of the oil reservoir with surfactant reduced the surface tension to 37 mN/m at the interface between the air and liquid interface. Calculate the thickness of the oil layer remaining on the soil surface after recovery of the oil. Assume constant, A for Tanner's law = 30.

5.6 Ten liquid plugs of length equal to 1 mm separated by air with a pitch of 1 mm between the two liquid plugs occupy a microcapillary of diameter 500 μm. The static contact angle of water with the capillary surface is equal to 60°. The surface tension between the liquid and air interface is equal to 73 mN/m. Using properties of air and water at standard atmospheric pressure and temperature, estimate the pressure drop due to friction and the interface at an average plug velocity equal to 1 cm/s.

5.7 Water droplet on a platinum substrate has contact angle equal to 40.0°. What will be the contact angle of water droplet with NaCl concentration, $C = 0.1$ mol/m^3? The surface tension of NaCl solution with air is given as

$$\sigma_{\text{water–air}} = \sigma_0 - 100 \times C$$

where $\sigma_0 = 72.9$ mJ/m^2 and C is concentration in mol/m^3.

5.8 Consider a 5 mm long mercury droplet inside a microcapillary of 200 μm diameter. The temperature, T, of the contact line at the heater end is equal to 35 °C, and room temperature of the other end is equal to 25 °C. Surface tension of mercury in air at 25 °C is equal to 486.5 mJ/m^2. The surface tension of mercury decreases with temperature as per the following relation:

$$\sigma = 486.5 - 10\Delta T$$

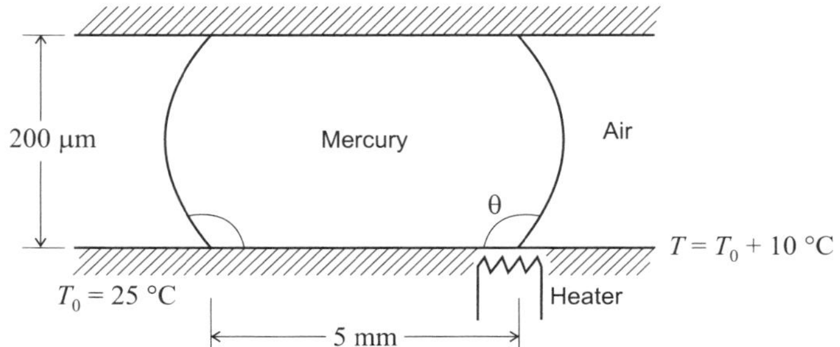

where ΔT is equal to $(T - 25)$ °C and T is the temperature in degree Celsius. The static contact angle of mercury is equal to 140°. Density of mercury is 13,600 kg/m^3. Viscosity of mercury is equal to 1.526×10^{-3} Pa-s.
(a) Calculate the speed of the mercury droplet in meter per second.
(b) Calculate the speed of the droplet while moving upward in the channel oriented in upward direction at an angle, $\theta = 5°$.

5.9 Teflon coating is carried out on a glass cover slip to obtain a hydrophobic surface with contact angle of water (aqueous solution) equal to 110°. The cover slip is cleaned before it is used, and 30% of the coating material got lost resulting in surface inhomogeneities. What will be the contact angle of aqueous solution after cleaning? The contact angle of uncoated cover slip is equal to 40°.

References

Aussilous P and Quere D 2000 Quick deposition of a fluid on the wall of a tube. *Phys. Fluids*, **12**(10), pp. 2367–2371.
Bretherton FP 1961 The motion of long bubbles in tubes. *J. Fluid Mech.*, **10**, pp. 166–188.
Broekhuis RR, Machado RM, and Nordquist AF 2004 Monolith catalytic process for producing sorbitol: catalyst development & evaluation. *Ind. Eng. Chem. Res.*, **43**(17), pp. 5146–5155.
Cho SK and Kim CJ 2003 Particle separation and concentration control for digital microfluidic systems. *Proceedings of the IEEE Micro Electro Mechanical Systems (MEMS)*, pp. 686–689.
Guenther A, Khan SA, Thalmann M, Trachsel F, and Jensen KF 2004 Transport and reaction in microscale segmented gas liquid flow. *Lab Chip*, **4**, pp. 278–286.
Khan SA, Guenther A, Schmidt MA, and Jensen KF 2004 Microfluidic synthesis of colloidal silica. *Langmuir*, **20**, pp. 8604–8611.
Laborie S, Cabassud C, Durand-Bourlier L, and Laine JM 1998 Fouling control by air sparging inside hollow fiber membrane-effects on energy consumption. *Desalination*, **118**, pp. 189–196.
Levich V G 1962 *Physicochemical Hydrodynamics*. Prentice-Hall, Englewood Cliffs, NJ.
Machado RM, Parrillo DJ, Boehme RP, and Broekhuis RR 1999 Use of monolith catalyst for the hydrogenation of dinitrotoluene to toluenediamine. US patent 6005143.
Taylor GI 1961 Deposition of a viscous fluid on the wall of a tube. *J. Fluid Mech.*, **10**, pp. 161–165.
Thiers RE, Reed AH, and Delander K 1971 Origin of the lag phase of continuous-flow analysis curves. *Clin. Chem.*, **17**(1), pp. 42–48.
Young T 1805 An essay on the cohesion of fluids. *Philos. Trans. R. Soc. London*, **95**(1805), pp. 65–87.
Laplace PS 1806 Méchanique Céleste, Supplement 10th vol.

Supplemental Reading

Baroud CN and Willaime H 2004 Multiphase flows in microfluidics. *C.R. Phys.*, **5**, pp. 547–555.
Berthier J and Silberzan P 2006 *Microfluidics for Biotechnology*. Artech House, Inc.
Bruus H 2007 *Theoretical Microfluidics*. Oxford University Press.
Cui ZF, Chang S, and Fane AG 2003 The use of gas bubbling to enhance membrane processes. A review. *J. Membr. Sci.*, **221**, pp. 1–35.
Dreyfus R, Tabeling P, and Willaime H 2003 Ordered and disordered patterns in two phase flows in microchannels. *Phy. Rev. Lett.*, **90**, p. 144505.
Fair RB 2007 Digital microfluidics: is a true lab-on-a-chip possible? *Microfluid. Nanofluid.*, **3**, pp. 245–281.

Fairbrother F and Stubbs AE 1935 The bubble-tube method of measurement. *J. Chem. Soc.*, **1**, pp. 527–529.

Huebner A, Sharma S, Srisa-Art M, Hollfelder F, Edel JB, and deMello AJ 2008 Microdroplets: a sea of application? *Lab Chip*, **8**, pp. 1244–1254.

Kreutzer MT, Kapteijn F, Moulijn JA, and Heiszwolf JJ 2005 Multiphase monolith reactors: chemical reaction engineering of segmented flow in microchannels. *Chem. Eng. Sci.*, **60**, pp. 5895–5916.

Probstein RF 1994 *Physicochemical Hydrodynamics: An Introduction*. John Wiley & Sons, Inc.

Rana GR 2012 Hydrodynamics of segmented gas-liquid flow inside a square micro-channel. Thesis, Mechanical Engineering Department, IIT Kanpur.

Trplett KA, Ghiaasiaan SM, Abdel-Khalik SI, and Sadowski DL 1999 Gas-liquid two-phase flow patterns. *Int. J. Multiphase Flow*, **25**, pp. 377–394.

Yoon JY 2008 Open surface digital microfluidics. *Open Biotechnol. J.*, **2**, pp. 94–100.

6

Charged Species Flow

6.1 Introduction

The motion of the liquids or the solutes in many lab-on-a-chip applications are controlled electrically. Therefore, it is highly relevant to study electrohydrodynamics, that is, the coupling of electromagnetism and hydrodynamics. Electrohydrodynamics comprises a wide range of phenomena in addition to hydrodynamics such as electrochemistry and electrokinetics, which involves electrical properties of liquids.

We will deal with electromagnetic phenomena in the electrostatic regime, that is, we disregard any magnetic and radiative effects. In accordance with the continuum hypothesis, the governing equations for continuous media are Maxwell equation. Here, the electric field E, the electric displacement field D, the magnetic field B, the polarization field P, the electrical current density J_{el}, and the electrical potential ϕ are averaged locally over their microscopic counterparts. The fundamental equations are

$$\nabla \times E = -\frac{\partial B}{\partial t} \tag{6.1a}$$

$$\nabla \cdot D = \nabla \cdot (\epsilon E) = \rho_E \tag{6.1b}$$

$$D = \epsilon_0 E + P = \epsilon E \tag{6.1c}$$

$$J_{\text{el}} = K_{\text{el}} E \tag{6.1d}$$

Here, ϵ_0 is the permittivity of free space equal to 8.854×10^{-12} F/m or $C^2/N\text{–m}^3$ or $\frac{A^2 S^4}{\text{kg m}^3}$, ρ_E is the charge density of the medium, ϵ is the permittivity of the medium, and K_{el} is the conductivity of the material. For electrostatic induction, that is, the creation of electric charge in an object with another electrically charged object, $-\frac{\partial B}{\partial t} = 0$. This is also known as Faraday's law of induction. The right-hand side of equation (6.1a) becomes equal to zero in the absence of changing magnetic field. Thus, E-field can be written as (minus) the gradient of a potential ϕ, since the curl of the electric field is equal to zero. We write

$$E = -\nabla \phi \tag{6.2}$$

Transport Phenomena in Microfluidic Systems, First Edition. Pradipta Kumar Panigrahi.

Companion Website: www.wiley.com/go/panigrahi/microfluidic

Assuming the permittivity of a material, ϵ to be a constant, the equation (6.1b) can be written as

$$\nabla \cdot E = \frac{\rho_E}{\epsilon} \tag{6.3}$$

Substituting equation (6.3) in equation (6.2), we have

$$-\nabla^2 \phi = \frac{\rho_E}{\epsilon} \tag{6.4}$$

This equation is known as the Poisson equation for electrostatics. The knowledge of charge density distribution (ρ_E) is required for obtaining potential distribution. If the charge density is zero, we get Laplace equation. These equations will be used in the electrohydrodynamic analysis of microfluidics.

6.2 Electrical Conductivity and Charge Transport

Ohm's law named after the German scientist Georg Ohm is the most fundamental law describing the measurements of applied voltage and current through simple electric circuits containing various lengths of wire. Its most commonly used form is

$$I = \frac{V}{R} \tag{6.5}$$

Here, I is the current in ampere, V is the potential difference across the conductor in volts, and R is the resistance in ohms. Another generalized form of Ohm's law is

$$J_{\text{el}} = K_{\text{el}} E \tag{6.6}$$

where J_{el} is the current density in ampere/square meter, K_{el} is the electrical conductivity in siemens/meter, and E is the electric field in volts/meter.

Electrical conductivity (K_{el}) of a material is its ability to conduct an electric current. It is also defined as the reciprocal of electrical resistivity (ρ_{el}) which is a measure of how strongly a material opposes the flow of current:

$$K_{\text{el}} = \frac{1}{\rho_{\text{el}}} \tag{6.7}$$

The SI unit of electrical resistivity is Ω-m, and that of electrical conductivity is siemens per meter (S/m). We also can demonstrate the Ohm's law conversion from one form to another as follows. Using equations (6.7) and (6.2) in equation (6.6), we have

$$J_{\text{el}} = \frac{1}{\rho_{\text{el}}} \frac{V}{L} \tag{6.8}$$

Using the definition of electrical resistivity, $\rho_{\text{el}} = \frac{RA}{L}$, we have

$$\frac{I}{A} = \frac{L}{RA} \frac{V}{L}$$

$$I = \frac{V}{R}$$

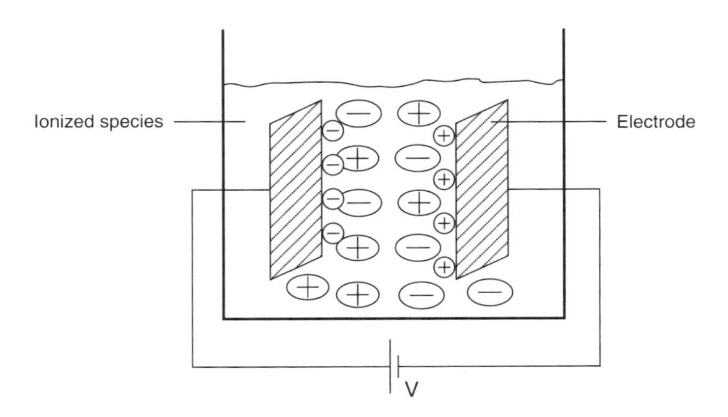

Figure 6.1 Schematic showing the influence of the electric field application on an ionized solution

Here, L and A corresponds to the length and area of the conductor/resistor, respectively. The aforementioned Ohm's law formation is valid for solid conductors or resistors. Let us define and discuss the electrical conductivity of a solution.

One example of an ionized solution is sodium chloride (NaCl) in water. A crystal of NaCl is made up entirely of ions, positively charged sodium ions and negatively charged chloride ions. Sodium ions are bound tightly to chloride ions by electrical forces. If the two ends of battery are attached to a large NaCl crystal, no electric current will flow, indicating that no ions are present. However, the situation changes when NaCl is added to water. Water molecules tear apart sodium ions and chloride ions. These ions are free to roam through the salt–water solution. Figure 6.1 shows the schematic of an ionized solution under the application of electric field. The positive species move toward the negative terminal, and the negative species move toward the positive terminal. We consider transfer of mass that takes place in a mixture of species under the action of an applied electric field. Unequal electrical forces act on different species as a result of the differences in the charge of species. This mode of mass transfer is termed as *electromigration*. It is similar to diffusion in a preferred direction and follows from the fact that electric field accelerates a charged particle which subsequently collides with other solute or solvent particles. The result is a migration rather than directed movement in the direction of the electric field. This is quite analogous to the particle diffusion that takes place in a concentration gradient. Let the solution be an unionized solvent containing ionized electrolytes. A gradient in electrostatic potential is applied to the solution. There will be an electric force exerted on the ion species, which is proportional to the potential gradient. The electric field (E) is the negative gradient of the electrostatic potential, $E = -\nabla\phi(N/c)$. Electric field is also defined as the force "F" experienced by a stationary positive point charge, q $(E = F/q)$. Force per mole exerted on the ionized species can be written as multiplication of sign of charge, magnitude of the particle charge, and electric field strength:

$$\text{Force per mole} = qE = -q\nabla\phi = z_i F\nabla\phi(N/\text{mol}) \tag{6.9}$$

where z_i is the charge number of species i and F is Faraday's constant, which is equal to charge of 1 mol of singly ionized molecules $= N_A \times e = 6.022 \times 10^{23}\text{mol}^{-1} \times 1.602 \times 10^{-19}\text{C} = 9.65 \times 10^4$ C/mol. Here, N_A and e are Avogadro's number and elementary charge, respectively.

Charge number or valence of an ion is a coefficient, which when multiplied by the elementary charge gives the charge of an ion. For example, chloride ion (Cl^{-1}) has a charge number equal to −1.

Similar to the constitutive relation for mass, heat, and momentum, the ionic flux due to migration in an electric field is proportional to the force acting on the particle multiplied by the particle concentration:

$$\text{Ionic flux} \propto \text{force} \times \text{concentration}$$

Using equation (6.9), the molar flux in stationary coordinate system can be written as

$$j_i^\star \propto (z_i F \nabla \phi) c_i \tag{6.10}$$

Introducing a proportionality constant, we write

$$j_i^\star = -\upsilon_i (z_i F \nabla \phi) c_i \left(\frac{\text{mol}}{\text{m}^2\text{s}}\right) \tag{6.11}$$

Here, υ_i is a transport property, known as *mobility*. It measures how mobile the charged particles are in an electric field. The mobility may be interpreted as the average velocity of a charged particle in a solution when acted upon by a force of one Newton per mole. The unit of mobility is $\frac{\text{mol-m}}{\text{N-s}}$. The mobility is related to diffusivity by Nernst–Einstein equation as

$$D_i = RT\upsilon_i \tag{6.12}$$

Electric current is defined as the flow of electric charge through a medium. This charge can be carried by moving electrons in a conductor or wire and carried by ions in an electrolyte. The motion of charged species gives rise to a current which expressed as current density for all species can be written as

$$J_{\text{el}} = -F\Sigma z_i j_i^\star \left(\frac{A}{\text{m}^2}\right) \tag{6.13}$$

Substituting equation (6.11) in equation (6.13), we have

$$J_{\text{el}} = F^2 \Sigma z_i^2 \upsilon_i c_i \nabla \phi \tag{6.14}$$

If there are no concentration gradients and no flow, the motion of the charge is only due to the applied electric field, and we write

$$J_{\text{el}} = K_{\text{el}} E = -K_{\text{el}} \nabla \phi \quad \text{(expression of Ohm's law)} \tag{6.15}$$

where K_{el} is the electrical conductivity of a solution $= F^2 \Sigma z_i^2 \upsilon_i c_i$ (S/m)

Note that conductivity is related to the total dissolved solid content of a solution. Typical drinking water has conductivity in the range of 5–50 mS/m. Seawater has conductivity about 5 S/m, and deionized water conductivity is about 5.5 μS/m. Conductivity measurement is traditionally used for monitoring the performance of water purification systems. It is also used as a fast inexpensive way to measure the ionic content of a solution in industrial and environmental applications.

When ordinary diffusion is present and there are concentration gradients, Ohm's law does not hold because there is contribution to the current from diffusion. Therefore, electrolyte

Table 6.1 Normalized molar conductivity of some ions in dilute solutions of water

Ion	Temperature [K]	Molar conductivity $[\text{m}^2\ \text{S}\ \text{mol}^{-1}] \times 10^3$
Na^+	288	3.98
	298	5.01
	308	6.15
Cl^{-1}	288	6.14
	298	7.63
	308	9.22
H^+	288	30.1
	298	35.0
	308	39.7
Ca^{2+}	298	11.9
Mg^{2+}	298	10.61

conductivities are also presented using normalized concentration as

$$K_{\text{el,N}} = \frac{K_{\text{el}}}{c_i} = F^2 \Sigma z_i^2 \upsilon_i \left(\frac{\text{m}^2\text{-S}}{\text{mol}} \right) \tag{6.16}$$

This is the conductivity of a solution when 1 mol of the substance is present in 1 m^3 of the solution. Normalized molar conductivity of some ions is presented in Table 6.1. Table 6.1 also shows the dependence of conductivity on the temperature of the solution. The conductivity of ions generally increases with increase in temperature.

6.3 Electrohydrodynamic Transport Theory

In this section, we attempt to develop appropriate equations of change governing the motion and behavior of species acted upon by an electric field. Let us assume the solution to be *dilute*. The criterion of diluteness is based on interparticle distance which is large enough such that motion of any particle is unaffected by the neighboring particles. For dilute solutions, the flux contributions from diffusion, electromigration, and convection can be linearly superposed. The molar flux (j_i^*) of the ith species is the superposition of all individual contributions:

$$j_i^* = j_{\text{el}}^* + j_{\text{diff}}^* + j_{\text{conv}}^* = -\upsilon_i z_i F c_i \vec{\nabla}\phi - D_i \vec{\nabla} c_i + c_i u \tag{6.17}$$

The first, second, and third term of the above equation, respectively, represents *electromigration, diffusion*, and *convection*. Here, the molar average velocity $u^* \left(= \frac{\Sigma c_i u_i}{c}\right)$ has been replaced by the mass average velocity $u \left(= \frac{1}{\rho}\Sigma \rho_i u_i\right)$, since in sufficiently dilute solutions $u^* = u$. The corresponding mass flux is

$$j_i = -\upsilon_i z_i F \rho_i \vec{\nabla}\phi - D_i \vec{\nabla}\rho_i + \rho_i u \tag{6.18}$$

The above two equations are called *Nernst–Planck* equations. The current density (J_{el}) can be written using equation (6.17) as

$$J_{el} = F\Sigma z_i j_i^* = -F^2\nabla\phi\Sigma z_i^2 v_i c_i - F\Sigma z_i D_i \vec{\nabla} c_i + Fu\Sigma z_i c_i \tag{6.19}$$

The above equation can also be written using electrical conductivity K_{el} similar to Ohm's law as

$$J_{el} = -K_{el}\nabla\phi - F\Sigma z_i D_i \vec{\nabla} c_i + Fu\Sigma z_i c_i \tag{6.20}$$

Here, K_{el} is the scalar electrical conductivity $= F^2\Sigma z_i^2 v_i c_i$. The first, second, and third term of the above equation represents contribution from electric field, concentration gradient, and convection of charge, respectively.

6.3.1 Transport Equation for Dilute Binary Electrolyte

Let us discuss about the governing equation for motion of dilute binary electrolyte with charged species. Dilute binary electrolyte is a combination of an unionized solvent and dilute fully ionized salt. The salt is composed of one negatively charged species and one positively charged species that do not enter into any reactions in the bulk of the fluid ($r_i = 0$). The solution is also assumed to be electrically neutral. For example, NaCl have Na^+ and Cl^{-1} species. For the solution to be electrically neutral, we have

$$\Sigma z_i c_i = 0 \tag{6.21}$$

This equation states that there is no accumulation of charge, that is, $\vec{\nabla} \cdot \vec{J}_{el} = 0$. If the positive ions are denoted by subscript + and the negative ions by subscript −, the electroneutrality condition is expressed by

$$z_+ c_+ + z_- c_- = 0 \tag{6.22}$$

We have described the species conservation equation for individual species in the previous chapter as

$$\frac{Dc_i}{Dt} + c_i \vec{\nabla}.\vec{u} = -\vec{\nabla} \cdot \vec{J}_i^* + r_i \tag{6.23}$$

Here, the mass average velocity is defined as $\rho u = \Sigma \rho_i u_i$ and molar flux with respect to the mass average velocity is defined as $J_i^* = c_i(u_i - u)$. The expression for molar flux with respect to stationary coordinate system has been expressed as (equation (6.17))

$$\vec{j}_i^* = -v_i z_i F c_i \vec{\nabla}\phi - D_i \vec{\nabla} c_i + c_i \vec{u}$$

The molar flux with respect to the reference velocity has been expressed as

$$\vec{J}_i^* = \vec{j}_i^* - c_i \vec{u}$$

Assuming incompressible flow ($\vec{\nabla} \cdot \vec{u} = 0$) and substituting the expression of molar flux (j_i^*) from above in the species conservation equation (6.23) for no reaction case ($r_i = 0$), we have

$$\frac{Dc_i}{Dt} = -\vec{\nabla} \cdot \left(-v_i z_i F c_i \vec{\nabla}\phi - D_i \vec{\nabla} c_i\right)$$

$$\frac{\partial c_i}{\partial t} + \vec{u} \cdot \vec{\nabla} c_i = v_i z_i F \vec{\nabla} \cdot (c_i \vec{\nabla}\phi) + D_i \nabla^2 c_i \tag{6.24}$$

Introducing a variable known as reduced concentration (c),

$$c = \frac{c_+}{\gamma_+} = \frac{c_-}{\gamma_-} \tag{6.25}$$

where γ_+ and γ_- are the number of positive and negative species produced by dissociation of one molecule of an electrolyte. Note that a *weak electrolyte* is defined as one which is not fully dissociated. *Strong electrolytes* are one which completely dissociate in a solution.

Using equations (6.24) and (6.25), we can write the general transport equation of species as

$$\frac{\partial c}{\partial t} + \vec{u} \cdot \vec{\nabla} c = z_{\pm} v_{\pm} F \vec{\nabla} \cdot (c \vec{\nabla} \phi) + D_{\pm} \nabla^2 c \tag{6.26}$$

Here, the mobility and diffusion coefficients are assumed constant. Subtracting the equation for the negative ions from that of the positive ions (equation (6.26)), we get

$$(z_+ v_+ - z_- v_-) F \vec{\nabla} \cdot (c \vec{\nabla} \phi) + (D_+ - D_-) \nabla^2 c = 0$$

$$F \vec{\nabla} \cdot (c \vec{\nabla} \phi) = -\frac{(D_+ - D_-) \nabla^2 c}{z_+ v_+ - z_- v_-} \tag{6.27}$$

Writing the ion concentration equation for positive ion,

$$\frac{\partial c}{\partial t} + \vec{u} \cdot \vec{\nabla} c = z_+ v_+ F \vec{\nabla} \cdot (c \vec{\nabla} \phi) + D_+ \nabla^2 c \tag{6.28}$$

Substituting $F \vec{\nabla} \cdot (c \vec{\nabla} \phi)$ from equation (6.27), in equation (6.28), we have

$$\frac{\partial c}{\partial t} + \vec{u} \cdot \vec{\nabla} c = -(z_+ v_+) \left(\frac{(D_+ - D_-) \nabla^2 c}{z_+ v_+ - z_- v_-} \right) + D_+ \nabla^2 c \tag{6.29}$$

$$\frac{\partial c}{\partial t} + \vec{u} \cdot \vec{\nabla} c = \frac{\nabla^2 c [-D_+ z_+ v_+ + D_- z_+ v_+ + D_+ z_+ v_+ - D_+ z_- v_-]}{z_+ v_+ - z_- v_-} \tag{6.30}$$

$$\frac{\partial c}{\partial t} + \vec{u} \cdot \vec{\nabla} c = D \nabla^2 c \tag{6.31}$$

where D is an effective diffusion coefficient defined by

$$D = \frac{z_+ v_+ D_- - z_- v_- D_+}{z_+ v_+ - z_- v_-} \tag{6.32}$$

The effective diffusion coefficient is related to the difference in charge and diffusion coefficients of the positive and negative ions. It should be noted that for $D_+ = D_-$, the effective diffusion coefficient, $D = D_+ = D_-$. Table 6.2 presents the diffusivity data for various ions.

Note: Equation (6.31) shows that the reduced concentration distribution in a dilute electrolyte is governed by the same convective diffusion equation as for neutral species even though there is a current flow.

If electroneutrality is assumed, the fluid mechanical conservation equations of mass, momentum, and energy remain unchanged. But if electroneutrality is not assumed with a net charge density of the liquid, the mass conservation equation remains unchanged, but the Lorentz body force must be added to the right-hand side of the Navier–Stokes (N–S) equation.

Table 6.2 Experimental values for ionic mobility and diffusivity for small ions in aqueous solutions at small concentrations

Ions at $T = 25°C$	H^+	K^+	Li^+	Na^+	Br^-	Cl^-	F^-	I^-	OH^-
Mobility, υ_{ion} $[10^{-8} m^2 (V\ s)^{-1}]$	36.2	7.62	4.01	5.19	8.09	7.91	5.70	7.96	20.6
Diffusivity, D_{ion} $[10^{-9} m^2\ s^{-1}]$	9.31	1.96	1.03	1.33	2.08	2.03	1.46	2.05	5.30

Remember that electric field is coupled with the fluid mechanics through the Lorentz relation for the force on a charged particle:

$$\vec{f}_E = \rho_E \vec{E} \tag{6.33}$$

where f_E is the electric force per unit volume, ρ_E is the electric charge density (c/m^3), and E is the electric field.

The coupling between the electromagnetism and hydrodynamics through the body force $\rho_E E$ leads to the modified N–S equation as

$$\rho\left(\frac{\partial}{\partial t}\vec{\upsilon}\right) + (\vec{\upsilon} \cdot \vec{\nabla})\vec{\upsilon} = -\vec{\nabla}p + \mu\nabla^2\vec{\upsilon} + \rho\vec{g} + \rho_E\vec{E} \tag{6.34}$$

Similarly, for energy equation, the contribution of $\rho_E\vec{E} \cdot \vec{u}$ is added to the contribution of work term. In the next section, an example of an electrolytic cell is presented to demonstrate the use of the above sets of governing equations.

6.4 Electrolytic Cell Example

Electrolytic cell is a device containing two electrodes in contact with electrolyte that brings about chemical reaction when connected to an outside source of electricity. Electrolytic cells have many practical applications, that is, recovery of pure metal from alloys, plating of one metal with another known as electroplating, etc.

A schematic of electroplating system is shown in Figure 6.2. Electroplating is primarily used for deposition of material of desired properties (abrasion, wear resistance, corrosion protection, lubricity, and aesthetic qualities) to a surface that otherwise lacks that property.

In Figure 6.2, the part to be plated is the cathode of the circuit. The electrolyte solution contains one or more dissolved salts. The application of direct current leads to dissolution of metal atoms from the anode and deposition on the cathode surface. The rate at which the anode is dissolved is equal to the rate at which the cathode is plated. Thus, the ions in the electrolyte bath are continuously replenished by the anode. The cations with positive charge coming from the anode associate with the ions in the solution. At the cathode, the cations are reduced to the metal form by gaining two electrons.

Figure 6.3 shows the schematic of an electrolytic cell. Two electrodes are submerged inside a cupric sulfate solution connected to an electric potential source. Electrode is a metal immersed in an electrolyte solution so that it makes contact with it. Here, copper in a solution of cupric sulfate is an example of an electrode. Electrochemical cell is a system consisting of two

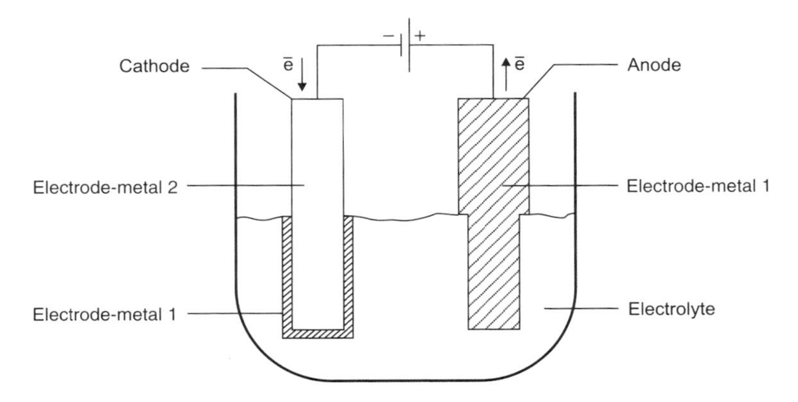

Figure 6.2 A schematic showing electroplating, where the electrode-metal 1 is consumed and plated on the surface of electrode-metal 2

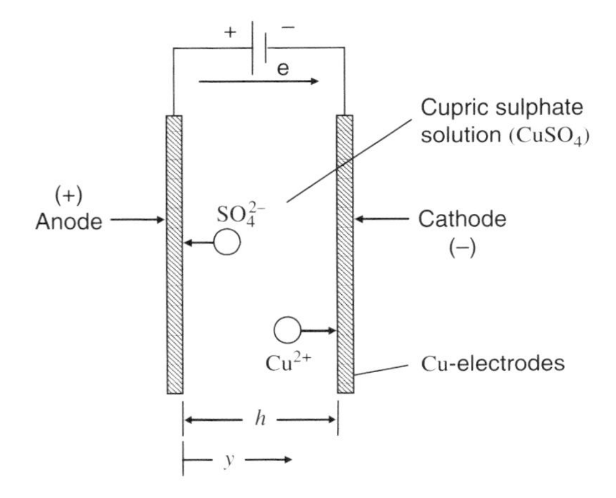

Figure 6.3 A schematic of an electrolytic cell

electrodes. If the cell generates an electromotive force (emf) by chemical reactions, it is termed as a *galvanic cell*. If an emf is imposed across the electrodes, it is an *electrolytic cell*. If a current is generated by the imposed emf, the electrochemical or electrolytic process that occurs is known as *electrolysis*. Whether or not a current flows, the electrolyte can be considered to be neutral except at the solution–electrode interface. At this interface, a thin layer termed as a *Debye sheath* or *electric double layer* forms that is composed predominantly of ions of charge opposite to the metal electrode. In Figure 6.3, the cupric sulfate dissociates into charged cupric ions Cu^{2+} and sulfate ions SO_4^{2-}. When a potential difference is imposed between the electrodes, there are a current flow and reactions at the electrode. The electric field drives the cupric ions (cations) toward the negative electrode (cathode) and the sulfate ions (anions) toward the positive electrode (anode). At the anode, there is a dissolution of copper, with an electron-producing reaction as

$$\mathrm{Cu} \rightarrow \mathrm{Cu}^{2+} + 2\mathrm{e}^- \tag{6.35}$$

At the cathode, there is a deposition of copper, that is, electrons are consumed:

$$\mathrm{Cu}^{2+} + 2\mathrm{e}^- \rightarrow \mathrm{Cu} \tag{6.36}$$

In the electrolytic cell, the cupric ions and sulfate ions both contribute to the conduction mechanisms. But only cupric ions enter into the electrode reaction and pass through the electrode–solution interface. The electrode therefore acts like a *semipermeable membrane* which is permeable to the Cu^{2+} ions but impermeable to the SO_4^{2-} ions. Anions accumulate near the anode and become depleted near the cathode, resulting in concentration gradients in the solution near the electrodes of both ions. This is termed as *concentration polarization*. Let us determine the current–voltage characteristic of the cell, that is, the concentration polarization. To do this, we must calculate the flux of metal ions (cations) arriving at the cathode and depositing on it. We assume that the overall rate of the electrode reaction is determined by this flux. Once the cation distribution is known, the potential drop can be calculated. Note that anions are effectively motionless and do not produce a current. Let us assume that electrodes of the electrolytic cell are infinite planes at the anode ($y = 0$) and cathode ($y = h$) (Figure 6.3). The electrolyte velocity is zero. The definition of the current densities is

$$i_{\pm} = F z_{\pm} j_{\pm}^{*} \tag{6.37}$$

We have derived before the molar flux of ith species as

$$j_i^* = -\upsilon_i z_i F c_i \nabla\phi - D_i \nabla c_i + c_i u \tag{6.38}$$

The last term on the right-hand side of the above equation is zero as the electrolyte is motionless ($u = 0$).

We know from the Nernst–Einstein equation the expression for mobility as

$$\upsilon_i = \frac{D_i}{RT}$$

Now, substituting j_i^* and υ_i in the equation of $i_{\pm}$ (equation (6.37)) for the motionless electrolyte ($u = 0$), we can write

$$i_+ = -D_+ F z_+ \frac{dc_+}{dy} - \frac{F^2 z_+^2 D_+ c_+}{RT} \frac{d\phi}{dy} \tag{6.39}$$

$$i_- = -D_- F z_- \frac{dc_-}{dy} - \frac{F^2 z_-^2 D_- c_-}{RT} \frac{d\phi}{dy} \tag{6.40}$$

In the simple electrolytic cell, only the cations deposit on the cathode. The flux of anions and therefore the anion current at the cathode must be zero. However, since the electrolyte is motionless, the anion current must be everywhere zero; that is,

$$i_- = 0$$

Thus, the current flowing in the cell due to electromigration and diffusion is thus due to only cation transport. It is the magnitude of this current which we wish to determine for a given value of applied voltage. From the definition of a reduced ion concentration,

$$c = \frac{c_+}{\gamma_+} = \frac{c_-}{\gamma_-} \tag{6.41}$$

where γ_+ and γ_- are the number of +ve and −ve ions produced by the dissociation of one molecule of electrolyte. The condition of electroneutrality is assumed to hold within the fluid, that is,

$$z_+\gamma_+ = -z_-\gamma_- \tag{6.42}$$

Now, the expression for current density can be written using equations (6.39)–(6.41) as

$$i_+ = -D_+Fz_+\gamma_+\frac{dc}{dy} - \frac{F^2z_+^2D_+\gamma_+c}{RT}\frac{d\phi}{dy} \tag{6.43}$$

$$0 = -D_-Fz_-\gamma_-\frac{dc}{dy} - \frac{F^2z_-^2D_-\gamma_-c}{RT}\frac{d\phi}{dy} \tag{6.44}$$

From equation (6.44), we can write

$$\frac{d\phi}{dy} = -\frac{RT}{Fz_-c}\frac{dc}{dy} \tag{6.45}$$

Substituting the above expression $\frac{d\phi}{dy}$ in the expression of i_+ (equation (6.43)) and replacing $\frac{z_+\gamma_+}{\gamma_-}$ by $-z_-$ (equation (6.42)), we have

$$i_+ = \_D_+F\gamma_-(z_+ - z_-)\frac{dc}{dy} \tag{6.46}$$

For a given applied voltage, the current is fixed. Therefore, equation (6.46) indicates that the "c" is linear across the cell, y.

Using the boundary condition $c = c_a$ at $y = 0$ and integrating the above equation (6.46), we get

$$c = c_a - \frac{i_+y}{D_+F\gamma_-(z_+ - z_-)} \tag{6.47}$$

Note that concentration, c_a, is not known in advance. It can be determined by overall conservation of species from the initial uniform species concentration and concentration drop across the wall.

The concentration at the cathode ($y = h$), denoted by c_c, is

$$c_c = c_a - \frac{i_+h}{D_+F\gamma_-(z_+ - z_-)} \tag{6.48}$$

When $c_c = 0$, the current approaches a limiting value

$$i_{\lim} = \frac{D_+F\gamma_-(z_+ - z_-)2c_0}{h} \tag{6.49}$$

We have used the condition $c_c + c_a = 2c_0$ from conservation of species where c_0 is the initial uniform reduced species concentration. This diffusion limiting current is reached when the cations at the cathode have been completely depleted by the electrode reaction. The limiting current density is based on the fact that metal ions cannot be deposited faster than they arrive at the cathode surface. When applied voltage is increased beyond this limit, other reactions start to occur such as the release of hydrogen. This should be avoided as it reduces the energy efficiency and the deposits during plating become rough. The potential distribution can be obtained by substitution of the concentration (c) distribution equation (6.47) to the i_+ equation (6.43), and upon integration, we get

$$\Delta\phi = \frac{RT}{z_+F}\frac{z_+ - z_-}{z_-}\ln\left(\frac{i_{\lim} - i_+}{i_{\lim} + i_+}\right) - \frac{RT}{z_+F}\ln\frac{c_a}{c_c} \tag{6.50}$$

Here,

$$\Delta\phi = \phi(y = 0) - \phi(y = h)$$

and $i_{\lim}$ is the limiting current density from equation (6.49).

The first term in the above equation is the *ohmic drop* due to flow of current through the electrolyte whose electrical conductivity varies because of the variation in ion concentration across the cell. The second term is the diffusion potential drop, which arises from the *concentration gradient* term, associated with the presence of the background immobile anions in equilibrium. This term does not disappear in the absence of current flow and leads to a differential electric field because of unequal rates of diffusion of charged particles. However, the total difference in potential between the anode and cathode is composed of three parts. The first part is the potential drop in the fluid where charge neutrality is assumed as given by $\Delta\phi$ expression (equation (6.49)). The second part is due to the difference in equilibrium potential between the anode and cathode. Because of the electrode reaction, even at equilibrium, that is, with no current flow, there will be potential drop. The formation of this metal–electrolyte potential difference arises from the transfer of metal ions from the metal into the electrolyte and vice versa. The amount of this transference is generally not equal in both directions and gives rise to the metal–electrolyte potential difference. This electrode potential is the potential difference that forms at the boundaries of the two phases. The standard electrode potential is a characteristic value of a metal, the solvent, the temperature, and the pressure. When rates of metal dissolution and deposition are equal, we get *equilibrium potential* (ϵ), which is expressed by *Nernst equation*. This is equivalent to chemical equilibrium with chemical reaction. For the example considered here, both electrodes are copper, and thus, we have the second potential drop, $\Delta\epsilon = 0$. The third part of the potential difference is called as *concentration over potential*. It is a consequence of current flow which leads to a lower ion concentration in the solution at the cathode and a higher concentration at the anode. Now, at an electrode, there is a change in concentration of the ions in solutions to the concentration of the adsorbed ions at the surface, a charge which is assumed to be discontinuous but which takes place over the thin *double layer* adjacent to the electrode in which charge neutrality does not hold good and in which a potential gradient exists. The concentration over potential is

$$\eta_{\text{conc}} = \frac{RT}{z_+F}\ln\frac{c_a}{c_c} \tag{6.51}$$

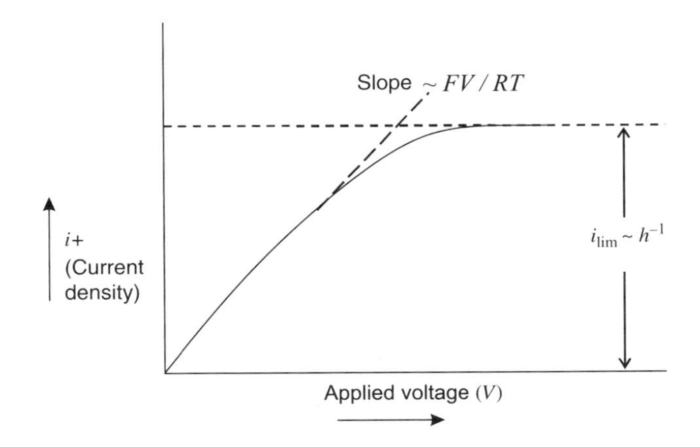

Figure 6.4 The current–voltage characteristic of the electrolytic cell

This equation is known as Nernst's equation. Thus, the potential boundary condition for the electrolytic cell is

$$V = \Delta\phi + \frac{RT}{z_+F}\ln\frac{c_a}{c_c} \tag{6.52}$$

Here, V is the applied voltage. Substituting $\Delta\phi$ from equation (6.50), we get the current voltage characteristic as

$$\frac{i_+}{i_{\text{lim}}} = \frac{1 - \exp[z_+z_-FV/(z_+ - z_-)RT]}{1 + \exp[z_+Z_-FV/(z_+ - z_-)RT]} \tag{6.53}$$

From equation (6.49), we have

$$i_{\text{lim}} \sim h^{-1}$$

We see that the limiting current is inversely proportional to the electrode spacing. At low values of FV/RT, the current is linear with applied voltage, and at sufficiently high voltage values, it approaches the limiting current exponentially (see Figure 6.4). Note that the previous example assumes no flow situation. From earlier study of concentration polarization discussed during reverse osmosis channel flow, it may be noted that the limiting flux (current density) can be increased by having the solution flow parallel to the walls (electrodes). With flow, the ions are carried with the solution, and the polarization at the electrode is thus reduced. It may be noted that this is the reason why the plating baths are kept agitated during plating.

The transport processes in fuel cell have many similarities with the electrolytic cell. It converts the chemical energy from a fuel through a chemical reaction with oxygen or another oxidizing agent. Fuel cell requires continuous source of fuel and oxidant to sustain the chemical reaction. This is contrary to a battery where the chemicals present in the battery react with each other to generate an emf. Similar issues of concentration polarization play an important role in the design of various fuel cells.

6.5 The Electric Double Layer and Electrokinetic Phenomena

Let us study the electric potential and charge distribution in an electrolyte, that is, an aqueous solution of ions, in equilibrium near a charged surface. This forms the basis of electrokinetic effect. Let an electrolyte be in contact with a solid surface, which may be the wall of a microfluidic channel. Depending on the chemical composition of the solid and the electrolyte, chemical processes at the surface result in a charge transfer between the electrolyte and the wall. As a result, the wall and the electrolyte get oppositely charged while maintaining global charge neutrality.

Generally, most substances acquire a surface electric charge when brought into contact with an aqueous (polar) medium. Some of the charging mechanisms include ionization, ion adsorption, and ion dissolution. Surface charge also spontaneously appears when a dielectric is immersed in an electrolyte. One example is a glass plate immersed in aqueous solution, which gets negatively charged if the pH of the solution is greater than 4. This is because the silane terminals Si–O–H localized on the glass surface lose hydrogen ions in the presence of the aqueous solution leaving Si–O^- terminals on the surface leading to negatively charged glass surface. The effect of any charged surface in an electrolyte solution will be to influence the distribution of nearby ions in the solution. Ions of opposite charge to that of the surface (counterions) are attracted toward the surface, while ions of like charges (co-ions) are repelled from the surface. This attraction and repulsion when combined with the mixing tendency resulting from the random thermal motion of the ions leads to the formation of an electric double layer. The electric double layer is a region close to the charged surface in which there is an excess of counterions over coions to neutralize the surface charge (see Figure 6.6). Evidently, there is no charge neutrality within the double layer because the number of counterions will be large compared with the number of coions. If there were no thermal motion, there would be just as many counterions in the electric double layer as needed to balance the charge of the surface. This is termed as *perfect shielding*, since all of the other ions are shielded from the surface charge. But because of the finite temperature and associated thermal motion of the ions, these ions have enough thermal energy to escape from the electrostatic potential wall at the edge of the "cloud" where the electric field is weak. Therefore, the edge of the double layer is at a position where the potential energy is approximately equal to the thermal energy of the counterions ($RT/2$ per mole per degree of freedom) and the shielding is not complete. Here, R is the universal gas constant (8.3145 J/mol-K), and T is the absolute temperature in kelvin. Let us estimate the approximate thickness of the double layer shown in Figure 6.5. The electric field is assumed to be parallel to the x-axis, that is, everywhere perpendicular to the plane charged surface. Let us consider a simple fully dissociated symmetrical salt in solution for which the number of +ve and −ve ions is equal:

$$z_+ = -z_- = z$$

Let us first make a rough estimate of the thickness of the double layer by assuming that there are no positive ions (coions) present (perfect shielding). Poisson's equation relating spatial variation in electric field to charge distribution for a medium of uniform dielectric constant is

$$\nabla^2\phi = -\frac{\rho_E}{\epsilon} \tag{6.54}$$

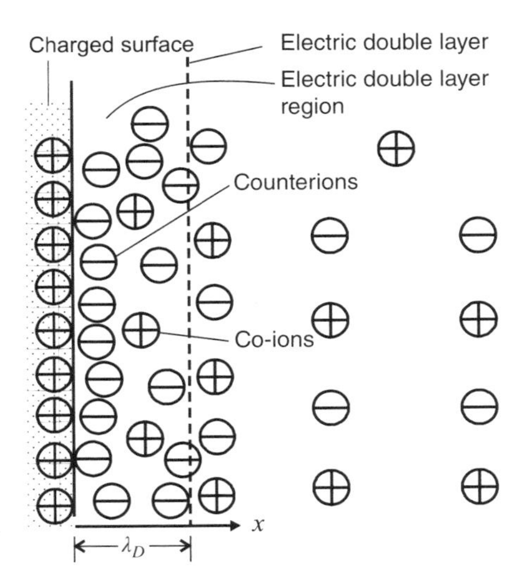

Figure 6.5 Schematic showing the electric double layer region

where ρ_E is the electric charge density (c/m^3) and ϵ is the permittivity of the medium ($c^2/\mathrm{N\text{-}m}^2$). We have earlier seen the electric change density expression as

$$\rho_E = F\Sigma z_i c_i = Fzc \tag{6.55}$$

Here, c is the average molar negative ion (counterion) concentration. Hence, the Poisson equation reduces to

$$\frac{d^2\phi}{dx^2} = \frac{Fzc}{\epsilon} \tag{6.56}$$

Integrating the above equation, we get

$$\frac{d\phi}{dx} = \frac{Fzcx}{\epsilon} + \text{Constant} \tag{6.57}$$

Assuming that electric field vanishes on one side of the plane layer, that is, we have the boundary condition as

$$\frac{d\phi}{dx} = 0 \quad \text{at} \quad x = \lambda_D$$

where λ_D is the thickness of the double layer (see Figure 6.5). Using the above boundary condition in equation (6.57), we get

$$\text{Constant} = -\frac{Fzc\lambda_D}{\epsilon}$$

Integrating equation (6.57) once again, we have

$$\phi(x) = \frac{Fzcx^2}{2\epsilon} - \frac{Fzc\lambda_D x}{\epsilon} + \text{Constant} \tag{6.58}$$

As electrical potential is the potential energy per charge, the electrical potential energy per mole of negative ion (counterion) can be expressed as

$$W = -Fz\phi$$

The change in W across a plane layer of width x is

$$\begin{aligned}\Delta W &= -Fz\Delta\phi = -Fz(\phi(x) - \phi(o)) \\ &= -Fz\left(\frac{Fzcx^2}{2\epsilon} - \frac{Fzc\lambda_D x}{\epsilon} + \text{Constant} - \text{Constant}\right) \\ &= -\frac{F^2z^2cx^2}{2\epsilon} + \frac{F^2z^2c\lambda_D x}{\epsilon}\end{aligned} \quad (6.59)$$

If we assume planar translational motion of the ions, the value of ΔW equals to $RT/2$ and we have

$$x = \lambda_D = \left(\frac{\epsilon RT}{F^2z^2c}\right)^{1/2} \quad (6.60)$$

We know that $c_c + c_a = 2c_0$, where c_0 is the initial uniform reduced species concentration, c_c is the reduced species concentration of counterions, and c_a is the reduced species concentration of co-ions. One of the species is present for perfect shielding. Therefore, we can write

$$\lambda_D = \left(\frac{\epsilon RT}{2F^2z^2c_0}\right)^{1/2}$$

For ionic concentration of 1 mol/m^3 and a dielectric constant equal to that of water, $\epsilon = 78\epsilon_0$, we have $\lambda_D \simeq 9.5$ nm at standard ambient condition where λ_D is termed as the Debye shielding distance or Debye length.

Note: (1) λ_D is inversely related to $\sqrt{c}$. This is because of the fact that there are more counterions per unit depth for balancing the surface charge.

(2) λ_D decreases with increase in valency (z) as fewer ions are required to equilibrate the surface charge.

(3) λ_D increases with $(RT)^{1/2}$, that is, without thermal agitation the double layer would collapse to an infinitely thin layer.

6.6 Debye Layer Potential Distribution

In the previous section, we have assumed perfect shielding, that is, presence of a single ion close to the charged surface. Let us use the spatial distribution of ion concentration for estimation of potential distribution inside a double layer. The concentration of ions in the sheath has Boltzmann distribution, which has been derived in the later section (see Figure 6.6) as

$$c_{\pm} = c_0 \exp\left(\mp\frac{zF\phi}{RT}\right) \quad (6.61)$$

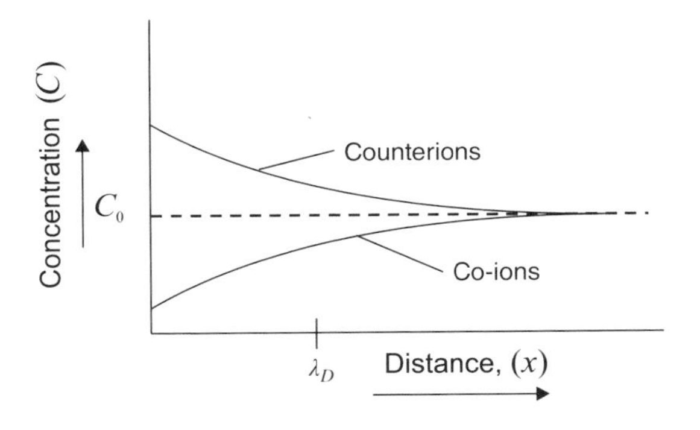

Figure 6.6 Concentration distribution inside a Debye layer

Here, the concentration far from the surface is $c \to c_0$ where $\phi \to 0$. The charge density can be written as

$$\begin{aligned}\rho_E &= F\Sigma z_i c_i = Fz_- c_- + Fz_+ c_+ \\ &= z_- Fc_0 \exp\left(\frac{zF\phi}{RT}\right) + z_+ Fc_0 \exp\left(\frac{-zF\phi}{RT}\right) \\ &= zFc_0 \left[\exp^{\frac{-zF\phi}{RT}} - \exp^{\frac{zF\phi}{RT}}\right] \\ &\qquad [\text{using} \quad z_+ = -z_- = z \quad \text{for symmetric salt}]\end{aligned} \tag{6.62}$$

The Poisson equation for potential distribution is

$$\nabla^2 \phi = \frac{-\rho_E}{\epsilon} \tag{6.63}$$

which can be expanded as

$$\frac{d^2\phi}{dx^2} = -\frac{zFc_0}{\epsilon}\left[\exp^{\frac{-zF\phi}{RT}} - \exp^{\frac{zF\phi}{RT}}\right] = \frac{2Fzc_0}{\epsilon}\sinh\left(\frac{zF\phi}{RT}\right) \tag{6.64}$$

Equation (6.64) can be integrated explicitly. The other approach is to use small potential approximation, $zF\phi \ll RT$. This expression is valid when the electrical energy is small compared to the thermal energy. Using the power series expansion of exponential function in equation (6.64), we get

$$\begin{aligned}\frac{d^2\phi}{dx^2} &= -\frac{zFc_0}{\epsilon}\left[1 - \frac{zF\phi}{RT} + \frac{1}{\lfloor 2}\left(\frac{zF\phi}{RT}\right)^2 \cdots - 1 - \frac{zF\phi}{RT} - \frac{1}{\lfloor 2}\left(\frac{zF\phi}{RT}\right)^2 \cdots\right] \\ &= +\frac{2z^2F^2c_o\phi}{RT\epsilon} = \frac{\phi}{\lambda_D^2}\end{aligned} \tag{6.65}$$

The small potential approximation is termed as the *Debye–Hückel* approximation. The boundary conditions can be written as

$$\phi = \phi_w \quad \text{at} \quad x = 0 \quad \text{and}$$
$$\phi = 0 \quad \text{and} \quad \frac{d\phi}{dx} = 0 \quad \text{at} \quad x \to \infty$$

Solution of the governing equation (6.65) and use of above boundary conditions give

$$\phi = \phi_w \exp\left(-\frac{x}{\lambda_D}\right) \tag{6.66}$$

Note: At close to the charged surface, where the potential is relatively high, the *Debye–Hückel approximation* is inapplicable, and the potential decreases faster than the exponential falloff indicated above. The above treatment about the diffuse double layer is based on the assumption that the ions in the electrolyte are treated as point charges. The ions are however of finite size. This limits the inner boundary of the diffuse part of the double layer, since the center of an ion can only approach the surface to within its hydrated radius without becoming specifically adsorbed. The inner part of the double layer next to the surface, the outer boundary of which is approximately a hydrated ion radius, is called the stern layer (see Figure 6.7). The plane separating the inner layer and outer diffuse layer is called the *stern plane*. The *shear surface* is characterized as the plane at which the mobile portion of the diffuse layer can "slip" or flow past the charged surface. Electrokinetic potential or *zeta* (ζ) *potential* is the potential at the shear surface between the charged surface and the electrolyte solution.

6.6.1 Surface Charge and Debye Layer Capacitance

The Debye layer acts as a capacitor since it contains electric charges due to the potential difference ζ between the surface and the bulk. One way to obtain the capacitance of the Debye layer is by integrating the charge density $\rho_E(x)$ along the x-direction from the surface at $x = 0$ to infinity, $x = \infty$.

Using the Poisson equation and the potential distribution, ϕ (equations (6.65) and (6.66)), we have

$$\rho_E(x) = -\epsilon\frac{\partial^2\phi}{\partial x^2} = -\frac{\epsilon\phi}{\lambda_D^2} = -\frac{\epsilon\phi_w}{\lambda_D^2}\exp\left(-\frac{x}{\lambda_D}\right) \simeq -\frac{\epsilon\zeta}{\lambda_D^2}\exp\left[-\frac{x}{\lambda_D}\right] \tag{6.67}$$

The resulting q_{liq}, which is the charge per area $\mathcal{A}$ that is contained in the liquid in the direction perpendicular to any small area $\mathcal{A}$ on the surface, can be written as

$$q_{\text{liq}}(x) = \int_0^\infty dx\ \rho_E(x) = \int_0^\infty dx\left[-\frac{\epsilon\zeta}{\lambda_D^2}\exp\left[-\frac{x}{\lambda_D}\right]\right] = -\frac{\epsilon}{\lambda_D}\zeta \tag{6.68}$$

The basic definition of capacitance is that it is equal to the amount of charge, Q, stored in between two plates for a potential difference or voltage, V, across the two plates $\left(c = \frac{Q}{V}\right)$.

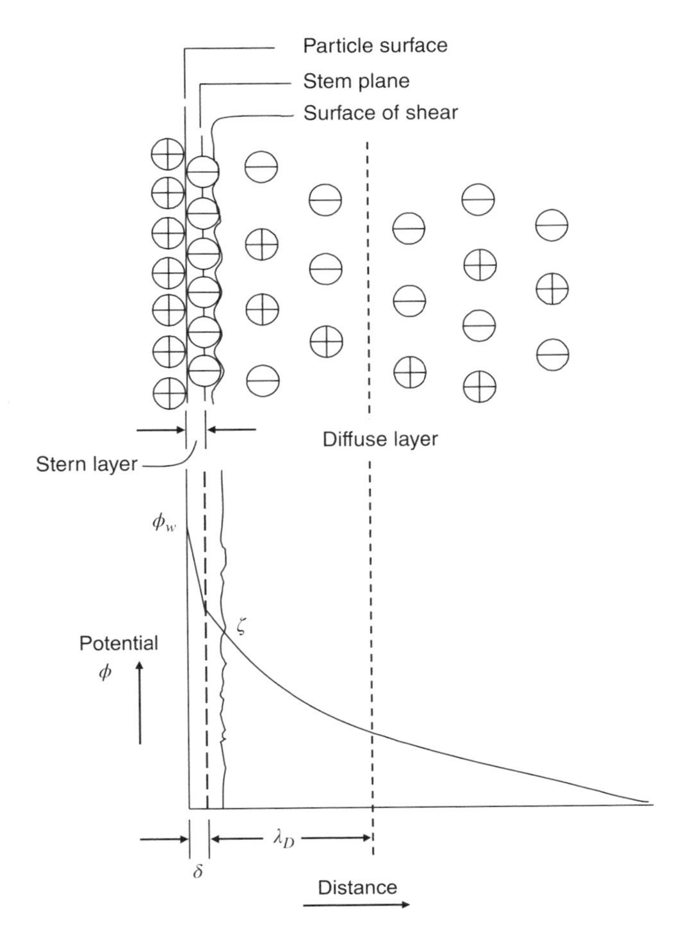

Figure 6.7 Detailed characterization of Debye layer

From this linear relation (equation (6.67)) between charge per area and applied potential difference, we can immediately read off the capacitance per area C of the Debye layer in thermal equilibrium as

$$C = \frac{\epsilon}{\lambda_D} \tag{6.69}$$

Using $\lambda_D = 9.6$ nm and $\epsilon = 78\epsilon_0$, we find the value of capacitance as

$$C \approx 0.073 \text{ F m}^{-2}$$

Debye layer acts as a charge-neutral capacitor, where the solid surface and the electrolyte are the two "plates" of the capacitor.

The Debye layer acts as a capacitor, and the electrolyte has a finite conductivity K_{el} or resistivity $1/K_{el}$. Therefore, one should be able to ascribe a characteristic RC time, τ_{RC}, of the system. The setup in Figure 6.8(a) shows two plane electrodes with an electrolyte sandwiched

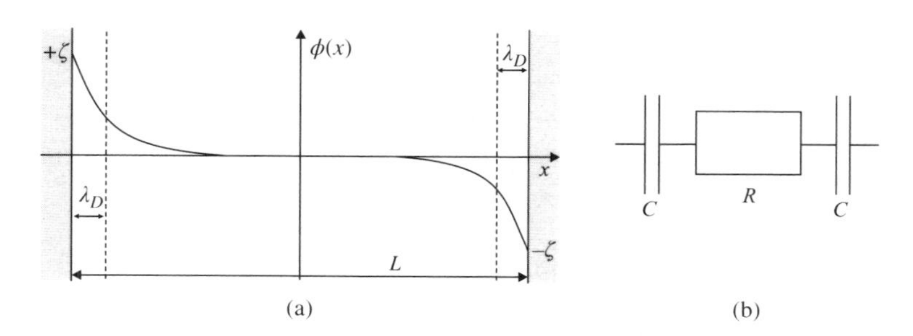

Figure 6.8 (a) An electrolyte occupying the space of width L between two parallel plane electrodes. Application of a voltage difference between the two electrodes develops a Debye layer of width λ_D on each of them. (b) The equivalent circuit diagram of panel (a) consisting of one capacitor C for each Debye layer and one resistor R for the bulk electrolyte

between the two. The distance between the electrodes is denoted by L. When one electrode is biased by the voltage $+\zeta$ and the other by $-\zeta$, a Debye layer builds up on each of them. The equivalent diagram of the system consists of a series coupling of one capacitor for each Debye layer and one resistor for the bulk electrolyte. The RC time for this system is

$$\tau_{\text{RC}} = \text{RC} = \left(\frac{L}{K_{\text{el}}\mathcal{A}}\right)\left(\frac{1}{2}\frac{\epsilon}{\lambda_D}\mathcal{A}\right) = \frac{\epsilon}{K_{\text{el}}}\frac{L}{\lambda_D} \tag{6.70}$$

If $\lambda_D = 9.6$ nm, $L = 100$ µm, $\epsilon = 78\epsilon_0$, and $K_{\text{el}} = 10^3$ S/m,

$$\tau_{\text{RC}} = 3.6 \text{ ms} \tag{6.71}$$

For processes with time period higher than 3.6 ms, enough time is available for establishing the Debye layer. However, the Debye layer cannot follow faster processes. In AC experiments, this translates into a characteristic frequency w_D of the Debye layer:

$$w_D = \frac{2\pi}{\tau_{\text{RC}}} = 1.7 \text{ kHz} \tag{6.72}$$

For frequencies higher than a few kilohertz, the Debye layer is not established.

6.7 Electrokinetic Phenomena Classification

Electrokinetic phenomenon arises when the mobile portion of the diffuse double layer and an external electric field interact in the viscous shear layer near the charged surface. If an electric field is applied tangentially along a charged surface, then the electric field exerts a force on the charge in the diffuse layer. This layer is part of the electrolyte solution, and migration of the mobile ions will carry the solvent with them and cause it to flow. On the other hand, an electric field is created if the charged surface and diffuse part of the double layer are made to move relative to each other. The four electrokinetic phenomena broadly classified are

1. *Electroosmosis*: The movement of liquid relative to a stationary charged surface (e.g., a capillary or porous plug) by an applied electric field (i.e., the complement of electrophoresis)

2. *Electrophoresis*: The movement of charged surface plus attached material (i.e., dissolved or suspended material) relative to stationary liquid by an applied electric field.
3. *Sedimentation potential*: The electric field created when charged particles move relative to stationary liquid (i.e., the opposite of electrophoresis).
4. *Streaming potential*: The electric field created when liquid is made to flow along a stationary charged surface (the opposite of elecroosmosis).

6.8 Electroosmosis

Electroosmosis has been used in a variety of applications, including the dewatering of soils for construction purposes, pumping and removal of contaminants from soils, and the dewatering of mine tailings and waste sludges. It is also used to understand the behavior of biological membranes. Under the influence of an applied electric field, water migrates through the porous clay diaphragms toward the cathode (see Figure 6.9). Clay, sand, and other mineral particles usually carry negative surface charges when in contact with water; the water normally contains small quantities of dissociated salts. The charged surface thus attracts positive ions present in the water and repels negative ions. The positive ions thus predominate in the Debye sheath next to the charged surface. Therefore, application of an external electric field results in a net migration of positive ions toward the cathode of ions in the surface water layer. Due to viscous drag, water in the pores is drawn by the ions and therefore flows through the porous medium.

6.8.1 *Electroosmotic Velocity*

The principle of electroosmotic (EO) flow is shown in Figure 6.10. Two metallic electrodes are situated at each end of a channel, in which charge separation at the walls has led to the formation of an equilibrium Debye layer. When a DC potential difference $\Delta V = \Delta\phi$ is applied over the electrodes the resulting electrical field, E is given by

$$E = -\nabla\phi \tag{6.73}$$

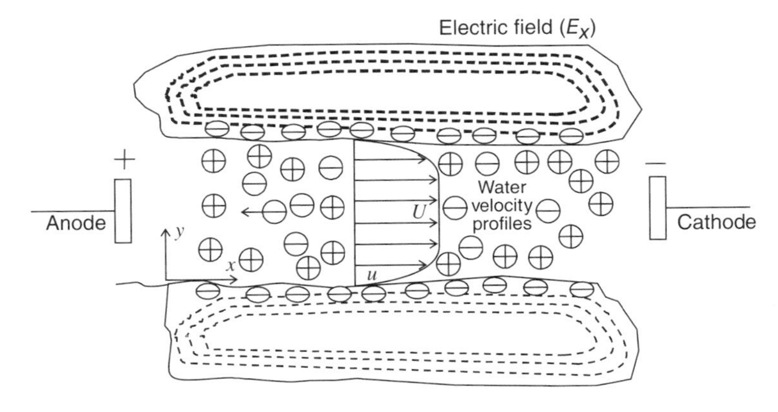

Figure 6.9 Electroosmotic flow of water in a porous charged medium

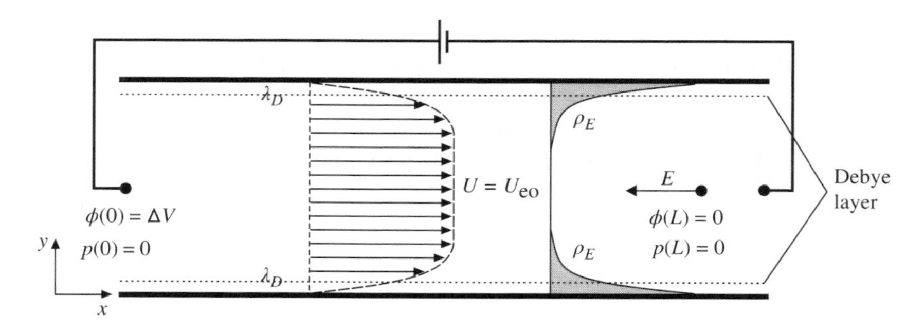

Figure 6.10 The velocity profile, U, and the negative Debye layer charge density profile, ρ_E, in an ideal electroosmotic (EO) flow inside a channel. The EO flow is induced by the external potential difference $\Delta\phi = \Delta V$, resulting in the homogeneous electric field E. The velocity profile reaches the constant value U_{eo} at a distance of a few times the Debye length λ_D from the walls

This leads to a body force $\rho_E E$ on the Debye layer, which begins to move and then by viscous drag pulls the charge neutral bulk liquid along. If no electrochemical processes occurs at the electrodes, the motion stops after a very short time (of the order of microseconds) when the electrodes are screened by the formation of a Debye layer around them. If, however, electrochemical processes, for example, electrolysis, take place at the electrodes, such a charge buildup is prevented, and electrical currents can flow in the system. We derive an expression for the resulting EO velocity field in the liquid below.

Let us estimate the EO velocity produced in a fine capillary by uniform electric field applied along the axis. It may be remembered from the previous section that for electrolyte flow, when electroneutrality is not satisfied, the Lorentz force should be added to the momentum equation. The momentum equation is

$$\rho\frac{Du}{Dt} = f_{\text{surf}} + f_{\text{body}} \tag{6.74}$$

Here, f_{surf} is the surface force, that is, $\vec{\nabla}\cdot\bar{\tau}$, and f_{body} is the body force, that is, gravitational (ρg) or electric body force per unit volume ($\rho_E E$). For incompressible flow of constant velocity, $f_{\text{surf}} = -\nabla P + \mu\nabla^2 u$. We write the momentum equation as

$$\frac{\rho Du}{Dt} = -\nabla p + \mu\nabla^2 u + \rho_E E \tag{6.75}$$

Assuming inertia $\left(\frac{\rho Du}{Dt}\right)$ free flow with no pressure gradient, the above equation simplifies to

$$\mu\nabla^2 u = -\rho_E E \tag{6.76}$$

For a long capillary, fully developed condition can be assumed, that is, $\frac{\partial u}{\partial x} = 0$ and $u = u(y)$. Therefore, we have

$$\mu\nabla^2 u = \mu\frac{\partial^2 u}{\partial y^2} \tag{6.77}$$

The Poisson equation is

$$\nabla^2\phi = -\frac{\rho_E}{\epsilon} \tag{6.78}$$

For small diffuse layer thickness, λ_D, compared to the channel dimension, the curvature terms can be neglected, and the problem reduces to 1D form appropriate for a long plane channel or infinite plane surface. Therefore, equation (6.78) can be written as

$$-\rho_E = \epsilon \frac{\partial^2 \phi}{\partial y^2} \tag{6.79}$$

Thus, the simplified momentum equation in x-direction can be written from equation (6.76) as

$$\mu \frac{\partial^2 u}{\partial y^2} = \epsilon \frac{\partial^2 \phi}{\partial y^2} E_x \tag{6.80}$$

Integrating the above equation, we have

$$\mu \frac{\partial u}{\partial y} = \epsilon \frac{\partial \phi}{\partial y} E_x + \text{Constant} \tag{6.81}$$

Here, E_x is the component of the electric field parallel to the surface in the positive x-direction.

At the edge of the diffuse layer ($y \to \lambda_D$), we have the boundary condition

$$\frac{\partial u}{\partial y} = \frac{\partial \phi}{\partial y} = 0 \tag{6.82}$$

Using above boundary condition, we get constant = 0. Integrating equation (6.81) again between the shear surface and diffuse layer at $y = \lambda_D$ with assumption of constant potential and velocity inside the Debye layer, we get

$$\begin{aligned} \mu u|_0^{\lambda_D} &= \epsilon E_x \phi|_0^{\lambda_D} \\ \mu(U - 0) &= \epsilon E_x (0 - \xi) \end{aligned} \tag{6.83}$$

We have used $\phi = \xi$ (zeta potential) and $u = 0$ at the shear surface.

Thus, we have EO velocity as

$$U_{\text{eo}} = -\frac{\epsilon \xi E_x}{\mu} \tag{6.84}$$

This equation is known as *Helmholtz–Smoluchowski* equation.

For the ideal EO flow, we obtain the simple velocity profile as

$$u \approx U_{\text{eo}} \quad \text{for} \quad \lambda_D \ll \frac{1}{2} h \tag{6.85}$$

The corresponding flow rate, so-called free EO flow rate Q_{eo}, for a channel section of the width w and height h is given by

$$Q_{\text{eo}} = \int_0^h dy \int_0^w dz\, u_x(y, z) = U_{\text{eo}} w h \quad \text{for} \quad \lambda_D \ll \frac{1}{2} h \tag{6.86}$$

The exact expression of EO flow valid for any value of λ_D can be obtained within the Debye–Hückel approximation. The next section presents the EO flow in a cylindrical channel.

6.8.2 Cylindrical Channel EO Flow

Analytical results for an ideal EO flow within the Debye–Hückel approximation can also be derived for a cylindrical channel of circular cross section with radius a. Similar to equation (6.65) for flat plate, the Poisson–Boltzmann equation, $\phi(r)$, for cylindrical geometry in the Debye–Hückel approximation is

$$\frac{\partial^2}{\partial r^2}\phi(r) + \frac{1}{r}\frac{\partial}{\partial r}\phi(r) = \frac{1}{\lambda_D^2}\phi(r) \tag{6.87}$$

with boundary condition $\phi(r = a) = \zeta$ and $\frac{\partial \phi}{\partial r}(r = 0) = 0$. This is the well-known modified Bessel equation. The equilibrium potential $\phi_{\text{eq}}(r)$ of the Debye layer in this case is given in terms of the modified Bessel function of the order 0 as

$$\phi_{\text{eq}}(r) = \zeta \frac{I_0\left(\frac{r}{\lambda_D}\right)}{I_0\left(\frac{a}{\lambda_D}\right)} \tag{6.88}$$

The fully developed velocity field $v = v_x(r)e_x$ with boundary conditions

$$\frac{\partial}{\partial r}(v_x(0)) = 0 \quad \text{(symmetry condition)}$$

$$v_x(a) = 0 \quad \text{(no-slip condition)}$$

The analysis can be carried out exactly as for the infinite parallel plate geometry (equations (6.80) and (6.81)), and we arrive at the result:

$$v_x(r) = \left[1 - \frac{I_0\left(\frac{r}{\lambda_D}\right)}{I_0\left(\frac{a}{\lambda_D}\right)}\right] U_{\text{eo}} \tag{6.89}$$

We note that $v \approx U_{\text{eo}} e_x$ for $\lambda_D \ll a$. Hence, the EO flow rate can be derived as

$$Q_{\text{eo}} = \int_0^{2\pi} d\theta \int_0^a r v_x(r, \theta) dr = U_{\text{eo}} \pi a^2, \quad \text{for} \quad \lambda_D \ll a \tag{6.90}$$

Note: (1) When the double layer thickness is very small compared to the characteristic length, say, $a/\lambda_D \gg 100$, the fluid moves as in plug flow. Thus, the velocity *slips* at the wall; that is, it goes from U_{eo} to zero discontinuously.

(2) For a finite thickness diffuse layer, the velocity drops continuously across the layer to zero at the wall and the constant EO velocity represents the velocity at the "edge" of the diffuse layer.

(3) A typical zeta potential is about 0.1 V. Thus, for $E = 10^3$ v/m, with viscosity that of water, the EO velocity $U_{\text{eo}} \sim 10^{-4}$ m/s, which is a very small value.

(4) In a capillary, the volume flow rate due to a fixed pressure gradient is proportional to the fourth power of radius $\left(Q = \frac{\pi a^4}{8\mu}\frac{dp}{dx}\right)$. We know that EO flow rate is equal to $U_{\text{eo}}\pi a^2$. Therefore, we can write

$$\frac{\text{Electroosmotic flow rate}}{\text{Hydraulic flow rate}} \propto a^{-2}$$

Thus, as the average pore size decreases, the EO becomes increasingly effective in driving the flow through the medium, compared to the pressure gradient provided $\lambda_D/a \ll 1$.

6.9 Exact Expression for Cylindrical Channel EO Flow

Here, we calculate the EO flow and potential change in a long capillary for different λ_D/a (Debye length to radius ratios). Let us assume that the mean velocity and tube radius are sufficiently small and inertia effects can be neglected. The dilute electrolyte solution in the capillary is assumed to be binary. The momentum equation is

$$\rho \frac{Du}{Dt} = -\nabla P + \mu \nabla^2 u + \rho_E E \tag{6.91}$$

The inertia force is assumed negligible, that is, $\frac{Du}{Dt} \simeq 0$.

The charge density can be written as

$$\rho_E = F\Sigma z_i C_i = F(z_+ C_+ + z_- C_-) \tag{6.92}$$

The electric field can be written as

$$E = -\nabla \phi \tag{6.93}$$

Substituting in equation (6.91), we have

$$0 = -\nabla P + \mu \nabla^2 u - F(z_+ C_+ + z_- C_-)\nabla \phi \tag{6.94}$$

Let us assume the electrolyte solution is of fully dissociated symmetrical salt with $z_+ = -z_- = z$ and the capillary is circular. We adopt cylindrical coordinate system (x, r) with x positive in the direction of flow and r the radial coordinate with origin at the axis of the symmetry. We have the momentum equation (6.94) in cylindrical coordinate as

$$\frac{\mu}{r}\frac{\partial}{\partial r}\left(r\frac{\partial u}{\partial r}\right) = \frac{\partial p}{\partial x} + Fz(C_+ - C_-)\frac{\partial \phi}{\partial x} \tag{6.95}$$

To define the interaction between the electric field and ion concentration, let us invoke the Poisson equation $\nabla^2\phi = -\frac{\rho_E}{\epsilon}$. For a capillary of length, L, larger compared with its radius, a, the term $\frac{\partial^2 \phi}{\partial x^2}$ may be neglected. Thus, using equation (6.92) in the Poisson equation, we get

$$\frac{1}{r}\frac{\partial}{\partial r}\left(r\frac{\partial \phi}{\partial r}\right) = -\frac{Fz}{\epsilon}(C_+ - C_-) \tag{6.96}$$

The above equation means that at any small segment of the capillary, the ion concentrations are in a local quasiequilibrium determined solely by the radial variation in ϕ. Because of the above behavior, it is convenient to divide the potential into two parts as

$$\phi(x, r) = \phi(x) + \psi(r) \tag{6.97}$$

Substituting the potential distribution from equation (6.97) in equation (6.96), we have

$$\frac{1}{r}\frac{\partial}{\partial r}\left(r\frac{\partial \psi}{\partial r}\right) = -\frac{Fz}{\epsilon}(C_+ - C_-) \tag{6.98}$$

We know that the Nernst–Planck equation for the molar flux of ith species is

$$j_i^* = -v_i z_i F C_i \nabla\phi - D_i \nabla C_i + C_i u$$

Since there is no radial flux or radial flow, the radial component of the above equation is

$$0 = \mp v_\pm z F C_\pm \nabla\phi - D_\pm \nabla C_\pm$$

$$\mp \frac{v_\pm z F \nabla\phi}{D_\pm} = \frac{\nabla C_\pm}{C_\pm}$$

Using $D = RTv$, we have

$$\mp \frac{zF\nabla\phi}{RT} = \frac{\nabla C_\pm}{C_\pm} \tag{6.99}$$

Integrating the above equation, we have

$$\mp \frac{zF\phi}{RT} = \ln C_\pm + \text{Constant}$$

$$C_\pm = \text{Constant}\left(\exp\left(\mp\frac{zF\phi}{RT}\right)\right) = C_0(x)\exp\left(\mp\frac{zF\psi}{RT}\right) \tag{6.100}$$

Note: This distribution (Boltzmann distribution) was used previously while deriving Debye–Hückel approximation. Note that in above, we have used $C_+^0 = C_-^0 \approx C_0$. Let us relook at the potential distribution equation derived above (equation (6.98)):

$$\begin{aligned}
\frac{1}{r}\frac{\partial}{\partial r}\left(r\frac{\partial\psi}{\partial r}\right) &= -\frac{Fz}{\epsilon}(C_+ - C_-) \\
&= -\frac{Fz}{\epsilon}C_0\left(\exp\left(-\frac{zF\psi}{RT}\right) - \exp\left(\frac{zF\psi}{RT}\right)\right) \\
&= \frac{2FzC_0}{\epsilon}\sinh\left(\frac{zF\psi}{RT}\right)
\end{aligned} \tag{6.101}$$

Using the dimensionless variables,

$$r^* = r/a,\ \lambda^* = \lambda_D/a, \quad \text{and} \quad \psi^* = \frac{zF\psi}{RT}$$

We can write the dimensionless form of equation (6.101) as

$$\lambda^{*2}\frac{1}{r^*}\frac{\partial}{\partial r^*}\left(r^*\frac{\partial\psi^*}{\partial r^*}\right) = \sinh\psi^*$$

Boundary conditions

$$\text{From symmetry:} \quad \frac{\partial\psi^*}{\partial r^*} = 0 \ \text{ at } \ r^* = 0$$

$$\text{At surface:} \quad \psi^* = \psi_w^* = \xi^* \ \text{ at } \ r^* = 1$$

Here, we assume that the solid wall is charged and its electrical nature affects the EO only through an effective surface potential, identified as the ξ-potential. The numerical solution of

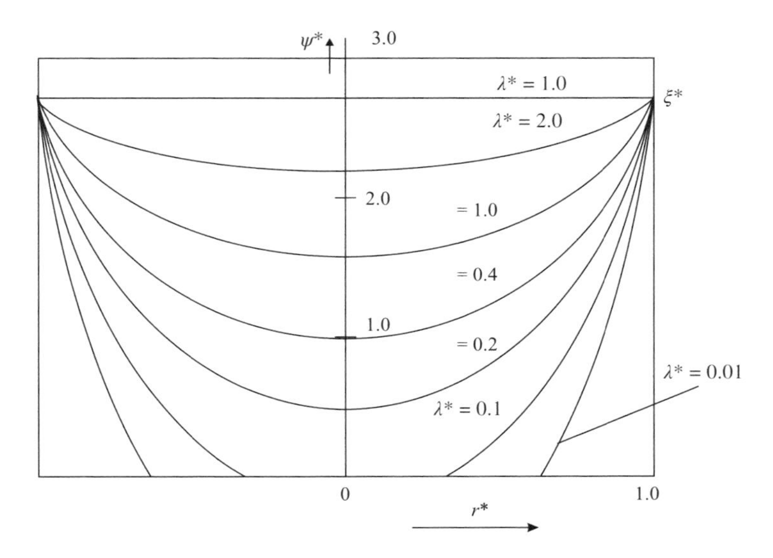

Figure 6.11 Dimensionless potential distribution across a cylindrical capillary obtained by the numerical solution of the potential distribution equation

the above equation is shown in Figure 6.11. This figure shows that the potential throughout the channel is uniform and equal to ξ-potential for small capillary diameter with respect to the size of the Debye layer, that is, for large $\lambda^\star$.

Eliminating $Fz(C_+ - C_-)$ from equation (6.98) in momentum equation (6.95), we have

$$\frac{\mu}{r}\frac{\partial}{\partial r}\left(r\frac{\partial u}{\partial r}\right) = -\frac{\epsilon}{r}\frac{\partial}{\partial r}\left(r\frac{\partial \psi}{\partial r}\right)\frac{d\phi}{dx} + \frac{dp}{dx} \tag{6.102}$$

The electrokinetic and flow boundary conditions are

$$\frac{\partial u}{\partial r} = 0, \frac{\partial \psi}{\partial r} = 0 \text{ at } r = 0$$

$$u = 0, \psi = \xi \text{ at } r = a$$

Integrating the governing equation (6.102) and using the above boundary conditions, we have

$$u(r) = -\frac{\epsilon(\psi - \xi)}{\mu}\frac{d\phi}{dx} + \frac{r^2 - a^2}{4\mu}\frac{dp}{dx} \tag{6.103}$$

The volume flow rate is obtained by integrating across the capillary cross section:

$$Q = -\int_0^a \frac{\epsilon(\psi - \xi)}{\mu}\frac{d\phi}{dx}2\pi r dr - \frac{\pi a^4}{8\mu}\frac{dp}{dx} \tag{6.104}$$

6.9.1 Small Debye Length

The distribution of ψ which was obtained numerically in Figure 6.11 can be obtained analytically for small Debye length. It should be noted from the plot of the numerical solution that

$\psi = 0$ for most of the capillary cross section in the limit of small Debye length, that is, for small $\lambda^\star$ value. Therefore, it is necessary to solve for ψ only near the pore wall, where $(r - a) \gg \lambda_D$. We can neglect the curvature effect, and the equation for ψ is exactly same as that for ϕ in 1D case:

$$\frac{\partial^2 \psi^*}{\partial y^{*2}} = \sinh \psi^* \tag{6.105}$$

Here, $y^* = y/\lambda_D$ and $y = a - r$. We write the above equation as

$$\frac{\partial^2 \psi^*}{\partial y^{*2}} = \frac{\psi^*}{\lambda_D^2} \tag{6.106}$$

With the Debye–Hückel approximation, the solution is

$$\psi = \xi e^{-(a-r)/\lambda_D} \tag{6.107}$$

Note that we have earlier derived

$$\phi = \phi_w \exp\left(\frac{-y}{\lambda_D}\right) \tag{6.108}$$

Now, the equation for ψ can be substituted to the equation of u (equation (6.103)) and Q, and we obtain

$$u = \frac{\epsilon\xi}{\mu}(1 - e^{-(a-r)/\lambda_D})\frac{d\phi}{dx} + \frac{r^2 - a^2}{4\mu}\frac{dp}{dx} \tag{6.109}$$

$$Q = \frac{\epsilon\xi}{\mu}\frac{d\phi}{dx}\pi a^2\left(1 - 2\frac{\lambda_D}{a}\right) - \frac{\pi a^4}{4\mu}\frac{dp}{dx} \tag{6.110}$$

Note: For a thin diffuse layer, that is, $a/\lambda_D \to \infty$, the velocity distribution equation reduces to

$$U_{\text{eo}} = -\frac{\epsilon\xi E}{\mu} \tag{6.111}$$

This velocity expression is identical to Helmholtz–Smoluchowski equation for zero pressure gradient derived previously.

6.9.2 *Large Debye Length*

Here, the entire capillary (pore) is within the double layer (see Figure 6.12). Therefore, $\psi = \xi$. In this case, the ion concentrations across the pore are given by the Boltzmann distribution

$$C_{\pm 2} = C_{\pm 3} = C_{\pm} = C_0 \exp\left(\mp\frac{zF\xi}{RT}\right) \tag{6.112}$$

Here, subscripts 2 and 3 correspond to the Debye layer at the entrance and exit of the cylindrical capillary. Substituting this in the momentum equation

$$\frac{\mu}{r}\frac{\partial}{\partial r}\left(r\frac{\partial u}{\partial r}\right) = \frac{\partial p}{\partial x} + Fz(C_+ - C_-)\frac{\partial \phi}{\partial x} \tag{6.113}$$

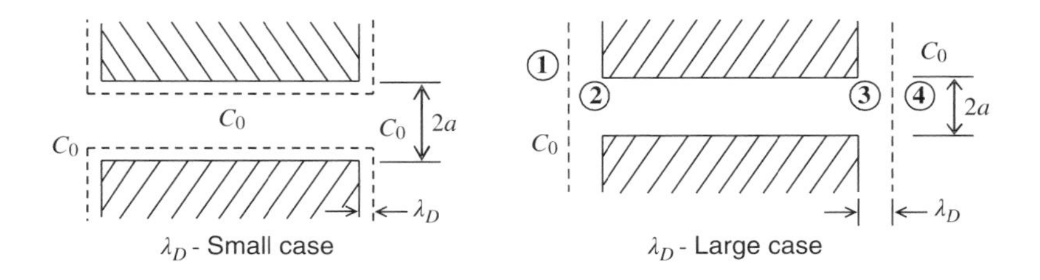

Figure 6.12 Sketch showing capillary with a different Debye length

we have

$$\frac{\mu}{r}\frac{\partial}{\partial r}\left(r\frac{\partial u}{\partial r}\right) = \frac{\partial p}{\partial x} + FzC_0 \exp\left(e^{-\frac{zF\xi}{RT}} - e^{\frac{zF\xi}{RT}}\right)\frac{\partial \phi}{\partial x}$$
$$= \frac{\partial}{\partial x}\left(P - 2FzC_0 \sinh\frac{zF\xi}{RT}\phi\right) \tag{6.114}$$

Integrating the above equation, we have

$$r\frac{\partial u}{\partial r} = \frac{\partial}{\partial x}\left(P - 2FzC_0\phi \sinh\frac{zF\xi}{RT}\right)\frac{r^2}{2\mu} + C_1 \tag{6.115}$$

Using symmetry condition $\frac{\partial u}{\partial r} = 0$ at $r = 0$, we have $C_1 = 0$. Integrating again, we have

$$u = \frac{\partial}{\partial x}\left(P - 2FzC_0\phi \sinh\frac{zF\xi}{RT}\right)\frac{r^2}{4\mu} + C_1 \tag{6.116}$$

Using B.C. $u = 0$ at $r = a$, we can obtain

$$u = -\frac{a^2 - r^2}{4\mu}\frac{d}{dx}\left(P - 2FzC_0\phi \sinh\frac{zF\xi}{RT}\right) \tag{6.117}$$

Integrating the velocity profile, u, for flow rate, Q, we can obtain

$$Q = -\frac{\pi a^4}{8\mu}\frac{d}{dx}\left(P - 2FzC_0\phi \sinh\frac{zF\xi}{RT}\right) \tag{6.118}$$

Note: Equation (6.118) shows that the ratio of EO flow to hydraulic flow is independent of the capillary radius. Hence, in terms of flow rates achievable, there is no particular advantage in using an electric field rather than a pressure gradient for large Debye length case.

6.9.3 Debye Layer Overlap

In this section, we discuss what happens if the Debye length λ_D becomes comparable with the transverse length scale a of the channel and the Debye layer from various parts of the wall overlaps at the center of the channel. For a standard electrolyte with concentration about

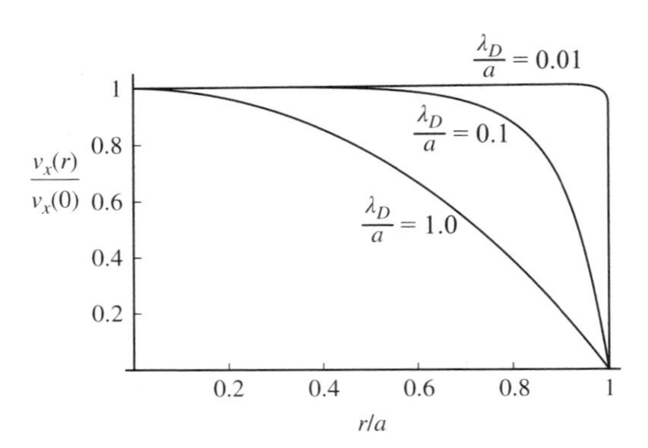

Figure 6.13 The normalized EO flow profile $v_x(r)/v_x(0)$ for a cylindrical channel of radius a, with three different values of the Debye length

1 mol/m^3 in water, this will happen for a cylindrical channel with radius a of the order 10 nm. With modern nanotechnology, it is in fact possible to make such channels intentionally. Such dimensions actually occur in nature for some porous materials.

Figure 6.13 shows three normalized EO flow profiles obtained by plotting $v_x(r)/v_{eo}$ of a cylindrical capillary with the value $\lambda_D/a = 0.01, 0.1$, and 1. It is seen that for small values of λ_D/a, a flat nearly constant velocity profile is obtained. As λ_D/a increases to 0.1, a rounded profile with nearly flat profile is observed at the center of the channel. When λ_D becomes comparable to the radius of the capillary, the profile changes to a paraboloid shape. The EO flow profile gets heavily suppressed as λ_D is increased beyond a.

In conclusion, when the Debye screening length is large compared to the transverse dimension of the channel, the screening of the charges on the wall becomes incomplete, and the electrical potential does not vary much across the channel. In this case, the Debye layers reaching out from the wall overlap in the center preventing the maintenance of a charge neutral bulk liquid. Together with the N–S equation, this further implies that the velocity profile likewise does not vary much, and as a result, the no-slip boundary condition is felt strongly even at the center of the channel. The EO flow is therefore strongly suppressed in the limit of Debye layer overlap, that is, when $\lambda_D \geq a$.

6.10 EO Pump

In the earlier section, we have seen how a nonequilibrium EO flow can be generated by applying an electrical potential difference, $\Delta\phi$, along a channel where an equilibrium Debye layer exists. For an ideal EO flow without any external pressure gradient, the expression for flow rate, Q_{eo}, has been derived. In this section, we study the flow rate variation as a function of externally applied pressure difference, Δp. This is analogous to studying the capability of an EO microchannel as a micropump.

The schematic of the flow condition is sketched in Figure 6.14. A cylindrical channel of radius, a, and length, L, is oriented along the x-axis between $x = 0$ and $x = L$. The walls and the Debye layer are negatively and positively charged, respectively. The external electrical

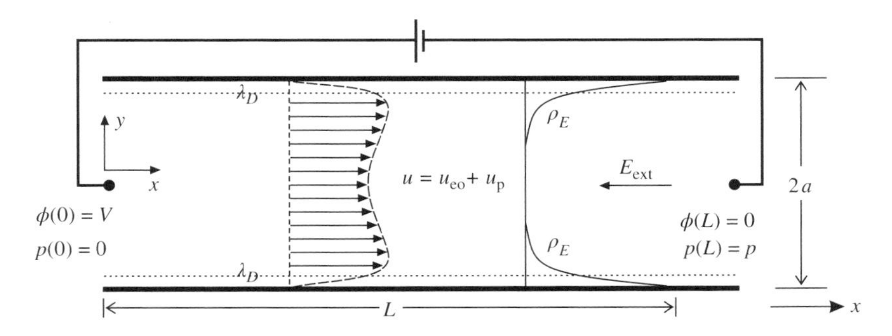

Figure 6.14 The velocity profile u and the Debye layer charge density profile ρ_E in an ideal electroosmotic (EO) flow with back pressure, p, inside a cylindrical channel of radius, a. The EO flow is induced by the external potential difference, $\Delta\phi = \Delta V$. It may be noted that the flat EO flow profile u_{eo} in Figure 6.10 now has a parabolic dent due to the superimposed back-pressure-driven Poiseuille flow profile u_p

potential V and pressure p lead to the following boundary condition:

$$\phi(x=0) = V, \quad \phi(x=L) = 0 \tag{6.119a}$$

$$p(x=0) = 0, \quad p(x=L) = p \tag{6.119b}$$

The N–S equation is

$$\rho\frac{D\vec{V}}{Dt} = \mu\nabla^2\vec{V} + \rho_E\vec{E} - \vec{\nabla}p \tag{6.120}$$

The x-component of the N–S equation can be written as

$$0 = \mu\nabla^2 u(r) + [\epsilon\nabla^2\phi(r)]\frac{V}{L} - \frac{p}{L} \tag{6.121}$$

The boundary conditions for the velocity profile are

$$u(a) = 0 \quad \text{(no-slip condition)} \tag{6.122}$$

$$\frac{\partial}{\partial r}u(0) = 0 \quad \text{(symmetry condition)} \tag{6.123}$$

The equation can also be solved by superimposing an EO flow $u_{eo}(r)$ and a standard Poiseuille flow $u_p(r)$, with opposite sign for Δp. Note that the superposition procedure works because of the linearity of the N–S equation due to negligible inertial term $(\vec{V}\cdot\vec{\nabla})\vec{V}$:

$$u(r) = u_p(r) + u_{eo}(r) \tag{6.124}$$

The governing equation for the EO flow case is

$$0 = \mu\nabla^2 u_{eo}(r) - \rho_E(r)\frac{V}{L} \tag{6.125}$$

The boundary conditions are

$$u_{eo}(a) = 0 \qquad \frac{\partial}{\partial r}u_{eo}(0) = 0$$

The governing equation for the pressure gradient-driven flow is

$$0 = \mu \nabla^2 u_p(r) - \frac{p}{L}$$

The boundary conditions are

$$u_p(a) = 0 \qquad \frac{\partial}{\partial r} u_p(0) = 0$$

Hence, for an ideal EO flow, that is, in the limit $\lambda_D \ll a$ with imposed pressure gradient, the flow rate is

$$Q = Q_{\text{eo}} + Q_p = \pi a^2 U_{\text{eo}} - \frac{\Delta p}{R_{\text{hyd}}} = \frac{\pi a^2 \epsilon \zeta}{\mu L} V - \frac{\pi a^4}{8 \mu L} p \tag{6.126}$$

Figure 6.15(a) shows the linear flow rate–pressure characteristic (Q–p) on the basis of equation (6.126). Two points on the Q–p graph characterize the capability of the EO microchannel to work as a micropump. One is the free-flow capability (Q_{eo}) obtained at zero back pressure, $P_{\text{ext}} = 0$. The other is the zero-flow pressure capability, or EO pressure, (p_{eo}), defined as the back pressure needed to exactly cancel the EO flow. At free flow, $p = 0$ and we have

$$Q_{\text{eo}} = \frac{\pi a^2 \epsilon \zeta V}{\mu L} \tag{6.127}$$

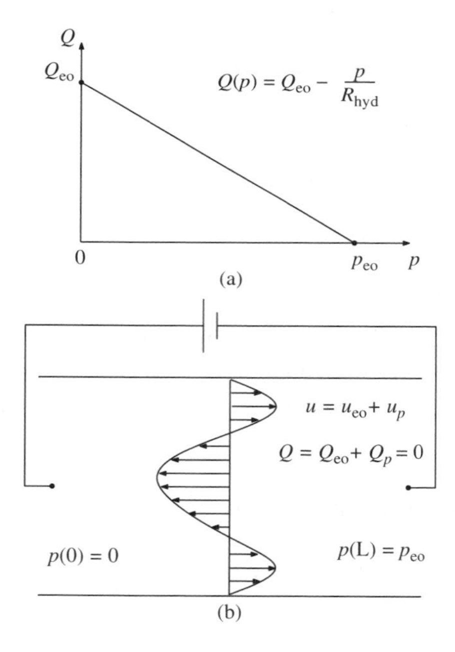

Figure 6.15 (a) The flow rate–pressure characteristic for an ideal EO flow with back pressure p. (b) The flow profile u in a cylindrical microchannel at the electroosmotic pressure P_{eo}, where the net flow rate is zero ($Q = 0$)

At zero flow, $Q = 0$ and we have

$$p_{eo} = \frac{8\epsilon\zeta V}{a^2} \tag{6.128}$$

The flow profile when running the EO micropump at zero flow is sketched in Figure 6.15(b).

Typical values for Q_{eo} and p_{eo} for $\zeta = 0.1$ V, $a = 10$ µm, $L = 100$ µm, $\mu = 1$ mPa-s, and $\epsilon = 78\epsilon_0$ are

$$\frac{Q_{eo}}{V} = 0.21 \text{ nL/s-V}$$
$$\frac{p_{eo}}{\Delta V} = 5.52 \text{ Pa/V} \tag{6.129}$$

These representative flow rates are not very high. Equation (6.127) shows that high flow rate in a single-channel EO micropump is obtained with a high voltage drop ΔV, a high zeta potential ζ, a large radius a, and a short channel length L. However, equation (6.128) reveals that the pressure capability of a single channel scales as a^{-2}. If a is chosen to be large to secure a decent flow rate, then the pressure capability is low and vice versa. A single EO channel can therefore mainly be used in the free-flow case, where the flat EO flow profile can be utilized to move the concentration profiles along a channel. More elaborate designs are needed if one is to construct a micropump based on EO flow. This is discussed in the following sections.

6.10.1 *Many-Channel EO Pump*

An EO-based micropump with decent flow rate and pressure capabilities can be constructed by using a large number of narrow channels in a parallel coupling. Let us consider a large number N of identical cylindrical channels of radius a, length L, and zeta potential ζ. All these channels experience the same external pressure P_{ext} when coupled in parallel. Here, the pressure capability $p_{eo,N}$ for the ensemble of channels is the same as for each individual channel, while the flow rate capability $Q_{eo,N}$ scales with the number of channels. At free flow, that is, $P = 0$, the total flow rate can be expressed as

$$Q_{eo,N} = NQ_{eo} = N\frac{\pi a^2 \epsilon\zeta V}{\mu L} \tag{6.130}$$

At zero total flow rate, the pressure developed can be expressed as

$$p_{eo,N} = p_{eo} = \frac{8\epsilon\zeta V}{a^2} \tag{6.131}$$

The above two equations indicate that the EO pressure $P_{eo,N}$ depends on the radius a, and the EO flow rate $Q_{eo,N}$ depends on the total open area $\mathcal{A}_{open} = N\pi a^2$. The area is of the same order of magnitude at the total cross-sectional area $\mathcal{A}_{tot}$ occupied with the channels including walls and voids between the channels.

In principle, it is possible to fabricate a many-channel EO pump by using microtechnology. However, it is not easy to fabricate sufficiently high number of parallel channels. There is a way around this problem by using the so-called frits. A frit consists of closely packed, sintered glass spheres with a diameter of the order 1 µm, which makes it a porous material due to the

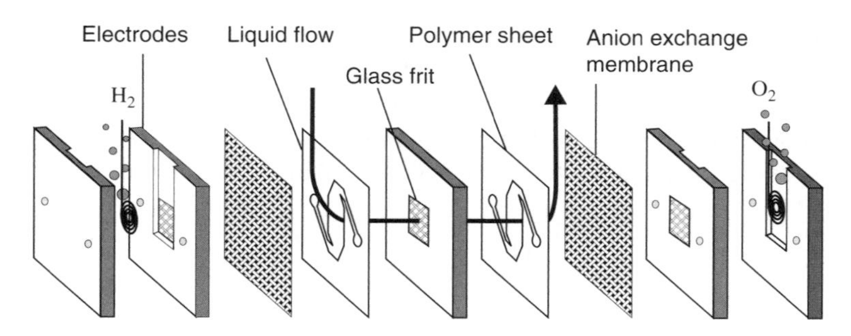

Figure 6.16 A frit-based electoosmotic micropump, where the glass frit (gray hatched square) is situated in the central layer between polymer sheets. The platinum electrodes, where gas bubbles are generated by electrolysis, are separated from the liquid flow by anion exchange membranes, which only allow the passage of OH^- ions

voids between the spheres. Frits are produced commercially and widely used as filters. The commercially available cylindrical frits have radius equal to 1.8 mm and length about 2 mm.

A frit can also be used as a many-channel EO pump. Instead of a regular array of cylindrical channels, the frit contains a high number of channels of irregular shape and size forming a percolation pattern through the frit. Despite the irregularities of the frit channels, they can still serve as EO flow channels.

Figure 6.16 shows a frit-based EO micropump. To stabilize the pump, the electrodes have been separated from the liquid flow by anion exchange membranes. These membranes only allow the passage of OH^- ions and thus prevent the electrolytic gases generated at the electrodes to interfere with the liquid flow. The pump can run steadily for hours and achieve $Q_{eo} \simeq 0.8$ μl/s and $P_{eo} \simeq 4$ kPa at voltage, $\Delta V = 30$ V.

6.10.2 Cascade EO Pump

The many-channel EO pump discussed in the previous section is capable of obtaining high flow rate. However, it has the drawback of low-pressure head. With a simple EO pump, this can be achieved by applying high voltages of the order of 1 kV. Using high voltage is doable but not very practical. A solution to this problem is a so-called cascade EO pump, which consists of several EO pump stages in series each with a zero voltage drop. A simple example of this principle is shown in Figure 6.17(a). It consists of several narrow channels in parallel connected to a wider channel. The narrow channels experience a potential gradient between the anode and cathode, while the wider channels experience potential gradient between the cathode and anode.

For a simple analogy of cascade EO pump, let us discuss the flow inside two cylindrical channels under zero potential difference. Consider two cylindrical channels of radius a_l and a_2, length L, and zeta potential ζ (Figure 6.17(b)). These two channels are joined together in series to form one long channel of length $2L$. All effects induced by the transition from the wide channel to the narrow channel may be neglected. The applied voltage $\phi_{ext}(x)$ is set from

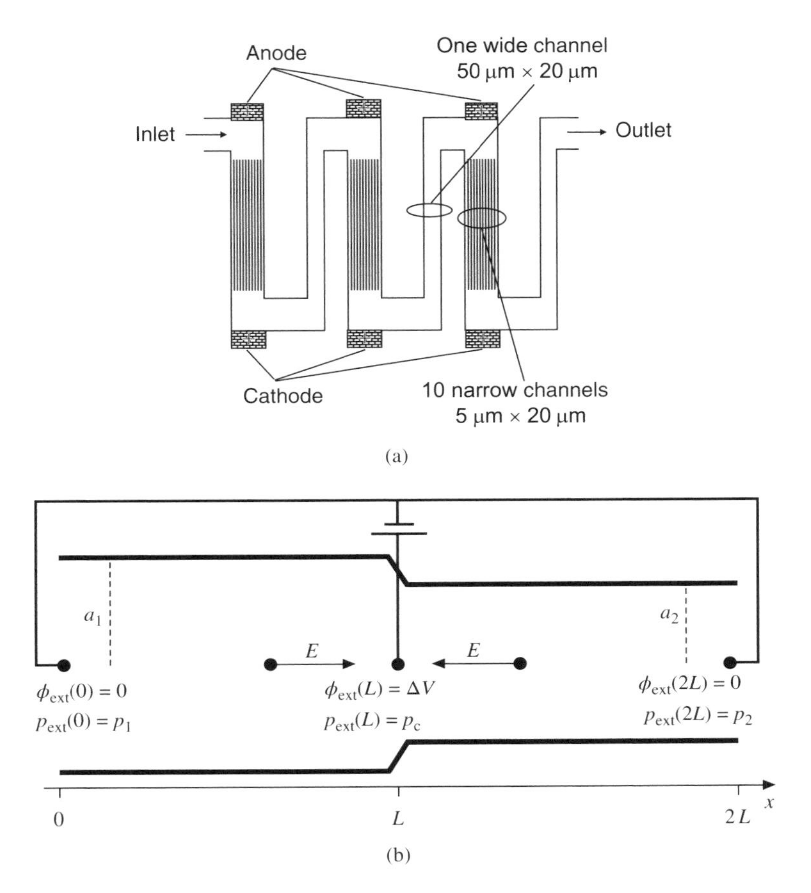

Figure 6.17 (a) The cascade EO pump with three identical stages. Each stage consists of 10 narrow channels in series with one wide channel, and the total voltage drop per stage is zero. (b) The schematic of a single stage of the cascade EO pump, which yields a finite EO flow with the total voltage drop along the channel equal to zero

0 at $x = 0$ to ΔV at $x = L$ and back to zero at $x = 2L$. The pressure setting is $p_{\text{ext}}(0) = p_1$, $p_{\text{ext}}(L) = p_c$, and $p_{\text{ext}}(2L) = p_2$.

Let us define size ratio, $\alpha = a_1/a_2$. The expression for hydraulic resistance and EO flow rates for two channels are

$$R_{\text{hyd},2} = \frac{8\mu L}{\pi a_2^4}, \qquad Q_{\text{eo},2} = \frac{\pi a_2^2 \epsilon \zeta}{\mu L} \Delta V \tag{6.132}$$

$$Q_{\text{eo},1} = \alpha^2 Q_{\text{eo},2}, \qquad R_{\text{hyd},1} = \alpha^{-4} R_{\text{hyd},2} \tag{6.133}$$

Note that the sign for $Q_{\text{eo},1}$ will be opposite to that of $Q_{\text{eo},2}$ due to the opposite direction of the voltage drop. The flow rate Q in the two sections must be identical due to mass conservation, and we can write the flow rate expression in each section, respectively, as

$$\text{Section 1}: \quad Q = Q_{\mathrm{eo},1} + \frac{p_1 - p_c}{R_{\mathrm{hyd},1}} \quad = \alpha^2 Q_{\mathrm{eo},2} + \alpha^4 \frac{p_1 - p_c}{R_{\mathrm{hyd},2}} \tag{6.134a}$$

$$\text{Section 2}: \quad Q = -Q_{\mathrm{eo},2} + \frac{p_c - p_2}{R_{\mathrm{hyd},2}} = -Q_{\mathrm{eo},2} + \frac{p_c - p_2}{R_{\mathrm{hyd},2}} \tag{6.134b}$$

Solving these equations and for $p_1 = 0$, we have

$$p_c = \frac{1+\alpha^2}{1+\alpha^4} R_{\mathrm{hyd},2} Q_{\mathrm{eo},2} + \frac{1}{1+\alpha^4} p_2 \tag{6.135a}$$

$$Q = \frac{\alpha^2 - \alpha^4}{1+\alpha^4} Q_{\mathrm{eo},2} - \frac{\alpha^4}{1+\alpha^4} \frac{p_2}{R_{\mathrm{hyd},2}} \tag{6.135b}$$

From equation (6.135b), we can derive the zero-flow ($Q = 0$) pressure capability, p_{eo}, and zero-pressure ($p_2 = 0$) free flow rate, Q_{eo}, as

$$p_{\mathrm{eo}} = \left(\frac{1}{\alpha^2} - 1\right) R_{\mathrm{hyd},2} Q_{\mathrm{eo},2} \tag{6.136a}$$

$$Q_{\mathrm{eo}} = \frac{\alpha^2 - \alpha^4}{1+\alpha^4} Q_{\mathrm{eo},2} \tag{6.136b}$$

These results demonstrate that despite the zero total voltage drop along the single-stage EO channel, the pump design shown in Figure 6.17(b) functions as an EO pump. The larger size difference between the two parts of the single-stage channel, that is, higher size ratio, α compared to unity, leads to larger EO effect.

In the special case, when $\alpha = 1$, the two parts of the single-stage EO pump are identical, and the pressure capability is zero simply because the EO flows in the two parts of the single-stage channel are equal but opposite in direction. In the limit $\alpha \gg 1$, the narrow part 2 dominates the pump characteristic with a reversal in flow direction due to the reverse voltage drop. In the opposite limit, $\alpha \ll 1$, the narrow part 1 dominates, but now without a reversal of flow direction. Thus, a zero-voltage EO pump stage can be constructed by letting channel 1 to be a many-channel EO pump and channel 2 a single-channel EO pump, as shown in Figure 6.17(a).

Let us take $N \times N$ narrow channels of radius, $a_1 = a_2/N$. Thus, we have

$$\alpha = \frac{a_1}{a_2} = \frac{1}{N} \tag{6.137}$$

With this choice of geometry, the two parts of the EO pump stage have the same area relationship available for flow, $\mathcal{A} = N^2 \pi a_1^2 = \pi a_2^2$. The hydraulic resistances and EO flow rates for small channel with respect to large channel can be written using equation (6.133) as

$$Q_{\mathrm{eo},1} = \frac{1}{N^2} Q_{\mathrm{eo},2}, \qquad R_{\mathrm{hyd},1} = N^4 R_{\mathrm{hyd},2} \tag{6.138}$$

The EO flow rate and hydraulic resistance for N^2 parallel channels in section 1 can be written as

$$Q_{\mathrm{eo},N} = Q_{\mathrm{eo},2}, \quad R_{\mathrm{hyd},N} = \frac{R_{\mathrm{hyd},1}}{N^2} = N^2 R_{\mathrm{hyd},2} \tag{6.139}$$

where subscript N refers to the ensemble of N^2 parallel channels in section 1, while a single channel here carries the subscript 1. The flow rate equation is written using the mass continuity, and the hydraulic resistance equation is written using the resistance in parallel analogy. The equation for flow rate in the two sections can be written on the basis of equations (6.134a) and (6.134b) as

$$\text{Section 1}: \quad Q = Q_{\text{eo},N} + \frac{0 - p_c}{R_{\text{hyd},N}} = Q_{\text{eo},2} - \frac{1}{N^2}\,\frac{p_c}{R_{\text{hyd},2}} \tag{6.140a}$$

$$\text{Section 2}: \quad Q = -Q_{\text{eo},2} + \frac{p_c - P_2}{R_{\text{hyd},2}} = -Q_{\text{eo},2} + \frac{p_c - p_2}{R_{\text{hyd},2}} \tag{6.140b}$$

In analogy with our previous derivation (see equations (6.135a) and (6.135b)), we can derive

$$Q = \frac{N^2 - 1}{N^2 + 1} Q_{\text{eo},2} - \frac{1}{N^2 + 1} \frac{p_2}{R_{\text{hyd},2}} \tag{6.141}$$

Thus, we find that the flow rate Q depends linearly on $Q_{\text{eo},2}$ the back pressure, p_2.

Using equation (6.133) and $\alpha = \frac{1}{N}$, the zero-flow pressure capability p_{eo} and the zero-pressure free flow rate Q_{eo} become (see equations (6.136a) and (6.136b))

$$p_{\text{eo}} = (N^2 - 1) R_{\text{hyd},2} Q_{\text{eo},2} = \left(1 - \frac{1}{N^2}\right) R_{\text{hyd},1} Q_{\text{eo},1} \tag{6.142a}$$

$$Q_{\text{eo}} = \frac{N^2 - 1}{N^2 + 1} Q_{\text{eo},2} = \frac{1 - \frac{1}{N^2}}{1 + \frac{1}{N^2}} Q_{\text{eo},2} \tag{6.142b}$$

It is thus seen that as the number of subchannels in section 1 grows, the pressure capability of the zero-voltage EO pump stage approaches that of a single narrow subchannel, while the flow rate capability approaches that of the wide channel.

Regardless of how a single zero-voltage EO pump stage with a pressure capability of p_2 is realized, it is possible to join several such stages in series leading to a cascade EO pump, where the pressure capability is augmented by p_2 for each stage added, while the flow rate capability remains unchanged. Hence, very efficient EO pump can be fabricated with the same low voltage applied across the center of each zero-voltage stage.

6.11 EO Flow in Parallel Plate Channel

This section reports the velocity distribution of EO flow inside parallel plate geometry. Figure 6.18 shows the parallel plate channel geometry undergoing EO flow through it. Let us find out the velocity distribution as a function of double layer thickness (λ_D). Let us assume that the flow field is inertia free flow $\left(\rho \frac{Du}{Dt} = 0\right)$ with no pressure gradient. The Debye layer thickness (λ_D) is very small compared to the channel dimension, and the velocity gradient in the x-direction can be neglected.

Therefore, the x-momentum equation is

$$\mu \frac{d^2 u}{dy^2} + \rho_E E = 0 \tag{6.143}$$

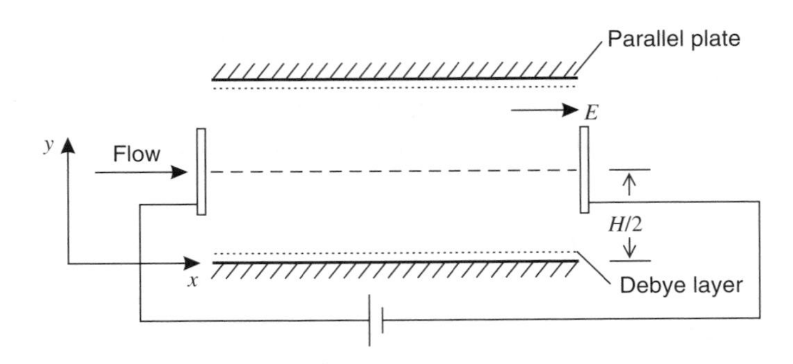

Figure 6.18 Schematic for electroosmotic flow inside parallel plate channel geometry

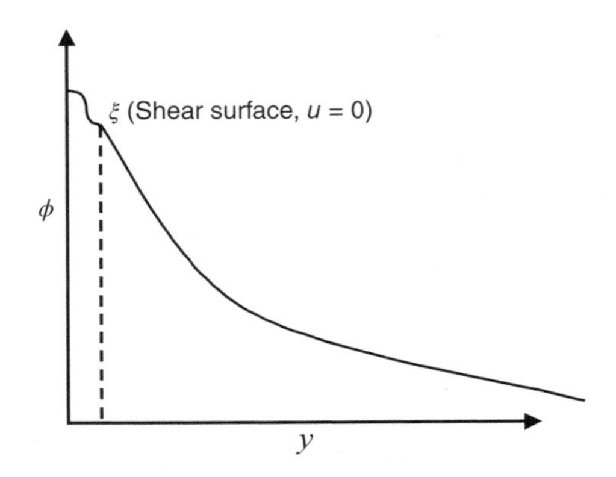

Figure 6.19 Distribution of electric potential inside the double layer

Using the Poisson equation, $\nabla^2\phi = -\frac{\rho_E}{\epsilon}$, that is, $\rho_E = -\epsilon\frac{d^2\phi}{dy^2}$, the above equation can be written as

$$\mu\frac{d^2u}{dy^2} - \epsilon\frac{d^2\phi}{dy^2}E = 0 \tag{6.144}$$

In the earlier section, we have assumed constant potential distribution of electric potential inside the electric double layer for the estimation of EO velocity. Let us use the Debye–Hückel approximation for potential distribution as

$$\phi(y) = \phi_w \exp(-y/\lambda_D) \tag{6.145}$$

Assuming the wall potential distribution, ϕ_w, is equal to the potential (ζ) at the shear surface (see Figure 6.19), we have

$$\phi(y) = \zeta \exp(-y/\lambda_D) \tag{6.146}$$

Thus, we can write from the above equation

$$\frac{d^2\phi}{dy^2} = \frac{\zeta}{\lambda_D^2}\exp(-y/\lambda_D) \tag{6.147}$$

Substituting equation (6.147) in the momentum equation (6.144), we have

$$\mu\frac{d^2u}{dy^2} - \frac{\epsilon E\zeta}{\lambda_D^2}\exp(-y/\lambda_D) = 0 \tag{6.148}$$

$$\lambda_D^2\frac{d^2u}{dy^2} + \left(-\frac{\epsilon E\zeta}{\mu}\right)\exp(-y/\lambda_D) = 0 \tag{6.149}$$

$$\frac{d^2u}{dy^2} = -\frac{U_{eo}}{\lambda_D^2}\exp(-y/\lambda_D) \tag{6.150}$$

where we have used EO velocity, $U_{eo} = -\frac{\epsilon E\zeta}{\mu}$.

Integrating equation (6.150), we get

$$\frac{du}{dy} = \frac{U_{eo}}{\lambda_D}\exp(-y/\lambda_D) + C_1$$

$$u = -U_{eo}\exp(-y/\lambda_D) + C_1 y + C_2 \tag{6.151}$$

Boundary conditions

B.C. 1: $u = 0$ at $y = 0$ (no-slip condition)

B.C. 2: $\frac{\partial u}{\partial y} = 0$ at $y = H/2$ (symmetry condition)

Using B.C. 1: $C_2 = U_{eo}$

Using B.C. 2: $0 = \frac{U_{eo}}{\lambda_D}\exp\left(-\frac{H}{2\lambda_D}\right) + C_1$

$$C_1 = -\frac{U_{eo}}{\lambda_D}\exp\left(-\frac{H}{2\lambda_D}\right) \tag{6.152}$$

Hence, the velocity profile inside the parallel plate geometry is

$$u = U_{eo} - \frac{U_{eo}}{\lambda_D}y\exp\left(-\frac{H}{2\lambda_D}\right) - U_{eo}\exp\left(-\frac{y}{\lambda_D}\right) \tag{6.153}$$

The dimensionless velocity profile can be written as

$$\frac{u}{U_{eo}} = 1 - \frac{y}{\lambda_D}\exp\left(-\frac{H}{2\lambda_D}\right) - \exp\left(-\frac{y}{\lambda_D}\right) \tag{6.154}$$

Figure 6.20 shows a representative velocity distribution inside the channel as a function of Debye number. It may be noted that the velocity gradient near the channel wall increases with increase in Debye number ($H/2\lambda_D$) and the velocity profile is similar to plug flow at high Debye number ($De = 300$).

The EO flow rate per unit width can be obtained by integrating the velocity profile as

$$\begin{aligned} Q_{eo} &= \int_0^H u dy \\ &= U_{eo}\int_0^H dy - \frac{U_{eo}}{\lambda_D}\exp\left(-\frac{H}{2\lambda_D}\right)\int_0^H y dy - U_{eo}\int_0^H \exp\left(-\frac{y}{\lambda_D}\right)dy \\ &= U_{eo}H - \frac{U_{eo}H^2}{2\lambda_D}\exp\left(-\frac{H}{2\lambda_D}\right) + U_{eo}\lambda_D\exp\left(-\frac{H}{\lambda_D}\right) - U_{eo}\lambda_D \end{aligned} \tag{6.155}$$

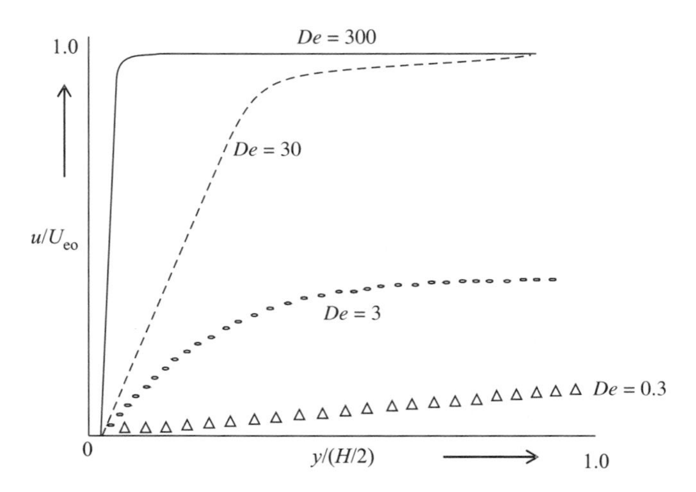

Figure 6.20 Velocity distribution as a function of Debye number ($De = H/2\lambda_D$) inside parallel plate channel geometry

Similar to equation (6.126), we can write the total flow rate under the combined influence of electric field and pressure gradient using

$$Q = Q_{\text{eo}} + Q_p = Q_{\text{eo}} - \frac{\Delta P}{R_{\text{hyd}}} \tag{6.156}$$

where Q_{eo} can be calculated using equation (6.155) and R_{hyd} is the hydraulic resistance of the parallel plate channel.

6.12 Electroosmosis and Forced Convection

When a liquid is forced through a microchannel under hydrostatic pressure, the ions in the mobile region of the electrodischarge layer (EDL) are carried toward one end. The motion of the ions in the diffuse mobile layer affects the bulk of the liquid flow via momentum transfer due to viscosity. In macroscale flows, the interfacial electrokinetic effects are negligible since the thickness of EDL is very small compared to the hydraulic diameter of the duct. However, in microscale flow, the EDL thickness is comparable to the hydraulic diameter, and the EDL effects must be considered during the analysis of fluid flow and heat transfer. The EDL effect on the flow depends on the *Debye number* defined as $De = \frac{L}{\lambda_D} = \frac{\text{Flow scale}}{\text{Debye length}}$. When $De \gg 1$, EDL effect can be neglected. When $De \sim 1$, EDL effect is significant. We can write the momentum equation for this case as

$$\mu \frac{d^2u}{dy^2} - \frac{dp}{dx} + \rho_E E = 0 \tag{6.157}$$

Using the Poisson equation,

$$\nabla^2 \phi = -\frac{\rho_E}{\epsilon} \tag{6.158}$$

where ρ_E is the electric charge density and ϵ is the permittivity of the medium.

Hence, the momentum equation (6.157) can be written as

$$\mu \frac{d^2u}{dy^2} - \frac{dp}{dx} - \epsilon \frac{d^2\phi}{dy^2} E = 0 \tag{6.159}$$

Integrating the above equation, we get

$$\mu \frac{du}{dy} - \frac{dp}{dx} y - \epsilon E \frac{d\phi}{dy} = C_1 \tag{6.160}$$

The boundary condition at the center of the microchannel is

$$\frac{du}{dy} = \frac{d\phi}{dy} = 0 \quad \text{at} \quad y = 0.$$

Substituting the above boundary condition in equation (6.160), we get $C_1 = 0$. Integrating equation (6.160) again, we get

$$\mu u - \frac{dp}{dx}\frac{y^2}{2} - \epsilon E \phi = C_2 \tag{6.161}$$

Using boundary condition at the surface of the microchannel (see Figure 6.21)

$$u = 0 \quad \text{and} \quad \phi = \zeta \quad \text{at} \quad y = \frac{H}{2}$$

we get

$$-\frac{dp}{dx}\frac{H^2}{8} - \epsilon E \zeta = C_2 \tag{6.162}$$

Substituting the value of "C_2" in equation (6.161), we have

$$-\mu u - \frac{dp}{dx}\frac{y^2}{2} - \epsilon E \phi = -\frac{dp}{dx}\frac{H^2}{8} - \epsilon E \zeta$$

$$\Rightarrow \mu u = \frac{dp}{dx}\frac{H^2}{8} - \frac{dp}{dx}\frac{y^2}{2} + \epsilon E \zeta - \epsilon E \phi$$

$$\Rightarrow u = \frac{1}{\mu}\frac{dp}{dx}\frac{H^2}{8}\left[1 - \left(\frac{y}{H/2}\right)^2\right] + \frac{\epsilon E \zeta}{\mu}\left[1 - \frac{\phi}{\zeta}\right] \tag{6.163}$$

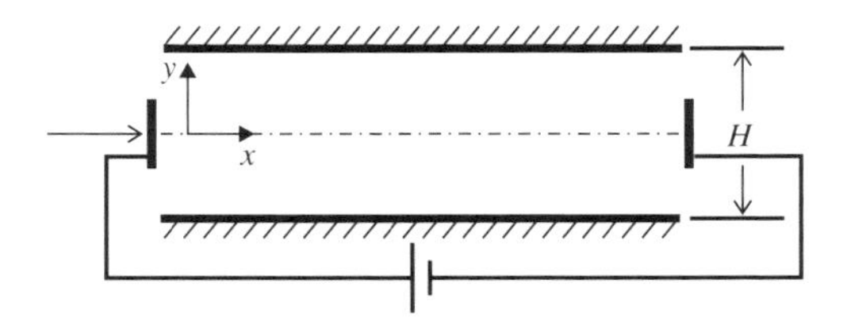

Figure 6.21 Fully developed liquid flow forced with convection in microchannel

Equation (6.163) gives the velocity distribution assuming constant potential inside the Debye layer. Let us derive the velocity profile assuming the Debye–Hückel linear approximation as

$$\frac{d^2\phi}{dy^2} = \frac{\phi}{\lambda_D^2}$$

Boundary condition

$$\phi = 0, \text{ at } y = 0$$

$$\phi = \zeta, \text{ at } y = \pm H/2$$

Let one of the solutions of the above governing equation be

$$\phi = e^{\lambda y} \quad \text{Hence} \quad \frac{d^2\phi}{dy^2} = \lambda^2 e^{\lambda y} \tag{6.164}$$

By substitution, we get

$$e^{\lambda y}\left(\lambda^2 - \frac{1}{\lambda_D^2}\right) = 0 \tag{6.165}$$

One of the possible solutions is

$$\lambda = \pm\frac{1}{\lambda_D} \tag{6.166}$$

Hence, the solution of the governing equation is

$$\phi = C_1 e^{y/\lambda_D} + C_2 e^{-y/\lambda_D} \tag{6.167}$$

The boundary conditions are

$$\phi = 0 \quad \text{at} \quad y = 0.$$

Hence, $C_1 = -C_2$

$$\phi = \zeta \quad \text{at} \quad y = \pm H/2,$$

Hence, $\zeta = C_1 e^{H/2\lambda_D} - C_1 e^{-H/2\lambda_D}$

$$\zeta = -C_1 2\sinh(H/2\lambda_D)$$

Hence, $C_1 = -\zeta(2\sinh(H/2\lambda_D))$.

Thus, we have

$$\phi = \frac{-\zeta}{2\sinh(H/2\lambda_D)}(e^{y/\lambda_D} - e^{-y/\lambda_D})$$

Hence,

$$\phi = \frac{\zeta \sinh(y/\lambda_D)}{\sinh(H/2\lambda_D)} \tag{6.168}$$

As $\phi = \zeta$ at $y = \pm H/2$, we can write

$$\phi = \frac{\zeta|\sinh(y/\lambda_D)|}{\sinh(H/2\lambda_D)}$$

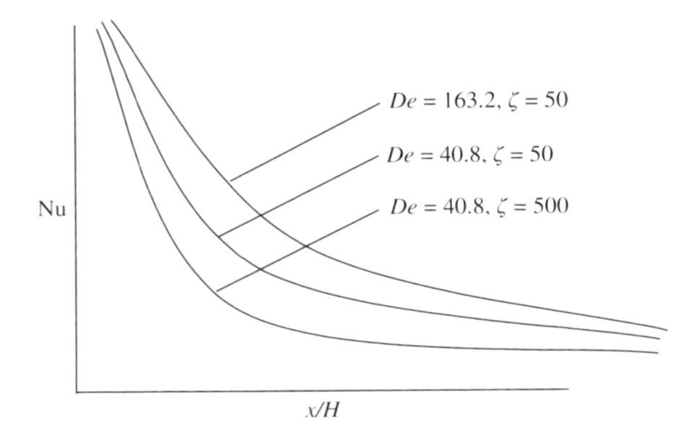

Figure 6.22 Microchannel heat transfer with EDL effect

Substituting the above expression for ϕ in equation (6.163), we have

$$u = \frac{1}{\mu}\frac{dp}{dx}\frac{H^2}{8}\left[1-\left(\frac{y}{H/2}\right)^2\right] + \frac{\epsilon E\zeta}{\mu}\left[1-\frac{|\sinh(y/\lambda_D)|}{\sinh De}\right] \tag{6.169}$$

where

$$\text{Debye number,} \quad De = H/2\lambda_D$$

The energy equation can be written as

$$\rho C_p\left(u\frac{\partial T}{\partial x}\right) = k\left[\frac{\partial^2 T}{\partial y^2}+\frac{\partial^2 T}{\partial x^2}\right] + \mu\left(\frac{\partial u}{\partial y}\right)^2 \tag{6.170}$$

Using the velocity field from equation (6.169), the energy equation (6.170) can be solved numerically for constant wall temperature boundary condition with given inlet liquid temperature. The representative Nusselt number distribution is shown in Figure 6.22.

Note: (1) Figure 6.22 shows that as ζ increases, Nu decreases.

(2) As Debye number (De) decreases, Nu decreases.

Hence, EDL results in a reduced flow velocity (higher apparent viscosity), thus affecting the temperature distribution and leading to smaller heat transfer rate.

6.13 Electrophoresis

Let us study how an applied electrical field E influences a spherical particle of charge Z_e and radius a in a stationary liquid of low electrical conductivity, say, deionized water (see Figure 6.23). This is a simple case of electrophoresis. The low conductivity of the liquid implies the lack of ions that otherwise would have accumulated around the charged particle and partly neutralized its charge, an effect known as *electrical screening*. Therefore, the electric force is

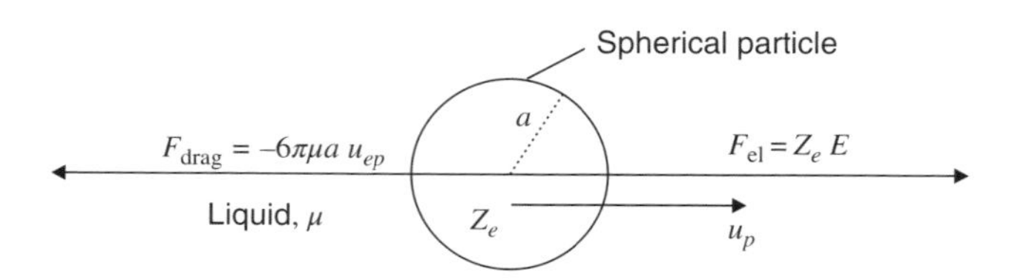

Figure 6.23 The principle of electrophoresis. A spherical particle of charge Z_e and radius a moves in a low-conductivity liquid with viscosity μ under the influence of an applied electrical field E

simply $F_{el} = Z_e E$. The charged particle reaches steady-state motion and electrophoretic velocity u_{ep}. The Stokes drag force is given by $F_{drag} = 6\pi\mu a\, u_{ep}$. At equilibrium, we have

$$F_{tot} = F_{el} + F_{drag} = 0 \tag{6.171a}$$

$$u_{ep} = \frac{Z_e}{6\pi\mu a}E \equiv \mu_{ion}E \tag{6.171b}$$

From equation (6.171), we see that the terminal velocity u_{ep} is proportional to the applied electrical field E. The proportionality constant is called the ionic mobility, μ_{ion}:

$$\mu_{ion} \equiv \frac{Z_e}{6\pi\mu a} \tag{6.172}$$

This simple theoretical estimate based on a macroscopic continuum model is in remarkable agreement with measured values of the ionic mobility of ions having a radius in the sub-nanometer range and moving in water. The radius a, however, is not the bare ionic radius $a \approx 0.05$ nm but instead the somewhat larger so-called hydrated radius $a \approx 0.2$ nm. This is due to the fact that ions in aqueous solutions accumulate approximately one atomic layer of water molecules. For $Z_e = 1$ C, $\mu = 1$ mPa s, and $a = 0.2$ nm, we get

$$\mu_{ion} \approx 4 \times 10^{-8}\ \text{m}^2(\text{V s})^{-1}$$

The dependence of the resulting drift velocity u_{ep} on particle charge and size makes electrophoresis usable in biochemistry for sorting of proteins and DNA fragments. The sample under consideration is dissolved in water and inserted in one end of a tube with electrodes at each end. A voltage difference is applied to the electrodes, and the part of the sample that arrives first at the other end of the tube contains the smallest and most charged particles.

6.13.1 Charged Particle in an Electrolyte

When the charged particle is immersed in an electrolyte, the mobile charges of the electrolyte redistribute themselves around the charged particle (see Figure 6.24). This charge distribution must be taken into account for the ionic mobility calculation.

A double layer is established near the charged surface. The thickness of the double layer, λ_D, is roughly 10 nm. Two cases are possible. The size of the charged particle may be larger than λ_D in the case of cells or beads. The particle size may be much smaller than λ_D in the case of smaller ions.

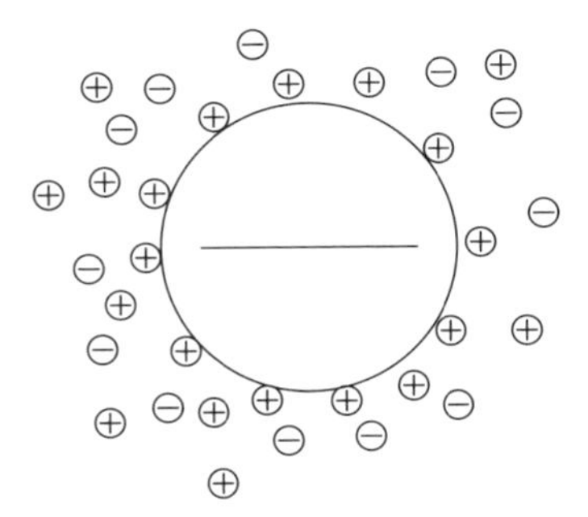

Figure 6.24 A negatively charged particle placed in an electrolyte forming a diffuse layer, which tends to screen the electric field produced by the particle

6.13.1.1 Small Particles

For small particles, the situation can be considered to be analogous to the particle immersed in an insulating medium. This situation corresponds to ions (Na^+, Ca^+, …), small ionized proteins, and small strands of DNA in solution. The ionic mobility for this case is similar to equation (6.172) as

$$\mu_{\text{ion}} = \frac{Z_e}{6\pi\mu a} \tag{6.173}$$

6.13.1.2 Large Particles

The particles in this case are enclosed in the Debye–Hückel double layer, which is much thinner than the particle itself. This case is analogous to the EO situation of the previous section with a change in the frame of reference, that is, the frame of reference is the resulting fluid. Thus, the migration velocity of the particle using equation (6.171b) is

$$U_{\text{eo}} \simeq -\frac{\epsilon\zeta E}{\mu} = \mu_{\text{ion}} E \tag{6.174}$$

$$\mu_{\text{ion}} \sim \frac{\epsilon\zeta}{\mu} \tag{6.175}$$

6.13.2 *Capillary Electrophoresis*

Capillary electrophoresis is an electrophoretic separation technique of ions confined inside a capillary or in a microchannel. A schematic of the system is shown in Figure 6.25. Two electrodes are submerged inside external reservoirs, which impose electric field along the capillary. The imposed electric field has two effects, that is, it induces an EO flow that displaces the whole fluid at some velocity and separates the charged particles that migrate at different velocities. Thus, the charged particles pass under the detector at different instants of time. Detection is more often based on fluorescence. The biological molecules we want to separate are marked by a fluorescent marker. If one wishes to separate simple inorganic ions like Na^+, Cl^-, etc., the nonfluorescent molecules are detected in a homogeneously fluorescent medium.

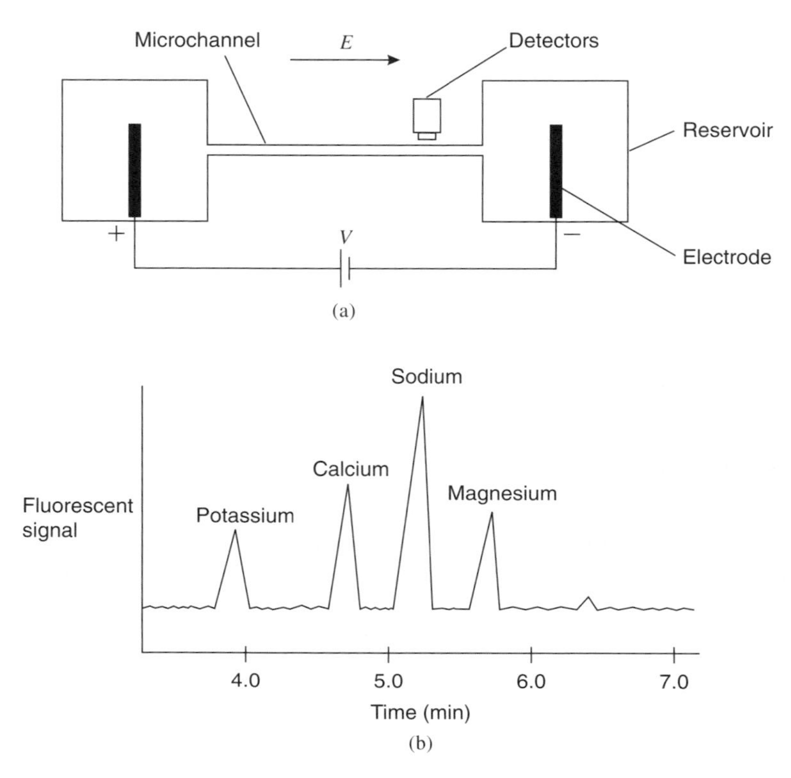

Figure 6.25 (a) A simple capillary electrophoresis system and (b) an example of temporal reading of species

Figure 6.25(b) shows peak corresponding to different groups of particles. With prior calibration, one can identify the different solutes present from the migration velocities. This figure shows the ions of magnesium, sodium, potassium, and calcium identified using this approach.

6.13.3 Debye Layer Screening

In the earlier section, we have studied electrophoresis in the case of a nonconducting liquid. This case was simple since the charge of the particle suspended in the liquid does not suffer any electrical screening. In the opposite limit, where the liquid is a highly conducting electrolyte, the particle charge is completely screened by the ions of the electrolyte within a distance of the Debye screening length λ_D. Since the effective charge of the particle in this case is zero, it cannot move by electrophoresis, but only by dielectric forces, a process called *dielectrophoresis* (DEP).

Let us solve the Debye layer problem for a charged spherical particle of radius a. The problem is spherical symmetric, so the potential due to the particle can only depend on the radial coordinate r, $\phi(r) = \phi(r)$. The boundary conditions for $\phi(r)$ are

$$\phi(a) = \zeta, \quad \phi(\infty) = 0 \tag{6.176}$$

If we again employ the Debye–Hückel approximation, the Poisson–Boltzmann equation becomes

$$\frac{1}{r^2}\frac{\partial}{\partial r}\left(r^2\frac{\partial}{\partial r}\phi(r)\right) = \frac{1}{\lambda_D^2}\phi \tag{6.177}$$

where the Laplace operator in spherical coordinates is simplified due to the lack of angular dependence in the problem. This differential equation is solved by the standard substitution $\psi(r) \equiv r\phi(r)$, and equation (6.177) is transformed into the simpler equation

$$\frac{\partial^2}{\partial r}\psi = \frac{1}{\lambda_D^2}\psi \tag{6.178}$$

with the straightforward exponential solutions $\psi(r) = \exp(\pm r/\lambda_D)$. Going back from $\psi(r)$ to $\phi(r)$ and employing the boundary condition equation (6.176) yield the solution

$$\phi(r) = \zeta\frac{a}{r}\exp\left[\frac{a-r}{\lambda_D}\right], \qquad \text{for} \quad r > a \tag{6.179}$$

This solution has the form of a modified or screened coulomb potential, which corresponds to a collection of counterions around the charged particle. This charge collection is known as a screening cloud. Just a few times the Debye length away from the particle, its charge cannot be observed, and it is completely screened. For strong electrolytes, where λ_D is very small, the originally charged particle becomes charge neutral for all practical purposes.

In the intermediate case of moderate Debye length, the electrophoresis problem becomes complicated. The motion of the particle distorts the screening cloud, which becomes asymmetric, resulting in very complex interactions between the electrolyte, the screening cloud, and the particle. In the case of very large Debye lengths, that is, for nonconducting liquids, we recover the simple unscreened charged particle.

6.14 Dielectrophoresis

For the neutral particle, the previously calculated coulomb force is equal to zero. However, a force appears when the particle is immersed in an electric field gradient. This force is called the dielectrophoretic force. *Dielectrophoresis* is the movement of a charge neutral particle in a dielectric fluid induced by an inhomogeneous electric field. The driving electric field can be either DC or AC.

A simple dipole consists of positive charge separated from a negative charge by a distance. A dielectric particle encloses dipoles that in the absence of exterior electric field are oriented randomly in space. In the presence of the electric field, the dipoles orient themselves parallel to the lines of the electric field, and the particle acquires a nonzero dipole. Let us consider a particle in a solvent subjected to an electric field. The application of an electric field leads to a charge accumulation at the interface with the surrounding medium. The nonuniform charge distribution creates a dipole, and this dipole interacts with the electric field. A net force acts on the particle when the electric field is different on both sides of the particle. The net force is toward the high electric field if the polarizability of the particle is higher than that of the medium, which is known as *positive dielectrophoresis*. This case is similar to a particle in vacuum. The direction of net force is reversed when the polarizability of the medium is higher than the

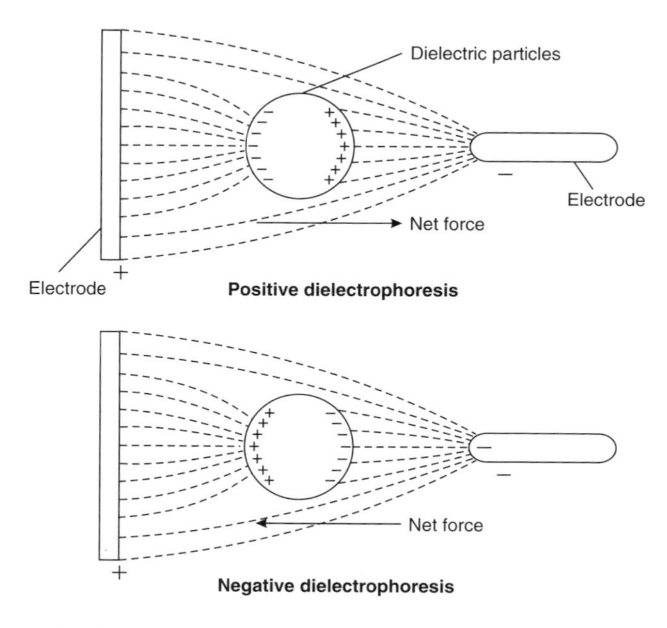

Figure 6.26 The schematic showing the positive and negative dielectrophoresis

particle, and this is known as *negative dielectrophoresis*. When the liquid is more polarizable than the particle, the dipoles localized in the fluid move the particle toward the minimum of the electric field. Figure 6.26 shows the schematic of positive and negative dielectrophoresis.

6.15 Polarization and Dipole Moments

Polarization effects play an important role in microfluidics. Electrical polarization corresponds to the rearrangement of bound electrons in the material. Consider a little particle, that is, a biological cell or a small part of some liquid, having the electric charge density ρ_E which occupies the region $\forall$ in space centered around the point r_0. General positions inside the particle are denoted $r_0 + r$. If all external electrical field E is imposed on the system, the ith component F_i of electrical force F_E acting on the particle is given by

$$F_i = \int_\forall dr \rho_E(r_0 + r) E(r_0 + r) \approx \int_\forall dr \rho_E(r_0 + r) \left[E_i(r_0) + r_j \frac{\delta}{\delta r_j} E_i(r_0) \right]$$
$$= QE(r_0) + p_j \frac{\delta}{\delta r_i} E_i(r_0) \tag{6.180}$$

Here, we have introduced the charge Q and electric dipole moment p of the particle as

$$Q \equiv \int_\forall dr \rho_E(r_0 + r) \tag{6.181a}$$

$$p \equiv \int_\forall dr \rho_E(r_0 + r) r \tag{6.181b}$$

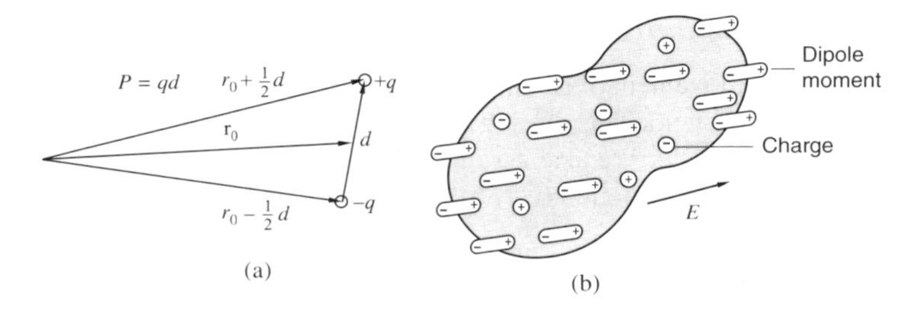

Figure 6.27 (a) The simple point dipole consisting of a charge $+q$ separated from a charge $-q$ by the distance d. (b) The dipole moments and external charges inside a body. Polarization charge is left behind in the body when the dipole moments stick out at the surface of the body

Equation (6.180) shows that there is an electrical force when the charge Q of the region $\forall$ is nonzero. A force is also present even when $Q = 0$ if both the dipole moment p and the electric field gradient ∇E are nonzero. The force in the latter case is denoted as *dielectric force*. The dielectric force plays the central role in dielectrophoresis.

A simple example of a dipole moment is the two-point-charge dipole, or in short the point dipole (Figure 6.27). It is defined by the charge distribution

$$\rho_E(r_0 + r) = +q\delta\left(+\frac{1}{2}d - r\right) - q\delta\left(-\frac{1}{2}d - r\right) \tag{6.182}$$

The polarization vector $P(r_0)$ appearing in equation (6.1b) is defined as the dipole moment density in a small region $\forall$ as the volume $\forall$ is taken to zero. Using equation (6.181b), we write

$$P(r_0) = \lim_{\forall \to 0}\left[\frac{1}{\forall}\int_{\forall} dr \rho_E(r_0 + r) r\right] \tag{6.183}$$

The divergence $\nabla \cdot P$ of the polarization can be interpreted as the polarization charge density. This is shown by considering the arbitrarily shaped body $\forall$ sketched in Figure 6.27(b), which contains a number of dipoles in the polarizable medium as well as some external charges, which are not part of the medium. In the bulk of the body, the charges from the dipole moments cancel each other, and at the surface part of the dipole, charges stick out at the surface of the body. There is a net amount of polarization charge Q_{pol} left behind in the body. The polarization charge density ρ_{pol} can be defined as

$$\rho_{\text{pol}} \equiv -\nabla \cdot P \tag{6.184}$$

We can write a simple expression for the density ρ_{ext} of the external charges as

$$\rho_{\text{ext}} = \rho_{\text{tot}} - \rho_{\text{pol}} = \epsilon_0 \nabla \cdot E + \nabla \cdot P = \nabla \cdot (\epsilon_0 E + P) \tag{6.185}$$

Thus, defining the displacement field as $D = \epsilon_0 E + P$ leads to equation (6.1a). We may note that ρ_E should not comprise the polarization charge density ρ_{pol}. For liquids and isotropic solids, the polarization is proportional to the electrical field, and the following expressions introducing the susceptibility χ and the relative dielectric constant ϵ can be used:

$$D = \epsilon_0 E + P = \epsilon_0 E + \epsilon_0 \chi E = \epsilon_0 (1 + \chi) E = \epsilon E \tag{6.186}$$

6.15.1 DC Dielectrophoresis

Dielectrophoresis is the movement of a charge neutral particle in a dielectric fluid induced by an inhomogeneous electric field. This driving field can be either DC or AC.

Let us consider a DC field E. We also assume linear media such that the polarization P of the dielectric fluid is given by equation (6.186)

$$P = \epsilon_0 \chi E \tag{6.187}$$

where χ is the susceptibility. The induced dipole moment p of the dielectric particle is

$$p = \alpha E \tag{6.188}$$

where α is the polarizability.

According to equation (6.180), a dielectric force F_{dip} acts on a dipole moment p situated in an inhomogeneous electric field E, that is, a field with a nonzero gradient tensor ΔE:

$$F_{\text{dip}} = (p \cdot \nabla)E \tag{6.189}$$

Let us consider a dielectric sphere with dielectric constant ϵ_p placed in a dielectric fluid with dielectric constant ϵ_l as shown in Figure 6.28. An inhomogeneous electric field E is imposed by charging a spherical electrode at the left side and a planar electrode at the right side. From equation (6.186),

$$D = \epsilon_0 E + P = \epsilon_0(1 + \chi)E = \epsilon E \tag{6.190}$$

We can observe that when an electric field E is applied to a medium with a large dielectric constant ϵ, the medium acquires a large polarization P and consequently contains many dipoles p. Figure 6.28(a) shows the medium with the smaller dielectric constant ϵ_l containing a few polarization charges at its surfaces compared to the sphere with larger dielectric constant $\epsilon_p > \epsilon_l$ containing more charges at its surface.

Figure 6.28(b) shows reverse trend. Here, the medium has larger dielectric constant ϵ_l and many polarization charges at its surfaces, while the sphere with the smaller dielectric constant $\epsilon_p < \epsilon_l$ has fewer polarization charges.

Figure 6.28(c) and (d) shows the unpaired surface charges of (a) and (b). The direction of the dipole moment p of the dielectric sphere and the direction of the dielectric force F_{dip} is also shown in the figure.

For $\epsilon_p > \epsilon_l$, the dielectric force pulls the dielectric particle toward the region of strong E-field (to the left), while for $\epsilon_p < \epsilon_l$, the particle is pushed away from this region (toward the right).

6.16 Point Dipole in a Dielectric Fluid

Let us determine the electrical potential $\phi_{\text{dip}}(r)$ arising from a point dipole $p = qd$ placed at the center of the coordinate system in a dielectric fluid with dielectric constant ϵ:

$$p = qd, \begin{cases} +q & \text{at} \quad +\frac{1}{2}d \\ -q & \text{at} \quad -\frac{1}{2}d \end{cases} \tag{6.191}$$

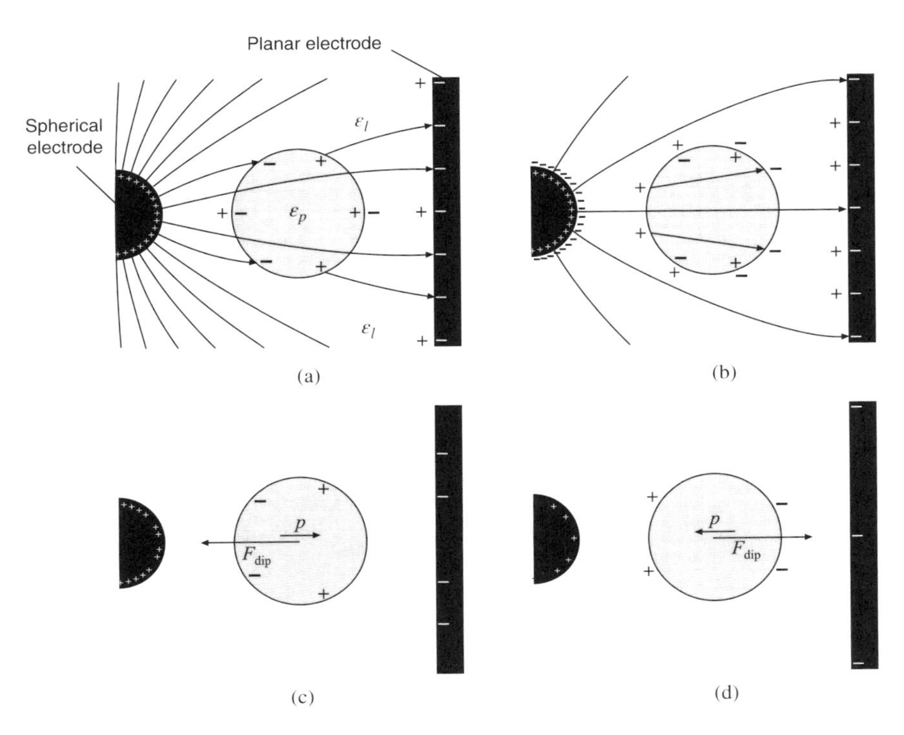

Figure 6.28 Sketch showing the direction of the electric dipole moment p induced in a dielectric sphere with dielectric constant ϵ_p inside a dielectric fluid having dielectric constant ϵ_l by the inhomogeneous electrical field E. (a) The particle is more polarizable than the fluid, that is, $\epsilon_p > \epsilon_l$. (b) The particle is less polarizable than the fluid, that is, $\epsilon_p < \epsilon_l$. (c) and (d) The effective charges and directions of p and F_{dip} corresponding to (a) and (b), respectively

The observation point is located at r_0 from the center of the dipole. The distances from the dipole charges $+q$ and $-q$ are $|r - d/2|$ and $|r + d/2|$, respectively (see Figure 6.27(a)). The potential $\phi_{\text{dip}}(r)$ from a point dipole, where $d \ll r$, can be written as

$$\phi_{\text{dip}}(r) = \frac{+q}{4\pi\epsilon}\frac{1}{|r_0 - d/2|} + \frac{-q}{4\pi\epsilon}\frac{1}{|r_0 + d/2|} \approx \frac{+1}{4\pi\epsilon}\frac{p.r}{r_0^3} = \frac{p}{4\pi\epsilon}\frac{\cos\theta}{r_0^2} \tag{6.192}$$

where θ is the angle between the dipole p and the observation point vector r. Based on the above equation, one may conclude that if a given potential $\phi_{\text{tot}}(r)$ contains a component of the form $B\cos\theta/r^2$, we can write

$$\phi_{\text{tot}}(r) = B\frac{\cos\theta}{r^2} + \phi_{\text{rest}}(r) \tag{6.193}$$

It implies that a dipole of strength

$$p = 4\pi\epsilon B \tag{6.194}$$

is located at the center of the coordinate system. This result will be used in the following section for calculating the induced dipole moment of a dielectric sphere placed in a dielectric fluid.

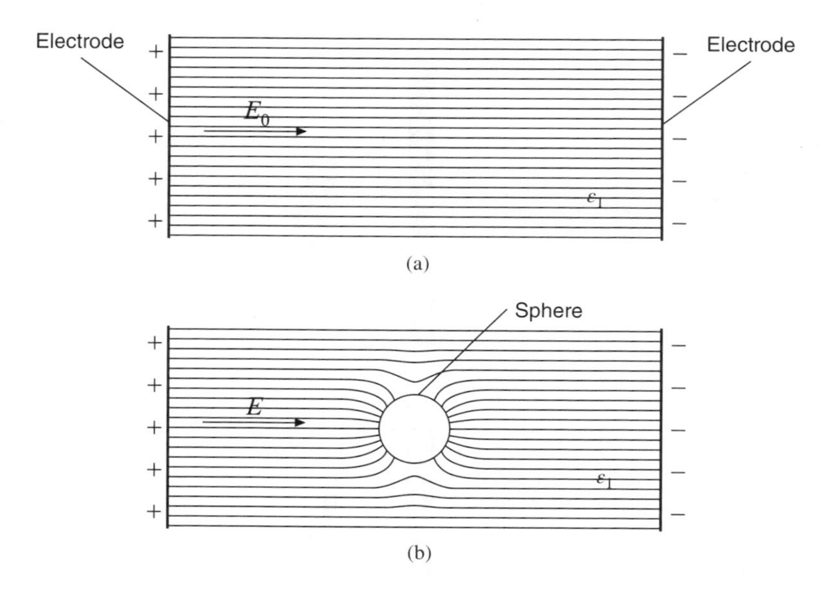

Figure 6.29 (a) A dielectric fluid with a dielectric constant ϵ_l penetrated by an unperturbed homogeneous electric field E_0. (b) A dielectric sphere of radius a and dielectric constant $\epsilon_p > \epsilon_l$ placed in the dielectric fluid. The electric field polarizes the sphere, resulting in a perturbed electric field, E

6.17 Dielectric Sphere in a Dielectric Fluid: Induced Dipole

Figure 6.29(a) shows a dielectric fluid with dielectric constant ϵ_l, which is penetrated by a homogeneouse electric field, $E_0 = -\nabla\phi_0$. In spherical coordinates (r, θ, φ), the unperturbed potential ϕ_0 is given by

$$\phi_0(r, \theta, \varphi) = -E_0 r \cos\theta \tag{6.195}$$

A dielectric sphere of radius, R, and dielectric constant, ϵ, is then placed in the fluid as shown in Figure 6.29(b). The electric field polarizes the sphere, resulting in a distortion of the electrical field, which now becomes $E = -\nabla\phi$, where the potential ϕ is given by one function ϕ_1 outside the sphere and another function ϕ_2 inside as

$$\phi(r, \theta, \varphi) = \begin{cases} \phi_1(r, \theta), & \text{for} \quad r > R \\ \phi_2(r, \theta), & \text{for} \quad r < R \end{cases} \tag{6.196}$$

Here, we have assumed the system to be rotation symmetric around the x-axis such that $\phi(r) = \phi(r, \theta)$ does not depend on the azimuthal angle φ.

The boundary conditions at the surface of the sphere, $r = R$, are same as electrostatics: the normal component $D \cdot e_r$ of D and the tangential component $E \cdot e_\theta$ of E must be continuous across the surface of the sphere at $r = R$. Hence, at $r = 0$, $r = R$, and $r = \infty$, we have a total of four boundary conditions:

$$\phi(0, \theta) \quad \text{is} \quad \text{finite} \tag{6.197a}$$

$$\phi_1(R, \theta) \quad = \quad \phi_2(R, \theta) \tag{6.197b}$$

$$\epsilon_l \frac{\delta}{\delta r}\phi_1(R,\theta) = \epsilon_p \frac{\delta}{\delta r}\phi_2(R,\theta) \tag{6.197c}$$

$$\phi_1(r,\theta) \underset{r\to\infty}{\rightarrow} -E_0 r\cos\theta \tag{6.197d}$$

Both the fluid and the sphere are dielectric media without external charges; hence, $\rho_E = 0$ in the equation $\nabla^2\phi(r) = \frac{-1}{\epsilon}\rho_E(r)$, and the potential obeys the Laplace equation

$$\nabla^2\phi(r) = 0 \tag{6.198}$$

Using the trial solution $\phi(r)$ in terms of Legendre polynomial, one can obtain the electrical potentials ϕ_1 and ϕ_2 as

$$\phi_1(r) = -E_0 r\cos\theta + \frac{\epsilon_p - \epsilon_l}{\epsilon_p + 2\epsilon_l}R^3 E_0 = \phi_1(r) + \phi_{\text{dip}}(r), \quad \text{for} \quad r > R \tag{6.199a}$$

$$\phi_2(r) = \frac{-3\epsilon_l}{\epsilon_p + 2\epsilon_l}E_0 r\cos\theta = \frac{-3\epsilon_l}{\epsilon_p + 2\epsilon_l}\phi_1(r), \quad \text{for} \quad r < R \tag{6.199b}$$

Equation (6.199b) shows that the potential, ϕ_2, inside the sphere is merely proportional to the unperturbed potential, ϕ_0. However, equation (6.199a) shows that the unperturbed potential is supplemented by a dipole potential, ϕ_{dip}. In an applied electric field, the dielectric sphere acquires an induced dipole moment p, which according to equations (6.193) and (6.194) has the value

$$p = 4\pi\epsilon_l \frac{\epsilon_p - \epsilon_l}{\epsilon_p + 2\epsilon_l}R^3 E_0 \tag{6.200}$$

The fraction in the prefactor plays a significant role. It is named as *Clausius–Mossotti factor* $K(\epsilon_l, \epsilon_p)$:

$$K(\epsilon_l, \epsilon_p) \equiv \frac{\epsilon_p - \epsilon_l}{\epsilon_p + 2\epsilon_l} \tag{6.201}$$

Note that the exact result in equation (6.200) indicates that when the sphere is more dielectric than the liquid, $\epsilon_p > \epsilon_l$, the induced dipole moment p and the unperturbed field E_0 are parallel. They become antiparallel when the sphere is less dielectric than the liquid, $\epsilon_p < \epsilon_l$. One can see that the induced dipole moment vanishes if the sphere and the fluid have the same dielectric constant $\epsilon_p = \epsilon_l$. Equation (6.200) provides us with a simple way to calculate the dielectric forces acting on a dielectric sphere immersed in a dielectric fluid.

6.18 Dielectrophoretic Force on a Dielectric Sphere

In this section, we present the dielectric force F_{dip} on a dielectric sphere of finite radius "a." It may be noted that the induced dipole moment reported in the previous section assumes homogeneous external electrical field E_0. However, the calculation becomes more involved for inhomogeneous case. If the radius R of the sphere is much smaller than the distance "l" over which the external electrical field varies, one can still use equation (6.200) for the induced dipole. Let us use the Taylor expansion (here just taken to first order) of the external electrical field $E_0(r)$ around the center coordinate r_0 of the sphere:

$$E_0(r) \approx E_0(r_0) + [(r - r_0).\nabla]E_0(r_0) + \mathcal{O}(R/l) \tag{6.202}$$

In this expression, the value of the gradient term is of the order of R/l since $|r - r_0| < R$ and $\nabla E_0(r_0) \approx (R/l)E_0(r_0)$. Equation (6.200) for the dipole moment can be generalized as

$$p \approx R^3 4\pi\epsilon_l K(\epsilon_l, \epsilon_p)E_0(r_0) + R^4[f_1(\epsilon_l, \epsilon_p).\nabla]E_0(r_0)$$
$$= R^3 4\pi\epsilon_l K(\epsilon_l, \epsilon_p)E_0(r_0) + \mathcal{O}(R/l) \qquad (6.203)$$

The vector function $f_1(\epsilon_l, \epsilon_p)$ appearing above is a generalized Clausius–Mossotti function. Combining equations (6.189), (6.202), and (6.203), we arrive at

$$F_{\text{dip}}(r_0) = [p(r_0) \cdot \nabla]E_0 + \mathcal{O}(R/l)$$
$$= 4\pi\epsilon_l \frac{\epsilon_p - \epsilon_l}{\epsilon_p + 2\epsilon_l} R^3 \cdot \nabla[E_0(r_0) \cdot \nabla]E_0(r_0) + \mathcal{O}(R/l)$$
$$= 2\pi\epsilon_l \frac{\epsilon_p - \epsilon_l}{\epsilon_p + 2\epsilon_l} R^3 \cdot \nabla[E_0(r_0)^2] + \mathcal{O}(R/l) \qquad (6.204)$$

In the last equality, we have used $\nabla[E^2] = 2E \cdot \nabla E$, which is valid in electrostatics where $\nabla \times E = 0$.

The dielectrophoretic force F_{DEP} can be written using the Clausius–Mossotti factor as

$$F_{\text{DEP}}(r_0) = 2\pi\epsilon_l K(\epsilon_l, \epsilon_p) R^3 \nabla[E_0(r_0)^2] \qquad (6.205)$$

This power of 2 of the electric field in equation (6.205) implies that the sign of the DEP force is independent of the sign of EO. The sign of the Clausius–Mossotti factor $K(\epsilon_l, \epsilon_p)$ determines the direction of the DEP force.

6.19 Dielectrophoretic Trapping of Particles

The dielectrophoretic force F_{DEP} can be used to trap dielectric particles suspended in microfluidic channel. An inhomogeneous electric field is created in a microchannel by charging carefully shaped metal electrodes at the walls of the channel. Dielectric particles suspended in the liquid flowing through the microchannel are attracted to the electrodes. The particles get trapped by the electrodes if the DEP force F_{DEP} is stronger than the viscous drag force F_{drag}:

$$|F_{\text{DEP}}| > |F_{\text{drag}}| \qquad (6.206)$$

We study a particular simple geometry as shown in Figure 6.30.

The microfluidic channel is rectangular with length L, width w, and height h. The channel in the flat and wide channel limit $h \ll w$ can be approximated by the infinite parallel plate channel. The origin of the coordinate system is placed at the center of the floor wall such that $-L/2 < x < L/2, -w/2 < y < w/2$, and $0 < z < h$. A pressure drop of Δp along the channel results in the flow profile given as

$$u(z) = \frac{\Delta p}{2\mu L}(h - z)z = 6\left(1 - \frac{z}{h}\right)\frac{z}{h}v_0 \qquad (6.207)$$

where v_0 is the average flow velocity $\left(v_0 = \frac{\Delta p h^2}{12\mu L}\right)$. The flow rate is given by $Q = v_0 wh$.

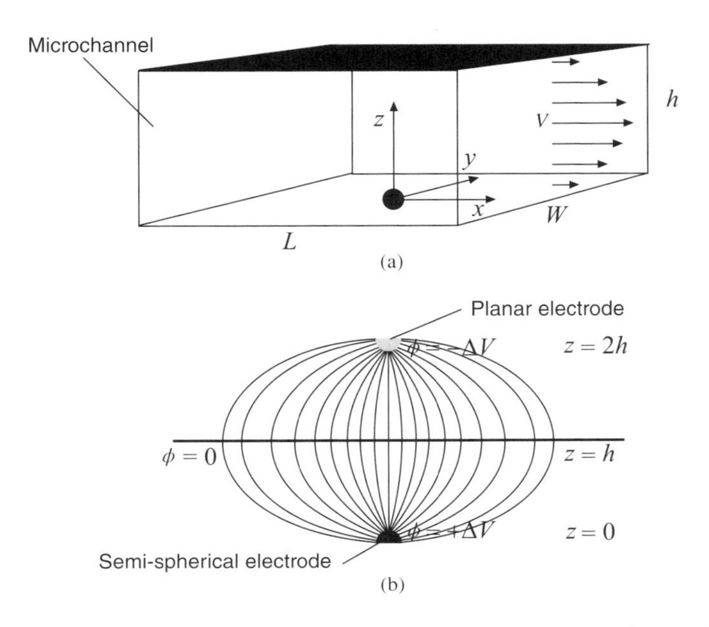

Figure 6.30 (a) An example of DEP trap in a rectangular microfluidic channel of dimensions $L \times w \times h$ to catch dielectric particle suspended in a liquid flow with velocity profile v. (b) Schematic showing the inhomogeneous electric field E created by applying a voltage difference ΔV between the semispherical electrode at the floor of the microchannel and the planar electrode covering the ceiling

The dielectric particles suspended in the liquid have radius R. One can approximate the drag force acting on a sphere trapped at the position "r" using the Stokes flow assumption as

$$F_{\text{drag}} \approx 6\pi \mu R v(r) \tag{6.208}$$

The inhomogeneous electric field is created by applying a potential $\phi = \Delta V$ to a semispherical metallic electrode of radius r_0 situated at the floor at $r = 0$ and the potential $\phi = 0$ to a planar metallic electrode covering the ceiling at the plane $r = he_z$ (see Figure 6.30). The liquid is assumed to have vanishing conductivity so that the formation of Debye screening layers near the electrodes can be discharged. By mirror image charge method, the electrical potential $\phi(r)$ of this configuration can be written as

$$\phi(r) = \frac{r_0}{|r|}\Delta V - \frac{r_0}{|r - 2he_z|}\Delta V \tag{6.209}$$

The planar top electrode with potential $\phi(he_z) = 0$ has been realized by placing a mirror electrode with the potential $\phi(2he_2) = -\Delta V$ at $r = 2he_2$. The above equation shows the potential to be equal to zero at $r = 2he_z$. The trapping of particles takes place close to the spherical electrode, that is, $|r| \ll h$. The electrical field is given approximately in this region as

$$E(r) = -\nabla\phi(r) \approx \frac{r_0 \Delta V}{r^2} e_r, \quad \text{for} \quad r_0 < |r| \ll h \tag{6.210}$$

where e_r is the radial vector pointing away from the spherical electrode.

Using the electrical field equation (6.210), one can derive an expression for the DEP force using equation (6.205) as

$$F_{\text{DEP}}(r) = 2\pi\epsilon_l \frac{\epsilon_p - \epsilon_l}{\epsilon_p + 2\epsilon_l} R^3 \nabla \left[\frac{(\Delta V)^2 r_0^2}{r^4} \right] e_r = -8\pi \frac{\epsilon_p - \epsilon_l}{\epsilon_p + 2\epsilon_l} \frac{R^3 r_0^2}{r^5} \epsilon_l (\Delta V)^2 e_r \tag{6.211}$$

The maximal DEP force $F_{\text{DEP}}^{\max}$ is achieved when the particle is as close to the spherical electrode as possible, $r = r_{\min} = r_0 + R$. If we denote the electrode radius by $r_0 = \Gamma R$, we obtain the minimal distance

$$r_{\min} = (1 + \Gamma)R, \quad \Gamma \equiv \frac{r_0}{R} \tag{6.212}$$

An estimate for the maximal DEP force can be written as

$$F_{\text{DEP}}^{\max} \equiv |F_{\text{DEP}}(r_{\min})| = 8\pi \frac{\epsilon_p - \epsilon_l}{\epsilon_p + 2\epsilon_l} \frac{\Gamma^2}{(1+\Gamma)^5} \epsilon_l (\Delta V)^2 \tag{6.213}$$

The average flow velocity at the position $z = r_0 + R = (1 + \Gamma)R$ follows from equation (6.207)

$$v_x(r_0 + R) = 6\left(1 - (1+\Gamma)\frac{R}{h}\right)(1+\Gamma)\frac{R}{h} v_0 \approx 6(1+\Gamma)\frac{R}{h} v_0 \tag{6.214}$$

The drag force on a particle placed at $r = r_{\min} = r_0 + R$ follows from equations (6.208) and (6.214)

$$|F_{\text{drag}}(r_0 + R)| \approx 6\pi\mu R v(r_0 + R) = 36\pi(1+\Gamma)\frac{\mu R^2}{h} v_0 \tag{6.215}$$

The largest average velocity $v_0^{\max}$ which still allows for the trapping of particles at the spherical electrode can be found from the condition $F_{\text{drag}} = F_{\text{DEP}}^{\max}$, which results in

$$v_0^{\max} = \frac{2}{9} \frac{\epsilon_p - \epsilon_l}{\epsilon_p + 2\epsilon_l} \frac{\Gamma^2}{(1+\Gamma)^6} \frac{h\epsilon_l (\Delta V)^2}{\mu R^2} \tag{6.216}$$

We need a liquid with a dielectric constant smaller than that of the particle to achieve trapping. Let the liquid be benzene with $\epsilon_l = 2.28\,\epsilon_0$ and $\mu = 0.65$ mPa s and pyrex glass particles have $\epsilon_p = 6.0\,\epsilon_0$. The length scales are set as $R = r_0 = 5\ \mu\text{m}$ and $h = 100\ \mu\text{m}$, while the applied voltage drop is $\Delta V = 10$ V. With these parameters, one can find

$$v_0^{\max} = 1.519 \times 10^{-3}\ \text{m/s} \tag{6.217}$$

The above example shows that DEP trap is also successful at high velocity. Actual DEP trap for biological cells is already available. In some DEP trap, spiral electrodes are used to trap the biological cells, that is, cloud of yeast cells. The yeast cells are caught at the center of the spiral electrode where the gradients are highest. The yeast cells are carried away by the liquid flow when the applied voltage to the electrodes is released.

6.20 AC Dielectrophoretic Force on a Dielectric Sphere

We have considered DC voltages driving the DEP trap in the previous section. However, there are many advantages in using AC voltage bias instead: (1) Any charge monopoles (ions) in the system will not change their mean position being influenced by AC electric fields. (2) The creation of permanent Debye screening layers at the electrodes is avoided. (3) In the AC mode, the DEP trap will also work even if the liquid and the particle have nonzero conductivities. The Clausius–Mossotti factor under AC field depends on the driving frequency w, and it can even change its sign. This allows to control in situ whether the DEP force should be attractive or repulsive.

Let us assume a simple harmonic time variation $\exp(-iwt)$ of the applied potential. The applied potential $\phi(r,t)$ and the associated electrical field $E(r,t) = -\nabla\phi(r,t)$ are denoted as

$$\phi(r,t) = \phi(r)e^{-i\omega t} \tag{6.218a}$$

$$E(r,t) = E(r)e^{-i\omega t} \tag{6.218b}$$

The general boundary condition for the radial component, $E_r(r,\theta) = -\frac{\delta}{\delta r}\phi(R,\theta)$, at the surface of the dielectric sphere is

$$\epsilon_l E_{r,1}(R,\theta,t) - \epsilon_p E_{r,2}(R,\theta,t) = q_{\text{surf}} \tag{6.219}$$

For perfect dielectric, the surface charge density q_{surf} is equal to zero, as stated in equation (6.197c). However, with nonzero conductivities and AC fields, it becomes nonzero and in fact time dependent. The time derivative of q_{surf} is given by charge conservation and Ohm's law:

$$\frac{\delta}{\delta r}q_{\text{surf}}(t) = J_{r,1}(R,\theta,t) - J_{r,2}(R,\theta,t) = \sigma_{\text{el},1}E_{r,1}(R,\theta,t) - \sigma_{\text{el},2}E_{r,2}(R,\theta,t) \tag{6.220}$$

where $\sigma_{\text{el},1}$ and $\sigma_{\text{el},2}$ are the electrical conductivities of liquid and particle, respectively.

Taking the time derivative of equation (6.219) using E-fields from equation (6.218b), substituting in equation (6.220), and multiplying with i/ω, we have

$$\left(\epsilon_l - i\frac{\sigma_{\text{el},1}}{\omega}\right)E_{r,1}(R,\theta) = \left(\epsilon_p - i\frac{\sigma_{\text{el},2}}{\omega}\right)E_{r,2}(R,\theta) \tag{6.221}$$

Let us define a complex dielectric function $\epsilon(\omega)$ as

$$\epsilon(\omega) = \epsilon - i\frac{\sigma}{\omega} = \epsilon_0\epsilon_r - i\frac{\sigma}{\omega} \tag{6.222}$$

Here, ϵ_r is the relative permittivity, σ is the conductivity, and ω is the frequency of the electric field. Now, we may observe that the boundary condition in the AC case equation (6.221) has the same mathematical form as the boundary condition equation (6.197c) in the DC case. We can therefore use the result of equation (6.205) directly using the complex dielectric functions in the Clausius–Mossotti factor, which is where the boundary condition has been used:

$$F_{\text{DEP}}(r_0,t) = 2\pi\epsilon_l\frac{\epsilon_p(\omega) - \epsilon_l(\omega)}{\epsilon_p(\omega) + 2\epsilon_l(\omega)}R^3\nabla[E(r_0,t)^2] \tag{6.223}$$

Note that the ϵ_l in the prefactor is the dielectric constant and not the dielectric function.

Let us define

$$A(t) = Re[A_0e^{-i\omega t}] \tag{6.224a}$$

$$B(t) = Re[B_0e^{-i\omega t}] \tag{6.224b}$$

where A_0 and B_0 are constant complex amplitudes. The time average $A(t)B(t)$ over one full period τ can be defined as

$$A(t)B(t) \equiv \frac{1}{\tau}\int_0^{\tau} dt \quad A(t)B(t) = \frac{1}{2}Re[A_0 B_0^*] \tag{6.225}$$

where B_0^* denotes the complex conjugate of B_0.

To obtain the real time-averaged DEP force F_{DEP} for equation (6.223), we use equation (6.225) with $A(t) = K[\epsilon_l(\omega), \epsilon_p(\omega)]E(r, t)$, and $B(t) = E(r, t)$. The result is

$$F_{\text{DEP}}(r_0, \omega) = 2\pi\epsilon_l \text{Re}\left[\frac{\epsilon_p(\omega) - \epsilon_l(\omega)}{\epsilon_p(\omega) + 2\epsilon_l(\omega)}\right] R^3 \nabla[E_{\text{rms}}(r_0)^2] \tag{6.226}$$

Here, we have introduced the usual root-mean-square value $E_{\text{rms}} = E/\sqrt{2}$.

6.20.1 Crossover Frequency

It is observed that $Re(f_{\text{CM}})$, that is, the real part of the Clausius–Mossoti factor determines the direction of the DEP force. Let us investigate the Clausius–Mossoti factor as a function of the electric field frequency.

The real part of the Clausius–Mossoti factor is written as

$$Re(f_{\text{CM}}) = \frac{(\epsilon_p - \epsilon_l)w^2\tau^2}{(\epsilon_p + 2\epsilon_l)(1 + w^2\tau^2)} + \frac{(\sigma_p - \sigma_l)}{(\sigma_p + 2\sigma_l)(1 + w^2\tau^2)} \tag{6.227}$$

$$\text{where} \quad \tau = \text{relaxation time} = \frac{\epsilon_p + 2\epsilon_l}{\sigma_p + 2\sigma_l} \tag{6.228}$$

Note that $Re(f_{\text{CM}})$ varies between two extremum values of frequency:

$$Re(f_{\text{CM}}) = \frac{\sigma_p - \sigma_l}{\sigma_p + 2\sigma_l} \quad (\text{for } w \to 0)$$

$$Re(f_{\text{CM}}) = \frac{\epsilon_p - \epsilon_l}{\epsilon_l + 2\epsilon_l} \quad (\text{for } w \to \infty) \tag{6.229}$$

Note that at low frequencies (<10 kHz), ϵ_p is the function of the frequency of the electric field. This is attributed to the relaxation of the double layer surrounding the particle, which depends on the radius of the particle with respect to the Debye length. This situation is complex, and no satisfactory model exists to describe this phenomenon. The counterions of the double layer do not have enough time to move in case of high frequencies. Thus, at high frequencies, the particle is nondispersive compared to that at low frequencies. Hence, ϵ_p and σ_p are independent of frequency at high frequencies. Here, the polarization is primarily due to the particle with respect to the surrounding medium. The behavior of the particles switches from positive to negative DEP when the frequency of the applied field varies. This frequency is easily calculated by setting

$$Re(f_{\mathrm{CM}}) = 0 \tag{6.230}$$

Its solution gives the crossover frequency as

$$f_0 = \frac{1}{2\pi}\sqrt{\frac{(\sigma_p + 2\sigma_l)(\sigma_p - \sigma_l)}{(\epsilon_p + 2\epsilon_l)(\epsilon_p - \epsilon_l)}} \tag{6.231}$$

The crossover frequencies from the above expression have shown some discrepancy with experimental observation. Therefore, some authors have considered this problem by assuming an infinitely thin conductive layer on the surface of the particle due to the double layer, leading to modification of the conductivity of the particle as

$$\sigma_p = \sigma_p + \frac{2\lambda_D}{a} \tag{6.232}$$

Figure 6.31 shows the real part of the Clausius–Mossoti factor as a function of the frequency clearly showing the crossover frequency. Figure 6.32 shows the crossover frequency of polystyrene latex beads in water as a function of its radius. The modified conductivity σ_p is used which compares well with the experimental data.

Let us calculate a characteristic value for ω_c of a biological cell, consisting mainly of the cytoplasm, in water. We use the following parameters: $\sigma_{\mathrm{el},2} = 0.1$ S/m and $\epsilon_p = 60.0\epsilon_0$ for the

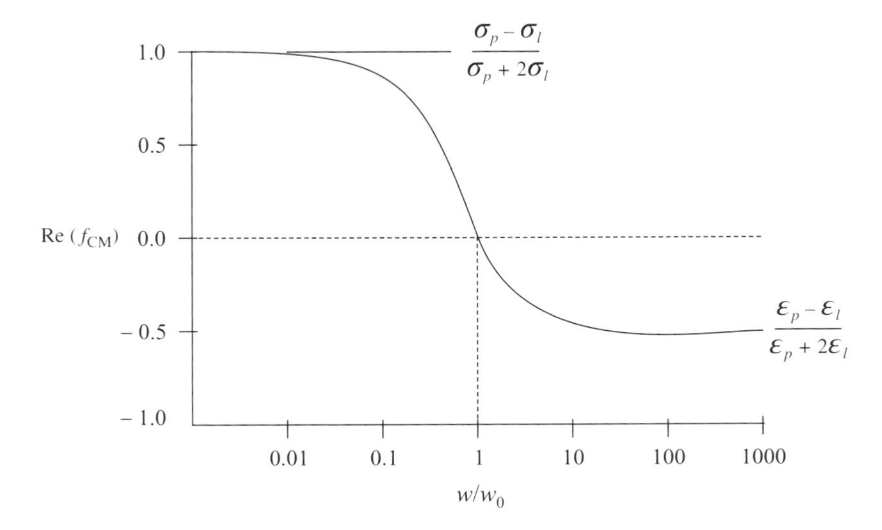

Figure 6.31 Real part of the Clausius–Mossoti factor as a function of frequency, where w_0 is the crossover frequency

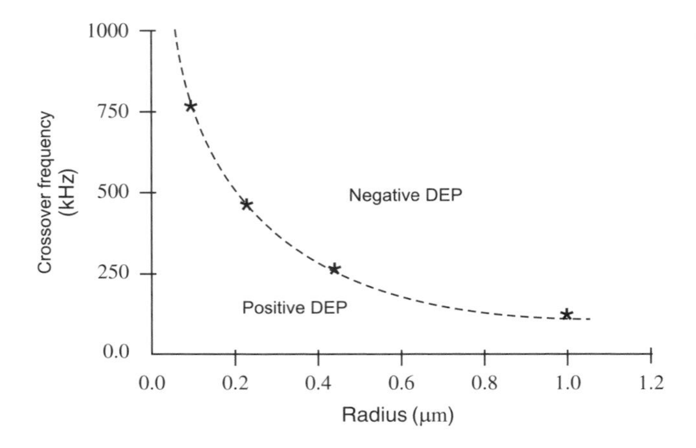

Figure 6.32 Crossover frequency of polystyrene bead in water as function of the radius. The symbols are expt-data

cell and $\sigma_{\mathrm{el},1} = 0.1$ S/m and $\epsilon_l = 78.0\epsilon_0$ for water. The crossover frequency value obtained is

$$\omega_c = 1.88 \times 10^8 \text{ Hz} \tag{6.233}$$

The frequency-dependent DEP can be used to separate living cells from dead cells and cancer cells from normal cells. Different cells have different electrical properties and consequently have different critical frequencies ω_c determining at which frequencies ω they are caught by the DEP electrode and at which they are expelled by it.

The neutral particle, subjected to dielectric force, is governed by the equation

$$m\frac{dv}{dt} \simeq 0 = F_{\text{DEP}} - F_{\text{Drag}}$$

where m is the mass of the particle, v is the velocity, and subscript DEP and drag corresponds to the dielectrophoretic and drag force, respectively.

Note: (1) The direction of the dielectric force depends on the sign of the real part of the Clausius–Mossoti factor, that is, if $\text{Re}(f_{CM}) > 0$, the particle is attracted to the region where the field is maximum (positive dielectrophoresis). In the other case, that is, negative dielectrophoresis, the particle is repelled.

(2) The dielectric force $F \sim \nabla E^2$ implies that the DEP arises both with AC and DC fields. In the case of DC, electrophoresis competes with DEP in the motion of the particle. The use of high-frequency AC fields suppresses electrochemistry at the surface of the electrodes, that is, electrolysis.

(3) The dielectrophoretic force depends on the gradient of the electric field intensity. Hence, a large electric field is necessary to transport particles over a large distance. This indicates high voltage requirement in the case of macroworld. However, large local gradients can more easily be created in the case of microstructures. Therefore, microfabrication techniques are

instrumental in rediscovering the capability of manipulation, characterization, and sorting of particles for life science.

Problems

6.1 (a) An electroosmotic μ-pump consisting of one microtube of diameter d = 10 μm and length l = 100 μm pumps against negligible pressure difference. Assuming small Debye layer with respect to the tube diameter, determine the flow rate (nano-liter/second) through the pump.

(b) The above pump is now required to pump against a pressure difference of 10.0 Pa. What will be the new flow rate?

(c) An additional microtube of the same length and diameter is now connected in series to the EO pump in (a) as shown below. What will be the new flow rate?

Use the following data:

Zeta potential (ζ) = 0.1 V
Applied electric potential difference (V) = 1.0 V
Viscosity of fluid (μ) = 0.001 Pa-s
Permittivity (ϵ) = 0.78×10^{-9} C^2/N-m^2

6.2 The annular gap between two infinitely long concentric cylinders of opposite charge is filled with water. Charged colloidal particles are to be moved from the inner cylinder (the source) to the outer cylinder (the collector) as a result of the voltage drop across the gap. The particle volume concentration is sufficiently small that each particle may be assumed to behave independently of the others. The width of the annular gap is 0.02 m, and there is an applied voltage drop of 2 V across the gap. The particles are spherical, have a radius a of 1 m and a density ρ twice that of water, and carry a charge q of 10^{-14} C. The temperature is constant at 300 K, and gravitational effects may be neglected. The drag force on a spherical particle moving in water at low speeds (low Reynolds number) is $F = -6\pi\mu aU$, where the viscosity of the water $\mu = 0.001$ Pa-s and U is the particle speed.

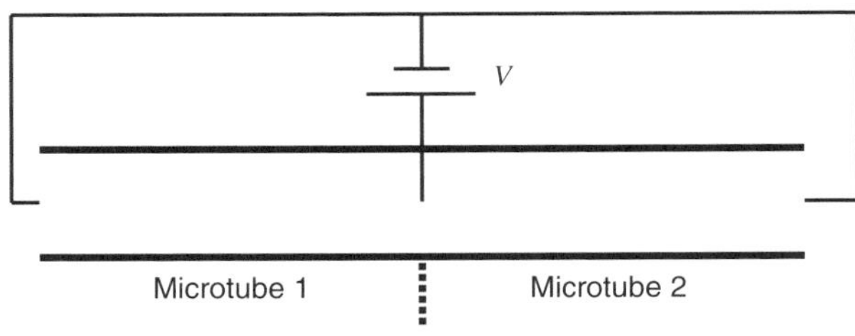

(a) Estimate the particle velocity at the collector surface.

(b) Neglecting the acceleration time of the particles to reach the steady-state velocity, estimate the time required to traverse the gap.

(c) Using an order of magnitude approach, estimate the ratio of the time required to reach the steady-state velocity to the time required to traverse the gap.

(d) Estimate the particle velocity at the collector surface. Assuming the particles start from rest at the source, estimate their time to traverse the gap.

6.3 Two infinite plates are held parallel to each other a distance $h = 0.02$ m apart with a dilute singly ionized electrolyte solution contained between them that has a uniform species concentration $c_0 = 1$ mol m^{-3}. The solution temperature is 300 K. A potential of 2 V is then applied across the gap between the plates.

(a) Assuming the electric field between the plates is constant and considering only the positive ions, write the governing differential equation for the steady-state concentration distribution in the gap subsequent to the application of the electric field.

Note that you can use for infinitely dilute solution, $Di = RTvi$ (Nernst–Einstein equation).

(b) Obtain an expression for the steady-state concentration distribution as a function of distance across the gap.

(c) Evaluate this expression in (b) using the numerical values given.

6.4 (a) A small spherical-shaped strand of human DNA with charge of 1×10^{-8} coulomb and radius of 0.2 nm inside deionized water medium ($\mu = 1$ mPa-s) is exposed to an external electric field of 1 V/m. Calculate the electrophoretic velocity of the human DNA strand.

(b) A spherical-shaped strand of mouse DNA inside the same water medium as above having radius equal to 0.4 nm and charge equal to 2×10^{-8} C is exposed to the same external electric field of 1 V/m. Calculate the ratio of electrophoretic velocity of the human DNA and mouse DNA.

6.5 An electroosmosis pump is made of glass channel of square cross section with edge $(a) = 10$ μm and length $(L) = 100$ μm. The permittivity of the working fluid $(\epsilon) = 45.0 \times 10^{-12}$ CV^{-1} m^{-1} and the viscosity of the fluid $(\mu) = 2$ mPa-s. The zeta potential of the glass surface $(\zeta) = 0.1$ V. A potential $(V) = 1$ V is applied across the channel, and the pressure difference between the inlet and outlet of the channel is equal to 1 Pa. Assume the size of the Debye layer to be small in comparison to the size of the channel.

(a) Calculate the total flow rate inside the channel in m^3/s.

(b) Calculate the free-flow capability and zero-pressure capability of the above electroosmotic pump.

6.6 An electroosmotic flow takes place inside a microchannel of cross section 20 μm × 400 μm (height × width) and 2 cm long under the applied potential of 30 V. The zeta potential of the wall is equal to 0.1 V, the viscosity of liquid is 1 m Pa-s, and the permittivity of the working fluid is equal to 45.0×10^{-12} C/V-m:

(a) Calculate the electroosmotic flow rate inside the channel assuming uniform potential inside the Debye layer.

(b) Calculate the electroosmotic flow rate through the channel assuming the potential inside the Debye layer to follow the Debye–Hückel approximation and the Debye layer size equal to 10 nm.

(c) Calculate the flow rate as in (b) with Debye layer size equal to 1000 nm.

(d) Calculate the pressure gradient for which the total flow rate in (c) will be equal to zero.

6.7 In a capillary electrophoretic system, potassium and sodium ions need to be separated. The mobility of potassium and sodium, respectively, are 7.62×10^{-8} m^2/s-V and 5.19×10^{-8} m^2/s-V. Calculate the electrophoretic velocity of potassium and sodium under an applied potential of 10 V in a 1 cm channel. If a photodetector is placed at 0.5 cm from inlet reservoir, calculate the time difference between the photodetector signature between sodium and potassium.

Supplemental Reading

Bruus H 2007 *Theoretical Microfluidics*. Oxford University Press.

Karniadakis G, Beskok A, and Aluru N 2005 *Microflows and Nanoflows, Fundamentals and Simulation*. Springer-Verlag.

Probstein RF 1994 *Physicochemical Hydrodynamics: An Introduction*. John Wiley & Sons, Inc.

7

Magnetism and Microfluidics

7.1 Introduction

Both magnetism and microfluidics are old concepts. However, the combined application of both concepts adds to the capability of microfluidic systems. In contrast to application of electric field, magnetic interactions are generally not affected by surface charges, pH, ionic concentrations, and temperature. Therefore, microfluidic applications with magnetic actuation have unique potential. This chapter reports and introduces few applications of using magnetic forces in microfluidic systems. Magnetic forces can be utilized to manipulated magnetic particles, magnetically labeled cells, plugs of ferrofluid, and pumping and mixing of fluids.

7.2 Magnetism Nomenclature

Figure 7.1 shows the magnetic field lines around a bar magnet. These magnetic field lines can be drawn using a bar magnet, graph paper, and magnetic compass. Another simpler way to generate magnetic line of force is by sprinkling iron powder around the bar magnet. When a magnetic material, such as soft iron, is placed in a magnetic field, the magnetic field lines are redirected through it (see Figure 7.1). This is attributed to the greater magnetic permeability of soft iron. Magnetic permeability is the ability of the material to permit the passage of magnetic lines of force through it. *Magnetic permeability*, μ, quantifies the density of magnetic field lines within a material. It represents the degree of magnetization attained by a material in response to an applied magnetic field. *Relative magnetic permeability* (μ_r) of a material is defined as the ratio of magnetic permeability of the material and magnetic permeability of the free space (μ_0)($\mu_r = \mu/\mu_0$). We know that $\mu_0 = 4\pi \times 10^{-7}$ Wb/A-m or H/m or N/m. *Magnetic flux density*, B (in tesla, T), is the number of magnetic field lines per unit area. *Relative permeability* is also defined as the magnetic flux density per unit area in that material to the magnetic flux density in vacuum ($\mu_r = \frac{B}{B_0}$). Magnetic field lines can be either homogeneous or inhomogeneous. There is no gradient in the flux density or the density of magnetic flux lines is constant over distance, x for *homogeneous magnetic field*. There is gradient in the density of flux lines over distance, x for the *inhomogeneous magnetic field*.

Transport Phenomena in Microfluidic Systems, First Edition. Pradipta Kumar Panigrahi.

Companion Website: www.wiley.com/go/panigrahi/microfluidic

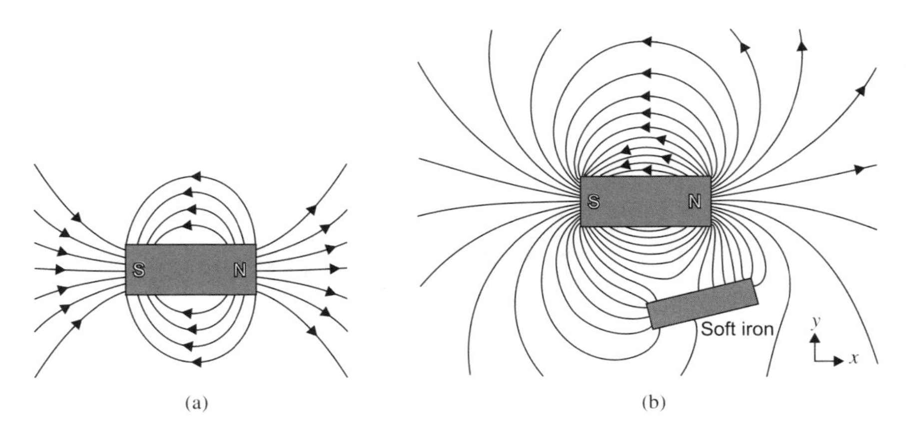

Figure 7.1 (a) Schematic of the magnetic field lines around a bar magnet. (b) Illustration of magnetic permeable material (soft iron) influence on the magnetic field lines

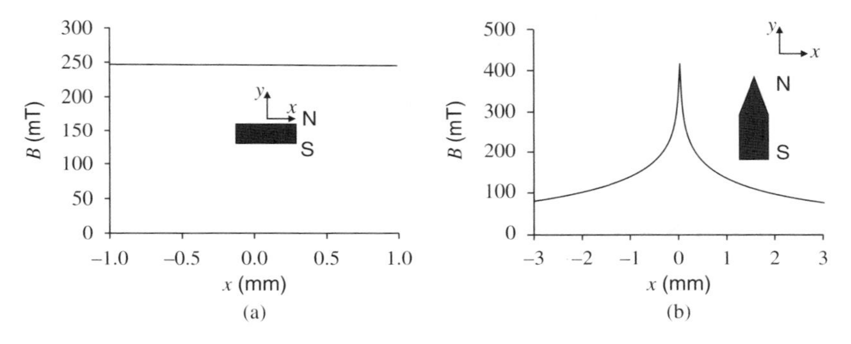

Figure 7.2 (a) Homogeneous magnetic field along the surface of a large NdFeB magnet and (b) inhomogeneous field above the surface of a tapered magnet

Magnetic field is defined as the space around a magnet in which magnetic effect, that is, force of attraction and repulsion on a magnet, can be experienced. Magnetic field strength is also defined as the force experienced by an isolated hypothetical unit north pole placed at that point. It is termed hypothetical as isolated poles or magnetic monopoles do not actually exist. They exist as dipoles. The unit of magnetic field strength is tesla (N/A-m)(Wb/m^2). Figure 7.2 shows the representative magnetic fields around a large magnet and a tapered magnet. The magnetic field around the permanent magnet approximates *homogeneous magnetic field*, and the magnetic field around the tapered magnet approximates *inhomogeneous magnetic field*. Magnetohydrodynamic (MHD) pumping and NMR spectroscopy use homogeneous magnetic field. Here, the magnets of large size with respect to the fluid volume are employed. Inhomogeneous magnetic fields are used for the trapping of particles and the transport of materials within a fluid volume.

The *magnetization* (M) of a block of *ferromagnetic material* is initially zero when the material is not magnetized. The application of external magnetic field (H_{ext}) leads to a net

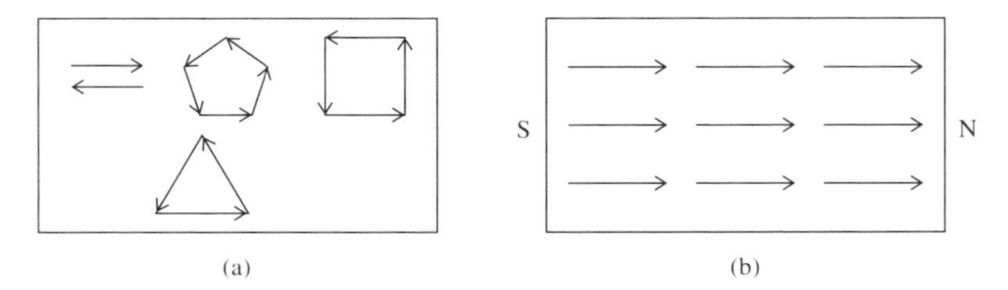

Figure 7.3 Molecular magnets alignment of a (a) unmagnetized substance and (b) magnetized substance

magnetization because of the alignment of the atomic moments with the applied field. Every molecule of a magnetic substance (whether magnetized or not) is a complete magnet in itself having a north pole and a south pole of equal strength. In an unmagnetized substance, the molecular magnets are randomly oriented as closed chains (see Figure 7.3(a)). The north pole of one molecular magnet cancels the south pole of the other so that the resultant magnetism of the unmagnetized specimen is zero. The molecular magnets are realigned for magnetizing the surface so that north poles of all molecular magnets point to one direction and south poles of all molecular magnets point to the opposite direction. The domains with parallel magnetic moments grow with an increase in external magnetization. It ultimately fills the entire block of the material and magnetization of the material cannot grow anymore. The whole piece of material attains its *saturation magnetization*. When the applied field is removed, the net magnetization reduces. However, it does not reach zero magnetization, and there is a *remnant magnetization*, $M_r \simeq B_r/\mu_0$. Ferromagnetic materials are divided into two classes, that is, soft and hard magnetic materials. Figure 7.4 shows the hysteresis loop for soft and hard magnetic materials. Soft magnetic materials have smaller hysteresis loop compared to hard magnetic materials and are easy to magnetize. Permanent magnetic materials are made of hard magnetic materials.

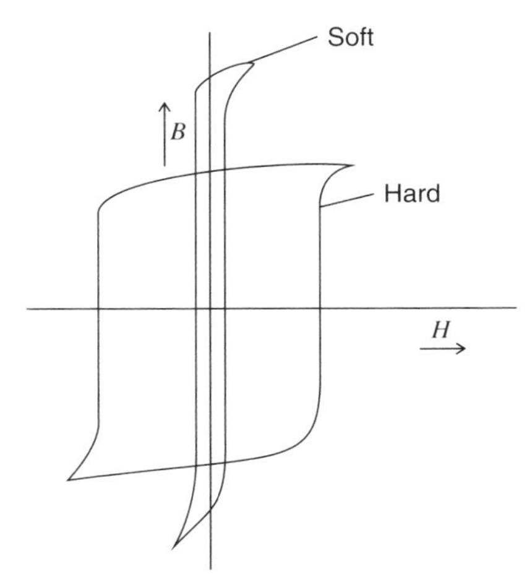

Figure 7.4 Hysteresis loops for soft and hard magnetic materials

Magnetic susceptibility, χ, is the ratio between H and M such that $M = \chi H$. Materials are classified as *diamagnetic*, *paramagnetic*, and *ferromagnetic* depending on the magnetic susceptibility, χ. Diamagnetic materials have $\chi < 0$, paramagnetic materials have $\chi > 0$, and ferromagnetic materials have $\chi \gg 0$. Diamagnetic materials are forced toward the minima of the magnetic field strength, that is, they are repelled from magnetic fields. Paramagnetic materials are attracted to magnetic fields, that is, they experience a small force toward magnetic field maxima. Ferromagnetic materials are strongly attracted to magnetic fields. Most materials, that is, water, proteins, DNA, cells, polymers, wood, and glass, are diamagnetic materials. Oxygen, platinum, and manganese salts are examples of paramagnetic materials. Iron, cobalt, and nickel are classified as ferromagnetic materials. Superparamagnetism is another special case of paramagnetism. *Superparamagnetism* materials have core of small iron oxide crystals encased by polymer shell. The particles are magnetized in a magnetic field. The particles redisperse and behave like nonmagnetic material after the removal of the external field.

7.3 Magnetic Beads

Magnetic beads with size in the range of 50 nm to 2 μm are commercially available. The smaller beads are used to displace smaller targets. For larger targets, larger beads or larger concentration of smaller beads attached to the target particle is used for magnetic actuation. Superparamagnetic beads having magnetization only during application of external field, which lose the magnetic field, are preferred as magnetic beads. These beads are obtained by embedding iron oxide (Fe_2O_3 or Fe_3O_4) paramagnetic micrograins of about 5 nm size in a matrix of latex or polystyrene. Superparamagnetic particles are preferred as the presence of remnant magnetic field leads to unwanted aggregates due to the absence of dispersion by Brownian motion when the external magnetic field is switched off.

For biodiagnostic devices, beads having both magnetic and fluorescence properties are preferred because magnetic fluorescence beads combine two functions, that is, displacement with external magnetic field and detection by fluorescence. Two approaches are available for obtaining combined magnetic and fluorescent effect. In the first approach, the fluorescent markers are incorporated inside the bead during the fabrication process. However, this process can influence the good properties, that is, sphericity, monodispersion, compactness, and so on, of the magnetic beads. In the second approach, the fluorescent particles are bonded to the magnetic bead externally. This approach is widely used due to its convenience of implementation.

One of the questions is the selection of proper magnetic bead size. Smaller beads are preferred as they are easily dispersed by Brownian motion after the removal of the external magnetic field.

After completing the job of carrying the target molecules, the smaller magnetic beads can easily be separated from them by thermal heating. However, it is difficult to free the targets from the aggregate of magnetic beads. The weak magnetic traction of smaller magnetic particle is not a serious drawback as more than one magnetic bead can be attached to the larger target particle (a cell, for instance) leading to the increase in resultant magnetic traction.

7.4 Magnetic Bead Characterization

The magnetic beads are characterized based on the average magnetic properties of the beads. Maxwell's equation can be used to determine the electromagnetic behavior of a material. The magnetic induction based on Maxwell's equation is

$$\bar{B} = \mu_0(\bar{H}_{\text{ext}} + \bar{M}) \tag{7.1}$$

where $\bar{B}$ is the average magnetic induction, $\bar{H}_{\text{ext}}$ is the average external magnetic field, $\bar{M}$ is the average magnetization, and μ_0 is the magnetic permeability of vacuum $= 4\pi \times 10^{-7}$ H/m(N/m^2).

The average magnetization can be expressed as

$$\bar{M} = \chi\bar{H}_{\text{ext}} + \bar{M}_r \tag{7.2}$$

where χ is the magnetic susceptibility of the bead and $\bar{M}_r$ is the average remnant magnetization. Substitution of the above equation in equation (7.1) leads to

$$\bar{B} = \mu_0\mu_r\bar{H}_{\text{ext}} + \mu_0\bar{M}_r \tag{7.3}$$

where the relative magnetic permeability $\mu_r = 1 + \chi$. The magnetic particles are defined based on the relation between the magnetization and external magnetic field. Diamagnetic substances are those in which the individual atoms/molecules do not posses any net magnetic moment on their own. When a specimen of a diamagnetic material is placed in a magnetic field, the field lines do not pass through them (see Figure 7.5). Magnetic permeability of diamagnetic material is less than unity ($\mu_r < 1$). Therefore, susceptibility of diamagnetic substance (χ) is negative. Table 7.1 presents the magnetic susceptibilities (χ) of some materials. Figure 7.6 shows the relation between M and H for different types of magnetic particles. Paramagnetic substances have net nonzero magnetic moment of its own. When a paramagnetic substance is placed in a magnetic field, the magnetic field lines pass through the specimen (see Figure 7.5(b)). The relative magnetic permeability of paramagnetic substance is greater than unity ($\mu_r > 1$). Hence, susceptibility of paramagnetic substance is positive. Ferromagnetic materials show all properties of a paramagnetic substance but to a greater degree ($\mu_r \gg 1$).

For ferromagnetic microparticles, there is a remnant magnetization (M_r) when the external field vanishes. For paramagnetic and superparamagnetic particles, there is no remnant magnetization. The average magnetization of supermagnetic microparticles is higher than that of paramagnetic particles, and the average magnetization is highest for the ferromagnetic particles.

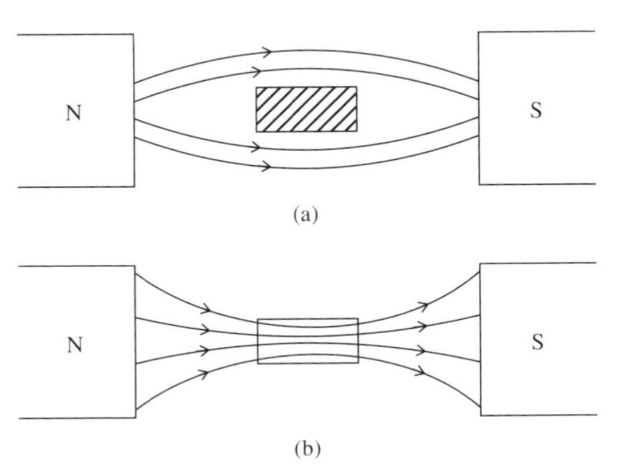

Figure 7.5 Magnetic field lines distribution inside (a) a diamagnetic material and (b) a paramagnetic material

Table 7.1 Magnetic susceptibilities (χ) of some materials at 20 °C

Material	χ
Vacuum	Zero
Copper	-9.8×10^{-6}
Diamond	-2.2×10^{-5}
Gold	-3.6×10^{-5}
Silver	-2.6×10^{-5}
Silicon	-4.2×10^{-5}
Sodium	-0.24×10^{-5}
Carbon (Graphite)	-9.9×10^{-5}
Aluminum	2.3×10^{-5}
Calcium	1.9×10^{-5}
Platinum	2.9×10^{-4}
Magnesium	1.2×10^{-5}
Tungsten	6.8×10^{-5}
Water	-9.035×10^{-6}

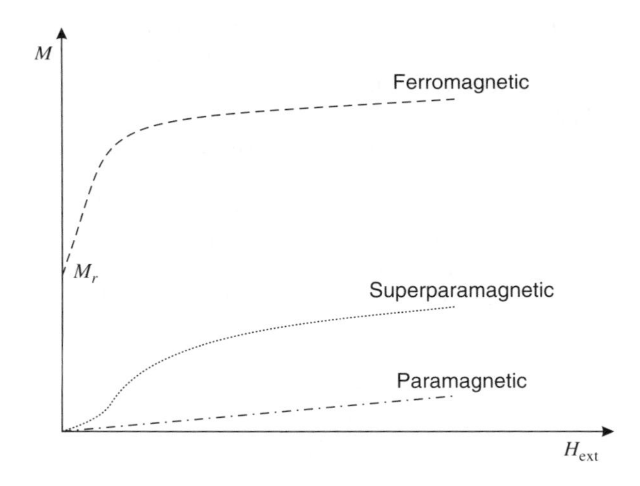

Figure 7.6 The relationship between magnetization (M) and the external magnetic field (H_{ext}) for different types of particles

7.5 Magnetostatics

The starting point for magnetism is the magnetostatic part of Maxwell's equations. Assuming only stationary current densities and neglecting all time derivatives, the two magnetostatic Maxwell's equations are

$$\vec{\nabla} \cdot \vec{B} = 0 \tag{7.4}$$

$$\vec{\nabla} \times \vec{B} = \mu_0 \vec{J}_{\text{tot}} \tag{7.5}$$

$$\vec{J}_{\text{tot}} = \vec{J}_{\text{ext}} + \vec{J}_{\text{mag}} \tag{7.6}$$

The $\vec{J}_{\text{tot}}$, $\vec{J}_{\text{ext}}$, and $\vec{J}_{\text{mag}}$ are total, external, and magnetic current densities, respectively. The current density due to magnetization is given by

$$\vec{J}_{\text{mag}} = \vec{\nabla} \times \vec{M} \tag{7.7}$$

The external current density is given by

$$\vec{J}_{\text{ext}} = \vec{\nabla} \times \vec{H}_{\text{ext}} \tag{7.8}$$

Here, $\vec{B}$ is the magnetic flux density and $\vec{J}$ is the current density. The unit of magnetic flux density is

$$\frac{\text{Newtons-seconds}}{\text{Coulomb-meter}} = \frac{\text{Weber}}{\text{Meter}^2} = \text{Tesla}$$

It may be noted that similar to electric flux density D with unit coulombs per meter square, the unit of magnetic flux density is webers per meter square. Therefore, coulombs and webers refer to electric flux and magnetic flux, respectively. However, it may be noted that similar to electrical charge as coulomb, there is no concept of magnetic charge.

Equations (7.4) and (7.5) specify the divergence and curl of magnetic flux density, $\vec{B}$. Equation (7.4) is also referred as Gauss's law for magnetics, similar to Gauss's law for electrostatics. It states that magnetic flux density neither converges nor diverges from a point. In other words, it says that there is no magnetic charge. Thus, the vector field, $\vec{B}$, rotates about a point but not diverge from it, which is described by equation (7.5). The equation (7.5) known as Ampere's law indicates that the magnetic flux density rotates around the current density, $\vec{J}$, which is a source of magnetic flux density.

7.6 Magnetophoresis

In biotechnology, there is a need to transport a biological target to a specific location of the biochip. The use of carrier fluid to transport these biological samples lacks specificity. Magnetic particles offer a second complementary carrier for the target particles.

Particles possessing either induced or permanent magnetization, M, can be moved by the application of external magnetic field, H. This is known as *magnetophoresis*. It is the magnetic analog of dielectrophoresis. In case of dielectrophoresis, most of the materials have enough dielectric responses to provide a significant effect. In contrast, most materials provide weak magnetic responses in magnetophoresis. Therefore, in case of dielectrophoresis, both target and auxiliary particles are influenced leading to the difficulty in operation of the device and cluttering its functionality. However, this is the advantage for magnetophoresis where full control of target particles is ensured. Magnetophoresis is used for in vitro applications, that is, biodiagnostics, biorecognition, cancer treatment, and flow fractionation.

When considering the use of magnetic actuation for biological application, it is important to investigate the magnetic properties of biological microsystems. DNA, proteins, cells, and antibodies are not magnetic. All bacteria except *Magnetospirillum magnetotacticum* are nonmagnetic. *M. magnetotacticum* has magnetic microreceptors, which uses the Earth's magnetic field for orientation. However, it may be noted that the magnetic actuation of the biological microsystem is possible by binding them to magnetic microparticles.

7.6.1 Magnetophoresis for Biodetection

Figure 7.7 shows a typical schematic for biodiagnostics using magnetophoresis. The objective is the detection of target molecules like DNA, protein cells, and so on. Direct detection of these targets is not very effective due to their lower concentration in the sample volume. Hence, it is necessary to concentrate the target in a small detection chamber. Functional magnetic beads are added to the carrier fluid with targets. The magnetic beads diffuse in the sample volume and bind to the targets. After the binding time, the beads are moved to the small diagnostic chamber by the movement of the external magnet.

The chemical linking between the magnetic beads and target is broken by increasing the temperature with heating the diagnostic chamber. Subsequently, magnetic beads are removed by the external magnet and the targets are concentrated in the detection chamber.

7.6.2 Magnetophoresis for Bioseparation

Magnetophoresis is a strong tool for bioanalysis as the biological samples are not affected or destroyed by the relatively weak magnetic field. Figure 7.8 shows a typical magnetic

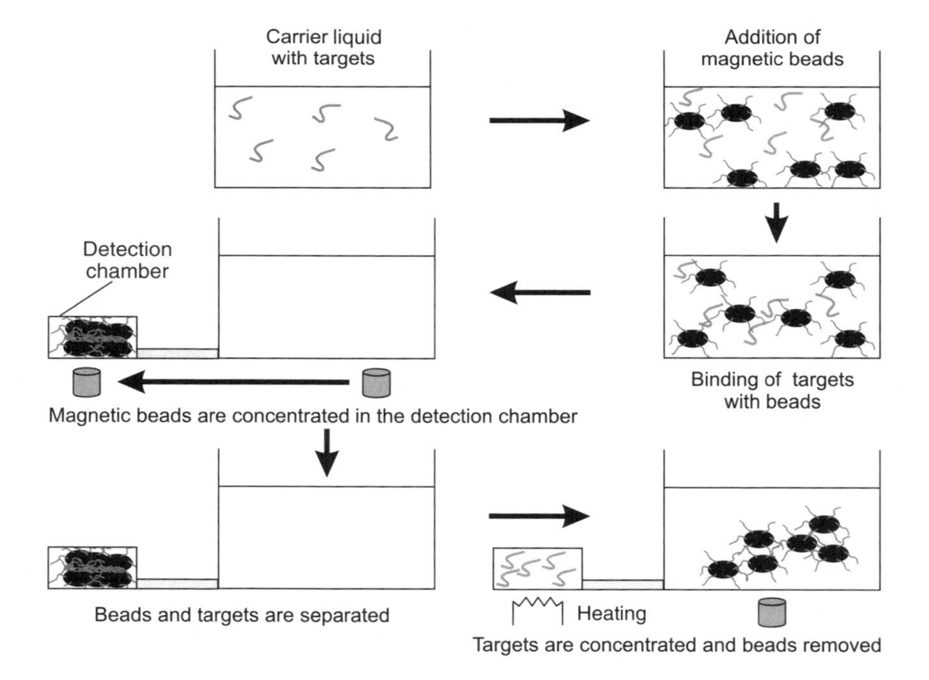

Figure 7.7 Schematic showing the magnetophoresis for the separation of the target molecule

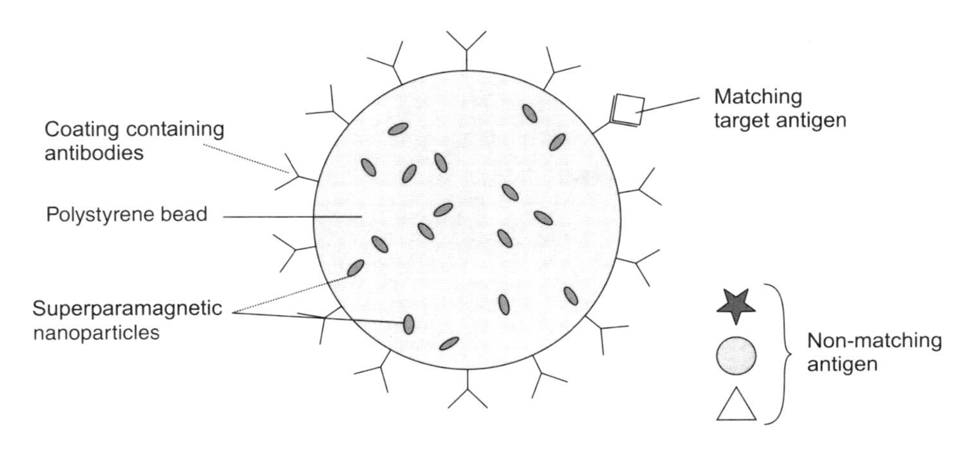

Figure 7.8 A schematic of the microbead for magnetic separation in a lab-on-a-chip (LOC) system

microbead. In Figure 7.8, the surface is coated with a specific antibody for capturing a given antigen (square shape). Antigen in immunology is the abbreviation of antibody generator. It may be a foreign substance from the environment, that is, pollens, viruses, bacteria, and so on. There is no interaction with other nonmatching antigens (triangle, circle, and pentagram). Its main body is a polymer sphere of the order of 1 μm diameter containing a large number of magnetic nanoparticles (iron oxide crystals) with diameters of about 10 nm embedded inside it. In these small particles, atomic magnetic moments are aligned, and the total magnetic moment can rotate freely under the influence of thermal fluctuations. This is also known as superparamagnetism. These particles have vanishing average magnetic moments in the absence of external magnetic field. They acquire large magnetic moments after placing in an external magnetic field. Hence, these particles can be captured by turning on the external magnetic field and released by turning them off. The advantage of polymer is that it allows to coat the microbead with specific biomolecules. Their surface can be coated with carefully chosen antibodies or DNA or RNA strings using a well-controlled biochemical process. After coating with such biomolecules, it is possible to have specific capture of target molecules.

Figure 7.9 shows the implementation procedure for the separation of biomolecules in an LOC system. The microfluidics channel surface has magnetic structures placed at the bottom wall. These magnetic structures can be either electromagnets or a magnetic material, which can be magnetized by electric currents and a magnetic field, respectively. The biocoated microbeads are flushed through the channel first (Figure 7.9(a)). These microbeads are attracted to the magnetic structures when these are turned on. After being captured on the magnetic structures, the antibodies form a layer of capture probes (Figure 7.9(b)). Next, the samples containing many antigens flow through the microchannel (Figure 7.9(c)). The specific antigens matching the antibody on the microbeads are captured (Figure 7.9(d)). The microchannel is rinsed next to flush out nonmatching antigens (Figure 7.9(e)). These target samples can be replaced by turning off the magnetic field, flushing and collecting at the outlet (Figure 7.9(f)). Overall, the square-shaped antigen is the target antigen, which is captured by the antibodies, and others of triangular-, circular-, and pentagonal-shaped antigens are flushed out.

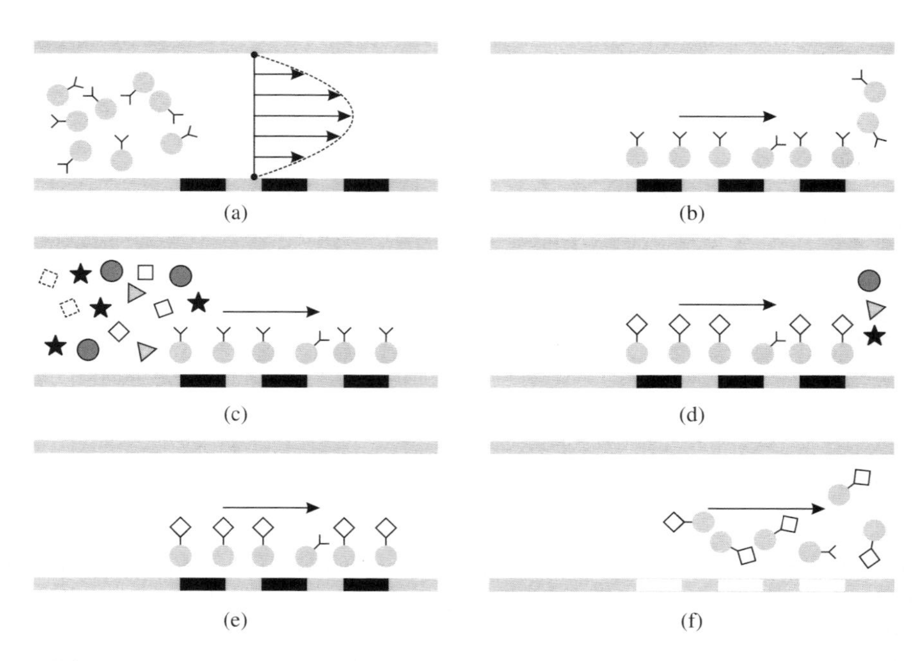

Figure 7.9 The principle of magnetic separation of biomolecules. (a) Flow carrying magnetic microbeads, (b) immobilization of the magnetic microbeads, (c) introduction of samples containing antigens, (d) capture of antigen by the immobilized antibody beads, (e) rinsing of the microchannel, and (f) release of the target sample by deactivating the magnets

7.7 Magnetic Force on Particles

Let us discuss about the magnetic forces acting on a magnetizable body placed in an external magnetic field. The external magnetic field is characterized by H_{ext} before the magnetizable object enters. The object acquires magnetization when placed in a magnetic field. This magnetization, M, generates extra contributions to the magnetic field. Hence, the magnetic field now changes from H_{ext} to H. The derivation of the magnetic force, F, on the magnetizable object is based on the magnetic energy of the system. In magneto-hydrodynamics (MHD), the energy density is written as

$$U_m = \frac{1}{2}(\vec{H} \cdot \vec{B}) \tag{7.9}$$

where B is the magnetic induction. The general expression of the magnetic energy of a particle immersed in a magnetic field H can be derived using equation (7.1) as

$$U_m = -\frac{1}{2}\mu_0 \int_{\forall} \vec{M} \cdot \vec{H} d\forall \tag{7.10}$$

where U_m is the magnetic energy and the integration is carried over the particle volume. The magnetic force on the particle is the gradient of the interaction energy as

$$\vec{F}_m = -\vec{\nabla} U_m \tag{7.11}$$

For small-size magnetic particles, the integration (equation (7.10)) can be replaced by the value of the field at the center of the particle multiplied by the particle volume, and we have

$$U_m = -\frac{1}{2}\mu_0 \forall_p \vec{M} \cdot \vec{H} \tag{7.12}$$

For paramagnetic particles ($M_r = 0$), $M = \chi H$, the magnetic energy is given by

$$U_m = -\frac{1}{2}\mu_0 \forall_p \chi_p |H|^2 \tag{7.13}$$

Considering the interaction between the fluid and the particle, the net magnetic energy of interaction is given as

$$U_{m,net} = -\frac{1}{2}\mu_0 \forall_p (\chi_p - \chi_f)|H|^2 \tag{7.14}$$

where χ_p and χ_f are the magnetic susceptibility of the particle and carrier fluid, respectively. Thus, the net magnetic force on the paramagnetic particle is given by

$$\vec{F}_{m,net} = \frac{1}{2}\mu_0 \forall_p (\chi_p - \chi_f)\nabla|H|^2 \tag{7.15}$$

$\nabla H = 0$ for a homogeneous field and therefore the force on the particle is equal to zero. The particle is magnetized in this case. However, it is not pulled into any direction. When diamagnetic objects ($\chi_p < 0$) are inside diamagnetic medium ($\chi_f < 0$), the particle can be either repelled from or attracted to the magnetic field. When a diamagnetic object ($\chi_p < 0$) is inside paramagnetic medium ($\chi_f > 0$), the diamagnetic particle is repelled from the magnetic field and pushed toward the field minima. The larger the χ_f is, the stronger the repelling force. A paramagnetic particle ($\chi_p > 0$) can be made to act like a diamagnetic material by placing it into a strongly paramagnetic medium ($\chi_f > \chi_p > 0$).

7.8 Magnetic Particle Motion

The magnetic bead motion needs to be known for understanding the performance of a microfluidic magnetic-bead-capture system. There are two possible approaches: (a) single-bead system (discrete approach) and (b) many-bead system (concentration approach). For a single-bead system, the influence of the bead on the liquid flow field is neglected. Hence, the liquid flow can be calculated before the addition of the bead and thus treated as a known field. Similarly, the influence of magnetic bead on the applied external magnetic field is also neglected. Thus, H_{ext} can be calculated without the presence of the bead, and it is a known field. In a many-bead system, the microbeads are modeled as continuous particle density $c(r, t)$ having an associated particle current density, J.

7.8.1 Single-Bead System

In macroscopic scale, the kinematics of a particle is governed by Newton's law as

$$m\frac{d\vec{V}}{dt} = \sum \vec{F}_{\text{ext}} \tag{7.16}$$

where m, $\vec{V}$, and $\vec{F}_{\text{ext}}$ are mass, velocity, and external force, respectively.

In microscopic scale, the effect of Brownian motion, that is, random hitting by other molecules, is more visible, and Newton's formula is replaced by Langevin's law as

$$m\frac{d\vec{V}}{dt} = \sum \vec{F}_{\text{ext}} + R(t) \tag{7.17}$$

where $R(t)$ is the white noise due to Brownian effect. If the size of the bead is larger than 1 μm, the Brownian motion can be neglected and the average trajectory calculated by Newton's law is sufficient for predicting the behavior of the microsystem. Considering the external force to be a combination of gravity, hydrodynamic drag, and magnetic force, Newton's law for a single-bead system is written as

$$m\frac{d\vec{V}_p}{dt} = \vec{F}_{\text{grav}} + \vec{F}_{\text{hyd}} + \vec{F}_{\text{mag}} \tag{7.18}$$

where V_p is the velocity of the particle.
The hydrodynamic drag for low Reynolds number and Stokes flow is

$$\vec{F}_{\text{hyd}} = -C_D(\vec{V}_p - \vec{V}_f) = 6\pi\mu r_h(\vec{V}_p - \vec{V}_f) \tag{7.19}$$

where C_D is the drag coefficient, r_h is the hydraulic diameter of the particle, and V_f is the velocity of the carrier fluid.

Here, it is assumed that the velocity field of the carrier fluid is not influenced by the magnetic bead motion and can be calculated by Navier–Stokes equation (N–S equation) prior to the introduction of the magnetic bead. Similarly, magnetic field is not influenced by the presence of the magnetic bead and can be calculated before attempting to calculate the trajectory of magnetic beads. The magnetic force can be calculated from equation (7.15) as

$$\vec{F}_{\text{mag}} = \mu_0 \forall_p \Delta\chi \nabla \left(\frac{1}{2}H^2\right) \tag{7.20}$$

where $\Delta\chi$ is the difference in magnetic susceptibility between the particle and the fluid. The gravity force is given by

$$\vec{F}_{\text{grav}} \equiv g\forall_p \Delta\rho\vec{j} \tag{7.21}$$

where g is the acceleration due to gravity, $\Delta\rho$ is the density difference between the particle and the liquid, $\vec{j}$ is the vertical unit vector oriented upward, and $\forall_p$ is the volume of the particle.

Thus, the fluid equation for the particle velocity is

$$m\frac{d\vec{V}_p}{dt} = \mu_0 \forall_p \Delta\chi \nabla \left(\frac{1}{2}H^2\right) - 6\pi\mu r_h \left(\vec{V}_p - \vec{V}_f\right) - g\forall_p \Delta\rho\vec{j} \tag{7.22}$$

The x and y coordinates of the particle can be obtained from the velocity field using

$$u_p = \frac{dx}{dt} \tag{7.23}$$

and

$$v_p = \frac{dy}{dt} \tag{7.24}$$

where u_p and v_p are the x and y components of the velocity ($\vec{V}_p$), respectively. The above equation can be solved either analytically or numerically depending on the complexity of the equation. For numerical approach, either Runge–Kutta or predictor–corrector method is used.

7.8.2 *Many-Bead System*

In the previous section, particles are considered as discrete entities. In this section, we model the particles as a continuous component in the carrier fluid. The presence of magnetic beads in the carrier fluid is defined using concentration, that is, the total number of particles per unit volume (ρ_α). Let the total density of the fluid be ρ. The dimensionless concentration can be defined as

$$C_\alpha = \frac{\rho_\alpha}{\rho} \tag{7.25}$$

The particle flux density (J_α) can be due to the combination of convection, diffusion, and magnetic force given as

$$J_\alpha = J_\alpha^{\text{Conv}} + J_\alpha^{\text{Diff}} + J_\alpha^{\text{Mag}} \tag{7.26}$$

The particle flux density, J_α^{Conv}, is due to the global velocity field of the solution given by

$$J_\alpha^{\text{Conv}} = \rho_\alpha V = C_\alpha \rho V \tag{7.27}$$

The diffusion flux density is due to the random motion of the solute relative to the solution given by Fick's law as

$$J_\alpha^{\text{Diff}} = -D_\alpha \rho \nabla C_\alpha \tag{7.28}$$

where D_α is the diffusion coefficient (meter square per second) defined by Einstein's relation as

$$D_\alpha = \frac{K_B T}{6\pi \mu r_h} \tag{7.29}$$

where K_B is the Boltzmann constant and T is the Kelvin temperature.

At a given instant, there will be balance between the magnetophoretic force (F_{mag}) and the viscous drag force (F_{drag}) of a particle.
The viscous drag force is assumed to follow Stokes relation as

$$F_{\text{drag}} = 6\pi \mu r_h \vec{V}_0 \tag{7.30}$$

where V_0 is the velocity of the fluid on a static particle. For a particle moving with velocity, u_p, the above relation is rewritten as

$$\vec{F}_{\text{drag}} = 6\pi \mu r_h (\vec{V}_0 - \vec{u}_p) \tag{7.31}$$

Now, the balance between the two forces gives

$$\vec{F}_{\text{drag}} = \vec{F}_{\text{mag}} \tag{7.32}$$

The magnetic force flux density can be written as

$$J_\alpha^{\text{mag}} = C_\alpha \rho (V_0 - u_p) = C_\alpha \rho \frac{F_{\text{mag}}}{6\pi \mu r_h} \tag{7.33}$$

Hence, the governing equation is the species continuity equation:

$$\frac{\partial C_\alpha \rho}{\partial t} = -\vec{\nabla} \cdot \vec{J}_\alpha \tag{7.34}$$

Substituting the relation for J_α from equation (7.26) and expression of J_α^{Conv}, J_α^{Diff}, and J_α^{Mag} from equations (7.27), (7.28), and (7.33), we have

$$\frac{\partial C_\alpha \rho}{\partial t} = -\vec{\nabla} \cdot C_\alpha \rho \vec{V} + \vec{\nabla} \cdot D_\alpha \rho \vec{\nabla} C_\alpha - \vec{\nabla} \cdot C_\alpha \rho \frac{\vec{F}_{\text{mag}}}{6\pi \mu r_h} \tag{7.35}$$

Canceling mean density ρ from both sides, we have

$$\frac{\partial C_\alpha}{\partial t} = \vec{\nabla} \cdot D_\alpha \vec{\nabla} C_\alpha - \vec{\nabla} \cdot C_\alpha \vec{V} - \vec{\nabla} \cdot C_\alpha b \vec{F}_{\text{mag}} \tag{7.36}$$

where $b = 1/6\pi \mu r_h$ is the Stokes mobility.

The magnetophoretic force should influence the external force contribution while deriving the N–S equation and should be included as body force. Thus, the N–S equation is modified as

$$\rho \frac{\partial \vec{V}}{\partial t} + \rho(\vec{V} \cdot \vec{\nabla})\vec{V} = -\vec{\nabla} p + \mu \nabla^2 \vec{V} + C_\alpha \vec{F}_{\text{mag}} \tag{7.37}$$

Here, the concentration of the magnetic particle is determined by solving the mass conservation equation of the particles in the liquid, while taking into account the convection due to the flow and the magnetic forces.

7.9 Magnetic Field Flow Fractionation

In biotechnology, there is a need to separate particles based on their mass, magnetic properties, and charge, that is, for purification of proteins and obtaining monodisperse magnetic beads. Figure 7.10 shows the schematic of a magnetic field flow fractionation (MFFF) device. Here, the carrier fluid flows horizontally in a channel. A magnetic force field perpendicular to the flow is set up. The resultant force field depends on the nature of the particle, and therefore, the particles take different trajectories and accumulate at different regions of the channel. In Figure 7.11, the upper wall is called the depletion wall and the lower wall is called the accumulation wall.

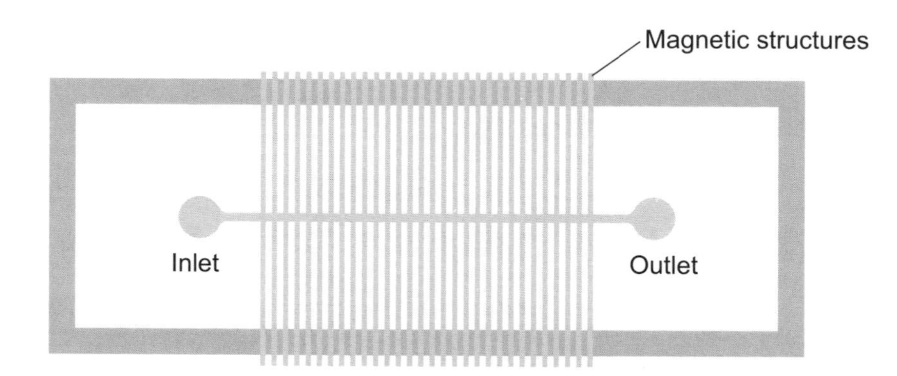

Figure 7.10 A schematic sketch of the magnetic field flow fractionation (MFFF)

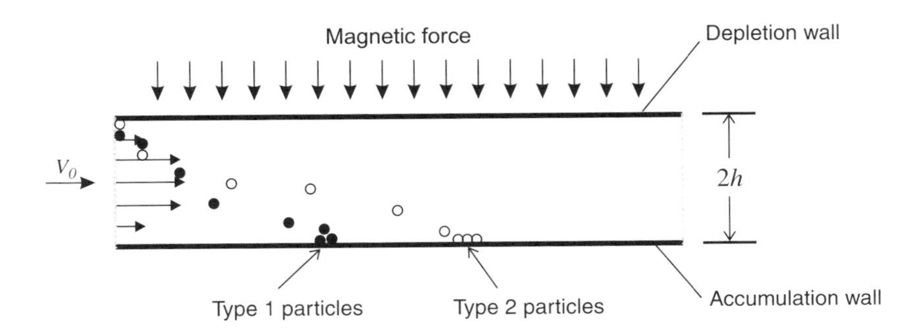

Figure 7.11 A schematic view of magnetic field flow fractionation (MFFF)

The velocity field between two horizontal parallel plates (Figure 7.11) is parabolic given by

$$V_f = \frac{3V_0}{2}\left(1 - \frac{y^2}{h^2}\right) \tag{7.38}$$

where V_0 is the average velocity and h is the half distance between the channel walls. The gravity force and the magnetic force act in the vertical direction. The drag force has a component in both horizontal (x) and vertical (y) directions. The x and y components of the momentum equation can be written using equation (7.22) as

$$\frac{du_p}{dt} = -C_1(u_p - u_f) \tag{7.39}$$

$$\frac{dv_p}{dt} = -C_1 v_p + C_2 \tag{7.40}$$

where u_p and v_p are the horizontal and vertical components of particle velocity and C_1 and C_2 are given by

$$C_1 = \frac{6\pi\mu r_h}{m} \tag{7.41}$$

$$C_2 = \frac{\forall_p \Delta\chi \vec{\nabla}\left(\frac{1}{2}H^2\right)}{m} \cdot \vec{j} + \frac{\vec{g}\forall_p \Delta\rho}{m} \tag{7.42}$$

It may be noted that the vertical velocity of fluid (v_f) is equal to zero due to unidirectional nature of liquid flow. The two equations are uncoupled first-order differential equations. For a homogeneous fluid, C_1 is constant. When the dimension of the magnet is sufficiently large compared to the dimension of the channel, the magnetic gradient is uniform, and thus, C_2 is constant. For constants C_1 and C_2 and the subscript 0 corresponding to the initial values at $t = 0$, the closed form solution of the above equations is

$$u_p = u_{p,0}e^{-C_1 t} + u_f\left[1 - e^{-C_1 t}\right] \tag{7.43}$$

$$v_p = v_{p,0}e^{-C_1 t} + \frac{C_2}{C_1}\left[1 - e^{-C_1 t}\right] \tag{7.44}$$

where C_2/C_1 represents the ratio between the applied external vertical forces and the horizontal drag force.

$$\frac{C_2}{C_1} = \frac{\forall_p \Delta \chi \vec{\nabla}\left(\frac{1}{2}H^2\right)\cdot \vec{j} + g\forall_p \Delta \rho}{6\pi \mu r_h} \tag{7.45}$$

The trajectory of the particle (x_p, y_p) at any time t can be written using equations (7.23), (7.24), (7.38), (7.43), and (7.44) as

$$\frac{dx_p}{dt} = u_{p,0}e^{-C_1 t} + \frac{3V_0}{2}\left(1 - \frac{y_p^2}{h^2}\right)\left(1 - e^{-C_1 t}\right) \tag{7.46}$$

$$\frac{dy_p}{dt} = v_{p,0}e^{-C_1 t} + \frac{C_2}{C_1}\left[1 - e^{-C_1 t}\right] \tag{7.47}$$

The above two equations are coupled as the y coordinate appears in the first x_p equation (equation (7.46)) and the trajectory of the particle is nonlinear.

Let us consider the particle trajectory starting at the top of the channel, where $u_{p,0} = v_{p,0} = 0$. Eliminating time t in equations (7.46) and (7.47) for this case, we get

$$\frac{dx_p}{dy_p} = \frac{3V_0}{2}\left(1 - \frac{y_p^2}{h^2}\right)\frac{C_1}{C_2} \tag{7.48}$$

Taking $x_{p,0} = 0$ and $y_{p,0} = h$ and integrating equation (7.48), we have

$$x_p = -\frac{V_0}{2h^2}\frac{C_1}{C_2}\left[-y_p\left(y_p^2 - 3h^2\right) - 2h^3\right] \tag{7.49}$$

This is a cubic equation. It can be verified from the equation that at $y_p = h$, we get $x_p = 0$. By setting $y_p = -h$, one can find the distance from the inlet at which the particle meets the bottom wall as

$$x_d = 2hV_0\frac{C_1}{C_2} \tag{7.50}$$

The above equation states that the larger magnetic force results in smaller distance, x_d, and the larger hydrodynamic drag results in larger distance, x_d. Let us assume that the particles are small and the gravity force can be neglected compared to the magnetic force. For two different types of particles (1 and 2), a simplified expression for the distance traveled in the x direction by respective particles L_1 and L_2 can be obtained using equations (7.41), (7.42), and (7.50) as

$$\frac{L_2}{L_1} = \frac{\Delta\chi_1}{\Delta\chi_2}\frac{r_1^2}{r_2^2} \tag{7.51}$$

Figure 7.12 shows the trajectories of two types of beads starting from the same point. Particle 1 has a lower magnetic susceptibility and particle 2 has a higher magnetic susceptibility. The two types of magnetic beads are gathering in two distinct packets on the accumulation wall. Thus, Figure 7.12 demonstrates efficient separation between two populations of magnetic beads using MFFF approach.

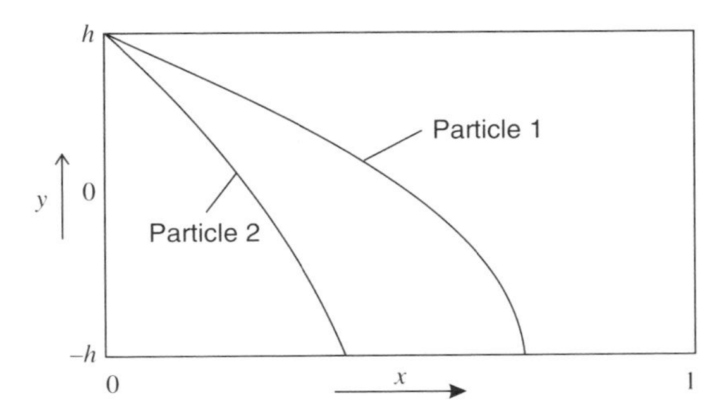

Figure 7.12 Two different trajectories followed by two different types of beads

7.10 Ferrofluidic Pumps

Ferrofluidics or magnetic fluids are another class of material used in microfluidic devices. Ferrofluid is a stable suspension of magnetic nanoparticles in a carrier fluid such as water or an organic solvent. The nanoparticles are coated with a surfactant to prevent aggregation. Ferrofluids retain fluidity in intense magnetic fields. Ferrofluid adopts to any geometry and can be moved through channels. Ferrofluids are superparamagnetic as they are suspensions consisting of nanometer-sized particles.

Ferrofluid plugs transported by magnetic fields can be used for pumping liquids through channels. These pumps can be cheap and easy to implement. The pumped liquid should be immiscible with the ferrofluid. Hydrophobic ferrofluids can be used for pumping aqueous solutions. Figure 7.13 shows the principle of a circular ferrofluid pump. It consists of two

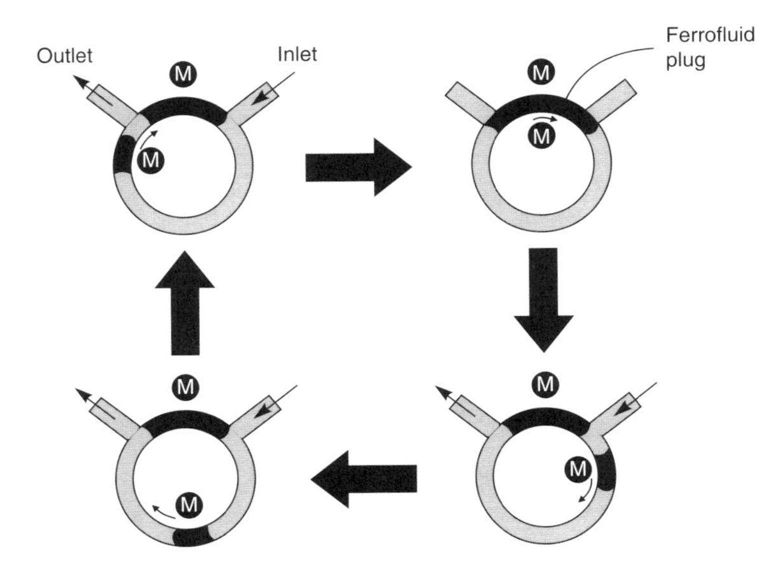

Figure 7.13 The principle of a circular ferrofluidic pump. Two magnets (M) are employed to manipulate ferrofluid plugs in a circular microchannel. One magnet is moving and another is stationary

permanent magnets and two plugs of ferrofluid. One plug is held in place between the inlet and outlet channels. The second plug is moved around a circle by a rotating external magnet. The moving plug pulls and pushes liquid into and out of the circular channel.

Several similar arrangements of magnet and ferrofluid plugs are possible for pumping both gases and liquids. For pumping water, the channel surface is coated with hydrophobic silane to prevent leakage around the ferrofluid plugs.

7.11 Magnetic Sorting and Separation

Figure 7.14 shows two possible arrangements for H-shaped separators. In the first arrangement, two electromagnets A and B are positioned at each end of the connecting channel. The magnets used are electromagnets. Electromagnetic fields are generated around any current-carrying wire. The electromagnetic fields can be switched on and off depending on the applied current. Strong magnetic fields can be achieved by using electromagnets with a highly permeable core material such as soft iron. In the first arrangement (Figure 7.14(a)), magnetic particles are pumped through one of the parallel channels with magnet A adjacent to one channel switched on. Subsequently, magnet A is switched off and magnet B adjacent to the other channel is switched on. The particles located at the channel junction are dragged through the connecting

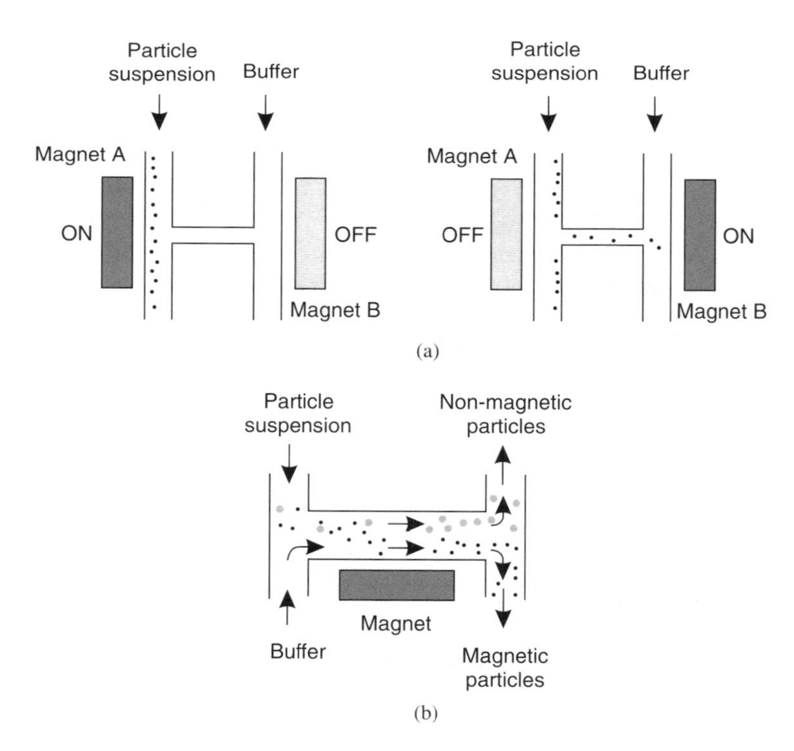

Figure 7.14 Principles of H-shaped separators: (a) from suspension of particles and (b) for continuous flow separation

channel into the neighboring channel. In the other arrangement (Figure 7.14(b)), a sample stream with particles is introduced into one of the inlets, and the buffer solution is introduced into another inlet. The two streams are unmixed and moved through the corresponding outlet channels. Application of magnetic field gradient over the middle channel drags the magnetic particles from the original stream into the buffer stream.

7.12 Magneto-Hydrodynamics

Magnetic force on a current-carrying wire is the primary basis for MHD flow. The straight wire carrying a current, I, when placed in a uniform magnetic field, $\vec{B}$, experiences magnetic force given as (see Figure 7.15)

$$\vec{F}_{\text{mag}} = \vec{I} \times \vec{B}$$

LOC systems integrate common laboratory procedures ranging from filtration, mixing, separation, and detection as a minute chemical processing plant within a single platform. The buffers and solutions used in these systems are many times electrically conductive. Therefore, electric current can be transmitted through these solutions. The electric current results in Lorentz body forces ($\vec{J} \times \vec{B}$) in the presence of an external magnetic field, which can be used to manipulate and propel the fluids. This domain is known as MHD which can be successfully applied to various applications such as pumps, fluidic networks, stirrers, and so on, using low conductivity solutions.

MHD devices can operate either under direct current (DC) or alternating current (AC) electric fields. The DC operation is limited by the electrochemistry of the electrodes leading to bubble formation and electrode corrosion. One must use sufficiently low potential differences between electrodes (less than 1.2 V) to avoid electrolysis of water. Electrolysis of water causes the accumulation of gas bubbles along the surface of the electrode leading to the formation of gas blanket. This gas blanket shields the electrode from current transmission through the solution. The electrode's chemistry in the absence of electrolysis may also lead to the accumulation of unwanted reaction products next to the electrodes reducing the useful life of the device due to electrode corrosion. One possible remedy is to operate with AC electric fields. AC electric field leads to a change in the direction of the flow due to a change in the direction of the electric field. Therefore, one must alternate the magnetic field in synchronous with the alternations in the electric field. AC operation requires the use of electromagnets instead of permanent magnets that are used in DC operation, which increases power consumption. AC

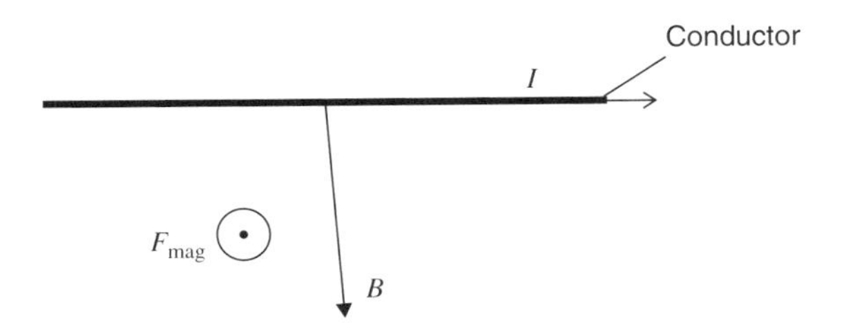

Figure 7.15 Schematic showing the magnetic force (F_{mag}) acting on a current-carrying conductor in a magnetic field

operation also induces parasitic eddy currents that may lead to excessive heating. The use of Redox species that undergo reversible electrochemical reactions also alleviates the disadvantage of DC MHD.

7.13 Governing Equations for MHD

The governing equations for MHD have two components: classical fluid dynamics and electromagnetics. The former includes mass continuity equation and N–S equation. The latter includes Maxwell's equation, current continuity equation, and constitutive equations. For an incompressible electrolyte solution of density, ρ, and viscosity, μ, the continuity and N–S equations are, respectively, described as

$$\vec{\nabla} \cdot \vec{V} = 0 \tag{7.52}$$

$$\rho\left(\frac{\partial \vec{V}}{\partial t} + \vec{V} \cdot \vec{\nabla}\vec{V}\right) = \vec{J} \times \vec{B} - \vec{\nabla}P + \mu\nabla^2\vec{V} \tag{7.53}$$

where $\vec{J}$ is the electric current flux, P is the pressure, and $\vec{B}$ is the magnetic field density. Here, we assume that the liquid's magnetic permeability is sufficiently small so that the magnetic field inside the fluid can be approximated as $\vec{B}$.

The equation for current flux can be written using Ohm's law as

$$\vec{J} = \sigma(\nabla\phi + \vec{V} \times \vec{B}) \tag{7.54}$$

where ϕ is the electric potential, σ is the electric conductivity, $\vec{V}$ is the fluid velocity, and B is the magnetic flux density. The second term in equation (7.54) represents the current induction caused by the motion of a conductor in a magnetic field similar to a dynamo or electric generator.

The current flux for various ionic species of electrolyte solutions can be obtained using the ionic flux density, N_K, of species, K, given as

$$N_K = \vec{V}C_K - D_K\vec{\nabla}C_K - Z_K\frac{D_K}{RT}FC_K(\nabla\phi + \vec{V} \times \vec{B}), K = 1, \dots, N \tag{7.55}$$

where C_K is the molar concentration, D_K is the diffusion coefficient, Z_K is the valance of the Kth ionic species, F is Faraday's constant ($F = 96,484.6$ C/mol), R is the universal gas constant, T is the absolute temperature, and N is the total number of species present in the electrolyte solution. The first, second, and third terms in the right-hand side of equation (7.55) correspond to motion of ionic species due to convection, diffusion, and electromigration, respectively. The current flux can be written as

$$\vec{J} = F\sum_{K=1}^{N} Z_K N_K \tag{7.56}$$

This equation is known as Nernst–Plank equation.
Under steady state conditions:

$$\vec{\nabla} \cdot \vec{N}_K = 0, \quad (K = 1, 2, \dots, N) \tag{7.57}$$

The potential in the electrolyte solution is governed by the local electroneutrality condition:

$$\sum_{K=1}^{N} Z_K C_K = 0 \tag{7.58}$$

The above electroneutrality assumption holds everywhere except in the thin Debye screening layer next to the solid surface. The potential drop across the Debye layer can be significant even though it is only a few nanometers in thickness. It may be noted that ionic mass transport affects the current density, J, which influences the flow field by Lorentz body force. Therefore, one needs to solve simultaneously the full mathematical model consisting of continuity, N–S equation, Nernst–Planck equation, and the local electroneutrality conditions with the appropriate boundary condition.
The energy equation for MHD flow is

$$\rho C_P \left(\frac{\partial T}{\partial t} + \vec{V} \cdot \vec{\nabla} T \right) = K \nabla^2 T + \frac{\vec{J} \cdot \vec{J}}{\sigma} + \phi \tag{7.59}$$

Here, K and C_P are, respectively, the thermal conductivity and specific heat capacity of the fluid. The viscous dissipation $\phi = 2\,\mu \vec{e} \cdot \vec{e}$, where $\vec{e}$ is the strain rate tensor, $e = \frac{1}{2}(\vec{\nabla} u + (\nabla u)^T)$. The second term of equation (7.59) is the ohmic heating due to the passage of the current.

7.13.1 Nondimensionalization

The nondimensionalization of the governing equations (7.52)–(7.54) for an isothermal flow inside a microchannel can be written as

$$J^* = E^* + \frac{1}{R_E}(V^* \times B^*) \tag{7.60}$$

$$\nabla^* \cdot V^* = 0 \tag{7.61}$$

$$\frac{\partial V^*}{\partial t^*} + V^* \cdot \nabla^* V^* = -\frac{1}{Re} \nabla^* P^* + N(J^* \times B^*) + \frac{1}{Re} \nabla^{*^2} V^* \tag{7.62}$$

Here, the characteristic velocity is U, characteristic length is L, characteristic pressure is $\mu U/L$, and characteristic time is L/U. The characteristic scale of magnetic field intensity is the maximum field density, B. The characteristic scale of electric field is the maximum electric field, E. The dimensionless numbers appearing in the above equations are $R_E = \frac{E}{UB}$; Reynolds number, $Re = \frac{\rho UL}{\mu}$; and interaction parameter, $N = \frac{\sigma EBL}{\rho U^2}$. In case of microchannel, R_E has large value due to high electric field. This indicates that J is little dependent on V. The interaction parameter, N, measures the ratio of electromagnetic force to inertia force. For low Reynolds number and high interaction parameter, the momentum equation can be written as

$$0 = -\vec{\nabla} P + \mu \nabla^2 \vec{V} + \vec{J} \times \vec{B} \tag{7.63}$$

7.13.2 DC MHD Micropump

Figure 7.16 shows the schematic of an MHD micropumping system for a microchannel of length, L; height, H; and width, W. A planar electrode of length, L_E, and height, H, is placed on

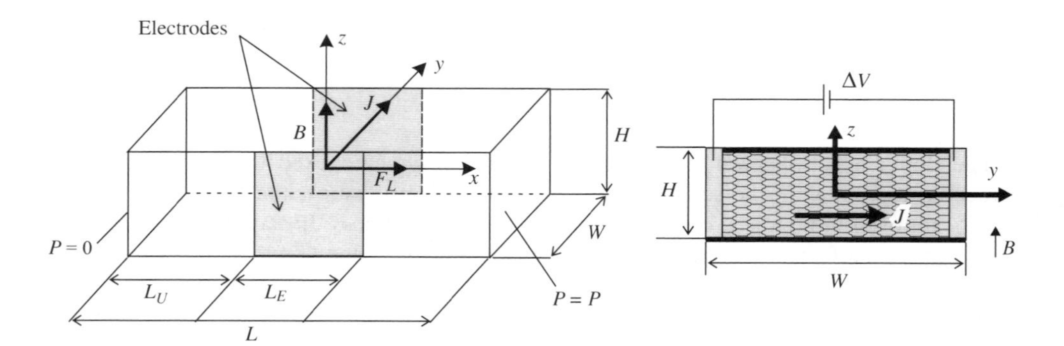

Figure 7.16 A schematic diagram of the MHD pump. Two electrodes with a potential difference ΔV are deposited along the opposing walls of the conduit. The right figure depicts a cross section of the conduit. The conduit is filled with an electrolyte solution and exposed to a uniform magnetic field of intensity B

each vertical surface of the channel. The leading edge of the electrode is located at a distance, L_U, downstream of the channel inlet. The channel is made of a dielectric material. The origin of the Cartesian coordinate system is positioned at the center of the channel. An electric current or potential difference is applied to two planar electrodes. The microchannel is filled with a weakly conductive electrolyte solution. The application of an electric potential difference across the two planar electrodes leads to a current density, J, through the electrolyte solution, $\vec{J} = J\vec{e}_y$, in an undisturbed state. Here, e_i $(i = x, y, \text{and } z)$ represents unit vector in the x, y, and z directions, respectively. A magnetic field with the magnetic flux density, B, is applied in the z direction by a permanent magnet or an electromagnet given by $\vec{B} = B\vec{e}_z$. The interaction between the current density, J, and the magnetic field, B, induces the Lorentz force, $\vec{J} \times \vec{B}$, which provides the driving force for the fluid transport along the channel. If the magnet is much larger than that of the microfluidic device, one can assume the magnetic field to be uniform in the z direction. Let us assume the natural convection generated by the density variation due to electrochemical reactions on the surfaces of the electrode to be negligible. The paramagnetic force induced by the concentration gradients of the paramagnetic species is also assumed to be negligible. The magnetic Reynolds number, $Re_m = \sigma\mu_e UL$, is small for a microchannel flow, where σ is the electrical conductivity, μ_e is the magnetic permeability, U is the characteristic velocity, and L is the characteristic length. This implies that the magnetic field is independent of the flow. There is no bubble generation leading to single phase flow. The magnetic, electric, and fluid properties are assumed as constant. The current density, J, is related to the ionic mass transport and the electrochemical reaction along the surface of the electrode. Even with a constant imposed electric current, the current density varies within the electrolyte solution. However, this complication is excluded, and the current density is assumed as unidirectional, $\vec{J} = J\vec{e}_y$.

The conservation laws for mass and momentum is described by

$$\vec{\nabla} \cdot \vec{V} = 0 \tag{7.64}$$

$$\rho\vec{V} \cdot \vec{\nabla}V = -\vec{\nabla}P + \mu\nabla^2\vec{V} + \vec{J} \times \vec{B} \tag{7.65}$$

The x component of the momentum equation under negligible inertia force assumption is given by

$$-\frac{\partial P}{\partial x}+\mu\left(\frac{\partial^2 u}{\partial y^2}+\frac{\partial^2 u}{\partial z^2}\right)+JB=0 \tag{7.66}$$

This equation describes the balance between the pressure gradient, the viscous diffusion, and the Lorentz force.

Using the fully developed flow solution in a rectangular cross-sectional channel (White, 2006), the following analytical solutions for MHD flow may be assumed as

$$u(y,z)=A_1\sum_{i=1,3,5,\ldots}^{\infty}(-1)^{(i-1)/2}\left[1-\frac{\cos h\left(\frac{i\pi z}{2a}\right)}{\cos h\left(\frac{i\pi b}{2a}\right)}\right]\frac{\cos\left(\frac{i\pi y}{2a}\right)}{i^3} \tag{7.67}$$

$$A_1=A_2\bar{U} \tag{7.68}$$

$$A_2=\frac{48}{\pi^3\left[1-\frac{192a}{\pi^5 b}\sum_{j=1,3,5,\ldots}^{\infty}\frac{\tan h(j\pi b/2a)}{j^5}\right]} \tag{7.69}$$

Here, $a=\min(H/2,W/2)$, $b=\max(H/2,W/2)$, and $\bar{U}$ is the cross-sectional average velocity. It may be noted that $\tan h(\pi/2)=0.9172$ and $\tan h(3\pi/2)=0.9998$. Thus, it may be approximated that $\tan h(j\pi b/2a)=1$ for $j\geq 3$. The above equation for A_2 can be approximated as

$$A_2=\frac{48}{\left[\pi^3-\frac{192a}{\pi^2 b}\left(\tan h\left(\frac{\pi b}{2a}\right)+\frac{31}{32}\zeta(5)-1\right)\right]} \tag{7.70}$$

Here, $\zeta(x)$ is the Riemann zeta function given as

$$\zeta(x)=\sum_{n=1}^{\infty}\frac{1}{n^x} \tag{7.71}$$

It may be evaluated that $\zeta(5)=1.0369$. We can write

$$A_2=\frac{48}{\pi^3-\frac{192}{\pi^2}\frac{a}{b}\left(\tan h\left(\frac{\pi b}{2a}\right)+0.00045\right)} \tag{7.72}$$

Thus, A_2 is only a function of the aspect ratio b/a of the microchannel. Substituting the assumed velocity profile, $u(y,z)$, into the momentum equation and taking the volume integration of the equation for the domain, $-L/2\leq x\leq L/2, -a\leq y\leq a$, and $-b\leq z\leq b$, we get

$$4\Delta Pab-2A_2\bar{U}\mu L\pi\frac{b}{a}\sum_{i=1,3,\ldots}^{\infty}(-1)^{(i-1)/2}\frac{\sin\left(\frac{i\pi}{2}\right)}{i^2}+IBW=0 \tag{7.73}$$

Here, $I=J\times L_E\times H$ is the electric current. The infinite series in the above equation can be simplified as

$$\sum_{i=1,3,\ldots}^{\infty}(-1)^{(i-1)/2}\frac{\sin\left(\frac{i\pi}{2}\right)}{i^2}=\sum_{j=1,2,\ldots}^{\infty}(-1)^j\frac{\sin\left((2j+1)\frac{\pi}{2}\right)}{(2j+1)^2}=\frac{\pi^2}{8} \tag{7.74}$$

Using the above simplification, the integral momentum equation (7.73) gives

$$\bar{U} = \frac{16a^2 \Delta P}{\pi^3 A_2 \mu L} + \frac{4aWB}{\pi^3 b A_2 \mu L} I$$

The volumetric flow rate is

$$Q = 4\bar{U}ab = \frac{64a^3 b}{\pi^3 A_2 \mu} \frac{\Delta P}{L} + \frac{16a^2 WB}{\pi^3 A_2 \mu L} I \tag{7.75}$$

Rewriting in terms of hydraulic conductivity (H_P) and MHD conductivity (H_{MHD}), we have

$$Q = H_P \Delta P + H_{\mathrm{MHD}} I \tag{7.76}$$

where $H_P = \frac{64a^3 b}{\pi^3 A_2 \mu L}$ and $H_{\mathrm{MHD}} = \frac{16a^2 WB}{\pi^3 A_2 \mu L}$.

From equation (7.75), the pressure head $\Delta P_{M,0}$ for zero flow ($Q = 0$) can be written as

$$\Delta P_{M,0} = \frac{16a^2 WBI}{\pi^3 A_2 \mu L} \times \frac{\pi^3 A_2 \mu L}{64a^3 b} = \frac{BIW}{4ab} = \frac{BI}{H}$$

Thus, the pressure capability of an MHD pump can be increased by reducing the height of the microchannel.

7.13.3 AC MHD Micropump

A square-shaped microchannel with $H = W = 800$ µm, $L = 20$ mm having electrolyte of viscosity, $\mu = 6 \times 10^{-4}$ Pa-s will develop flow rate $Q = 599$ µl/s under zero pressure gradient and with Lorentz force per unit length $IB = 0.4$ N/m. Higher flow rate can be obtained by increasing BI, which is limited by electrolysis problem. The use of AC electric field circumvents the problem of electrolysis. However, change in the direction of electric field results in change in the direction of the flow. One must alternate the magnetic field in synch with alternations in the electric field. For AC MHD, the magnetic flux density and current density are, respectively, $B\sin(wt)$ and $J\sin(wt + \phi)$, where w is the angular frequency of the fields and ϕ is the phase angle between the electric and magnetic fields. The low Reynolds number momentum equation for the fully developed AC MHD flow is

$$\rho \frac{\partial u}{\partial t} = -\frac{\partial P}{\partial x} + \mu \left(\frac{\partial^2 u}{\partial y^2} + \frac{\partial^2 u}{\partial z^2} \right) + JB\sin(wt)\sin(wt + \phi) \tag{7.77}$$

Integrating the above equation with respect to wt from $wt = 0$ to 2π, we obtain

$$-\frac{\partial < P >}{\partial x} + \mu \left(\frac{\partial^2 < u >}{\partial y^2} + \frac{\partial^2 < u >}{\partial z^2} \right) + JB\frac{\cos\phi}{2} = 0 \tag{7.78}$$

where $< u >= \int_0^{2\pi} u d(wt)$ and $< P >= \int_0^{2\pi} P d(wt)$.

The difference between the above equation and that of DC MHD equations is the factor $\cos\phi/2$ in the last term on the left-hand side of the above equation. Hence, similar to the DC MHD case, we can write for AC MHD the average velocity and flow rate as

$$<\bar{u}> = \frac{16a^2}{\pi^3 A_2 \mu}\frac{\Delta P}{L} + \frac{2aWB}{\pi^3 b A_2 \mu L} I \cos\phi \tag{7.79}$$

$$<\dot{Q}> = \frac{64a^3 b}{\pi^3 A_2 \mu}\frac{\Delta P}{L} + \frac{8a^2 WB}{\pi^3 A_2 \mu L} I \cos\phi \tag{7.80}$$

The time average flow rate in AC MHD in the absence of the imposed pressure gradient is given by

$$<Q> = \frac{8a^2 WB}{\pi^3 A_2 \mu L} I \cos\phi \tag{7.81}$$

Maximum flow rate for AC MHD is observed when $\phi = 0$ and the maximum flow rate for AC MHD is

$$Q_{\max}(\text{AC MHD}) = \frac{8a^2 WIB}{\pi^3 A_2 \mu \text{L}} \tag{7.82}$$

Comparing the above equation with DC MHD (equation (7.75)), we see that the flow rate in DC MHD is twice that of AC MHD for the same current flow. However, AC MHD can operate at a higher current rating without the problem of electrolysis.

Problems

7.1 A channel of 8 cm long has depth of 7 mm. It is filled with NaCl solution of viscosity equal to 6×10^{-4} Pa-s and density equal to 1058 kg/m^3. The electrode length is equal to 35 mm. The magnetic field is kept at 0.02 T. The input current is set at 0.7 A. Calculate the flow rate in milliliters per second for channel width of (a) 500 μm and (b) 1000 μm. Assume, dP/dx = 0.0.

7.2 A channel dimension is given by $L \times W \times H =$ 80 mm × 3 mm × 7 mm. It is filled with NaCl solution of viscosity, $\mu = 6 \times 10^{-4}$ Pa-s, and density, $\rho =$ 1058 kg/m^3. The magnetic field is kept at 0.02 T and the electrode length is equal to 35 mm. Calculate the average fluid velocity in the channel in centimeters per second for current flows of (a) 0.4 A and (b) 0.8 A. Assume, dP/dx = 0.0.

7.3 An AC MHD micropump uses a microchannel of 20 mm long, 800 μm wide, and 380 μm tall. The electrode length is equal to 4 mm. The magnetic field used is 13 mT. Calculate the average velocity (millimeters per second) in the microchannel for (1) phosphate buffered saline (PBS) (pH 7.2) with 12 mA having viscosity 5×10^{-4} Pa-s and (2) lambda DNA is 5 mm NaCl with 10 mA having viscosity 6×10^{-4} Pa-s.

7.4 For the above problem (3), calculate the time-averaged velocity of PBS with magnetic field of 2.51 mT and applied current of 75 mA at phase angle, ϕ, equal to (a) 45° and (b) 120°.

7.5 Nondimensionalize the energy equation and show the dimensionless form of the energy equation as

$$\frac{\partial T^*}{\partial t^*} + V^* \cdot \nabla^* T^* = \frac{1}{Pe}\nabla^{*^2} T^* + J^* \cdot J^* + \frac{Ec}{Pe}\nabla^* V^* \cdot (\nabla^* V^* + \nabla^* V^*)$$

where the Peclet number $Pe = \frac{UH\rho C_P}{K}$, the Eckert number $Ec = \frac{\mu U^2}{K\nabla T}$, and the height of the conduit is equal to H.

7.6 A magnetic field flow fractionation unit consisting of a square channel of 1 mm × 1 mm cross section is used to separate magnetic beads of different sizes. The magnetic susceptibility of magnetic bead is equal to 3500. The magnetic beads of 1 μm and 10 μm diameter are submerged in water with susceptibility of -9.035×10^{-6}. The 1 μm particle is deposited at a mean distance of 500 mm from the entrance of the channel. Calculate the distance at which the 10 μm particle will be deposited.

Reference

White FM 2006 *Viscous Flow*. 3rd Ed., McGraw-Hill, New York.

Supplemental Reading

Hartshorne H, Backhouse CJ, and Lee WE 2004 Ferrofluid-based microchip pump and valve. *Sens. Actuators, B*, **99**, pp. 592–600.

Hatch A, Kamholz AE, Halman G, Yager P, and Bohringer KF 2001 A ferrofluidic magnetic micropump. *J. Microelectromech. Syst.*, **10**, pp. 215–221.

Jang J and Lee SS 2000 Theoretical and experimental study of MHD (Magnetohydrodynamics) micropump. *Sens. Actuators*, **80**, pp. 84–89.

Kabbani HS, Mack MJ, Joo SW, and Qian S 2008 Analytical prediction of flow field in magnetohydrodynamics based microfluidic devices. *J. Fluids Eng.*, **130**, pp. 091204–091215.

Ostergaard S, Blankenstein G, Dirac H, and Leistiko O 1999 A Novel approach to the automation of clinical chemistry by controlled manipulation of magnetic particles. *J. Magn. Magn. Mater.*, **194**, pp. 156–162.

Pamme N 2006 Magnetism and microfluidics. *Lab Chip*, **6**, pp. 24–38.

Pamme N and Manz A 2004 On-chip free-flow magnetophoresis: continuous flow separation of magnetic particles and agglomerates. *Anal. Chem.*, **76**, pp. 7250–7256.

Qian S and Bau HH 2009 Magneto-hydrodynamics based microfluidics. *Mech. Res. Commun.*, **36**(1), pp. 10–21.

Weddermann A, Wittbracht F, Auge A, and Huetten A 2009 A hydrodynamics switch: microfluidic separation system for magnetic beads. *Appl. Phys. Lett.*, **94**, pp. 173501.

8

Microscale Conduction

8.1 Introduction

Microelectronics technology has become the driving force for miniaturization during the last several decades due to faster operational speeds and more compact systems. Some of the examples are diode lasers, photovoltaic cells, and microelectromechanical systems (MEMS). Most of the studies have focused on electrical and/or microstructural properties of these devices. However, thermal issues which limit the performance of these devices have been overlooked to a great extent. The thermal properties of these materials and associated thermal behavior of these devices are of critical importance for the continuous development of these high-technology systems. The need for increased understanding of this new field of study has given rise to microscale heat transfer. Microscale heat transfer is the study of thermal energy transfer when the continuum models break down and individual carriers are considered.

Microscale conduction phenomena can be introduced using the experimental results of Bertman and Sandiford (1970). Bertman and Sandiford (1970) in their experiment supplied an electric pulse from a generator providing a pulse of heat to the superfluid liquid helium sample at about 1 K. The electrical pulse simultaneously triggers the oscilloscope, and the temperature at the fixed measuring point in the sample was recorded by a resistance thermometer. The oscilloscope trace recorded in their experiment is shown in Figure 8.1. The temperature at the fixed point rises and falls sharply at a certain time after the heat pulse is applied to one end of the sample. A clear time lag between the application of electric pulse and the temperature rise is observed in Figure 8.1.

The classical approach of conduction does not include the above time lag concept or the wavelike response. It expects a delta function-like response of temperature without any time lag with respect to the applied heat pulse. In the classical approach, we use phenomenological models which do not require any knowledge of the mechanism of energy transport or the microstructure of the solids. Fourier's law of heat conduction uses thermal conductivity as a material property which is a function of temperature. This thermal conductivity depends on the microstructure of the solids, which the thermal conductivity data does not show. For example, thermal conductivity of diamond can span an order of magnitude depending on the type of microstructure obtained by chemical vapor deposition. Thermal conductivity of natural

Transport Phenomena in Microfluidic Systems, First Edition. Pradipta Kumar Panigrahi.

Companion Website: www.wiley.com/go/panigrahi/microfluidic

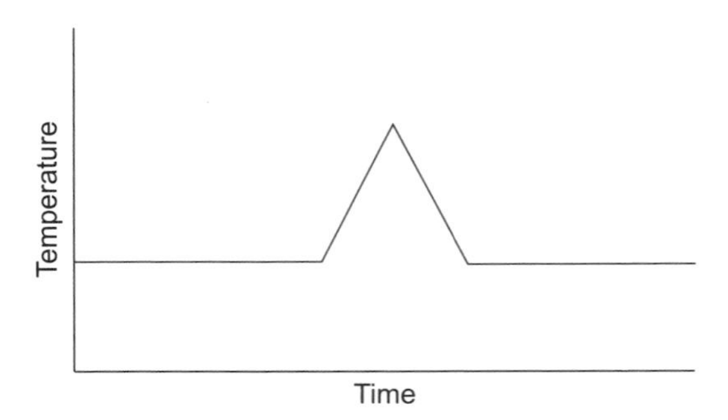

Figure 8.1 Temperature–time trace from oscilloscope during the experiment of Bertman and Sandiford (1970). Note that no temperature and time scale was given

diamond is about 22 W/cm-K which is five times more than copper. Monocrystalline synthetic diamond has thermal conductivity equal to 33.2 W/cm-K. At low temperature (104 K), the conductivity is higher, that is, about 410 W/cm-K.

The classical approach of conduction is sufficient for large-scale systems. However, it is not sufficient for miniaturized engineering systems. Laser heating of semiconductors during fabrication of microscale electronic devices requires understanding of the time scale and length scale of thermal conduction phenomena. Heating and cooling of microelectronic elements involve a time duration of nanosecond or picosecond in which energy is absorbed within a distance of about micron from the surface.

The methodology for microscale heat transfer study can be classified into *three categories*. The first method attempts to modify the *continuum model* in such a way that microscale consideration is taken into account. The second method is application of the *Boltzmann transport equation* (BTE). The third approach is computationally exhaustive *molecular dynamics approach*, which is typically used when the first two methods fail. The present chapter discusses the introductory concepts of microscale conduction.

8.2 Energy Carriers

Heat transfer is the transfer of energy from one region to another by energy carriers. Let us look into the energy carriers for different media, that is, solid, liquid, and gas:

Gas: Gas molecules carry energy either by random molecular motion (diffusion) or an overall drift of molecules in certain direction (convection).

Liquid: In liquids, energy can be transported by diffusion and advection of molecules.

Solid: Energy is transported by phonons, electrons, and photons in solid. *Phonon* is a quantum of crystal vibration energy, which dominates the heat conduction in insulators and semiconductors. Any solid crystal consists of atoms bound into a specific repeating three-dimensional spatial pattern called a lattice. These atoms behave as if they are connected by springs. The thermal energy of the atoms or outside forces makes the lattice vibrate. The atoms are tied together with bonds and they cannot vibrate independently. The vibration of the lattice propagates through the material in the form of collective modes. The propagation speed is equal to the speed of sound in the material. The waves are treated as a particle, called a "phonon," in definite discrete unit or quantum of vibration mechanical energy. *Electrons* dominate energy transport in metals. *Photons* are quanta of electromagnetic energy as phonon is quantum of vibrational mechanical energy. One mode of energy transport in vacuum is by photons. Photons can interact with photons and phonons to render radiative properties of solids.

Note: It is interesting to note that even though the physics of energy carriers are different, they obey Fourier's law of heat conduction for solids, liquids, and gas at macroscopic/continuum scales. The question is: does this universality breaks down for small scales?

8.3 Scattering Mechanism

Photons, phonons, and electrons transport the energy in solid. Electrons are the primary heat carriers in metallic film. Phonons are the primary heat carriers in insulating materials and semiconductors. A different scattering mechanism takes place during the transport process. Figure 8.2 illustrates four different scattering mechanisms of electrons: (1) electron–electron scattering, (2) electron–lattice scattering, (3) boundary scattering, and (4) defect scattering. When the material is heated by ultrashort pulses, the electron system becomes very hot, and electron–electron scattering becomes significant. The electron–lattice scattering becomes important inside the bulk metal. Boundary scattering becomes important when the film thickness is on the order of the mean free path. This is known as size effect, where the physical size of the film influences the transport properties. The defect and grain boundary

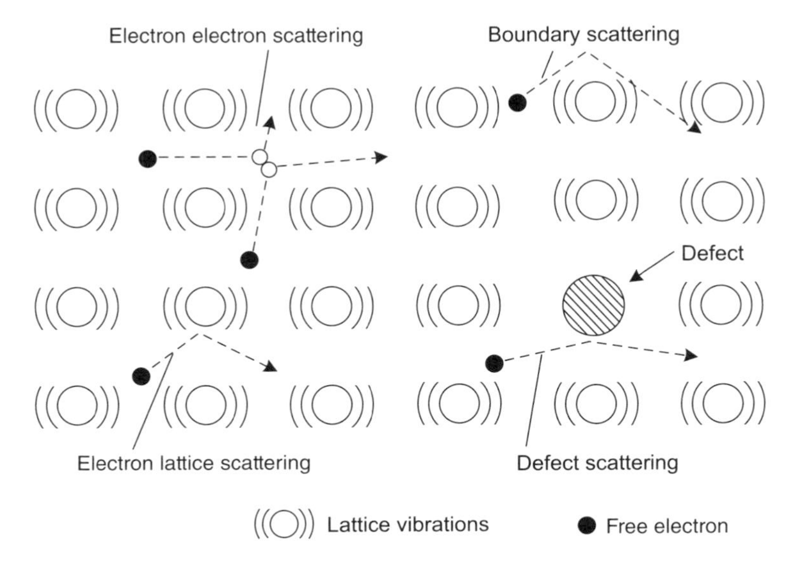

Figure 8.2 Different scattering mechanisms of free electrons within a metal

scattering is another mode depending on the microstructure of the material influenced by the manufacturing conditions.

8.4 Nonequilibrium Conditions

Temperature is the measurement of the average kinetic energy of the molecules. When we write Fourier's law, we assume that one can define the temperature at any point in space. This can be a severe assumption because temperature can be defined only under thermodynamic equilibrium. Someone may ask: if there is thermodynamic equilibrium in the system, why there should be transport of energy? To explain this, we have to resort to the concept of local thermodynamic equilibrium, where we assume thermodynamic equilibrium to be valid over a volume which is much smaller than the overall size of the system. What happens when the size of the object becomes the order of this volume? Hence, the macroscopic or continuum theories break down and new laws based on nonequilibrium thermodynamics need to be formulated.

Note: Nonequilibrium conditions can arise from both size restrictions and also at short time scales. For example, in metals, the response time for electrons is much shorter than that of the crystal vibrations or phonons. If we heat a metal by sufficiently short pulse of energy, only the electrons are energized, leaving the phonons relatively untouched. This creates nonequilibrium between the electrons and phonons and leads to nonequilibrium phenomena. The following section describes various length and time scales required for explanation of nonequilibrium phenomena.

8.5 Time and Length Scales

The smallest length scale for each of energy carriers in solid, that is, phonon, electron, or photon, is its wavelength, λ. The wavelength ranges for these carriers are the following:

Electrons: $\lambda \simeq$ 1–10 nm (in metals)
Phonons: $\lambda \simeq$ 3–20 nm (in semiconductors and insulators)
Photons: $\lambda \simeq$ 0.4–0.7 μm (in visible spectrum), 0.7–50 μm (for infrared range)

The scattering of energy carriers results in resistance to energy transport. In the absence of scattering, the conductivity will be infinite. There are several time and length scales associated with the scattering process, which are discussed in the succeeding text.

Collision Time (τ_c)

Collision time is the duration of collision. In classical physics, collisions are considered instantaneous. However, there is a finite collision time during wave scattering. The collision time is defined as the ratio of wavelength of the carrier and the propagation speed of the carrier. The typical values of collision time (τ_c) are as follows:

Electrons: $\tau_c \simeq 10^{-15}$ s (for metals)
Phonons: $\tau_c \simeq 10^{-13}$ s

Mean Free Time (τ)

Mean free time, τ is the average time between collisions. Generally, $\tau \gg \tau_c$. The typical values of mean free time are:

Electrons: $\tau \simeq 10^{-14}$ s (for metals)
Phonons: $\tau \simeq 10^{-11}$ s

Relaxation Time (τ_r)

Relaxation time is the statistical time lag value (a nonnegative constant) needed to establish steady-state heat flow conditions in a small elemental volume of material when a temperature gradient is suddenly imposed on the boundary. Chandrasekharaiah (1986) has reported relaxation time for different types of materials. The relaxation time for gases is in the range of 10^{-10} s, and the relaxation time for metals is in the range of 10^{-14} s. Relaxation time, τ_r, is associated with local thermodynamic equilibrium. Equilibrium is achieved in 5–20 collisions, $\tau_r > \tau_c$. **Note:** A system may have different momentum (τ_{rm}) and energy ($\tau_{r\varepsilon}$) relaxation times. For example, when the collision of an electron is elastic, the momentum is changed but not its energy. Therefore, typically, $\tau_{r\varepsilon} > \tau_{rm}$.

Diffusion Time (τ_d)

The diffusion time (τ_d) depends on the characteristic size of an object (L) and is equal to L^2/α, where "L" is the size of the object and "α" is the thermal diffusivity.

Length Scale

Analogous to above time scales, we have the corresponding length scales, that is, *wavelength* (λ) and *mean free path* ($l = v\tau$), where v is the particle speed. *Relaxation length* (l_r) is the characteristic size of volume over which the local thermodynamic equilibrium can be defined.

8.6 Scale Effects

Fourier's model of heat conduction applies to certain regimes or scales of dimensions, times, and temperature regimes. The applicability of microscopic conduction model can be described based on the following criteria: (a) space scale, (b) time scales, and (c) temperature regimes.

Space Scale

Tien and Chen (1994) identified two microscale heat transfer regimes. The first microscale regime represents the classical size effect, which is defined as follows:

(i) *First regime*

$$\frac{L}{l} < 0(1)$$

or

$$\frac{d_r}{l} < 0(1) \tag{8.1}$$

and

$$\frac{L}{\lambda_c} > 0(1) \tag{8.2}$$

where L is the characteristic device dimension, l is the heat carrier mean free path, d_r is the temperature penetration depth, and λ_c is the characteristic wavelength of electrons or phonons. The penetration depth is the ratio of characteristic temperature divided by the maximum temperature gradient defined as

$$d_r = \frac{T}{\left(\frac{dT}{dx}\right)_{\max}} \tag{8.3}$$

(ii) *Second regime*

In the second microscale heat transfer regime, the quantum size effects are important, which is defined as

$$\frac{L}{\lambda_c} = 0(1) \tag{8.4}$$

Figure 8.3 shows the approximate regimes of applicability of Fourier's diffusion, BTE approach, molecular dynamics, and quantum BTE approach for the simulation of thermal transport in silicon at room temperature. The dominant wavelength λ_c and mean free path l for silicon at 300 K are shown in the figure for reference. Fourier's law is valid for devices with length scale greater than 0.5 μm. Quantum approach is essential for devices with length scale of the order of nanometers, that is, nanotubes. Molecular dynamics and Boltzmann transport approach are applicable for devices with length scale greater than nanometers. The computational requirement poses practical limitation for the use of molecular dynamics simulation. Computational requirement for molecular dynamics simulation approach is computationally prohibitive for larger systems with boundaries that cannot be approximated

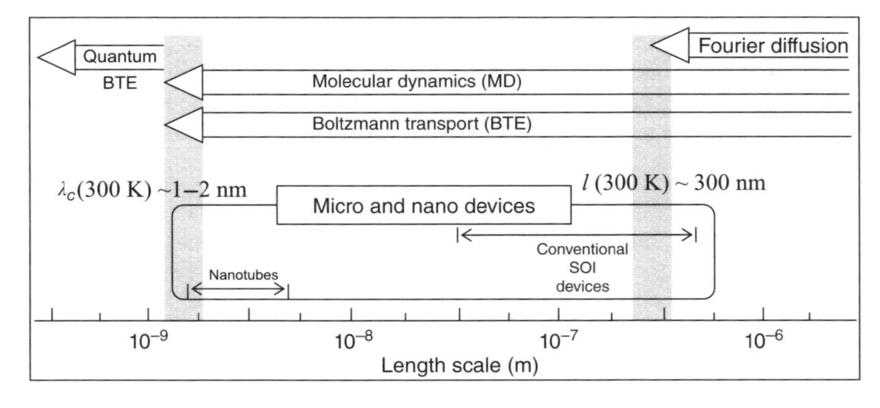

Figure 8.3 Range of applicability for different heat transfer modeling approach of silicon at ambient temperature as a function of the system length

by boundary conditions. The advantage of molecular dynamics simulation is that no prior assumption is needed about physical properties of phonons.

Temperature Regimes

Extreme temperature regime can contribute to the importance of microscale transport phenomena. Peshkov (1944) experimentally observed the wave type of temperature propagation in liquid helium at temperature less than 2.2 K. Overall, this experimental evidence suggests that the wave type of temperature propagation may become important for certain material under certain extreme conditions, that is, at very low temperature near absolute zero. The wave-type propagation of heat is also called second sound. This concept was introduced by Landau (1941), which predicts the speed of second sound between $\frac{C_l}{\sqrt{3}}$ and zero at 2.2 K, where C_l is the ordinary first speed of sound. Peshkov (1944) measured the velocity of temperature as 1.9 m/s at 1.4 K.

8.6.1 Approach Details (Methodology)

Depending on the time and length scales, different transport laws can be used. When the objects have comparable size to the wavelength of energy carrier, wave phenomena, that is, reflection, refraction, diffraction, etc., dominate the energy transport mechanism. When the time scale of interest (t) is of the order of collision time scale (τ_c), time-dependent wave mechanics must be used. Schrodinger's equation must be used for electrons and phonons. Maxwell's equation must be used for photons:

Case I [$L \gg l, l_r$ and $t \gg \tau, \tau_r$]. Local thermodynamic equilibrium can be applied over both space and time leading to validity of Fourier's law of heat conduction.

Case II [$L \gg l, l_r$ and $t \simeq \tau, \tau_r$]. Local thermodynamic equilibrium can be assumed over space and time-dependent terms cannot be averaged.

Case III [$L \simeq l, l_r$ and $t \gg \tau, \tau_r$]. Time-averaged statistical transport equations need to be used.

Case IV [$L \simeq l, l_r$ and $t \simeq \tau, \tau_r$] Statistical transport equations without any spatial or temperature averaging need to be used.

8.7 Fourier's Law

Most of the engineering applications are analyzed by Fourier's law, which is based on the phenomenological models. Heat transfer by conduction takes place by molecular motion including the transport by heat carriers (free electrons) and vibration of lattices (phonons) from a high-temperature region to low-temperature region by diffusion inside the medium. The heat conduction deals with random movement of large number of heat carriers. The movement of individual heat carriers is mostly considered not to significantly affect the heat conduction phenomena. Therefore, continuum assumption is used in macroscopic sense ruling out the detail movement of heat carriers on heat conduction. This assumption neglects the scales of the media and microstructures of the media. The Fourier law is assumed to be valid for the following regime:

$$\frac{L}{l} \gg 0(1)$$

$$\frac{t}{\tau_r} \gg 0(1)$$

$$T \gg 0°(K)$$

where L is the physical characteristic length of the system, l is the heat carriers' mean free path, t is the physical time, τ_r is the mean relaxation time of the heat carriers, and T is the temperature in kelvin temperature scale.

The Fourier heat flux model under macroscopic continuum formulation is given as

$$\vec{q}(\vec{r},t) = -K\nabla T(\vec{r},t) \tag{8.5}$$

Substituting this to the first law of thermodynamics of a given system gives the general heat conduction equation as

$$\vec{\nabla} \cdot K\nabla T(\vec{r},t) + S(\vec{r},t) = \rho C_p \frac{\partial T(\vec{r},t)}{\partial t} \tag{8.6}$$

where K is the thermal conductivity tensor, S is the internal heat source, ρ is the density, and c_P is the specific heat.

Assuming constant thermophysical property, the above equation reduces to

$$\nabla^2 T(\vec{r},t) + \frac{S(\vec{r},t)}{K} = \frac{1}{\alpha}\frac{\partial T(\vec{r},t)}{\partial t} \tag{8.7}$$

where α is the thermal diffusivity (for isotropic material) equal to $K/\rho C_p$.

The heat conduction equation in an isotropic material with no internal source is given as

$$\nabla^2 T(\vec{r},t) = \frac{1}{\alpha}\frac{\partial T(\vec{r},t)}{\partial t} \tag{8.8}$$

For steady-state and constant thermal conductivity, the heat conduction equation is

$$\nabla^2 T(\vec{r},t) + \frac{S(\vec{r},t)}{K} = 0 \tag{8.9}$$

which is a Poisson equation.

For steady state with no internal heat source and constant thermal conductivity, the heat conduction equation is given as

$$\nabla^2 T = 0 \tag{8.10}$$

which is the Laplace equation.

8.8 Hyperbolic Heat Conduction Equation

The lagging response in time between the heat flux vector and the temperature gradient observed by Bertman and Sandiford (1970) requires a new look into the macroscopic relationship between the heat flux and the temperature gradient. The following constitutive relationship is proposed for heat flux:

$$\vec{q}(\vec{r},t+\tau) = -K\vec{\nabla}T(\vec{r},t) \tag{8.11}$$

The above equation indicates that the temperature gradient established at time t results in a heat flux vector at a later time $t + \tau$.

The constitutive relationship proposed in equation (8.11) can be expanded using Taylor's expansion as

$$\vec{q}(\vec{r}, t + \tau) = \vec{q}(\vec{r}, t) + \tau \frac{\partial \vec{q}(\vec{r}, t)}{\partial t} + 0(\tau^2) = -K\vec{\nabla}T(\vec{r}, t) \quad (8.12)$$

When τ is small, second-order term can be neglected. Thus, we have

$$\vec{q}(\vec{r}, t) + \tau \frac{\partial \vec{q}(\vec{r}, t)}{\partial t} = -K\vec{\nabla}T(\vec{r}, t) \quad (8.13)$$

This is same as the Cattaneo type of model originally developed for gases. This assumes finite speed of heat propagation contrary to the infinite propagation speed assumption of Fourier's model (zero relaxation time).

Let us take the divergence of the above equation:

$$\vec{\nabla} \cdot \vec{q}(\vec{r}, t) + \tau \vec{\nabla} \cdot \frac{\partial \vec{q}(\vec{r}, t)}{\partial t} = -\vec{\nabla} \cdot K\vec{\nabla}T(\vec{r}, t) \quad (8.14)$$

The energy equation is

$$-\vec{\nabla} \cdot \vec{q}(\vec{r}, t) + S(\vec{r}, t) = \rho C_p \frac{\partial T(\vec{r}, t)}{\partial t} \quad (8.15)$$

where S is the volumetric heat generation rate.

Combining the above two equations, we have

$$S(\vec{r}, t) - \rho C_p \frac{\partial T(\vec{r}, t)}{\partial t} + \tau \frac{\partial}{\partial t}\left(S(\vec{r}, t) - \rho C_p \frac{\partial T(\vec{r}, t)}{\partial t}\right) = -\vec{\nabla} \cdot K\vec{\nabla}T(\vec{r}, t) \quad (8.16)$$

Rearranging the above equation, we have

$$S(\vec{r}, t) + \tau \frac{\delta S(\vec{r}, t)}{\delta t} - \rho C_p \frac{\delta T(\vec{r}, t)}{\delta t} - \rho C_p \tau \frac{\delta^2 T(\vec{r}, t)}{\delta t^2} = -\vec{\nabla} \cdot K\vec{\nabla}T(\vec{r}, t) \quad (8.17)$$

At constant thermal conductivity, equation (8.17) can be simplified as

$$K\nabla^2 T(\vec{r}, t) + S(\vec{r}, t) + \tau \frac{\delta S(\vec{r}, t)}{\delta t} = \rho C_p \frac{\delta T(\vec{r}, t)}{\delta t} + \rho C_p \tau \frac{\delta^2 T(\vec{r}, t)}{\delta t^2} \quad (8.18)$$

The rearrangement of the above equation gives

$$\nabla^2 T + \frac{1}{K}\left(S(\vec{r}, t) + \tau \frac{\delta S(\vec{r}, t)}{\delta t}\right) = \frac{1}{\alpha}\left(\frac{\delta T(\vec{r}, t)}{\delta t} + \tau \frac{\delta^2 T(\vec{r}, t)}{\delta t^2}\right) \quad (8.19)$$

When τ is small, the above equation reduces to classical diffusion equation.

A wave term represented by $\frac{\tau}{\alpha}\frac{\delta^2 T}{\delta t^2}$ is present in equation (8.19). Here, $\sqrt{\alpha/\tau}$ is a velocity-like quantity known as *thermal wave speed* C. Here, α and τ are thermal properties of the medium. At steady state, the conduction equation reverts back to the Fourier model, even though the relaxation parameter, $\tau \neq 0$. Therefore, temperature results for two models differ during the transient state only.

The application of Fourier's conduction equation and hyperbolic heat conduction equation to the transient heating of semi-infinite solid is discussed in the following section.

8.8.1 Fourier's Conduction in Semi-Infinite Solid

The application of hyperbolic heat conduction equation is shown here using a pulse heating example of semi-infinite solid. The wall at $x = 0$ is impulsively stepped to a temperature T_w (Figure 8.4).

The initial and boundary conditions for the problem are

$$T = T_0 \quad \text{at} \quad t = 0, \quad x > 0$$
$$T = T_w \quad \text{at} \quad x = 0, \quad t > 0$$
$$T \to T_0 \quad \text{at} \quad x \to \infty, \quad t > 0$$

The energy equation based on Fourier's law of conduction for the above problem is

$$\frac{\partial^2 T}{\partial x^2} = \frac{1}{\alpha}\frac{\partial T}{\partial t} \tag{8.20}$$

It may be noted that $\sqrt{\alpha t}$ has the dimension of length. Let us define a similarity variable, ζ, as

$$\zeta = \frac{x}{\sqrt{\alpha t}} \tag{8.21}$$

Using chain rule,

$$\begin{aligned}
\frac{\partial T}{\partial x} &= \frac{dT}{d\zeta}\frac{\partial \zeta}{\partial x} = \frac{1}{\sqrt{\alpha t}}\frac{dT}{d\zeta} \\
\frac{\partial^2 T}{\partial x^2} &= \frac{1}{\alpha t}\frac{d^2 T}{d\zeta^2} \\
\frac{\partial T}{\partial t} &= \frac{dT}{d\zeta}\frac{\partial \zeta}{\partial t} = -\frac{x}{2t\sqrt{\alpha t}}\frac{dT}{d\zeta}
\end{aligned} \tag{8.22}$$

Now, the modified governing equation (equation (8.20)) using the similarity variable is

$$\frac{d^2 T}{\partial \zeta^2} + \frac{\zeta}{2}\frac{dT}{d\zeta} = 0 \tag{8.23}$$

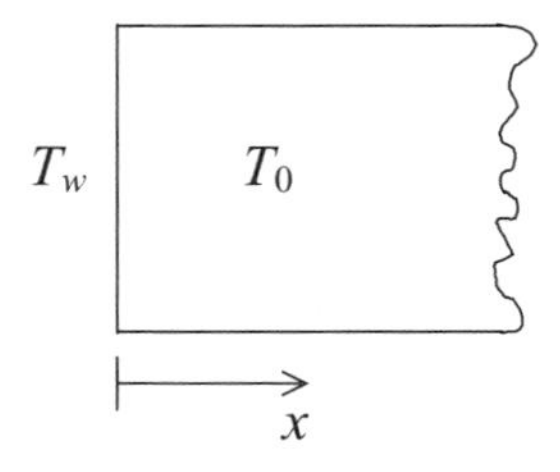

Figure 8.4 A schematic of semi-infinite solid exposed to a constant temperature at one end

The integration of the above equation gives

$$\frac{d\left(\frac{dT}{d\zeta}\right)}{\left(\frac{dT}{d\zeta}\right)} = -\frac{\zeta}{2}d\zeta$$

$$\ln\frac{dT}{d\zeta} = \frac{-\zeta^2}{4}d\zeta$$

$$\frac{dT}{d\zeta} = C_1\exp\left(\frac{-\zeta^2}{4}\right)$$

$$T = C_1\int_0^{\zeta}\exp\left(\frac{-\zeta^2}{4}\right)d\zeta + C_2 \tag{8.24}$$

Imposing the B.C.s, we have

$$\frac{T(x,t)-T_0}{T_w - T_0} = erfc\left(\frac{x}{2\sqrt{\alpha t}}\right) \tag{8.25}$$

8.8.2 Hyperbolic Conduction in Semi-Infinite Solid

The hyperbolic heat conduction governing equation for the ID problem is

$$\frac{\partial^2 T}{\partial x^2} = \frac{1}{\alpha}\frac{\partial T}{\partial t} + \frac{\tau}{\alpha}\frac{\partial^2 T}{\partial t^2}$$

$$\frac{\partial^2 T}{\partial x^2} = \frac{1}{\alpha}\frac{\partial T}{\partial t} + \frac{1}{c^2}\frac{\partial^2 T}{\partial t^2} \tag{8.26}$$

where c is the thermal wave speed. The presence of second-order time derivative in hyperbolic heat conduction equation requires one additional boundary condition for solution compared to the Fourier conduction equation.

The boundary conditions are

$$T = T_0 \text{ and } \frac{\delta T}{\delta t} = 0 \text{ at } t = 0, x > 0$$
$$T = T_w \text{ at } x = 0, t > 0$$
$$T \to T_0 \text{ at } x \to \infty, t > 0$$

Let us define the dimensionless variables as

$$\theta = \frac{T(x,t)-T_0}{T_w - T_0} \quad \text{(normalized temperature)}$$
$$\beta = \frac{c^2 t}{2\alpha} \quad \text{(dimensionless time)}$$
$$\delta = \frac{cx}{2\alpha} \quad \text{(dimensionless distance)}$$

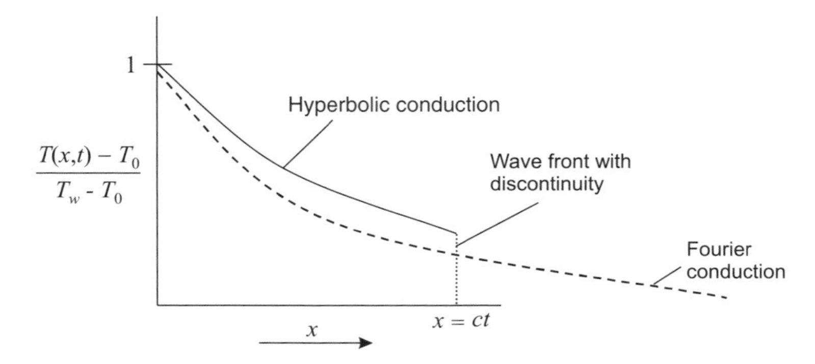

Figure 8.5 The comparison of temperature distribution in a semi-infinite solid based on Fourier's conduction and hyperbolic conduction approach

Substituting the dimensionless variables in governing equation (8.26), we have

$$\frac{\partial \theta}{\partial \beta^2} + 2\frac{\partial \theta}{\partial \beta} = \frac{\partial^2 \theta}{\partial \delta^2} \tag{8.27}$$

The boundary conditions in dimensionless form are

$$\begin{aligned} \theta = 0, \frac{\delta\theta}{\delta\beta} = 0 \ \text{ at } \ \beta = 0, \delta > 0 \\ \theta = 1 \ \text{ at } \ \delta = 0, \beta > 0 \\ \theta \to 0 \ \text{ at } \ \delta \to \infty, \beta > 0 \end{aligned}$$

Either the Laplace transform or the numerical solution procedure can be used to obtain the solution of the above problem (equation (8.27)).

Figure 8.5 shows the normalized temperature distribution inside the semi-infinite solid after imposition of the temperature pulse from both Fourier's conduction and hyperbolic conduction approach. There is propagation of a weak shock wave through the medium at speed C based on the hyperbolic conduction approach. The above comparison explains the observation from Bertman and Sandiford's (1970) experiment. The sensor in their experiment senses the heat pulse at a later time after initiation of the temperature pulse, that is, when the shock wave reaches it. Figure 8.6 shows the thermal wave passing through the semi-infinite solid. It divides the solid into two parts, that is, undisturbed region and disturbed region through which the thermal wave has passed. In the disturbed region, the temperature variation shows a similar trend as that from diffusion solution. Figure 8.7 shows the schematic representing the transport process of particles leading to a net energy flux through a plane.

8.9 Kinetic Theory

Kinetic theory of transport phenomena is the first step toward understanding complex transport processes in a medium. The movement of energy-carrying particles is modeled inside the medium. Figure 8.7 shows the schematic representing the transport process of particles leading to a net energy flux through a plane.

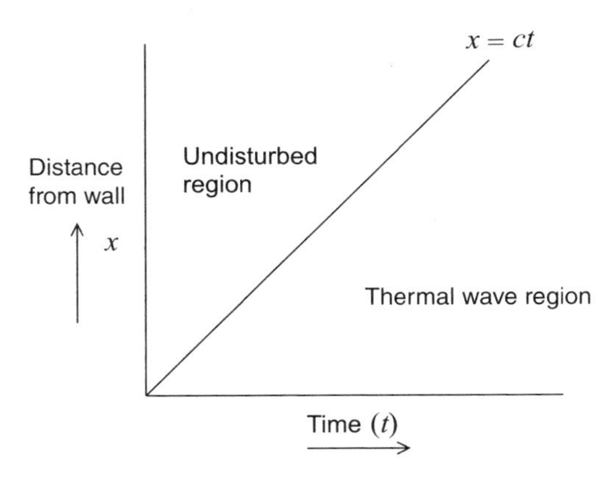

Figure 8.6 Schematic showing the traveling of thermal wave

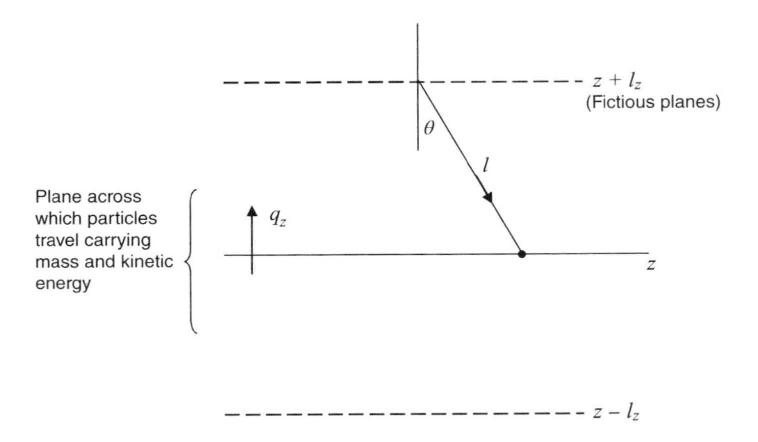

Figure 8.7 The schematic showing the transport phenomena based on kinetic theory

The terminology with reference to Figure 8.7 can be defined as follows:

l_z: z component of mean free path l, which makes an angle, θ
$u(z + l_z)$: The energy density of particles moving down from plane, $z + l_z$
$u(z - l_z)$: The energy density of particles moving up from plane, $z - l_z$
v: velocity of particles

The net energy flux of particle in the +ve z-direction is

$$q_z = \frac{1}{2} v_z [u(z - l_z) - u(z + l_z)] \tag{8.28}$$

Note: Here, 1/2 is used because it is assumed that only 1/2 of the total number of particles at each location move up from $z - l_z$ plane or down from plane $z + l_z$.

Using

$$\frac{du}{dz} = \frac{u(z + l_z) - u(z - l_z)}{2l_z} \tag{8.29}$$

we rewrite equation (8.28) as

$$q_z = -v_z l_z \frac{du}{dz} = -\cos^2\theta v l \frac{du}{dz} \quad \text{(using:} \quad v_z = v\cos\theta, l_z = l\cos\theta) \tag{8.30}$$

Averaging over the whole hemisphere of solid angle 2π,

$$q_z = -vl\frac{du}{dz}\left[\frac{1}{2\pi}\int_{\psi=0}^{2\pi}\int_{\theta=0}^{\pi/2}\cos^2\theta\sin\theta d\theta d\psi\right] = -\frac{1}{3}vl\frac{du}{dz} \tag{8.31}$$

where ψ is the azimuthal angle, θ is the polar angle, and $\sin\theta\, d\theta\, d\psi$ is the elemental solid angle. Let us assume local thermodynamic equilibrium such that "u" is a function of temperature. Using the chain rule, we can write equation (8.31) as

$$q_z = -\frac{1}{3}vl\frac{du}{dT}\frac{dT}{dz}$$

$$q_z = -\frac{1}{3}cvl\frac{dT}{dZ} \tag{8.32}$$

where c = heat capacity $= \frac{du}{dT}$. Alternatively, equation (8.32) can be written as

$$q_z = -K\frac{dT}{dz} \tag{8.33}$$

where $K = \frac{cvl}{3}$ for Fourier's law.

Note: We have not made any assumption on the type of energy carriers. Hence, this is a universal law for all energy carriers. Only assumption is made about local thermodynamic equilibrium. Neglecting photon contribution, the thermal conductivity can be written as

$$K = \frac{1}{3}\left[(cvl)_{\text{lattice}} + (cvl)_{\text{electron}}\right] \tag{8.34}$$

where the lattice contribution is dominant for semiconductors and insulators and the electron contribution is dominant for metals.

Note: For copper, $C \propto T$ at low temperature while l is constant. Therefore, $K \propto T$ for copper at low temperature. However, at high temperature, for copper, $l \propto T^{-n} (n > 1)$. Therefore, $K \propto \frac{1}{T}$ at high temperature for copper. Figure 8.9 has shown the thermal conductivity variation of copper as a function of temperature.

8.10 Heat Capacity

Heat capacity of a material is defined as the change in internal energy resulting from a change in temperature. The energy within a crystalline material is stored in the free electrons of a metal and within the lattice in the form of vibrational energy.

8.10.1 Electron Heat Capacity

Electron heat capacity of a free electron metal, C_e, is defined as

$$C_e = \frac{\partial u_e}{\partial T} = \frac{\partial}{\partial T} \int_0^{\infty} \varepsilon D(\varepsilon) f(\varepsilon) d\varepsilon \tag{8.35}$$

where $D(\varepsilon)$ is the number of states of particle between energy ε and $(\varepsilon + d\varepsilon)$ and $f(\varepsilon)$ is the Fermi–Dirac function (occupational probability of free electron gas):

$$f(\varepsilon) = \frac{1}{e^{\frac{(\varepsilon - \mu)}{K_B T}} + 1} \tag{8.36}$$

where μ is the thermodynamic potential, K_B is the Boltzmann constant, and T is the temperature of the electron gas. The chemical potential μ is a function of temperature, which can be approximated by the Fermi energy for temperatures at or below room temperature (Kittel, 1996). The Fermi energy can be calculated using the relation

$$\varepsilon_F = \frac{h^2}{2m}\left(\frac{3\pi^2 N_e}{V}\right)^{2/3} \tag{8.37}$$

where m is the effective mass of an electron, h is Planck's constant, N_e is the number of free electrons, and V is the volume of electron wavevectors. It may be noted that Fermi energy, ε_F, is visualized as a sphere plotted as a function of wavevector, where the radius is given by the Fermi wavevector, which is the wavevector of the highest occupied energy level. No two electrons can occupy the same energy state according to the Pauli exclusion principle. The electrons fill the energy levels beginning with the lowest energy level. The energy of the highest level occupied at zero temperature is called Fermi energy. The integral for free electron capacity calculation can be simplified as

$$C_e = \frac{\pi^2 K_B^2 n_e}{2\varepsilon_F} T \tag{8.38}$$

Here, n_e is the electron number density. The above expression indicates electron capacity C_e to be a linear function of temperature.

8.10.2 Phonon Heat Capacity

The phonon heat capacity, C_l, is defined as the derivative of the internal energy, u_l, stored within the vibration lattice.

$$C_l = \frac{\partial u_l}{\partial T} = \frac{\partial}{\partial T}\left[\sum_s \int D_s(w) n_s(w) h w_s \partial w\right]$$

where $D_s(w)$ is the phonon density of states, which is the number of phonon states with frequency between w and $(w + dw)$ for each phonon branch designated by s. The integer quantity $n_s(w)$ is the number of phonons with energy hw_s. The number of phonons at a particular

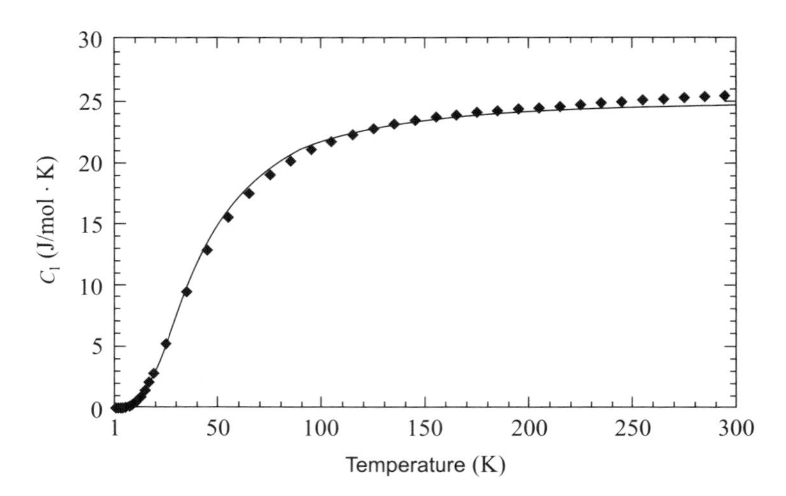

Figure 8.8 Molar specific heat of Au compared to the Debye model using 170 K for the Debye temperature (Weast *et al.*, 1985)

frequency represents the amplitude to which that vibrational mode is excited. The number of phonons with a particular frequency w at an equilibrium temperature is given by the relation (Bose–Einstein statistical distribution)

$$n_s(w) = \frac{1}{e^{hw/K_BT} - 1} \tag{8.39}$$

An expression for the phonon density of states is required to calculate the lattice heat capacity, C_l. Two common models for the calculation of density of phonon states are (a) Debye model and (b) Einstein model.

Figure 8.8 shows the lattice thermal conductivity of Au and comparison with the Debye model. It shows T^3 temperature dependence of thermal conductivity at low temperature.

8.10.3 Electron Thermal Conductivity in Metals

The kinetic theory of gas predicts that three factors govern the thermal conduction in metals: (1) heat capacity of energy carrier, (2) average velocity, and (3) the mean free path. We have seen earlier that the electron heat capacity is linearly related to temperature. Fermi–Dirac distribution dictates that electrons located at energy levels near the Fermi energy undergo transitions and transport energy. The electrons contain purely kinetic energy. As all electrons involved in the transport of energy have kinetic energy close to the Fermi energy, they travel at velocities near the Fermi velocity. The Fermi velocity is given by

$$v_F = \sqrt{\frac{2}{m}\varepsilon_F} \tag{8.40}$$

The mean free path of electron is a direct function of the electron collision frequency. Electrons can collide with other electrons, lattice, defects, grain boundaries, and surfaces.

Assuming that each scattering mechanism is independent, collisional rate is the sum of the individual scattering mechanism:

$$\upsilon_{\text{tot}} = \upsilon_{\text{ee}} + \upsilon_{\text{el}} + \upsilon_d + \upsilon_b \tag{8.41}$$

where υ_{ee} is the electron–electron collisional frequency, υ_{el} is the electron–lattice collisional frequency, υ_d is the electron-defect collisional frequency, and υ_b is the electron-boundary collisional frequency. All these scattering mechanisms are important in the area of microscale heat transfer.

The collisional frequency can also be dependent on temperature. Electron-defect and electron-boundary scattering are typically independent of temperature. Debye temperature is a characteristic temperature arising in the computation of Debye specific heat. It is defined as

$$T_D = \frac{hw_{\max}}{K_B} \tag{8.42}$$

where h is Planck's constant, $w_{\max}$ is the maximum frequency of oscillation (Debye frequency), and K_B is the Boltzmann constant. It is also defined as the temperature at which the wavelength of vibration of the atoms in a crystal is equal to the length of the unit cell. The Debye temperature of copper and aluminum is, respectively, 343.5 K and 428 K. Above the Debye temperature, electron–lattice collisional frequency is proportional to the lattice temperature, and electron–electron scattering is proportional to the square of the electron temperature:

$$\upsilon_{\text{ee}} \simeq A\,T_e^2 \tag{8.43}$$

$$\upsilon_{\text{el}} \simeq B\,T_l \tag{8.44}$$

Here, A and B are constants and T_e and T_l are the electron and lattice temperature, respectively.

The temperature dependence of the thermal conductivity isolates different mechanisms affecting the thermal conductivity. Figure 8.9 shows thermal conductivity of three materials

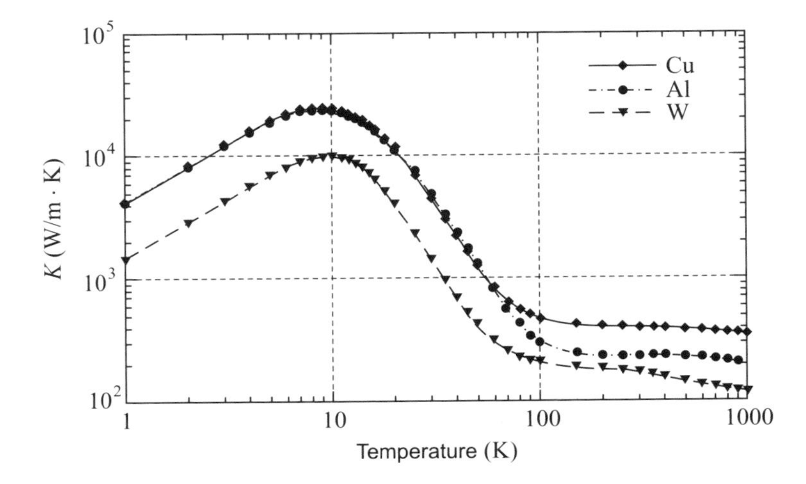

Figure 8.9 Thermal conductivity of Cu, Al, and W plotted as function of temperature

(Cu, Al, and W) which are commonly used in the electronics industry. There is general similarity in temperature dependence of the three metals. The primary scattering mechanism at very low temperature (below 10 K) is due to either defect or boundary scattering. Both these mechanisms are independent of temperature. However, the linear temperature dependence of the electron heat capacity is responsible for the linear relationship between the thermal conductivity and temperature in this regime. The thermal conductivity is roughly independent of temperature at temperatures above the Debye temperature due to competing temperature effects. In this regime, electron heat capacity increases with temperature. The mean free path is inversely proportional to the temperature due to increased electron–lattice collisions. Therefore, the thermal conductivity decreases above the Debye temperature.

8.10.4 Lattice Thermal Conductivity

Phonons are primarily responsible for thermal conduction within the crystalline lattice. Phonons are originally defined as the amplitude of a particular vibrational mode with finite energy content. Phonons are also assumed as particles analogous to a localized wave packet. Debye model is generally adopted when modeling the thermal transport properties. Group velocity is assumed equal to the speed of sound within the material, which is independent of temperature. The phonon heat capacity is proportional to T^3 at very low temperatures, and the heat capacity is nearly constant at temperatures above Debye temperature. Only long-wavelength phonons are excited at low temperatures, which have small wavevectors. Therefore, only normal scattering processes occur at low temperature, which do not contribute to thermal resistance. Therefore, phonon–phonon collisions do not contribute to low-temperature thermal conductivity. All allowable modes of vibration are excited at higher temperatures, that is, above the Debye temperature, and overall phonon population increases with temperature. The mean free path, l_l, is inversely proportional to temperature:

$$l_l \propto \frac{1}{T} \tag{8.45}$$

Figure 8.10 shows thermal conductivity of three elements (C, Si, and Ge) as a function of temperature. The general trend of thermal conductivity is similar for all materials. At low temperature, normal scattering processes do not affect the thermal conductivity. Defect and boundary scattering are independent of temperature. Therefore, the temperature dependence of thermal conductivity only arises from the heat capacity and follows the expected T^3 behavior. The heat capacity becomes constant at higher temperature and mean free path decreases. Therefore, the thermal conductivity at higher temperature shows approximately T^{-1} behavior.

8.10.5 Scale Effects on Thermal Conductivity

The characteristic dimensions in several areas of technology are on the order of micrometers or nanometers. The thermal conductivity, K, defined for large-scale applications need to be modified when the physical dimensions become small. Figure 8.11 shows the electron and phonon trajectories for films of the same material having thickness L_1 and L_2 where one thickness (L_1) is larger than the other thickness (L_2). The physical boundaries of the film also

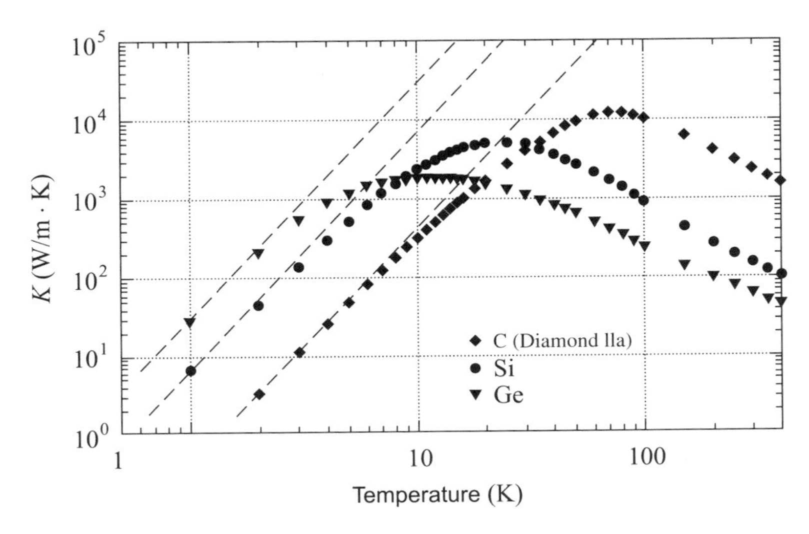

Figure 8.10 Thermal conductivity of three elements having diamond structure as a function of temperature

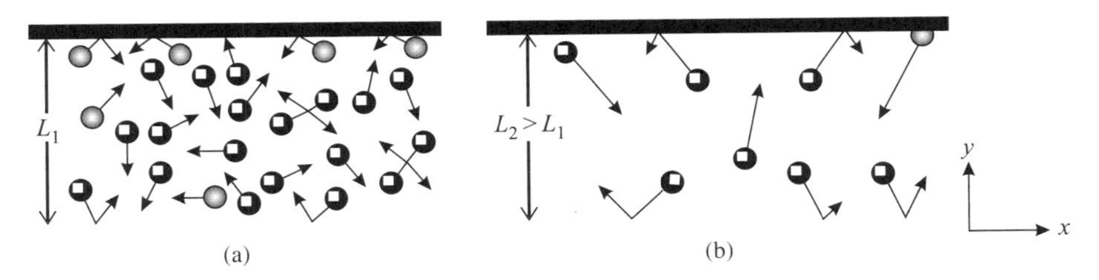

Figure 8.11 Schematic of electron and phonon trajectories in (a) relatively thick film and (b) relatively thin film

scatter the energy carrier and redirect their propagation. When L/λ_{mfp} is larger, the influence of boundaries on reducing the average energy carrier path length is minor. The conduction heat transfer occurs similar to that of bulk materials. The physical boundaries of the material can decrease the average net distance traveled by the energy carriers for the thin film thickness case. Electrons and phonons moving in the thin y-direction (representing conduction in the y-direction) are affected by the boundaries to a more significant degree than energy carriers moving in the x-direction. Therefore, for thin film characterized by small L/λ_{mfp}, we have $K_y < K_x < K$, where K is the bulk thermal conductivity of the film material.

Flik *et al.* (1992) proposed the following relation for K_x and K_y when $L/\lambda_{\text{mfp}} > 1$:

$$\frac{K_x}{K} = \frac{1 - 2\lambda_{\text{mfp}}}{(3\pi L)} \tag{8.46}$$

$$\frac{K_y}{K} = \frac{1 - \lambda_{\text{mfp}}}{(3L)} \tag{8.47}$$

Table 8.1 Mean free path and critical film thickness for various materials at $T \simeq 300$ K

Material	λ_{mfp} (nm)	$L_{\mathrm{crit},y}$ (nm)	$L_{\mathrm{crit},x}$ (nm)
Aluminum oxide	5.08	36	22
Diamond (IIa)	315	2200	1400
Gallium arsenide	23	160	100
Gold	31	220	140
Silicon	43	290	180
Silicon dioxide	0.6	4	3
Yttria-stabilized zirconia	25	170	110

The above relation predicts that the thermal conductivity within 20% value of the mean free path and critical film thickness below which microscale effects must be considered is tabulated in Table 8.1.

The above equations indicate that values of K_x and K_y are within approximately 5% of the bulk thermal conductivity if $L/\lambda_{\mathrm{mfp}} > 7$ (for K_y) and $L/\lambda_{\mathrm{mfp}} > 4.5$ (for K_x). Therefore, the above relations can be used for determination of thermal conductivity when $\lambda_{\mathrm{mfp}} < L < L_{\mathrm{crit}}$. No general guidelines exist for thermal conductivity calculation when $L/\lambda_{\mathrm{mfp}} < 1$.

The above example shows the effect of scattering from physical boundaries. Chemical dopants embedded within a material or by grain boundaries may also redirect the energy carriers. Nanostructured materials may provide very small grain size and increase the scattering and reflection of energy carriers at the grain boundaries. The measured thermal conductivity values of bulk, nanostructured yttria-stabilized zirconia material is shown in Figure 8.12. High-temperature combustion devices such as gas turbine engines use this ceramic for insulation purpose. The mean free path of the energy carrier for yttria-stabilized zirconia is $\lambda_{\mathrm{mpf}} = 25$ nm. The reduction in grain size to less than 25 nm and introduction of more grain boundaries in the material per unit volume significantly reduce the thermal conductivity and works as an efficient insulator.

8.11 Boltzmann Transport Theory

Kinetic theory formulation assumes local thermodynamic equilibrium in space and time. It cannot be applied when length scale $L \simeq l$ or l_r and time scale $t \simeq \tau$ or τ_r. The Boltzmann transport theory is more fundamental and can be applied at these situations. The BTE applies to all particles, that is, electrons, ions, photons, phonons, gas molecules, and so on.

The BTE for gas molecules has been introduced in earlier chapter as

$$\frac{\partial(nf)}{\partial t} + \zeta_j \frac{\partial(nf)}{\partial x_j} + F_j \frac{\partial(nf)}{\partial \zeta_j} = J(f, f^*) \tag{8.48}$$

$$\frac{\partial f}{\partial t} + \vec{v} \cdot \vec{\nabla} f + \vec{F} \cdot \frac{\partial f}{\partial p} = \left(\frac{\partial f}{\partial t} \right)_{\mathrm{Scat}} \tag{8.49}$$

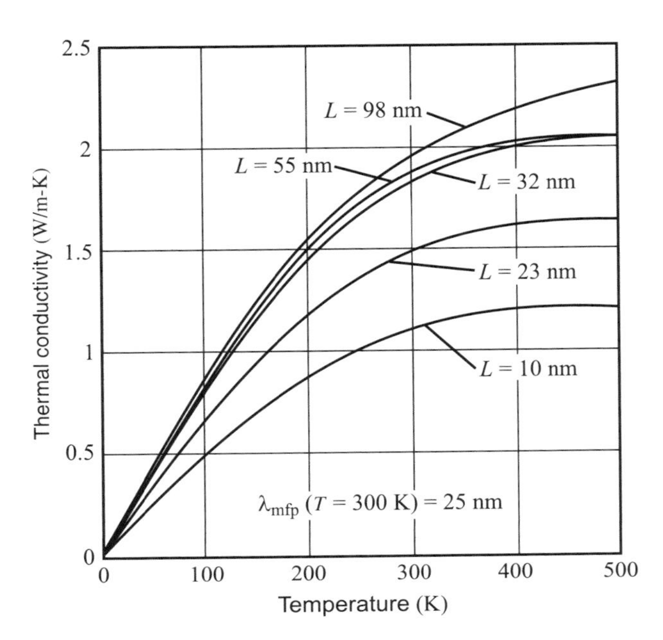

Figure 8.12 Measured thermal conductivity of yttria-stabilized zirconia as a function of temperature and mean grain size (Yang *et al.*, 2002).

Here, $f(\vec{r}, \vec{p}, t)$ is the statistical distribution of ensemble of particles, $\vec{r}$ is the position vector, $\vec{p}$ is the momentum vector, υ is the average heat carrier velocity, and F is the force applied to the particle. The term $\frac{df}{dt}|_{\text{scattering}}$ is the time rate of change of the scattering term or the so-called collision term. The Boltzmann equation describes the time rate of change of the heat carrier distribution due to diffusion. The Boltzmann equation in one dimension without any external force in reduced form is

$$\frac{\partial f}{\partial t} + \vec{\upsilon} \cdot \frac{\partial f}{\partial x} = \left[\frac{\partial f}{\partial t}\right]_{\text{scattering}} \tag{8.50}$$

Note that among all particles, only electrons and ions can experience appreciable force due to electric and magnetic fields. As discussed in the earlier chapter, BGK model can be used for simplifying collision integral as

$$J(f, f^*) = \gamma_*(f^{\text{o}} - f^*) \tag{8.51}$$

where γ_* is the collision frequency $(\frac{\upsilon}{l})$. Similar to the above model, we can use

$$\left(\frac{\partial f}{\partial t}\right)_{\text{Scat}} = \frac{f_0 - f}{\tau(\vec{r}, \vec{p})} \tag{8.52}$$

where f_0 is the equilibrium distribution and $\tau(\vec{r}, \vec{p})$ is the mean relaxation time of the heat carriers at velocity, υ. The next step is to solve BTE to obtain $f(\vec{r}, \vec{p}, t)$. Then, the rate of energy flow per unit area or the energy flux can be written as

$$\vec{q}(\vec{r},t) = \Sigma_p f(\vec{r},\vec{p},t)\vec{v}(\vec{r},t)\varepsilon(\vec{p}) \tag{8.53}$$

where $\vec{v}$ is the velocity vector and $\varepsilon(\vec{p})$ is the particle energy (integral over momentum). Introducing the density of states $D(\varepsilon)$), the energy flux equation can also be written as

$$\vec{q}(\vec{r},t) = \int f(\vec{r},\vec{p},t)\vec{v}(\vec{r},t)\varepsilon(\vec{p})d^3p \tag{8.54}$$

$$\vec{q}(\vec{r},t) = \int \vec{v}(r,t)f(\vec{r},\vec{p},t)\varepsilon D(\varepsilon)d\varepsilon \tag{8.55}$$

The density of state $D(\varepsilon)$ is the number of states of particle between energy ε and $(\varepsilon + d\varepsilon)$.

Fourier's law, Ohm's law, Fick's law, hyperbolic heat conduction equation, and mass, momentum, and energy equation can be derived from the BTE.

8.11.1 Fourier's Heat Conduction Equation

The BTE in the absence of any external force applied to the particle can be written as

$$\frac{\partial f}{\partial t} + \vec{v}.\vec{\nabla} f = \frac{f_0 - f}{\tau} \tag{8.56}$$

Let $t > \tau$ (mean free time) and τ_r (relaxation time). We can drop the time-varying term in the Boltzmann transport if $t > \tau$, τ_r. We also can assume $\nabla f \simeq \nabla f_0$ (local thermodynamic equilibrium) if $L > l$, l_r.

The first term in the left-hand side of the above equation is 0 and $\vec{\nabla} f \simeq \vec{\nabla} f_0$, for Fourier's condition case.

For 1D case, we have

$$v_x \frac{\partial f_0}{\partial x} = \frac{f_0 - f}{\tau} \tag{8.57}$$

$$f = f_0 - \tau v_x \frac{\partial f_0}{\partial x} \tag{8.58}$$

As equilibrium, distribution (f_0) is a function of temperature, and we can write

$$\frac{\partial f_0}{\partial x} = \frac{df_0}{dT}\frac{\partial T}{\partial x} \tag{8.59}$$

We can write the heat flux from equation (8.55) as

$$\vec{q}(\vec{r},t) = \int \vec{v}(\vec{r},t)f(\vec{r},\vec{p},t)\varepsilon D(\varepsilon)d\varepsilon \tag{8.60}$$

Using equation (8.58), we have

$$\begin{aligned}\vec{q}(\vec{r},t) &= \int v_x\left(f_0 - \tau v_x \frac{\partial f_0}{\partial x}\right)\varepsilon D(\varepsilon)d\varepsilon \\ &= \int v_x f_0 \varepsilon D(\varepsilon)d\varepsilon - \int v_x \tau v_x \frac{\partial f_0}{\partial x}\varepsilon D(\varepsilon)d\varepsilon \end{aligned} \tag{8.61}$$

The first term of the above heat flux expression $\vec{q}(\vec{r}, t)$ is equal to zero as the integral over all directions is equal to zero due to the local thermal equilibrium. Using equation (8.59), we can write

$$q_x(x) = -\frac{\partial T}{\partial x} \int v_x^2 \tau \frac{df_0}{dT} \varepsilon D(\varepsilon) d\varepsilon \tag{8.62}$$

Assuming τ and v_x independent of particle energy, the thermal conductivity K can be expressed as

$$K = v_x^2 \tau \int \frac{df_0}{dT} \varepsilon D(\varepsilon) d\varepsilon \tag{8.63}$$

The thermal conductivity, K, can be expressed as

$$K = \frac{1}{3} C v^2 \tau \tag{8.64}$$

where $v^2 = v_x^2 + v_y^2 + v_z^2$ where the heat capacity C is

$$C = \int \frac{df_0}{dT} \varepsilon D(\varepsilon) d\varepsilon \tag{8.65}$$

Using $v\tau = l$ in equation (8.64), we can write

$$K = \frac{1}{3} c v l$$

It may be noted that kinetic theory has derived same expression of K. We thus have derived Fourier's law as

$$q_x = -K \frac{\partial T}{\partial x}$$

8.11.2 Hyperbolic Heat Conduction Equation

The simplified BTE in the absence of external force is written as

$$\frac{\partial f}{\partial t} + \vec{v} \cdot \nabla f = \frac{f_0 - f}{\tau} \tag{8.66}$$

Multiplying both sides of the above equation by $v_x \varepsilon D(\varepsilon) d\varepsilon$, we have

$$\left(\frac{\partial f}{\partial t} + \vec{v} \cdot \nabla f = \frac{f_0 - f}{\tau} \right) v_x \varepsilon D(\varepsilon) d\varepsilon \tag{8.67}$$

Integrating over energy,

$$\int \frac{\partial f}{\partial t} v_x \varepsilon D(\varepsilon) d\varepsilon + \int v_x \frac{\partial f}{\partial x} v_x \varepsilon D(\varepsilon) d\varepsilon = \int \left(\frac{f_0 - f}{\tau} \right) v_x \varepsilon D(\varepsilon) d\varepsilon \tag{8.68}$$

Using equation (8.55) for heat flux, we have

$$\frac{\partial q_x}{\partial t} + \int v_x^2 \frac{\partial f}{\partial x} \varepsilon D(\varepsilon) d\varepsilon = \int \frac{f_0}{\tau} v_x \varepsilon D(\varepsilon) d\varepsilon - \int \frac{f}{\tau} v_x \varepsilon D(\varepsilon) d\varepsilon \tag{8.69}$$

The first term in the right-hand side of the above expression is equal to "0" due to integration in all directions.

Let $L \gg l, l_r$ and $t \simeq \tau, \tau_r$.

It may be noted that $\tau =$ constant as the relaxation time is independent of particle energy.

From quasiequilibrium assumption, we have

$$\frac{\partial f}{\partial x} = \frac{df_0}{dT}\frac{\partial T}{\partial x} \tag{8.70}$$

Equation (8.69) can be written as

$$\frac{\partial q_x}{\partial t} + \frac{\partial T}{\partial x}\int v_x^2 \frac{df_0}{dT}\varepsilon D(\varepsilon)d\varepsilon = -\frac{q_x}{\tau}$$

$$\frac{\partial q_x}{\partial t} + \frac{q_x}{\tau} = -\frac{K}{\tau}\frac{\partial T}{\partial x}$$

$$q_x + \tau\frac{\partial q_x}{\partial t} = -K\frac{\partial T}{\partial x} \tag{8.71}$$

The above equation is known as *Cattaneo equation* similar to equation (8.13). The 1D energy equation is written as

$$\rho C\frac{\partial T}{\partial t} + \frac{\partial q_x}{\partial x} = 0 \tag{8.72}$$

Combining Cattaneo equation with energy equation, we get

$$\frac{\partial T}{\partial t} + \tau\frac{\partial^2 T}{\partial t^2} = \frac{K}{\rho C}\frac{\partial^2 T}{\partial x^2} \tag{8.73}$$

The above equation is the hyperbolic heat conduction equation.

Note: For a realistic rigorous solution, the BTE must be solved for electrons, phonons, and photons. Each of these functions depends on six variables, that is, three space and three momentum (energy). The time scales of electron–phonon and phonon–phonon interactions vary by two orders of magnitude. Monte Carlo solution is one approach to solve these problems, which is time consuming. Another approach (two-step model) has been to propose hydrodynamic-type equation for modeling electron and phonon transport, which is discussed in the following section.

8.12 Microscale Two-Step Models

Microscale thermal transport phenomenon involves complex transfer mechanism of free electrons and phonons. The molecular dynamics and processes are not significant in most of the microscopic engineering applications. However, scale effects become extremely important in system with sudden high heat flux irradiation by laser pulses and some other dimensionally space- and time-governed problems. Anisimov *et al.* (1974) proposed the first two-step model for microscale conduction as

$$C_e(T_e)\frac{\partial T_e}{\partial t} = -\frac{\partial q_e}{\partial r} - G(T_e - T_l) + f(r,t) \tag{8.74}$$

$$C_l \frac{\partial T_l}{\partial t} = G(T_e - T_l) \tag{8.75}$$

$$q_e = -K_e \Delta T_e \tag{8.76}$$

where C_e is the electron heat capacity, T_e is the electron temperature, T_l is the lattice temperature, G is the coupling constant (electron–lattice coupling factor) to be found from pulse experiments, $f(r, t)$ is the internal heating due to a laser pulse, and K_e is the electron thermal conductivity. The heat flux from Fourier's model is denoted as q_e. The aforementioned equations can be used to obtain lattice and electron temperature. The heat flux relation (equation (8.76)) was later substituted with Cattaneo-type constitutive relation as

$$q_e = -K_e \nabla T_e - \tau \frac{\partial q_e}{\partial t} \tag{8.77}$$

This transforms the two-step model (equations (8.74) and (8.76)) to hyperbolic in nature.

Qiu and Tien (1992, 1993) proposed the coupling factor G for radiation deposition in metals as

$$G = \frac{\pi^4}{18} \frac{(n_e u K_B)^2}{K_{eq}}$$

where K_B is the Boltzmann constant, n_e is the electron number density, u is the speed of sound, and K_{eq} is the electron equilibrium thermal conductivity.

8.13 Thin Film Conduction

The influence of thermal wave is more significant for thermal conduction in microscale systems compared to macrosystem. An example of thin film conduction is illustrated here to justify this point. Figures 8.13 and 8.14 compare the conduction through a thin film and thick film, respectively. Here, the film is initially at temperature T_0. There is a sudden change in the temperature of both sides to T_w. The nondimensional variables for the problem are defined as

$$\theta = \frac{T - T_0}{T_w - T_0}, \quad \eta = \frac{Cx}{2\alpha} \quad \text{and} \quad \zeta = \frac{C^2 t}{2\alpha}$$

where C is the thermal wave speed and α is the thermal diffusivity. The boundary conditions are

$$\theta = 1 \quad \text{at} \quad t > 0 \quad \text{and} \quad x = 0 \quad \text{and} \quad x_0$$

where x_0 is the thickness of the film. The hyperbolic heat conduction equation is solved for both thick and thin silicon films to observe the nature of thermal wave phenomena.

After raising the wall temperature in both sides, a sharp wave front develops and propagates toward the center of the physical domain which separates between the heat-affected zone from the thermally undisturbed zone (see Figure 8.13). The two thermal wavefronts from both sides collide with each other at the center of the film. After the collision, the center temperature is amplified, and reverse thermal wavefronts take place and travel toward both sides of the thin film. When thermal wave-front reaches both side walls, the film temperatures at the side walls exceed the imposed wall temperature, called temperature *overshoot*. For thick film, the wavefronts do not show reverse temperature waves after meeting the center of the thin film,

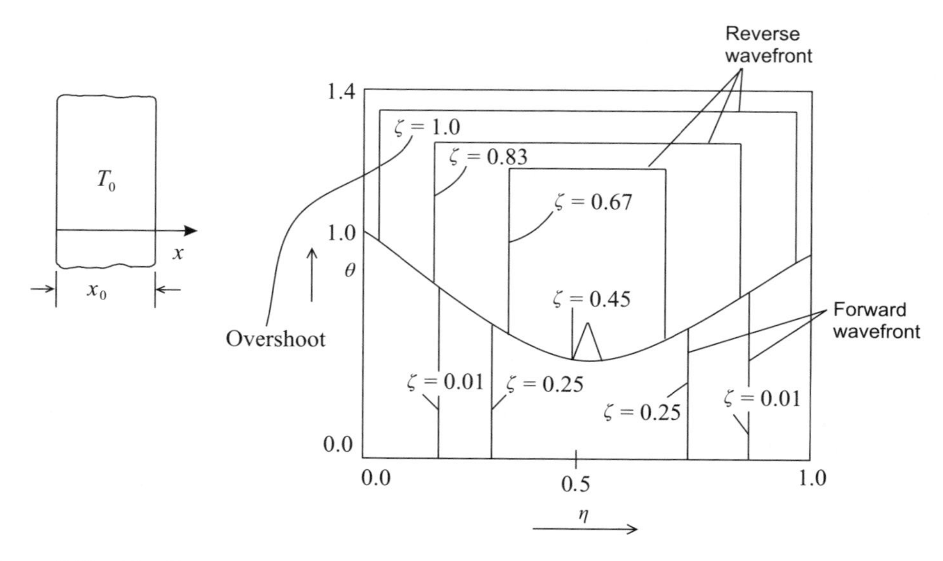

Figure 8.13 Nondimensional unsteady temperature distribution (θ) inside a thin film ($C_0x_0/\alpha = 1$) as a function of nondimensional time, ζ, and nondimensional distance, η

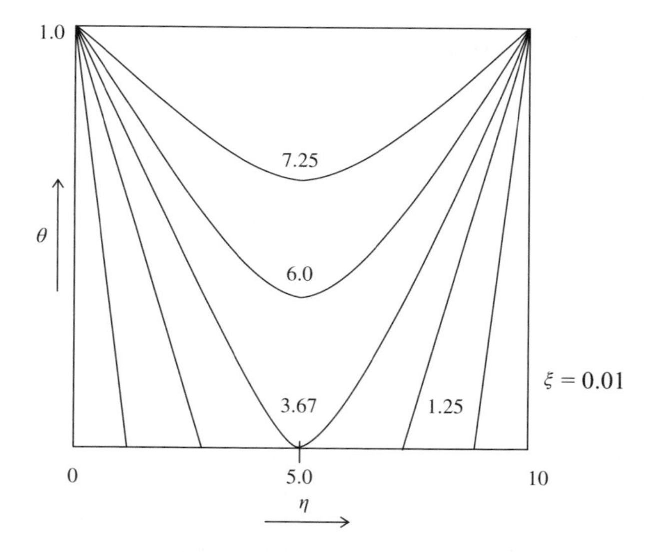

Figure 8.14 Nondimensional temperature distribution (θ) in a thick film ($C_0x_0/\alpha = 10$) as a function of nondimensional distance, η, and nondimensional time, ζ

and there is no overshoot (see Figure 8.14). Hence, the criterion for thermal wave domination is

$$Cx_0/\alpha < 10$$

For silicon,
Mean free time $\tau = 10^{-14}$ s and $\alpha = 93.4 \times 10^{-6}$ m^2/s

Therefore, using $C^2 = \alpha/\tau$ x_0 (thickness of film) = 0.01 micron for presence of thermal wave effect in silicon.

References

Anisimov SI, Kapeliovich BL, and Perelman TL 1974 Femtosecond electronic heat-transfer dynamics in thin gold film. *Sov. Phys. JETP*, **39**, p. 375.

Bertman B and Sandiford DJ 1970 Second sound in solid helium. *Sci. Am.*, **222**(5), pp. 92–101.

Chandrasekharaiah DS 1986 Thermoelasticity with second sound: a review. *Appl. Mech. Rev.*, **39**, p. 355.

Flik MI, Choi BI, and Goodson KE 1992 Heat transfer regimes in microstructures. *J. Heat Transfer*, **114**, pp. 666–674.

Kittel C 1996 *Introduction to Solid State Physics*. John Wiley & Sons, Inc., New York.

Landau L 1941 Theory of the superfluidity of Helium II. *Phys. Rev.*, **60**, p. 356.

Peshkov V 1944 Second sound in helium II. *J. Phys.*, **8**, p. 381.

Qiu TQ and Tien CL 1992 Short-pulse laser heating on metals. *Int. J. Heat Mass Transfer*, **35**, p. 719.

Qiu TQ and Tien CL 1993 Heat transfer mechanism during short-pulse laser heating of metals. *J. Heat Transfer*, **115**, p. 835.

Tien CL and Chen G 1994 Challenges in microscale conductive and radiative heat transfer. *J. Heat Transfer*, **116**, pp. 799–807.

Weast RC, Astle MJ, and Beyer WH 1985 *CRC Handbook of Chemistry and Physics*. CRC Press, Boca Raton, FL.

Yang HS, Bai GG, Thompson LJ, and Eastman JA 2002 Interfacial thermal resistance in nanocrystalline yttria-stabilized zirconia. *Acta Mater.*, **50**, pp. 2309–2313.

9

Microscale Convection

9.1 Introduction

The trend of miniaturization for computer technology and electronics has increased the overheating problem of integrated circuits (ICs). The development of efficient and effective cooling techniques for microchips has initiated extensive research in microchannel heat transfer. A microchannel heat sink is one of the successful approaches for removing high rates of heat. In a microchannel heat sink, many microscale channels are machined on the electrically inactive face of the microchip. There is a very high heat transfer per unit volume in these microchannel heat sinks. It is important to know the fluid flow and heat transfer characteristics of these channels for a better design of the systems. The literature related to microscale convection is emerging. Heat transfer in microchannels can be significantly different from conventional-sized channels due to the effect of velocity slip, temperature jump, and viscous dissipation. Similarly, different flow regimes are observed during boiling and condensation inside a microchannel. Therefore, different heat transfer correlations are required for boiling and condensation heat transfer in microchannels. All these aspects are discussed in this chapter.

9.2 Scaling Analysis

The present section discusses the convective heat transfer governing equation from scaling perspective to understand the importance of various contributions to heat transfer. This can be useful for the simplification of the governing equation. Thermal convection is governed by the Navier–Stokes (N–S) equation along with the buoyancy force. The buoyancy force can be estimated as

$$F_b \sim \rho g \Delta T l^3 \tag{9.1}$$

where ρ is the density, g is the gravity, ΔT is the temperature deviation, and l is the length scale or size of the system. The other force relevant for the convective flow is the viscous force, which can be estimated as

$$F_\mu \sim \mu A \frac{\partial u}{\partial y} \sim \mu v l \sim \mu l^2 \tag{9.2}$$

Transport Phenomena in Microfluidic Systems, First Edition. Pradipta Kumar Panigrahi.

Companion Website: www.wiley.com/go/panigrahi/microfluidic

Comparison of the above two estimates indicates that the buoyancy force is typically weaker than the viscous force in microsystems. The other physical significance is that convective motion is substantially damped by the effect of viscosity. There may be a heating source or sink due to exothermic and endothermic chemical reactions inside a microchannel. This volumetric heat source or sink can be estimated as

$$Q_v \sim l^3 \tag{9.3}$$

This other source of internal heating is the viscous heating produced by the viscosity of the flowing fluid. The viscous heating per unit volume is independent of the size of the system. Hence, the quantity of viscous heating produced by the system can be estimated as

$$Q_{\text{Diss}} \sim l^3 \tag{9.4}$$

The heat exchanged by conduction can be estimated as

$$Q_{\text{Cond}} \sim Kl\Delta T \tag{9.5}$$

where K is the thermal conductivity.

The viscous dissipation (Q_{Diss}) and heat generation can be neglected for a microsystem. The generalized convection equation for an incompressible Newtonian fluid governed by the Fourier law is given as

$$\rho C_P \left(\frac{\partial T}{\partial t} + u\nabla T \right) = \nabla \cdot K\nabla T + Q_{\text{Diss}} + Q_v \tag{9.6}$$

where the viscous dissipation is given as

$$Q_{\text{Diss}} = \frac{\mu}{2} \left(\frac{\partial u_i}{\partial x_j} + \frac{\partial u_j}{\partial x_i} \right)^2 = \mu\phi \tag{9.7}$$

where ϕ is the dissipation function, which is important in high-speed flows and for very viscous fluids. In microsystems, Q_v and Q_{Diss} can be neglected based on the scale analysis. The convective flow governing equation in that case for constant thermal conductivity can be written as

$$\frac{\partial T}{\partial t} + u\nabla T = \alpha \nabla^2 T \tag{9.8}$$

where $\alpha = \frac{K}{\rho C_P}$ is the thermal diffusivity of fluid.

The N–S equation with the inclusion of buoyancy force is written as

$$\rho \frac{Du}{Dt} = -\nabla P + \rho g + \mu \nabla^2 u \tag{9.9}$$

In nonisothermal flow, temperature variation induces density variation. Boussinesq assumed that density varies little with temperature and can be assumed as an average density with the exception of buoyancy term. However, the scaling analysis has shown that the buoyancy term can be neglected for a microsystem. Therefore, the simplified N–S equation is written as

$$\rho \frac{Du}{Dt} = -\nabla P + \mu \nabla^2 u \tag{9.10}$$

Let us define the following dimensionless variables:

$$u^* = \frac{u}{U}, p^* = \frac{\Delta P}{\rho U^2}, T^* = \frac{T - T_\infty}{T_S - T_\infty}$$
$$x^* = \frac{x}{L}, y^* = \frac{y}{L}, z^* = \frac{z}{L}, t^* = \frac{Ut}{L}$$

Substitution of these variables into the N–S equation gives

$$\frac{Du^*}{Dt^*} = -\nabla^* P^* + \frac{1}{Re}\nabla^{*2}u^*$$

where $\nabla = \frac{1}{L}\nabla^*$ and $\nabla^2 = \frac{1}{L^2}\nabla^{*2}$

$$Re = \frac{UL}{\nu}$$

Substitution of these variables in the energy equation gives

$$\frac{DT^*}{Dt^*} = \frac{1}{RePr}\nabla^{*2}T^*$$

where $Re \times Pr$ is also known as the Peclet number $\left(\frac{UL}{\alpha}\right)$.

If viscous dissipation is included, the nondimensional energy equation is written as

$$\frac{DT^*}{Dt^*} = \frac{1}{Pe_{\text{th}}}\nabla^{*2}T^* + \frac{Ec}{Re}\phi^*$$

where the Eckert number $Ec = \frac{U^2}{C_p(T_S - T_\infty)}$ and nondimensional dissipation function, ϕ^*, is written as

$$\phi^* = 2\left[\left(\frac{\partial u^*}{\partial x^*}\right)^2 + \left(\frac{\partial v^*}{\partial y^*}\right)^2 + \cdots\right]$$

Assuming the velocity to vary with l, we can write

$$Pe_{\text{th}} \sim \frac{Ul}{\alpha} \sim \frac{l^2}{\alpha} \tag{9.11}$$

For a microsystem, it is tempting to assume the Peclet number to be small due to the small value of "l" indicating the negligible role of convection in microsystems. Let us consider the typical size of a microsystem to be 100 μm with velocity of the order of meters per second. For kinematic viscosity of water at 30 °C ($\nu = 0.801 \times 10^{-6}\text{m}^2/\text{s}$), $Re = 124$. For water with thermal diffusivity, $\alpha = 0.143 \times 10^{-6}\text{m}^2/\text{s}$, $Pe_{\text{th}} \sim 700$. Thus, the thermal Peclet number and Reynolds number cannot generally be considered to be small in microsystem convective heat transfer.

9.2.1 Brinkman Number

The Brinkman number is a nondimensional number, which influences the convective heat transfer coefficient, the thermal development length, and laminar to turbulent transition phenomena. It is defined as

$$Br = \frac{\mu u_m^2}{K \Delta T} \tag{9.12}$$

where u_m is the mean velocity and ΔT is the wall-fluid temperature difference $(T_s - T_\infty)$. Br is a measure of relative importance of viscous heating (work done against viscous shear) to heat conduction in the fluid along the channel. For low-speed and low-viscosity flows in conventionally sized channels of short lengths, the viscous effect is neglected. For flows in microchannels, the length-to-diameter ratio is quite large, and therefore, the effect of the Brinkman number is significant. The Brinkman number is related to the Eckert number as $Br = Ec \times Pr$.

9.3 Laminar Fully Developed Nusselt Number

The Nusselt number solutions for various channels with noncircular cross-sections have been reported for fully developed laminar flow. Two surface conditions – (1) uniform heat flux and (2) uniform surface temperature – have been considered. The Nusselt number for noncircular cross-section is based on the hydraulic diameter, D_h, defined as

$$D_h = \frac{4A}{P}$$

where A is the flow area and P is the perimeter. Table 9.1 shows the fully developed Nusselt number for various channel cross-sections. It may be noted that the Nusselt number for uniform heat flux is greater than that for uniform surface temperature for all cases. The heat transfer coefficient for noncircular channels varies along the periphery. The Nusselt number in Table 9.1 is the average Nusselt number along the periphery. The surface temperature varies axially and along the periphery for the uniform surface heat flux case. The results in Table 9.1 are based on uniform periphery temperature.

9.4 Why Microchannel Heat Transfer

The Nusselt number for heat transfer inside a macrotube for a constant temperature fully developed case is equal to 3.657. Thus, the heat transfer coefficient for this case is expressed as

$$h = 3.657 \frac{K}{D}$$

The above expression indicates that the heat transfer coefficient is larger for smaller diameter cases. The dramatic increase in heat transfer coefficient with a decrease in diameter has motivated the use of microchannel for heat transfer applications specifically for MEMS and high heat flux applications. Figure 9.1 shows a typical example of heat transfer in microelectronics applications. Microsized grooves are machined in a sink to form microfins for enhancing heat transfer. The heat sink is attached to a substrate. Microchannels are typically rectangular or trapezoidal shaped due to the limitation in microfabrication technique. It may be noted that high heat transfer in microscale flows is associated with high pressure drop necessitating the optimization of flow geometries.

Table 9.1 Nusselt number for laminar fully developed flow in microchannels for various channel cross-sections

Geometry	$\frac{b}{a}$	Nu_∞	
		Uniform surface flux	Uniform surface temperature
Circle		4.36	3.66
Square (a × a)	1	3.41	2.98
Rectangle (a × b)	2	4.12	3.39
Rectangle (a × b)	8	6.5	5.6
Parallel plates	∞	8.24	7.54
Equilateral triangle (a, a, a)		3.1	2.46

9.5 Gases versus Liquid Flow in Microchannels

There is no distinction between gases and liquids for macroscale flow and heat transfer as long as the governing parameters like the Reynolds number, Prandtl number, Grashof number, and so on are the same for both. However, this is not true for microscale flows. The following observations can be made for microscale gas and liquid flows.

1. Mean free path of liquids are much smaller than those of gases. Therefore, the continuum assumption may hold for liquids but fail for gases.
2. The Knudsen number provides criterion for the validity of thermodynamic equilibrium and continuum model for gases. However, it does not hold for liquids.
3. The failure of thermodynamic equilibrium and continuum is not well defined for liquids. Therefore, the validity of no-slip, no-temperature-jump, linear stress–strain relation and Fourier heat flux–temperature relationship are unknown for liquids.

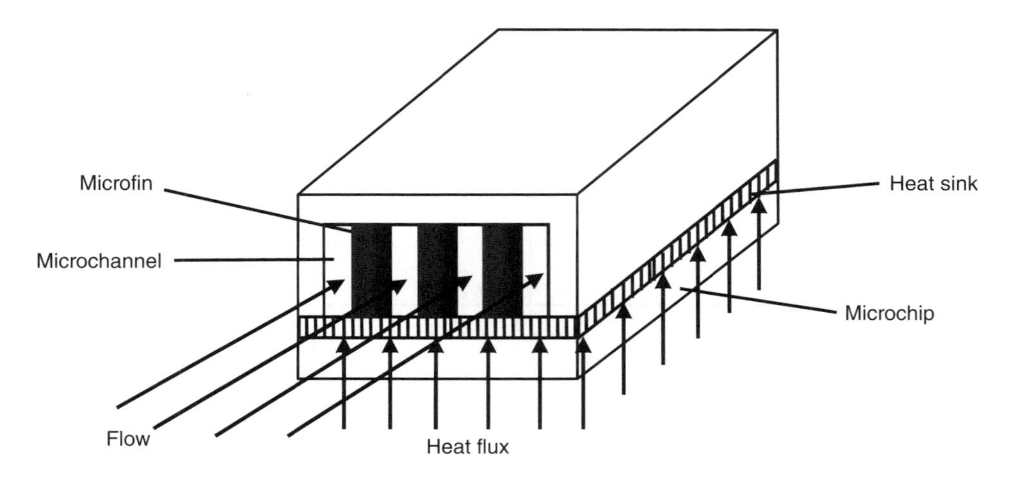

Figure 9.1 A schematic showing microchip heat transfer using heat sink with microchannels and microfins

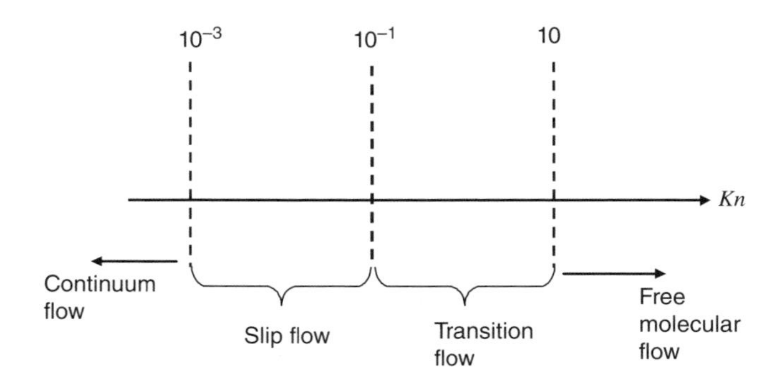

Figure 9.2 Different flow regimes of microscale convection

4. Surface forces become important for microchannel flow. The nature of surface forces in liquids is different from that of gases. Therefore, the boundary conditions for liquids are different from those of gases.
5. Liquid molecules are closer to each other than gas molecules. Therefore, liquids are almost incompressible and gases are compressible.

9.6 Temperature Jump

The flow regimes are classified based on "*Kn*" as shown in Figure 9.2. We have derived the expression for slip flow velocity B.C. based on kinetic theory of gas earlier as

$$u_s - u_w = \frac{2-\sigma_v}{\sigma_v}\left[\lambda\left(\frac{\partial u}{\partial n}\right)_s\right] + \frac{\lambda^2}{2}\left(\frac{\partial^2 u}{\partial n^2}\right)_s + \cdots \tag{9.13}$$

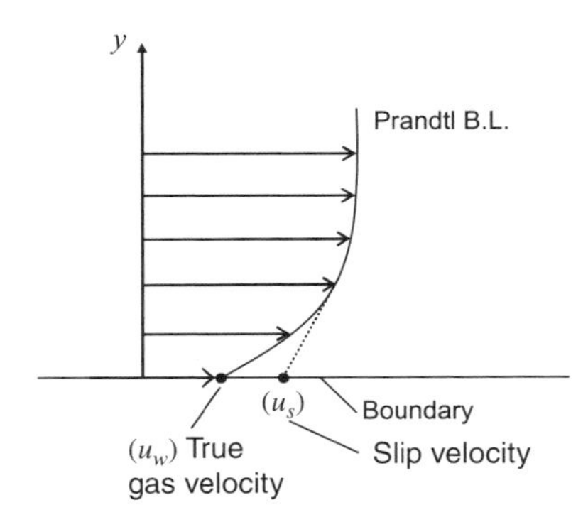

Figure 9.3 Representation of velocity slip condition

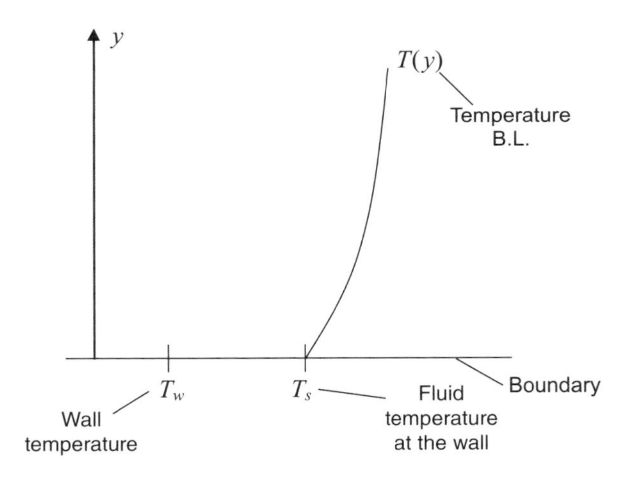

Figure 9.4 Temperature jump boundary condition near a boundary

Similar to velocity (Figure 9.3), we need to have appropriate boundary conditions for temperature. Figure 9.4 shows the temperature jump boundary condition near a boundary.

There is finite temperature between the fluid temperature at the wall and wall temperature, which is known as temperature jump.

$$T_s - T_w = C_{\text{jump}} \frac{\partial T}{\partial y} \tag{9.14}$$

We need to determine appropriate expressions for C_{jump} known as temperature jump coefficient. The thermal accommodation coefficient (σ_T) was defined earlier as

$$\sigma_T = \frac{Q_i - Q_r}{Q_i - Q_w} \tag{9.15}$$

where Q_i = energy of the impinging molecules, Q_r = energy carried by the reflected molecules, and Q_w = energy of the molecules leaving the surface at wall temperature.

Hence, the thermal accommodation coefficient (σ_T) is defined as the fraction of molecules reflected by the wall that accommodates their energy to the wall temperature. Let us assume the temperature of approaching molecules as T_s. The energy of these molecules is the combination of kinetic energy, internal energy, and the contribution of the incoming molecules to conduction. Energy of impinging molecules is given by

$$Q_i = m(2RT_s + U_s) + \frac{1}{2}K\left.\frac{\partial T_s}{\partial y}\right|_w \tag{9.16}$$

where m is the mass, R is the gas constant, U_s is the internal energy, and K is the thermal conductivity of fluid. For the calculation of conduction, $\frac{1}{2}$ is used as it is assumed that half of the molecules at the fictitious plane are impinging and the other half are reflected molecules. Energy of reflected molecules at $T_w =$

$$Q_w = m(2RT_w + U_w) + \frac{1}{2}K\left.\frac{\partial T_w}{\partial y}\right|_w \tag{9.17}$$

The second term of the above expression is approximately equal to zero as $\left.\frac{\partial T_w}{\partial y}\right|_w \simeq 0$ Subtracting equation (9.17) from equation (9.16), we have

$$Q_i - Q_w = m(2R(T_s - T_w) + (U_s - U_w)) + \frac{1}{2}K\left.\frac{\partial T_s}{\partial y}\right|_w \tag{9.18}$$

From kinetic theory, we can write

$$m = \frac{P}{\sqrt{2\pi RT}}, R = C_p - C_v, \gamma = \frac{C_p}{C_v}$$

$$C_v = \frac{3}{2}R + \frac{dU}{dT}$$

$$= \frac{3}{2}R + \frac{U_s - U_w}{T_s - T_w}$$

Rearranging the above equation, we have:

$$U_s - U_w = C_v(T_s - T_w) - \frac{3}{2}R(T_s - T_w) \tag{9.19}$$

Substituting the above equation in equation (9.18), we have

$$Q_i - Q_w = m\left(2R(T_s - T_w) + C_v(T_s - T_w) - \frac{3}{2}R(T_s - T_w)\right) + \frac{1}{2}K\left.\frac{\partial T_s}{\partial y}\right|_w$$

$$= m(T_s - T_w)\left(\frac{R}{2} + C_v\right) + \frac{1}{2}K\left.\frac{\partial T_s}{\partial y}\right|_w$$

$$= m(T_s - T_w)\left(\frac{C_p}{2} - \frac{C_v}{2} + C_v\right) + \frac{1}{2}K\left.\frac{\partial T_s}{\partial y}\right|_w$$

$$= m(T_s - T_w)\frac{C_p + C_v}{2} + \frac{1}{2}K\left.\frac{\partial T_s}{\partial y}\right|_w$$

$$= mC_v(T_s - T_w)\frac{(\gamma+1)}{2} + \frac{1}{2}K\left.\frac{\partial T_s}{\partial y}\right|_w$$

$$= \frac{C_v(T_s - T_w)(\gamma+1)P}{2\sqrt{2\pi RT}} + \frac{1}{2}K\left.\frac{\partial T_s}{\partial y}\right|_w \tag{9.20}$$

Noting that the net energy carried to the surface $Q_i - Q_r$ is equal to the heat flux at the wall.

$$Q_i - Q_r = K\left.\frac{\partial T}{\partial y}\right|_w \tag{9.21}$$

Using the definition of σ_T (equation (9.15)) and equation (9.20), we have

$$\sigma_T = -\frac{K\left.\frac{\partial T}{\partial y}\right|_w}{\frac{c_v(T_s - T_w)P(\gamma+1)}{2\sqrt{2\pi RT}} + \frac{1}{2}K\left.\frac{\partial T}{\partial y}\right|_w} \tag{9.22}$$

Rearranging equation (9.22), we can write

$$(T_s - T_w)\frac{C_v\sigma_T(\gamma+1)P}{2\sqrt{2\pi RT}} + \frac{\sigma_T}{2}K\left.\frac{\partial T}{\partial y}\right|_w = K\left.\frac{\partial T}{\partial y}\right|_w$$

$$T_s - T_w = \frac{K\left.\frac{\partial T}{\partial y}\right|_w\left(1 - \frac{\sigma_T}{2}\right)}{C_v\sigma_T P(\gamma+1)}2\sqrt{2\pi RT}$$

$$T_s - T_w = \frac{\frac{2-\sigma_T}{\sigma_T}\sqrt{2\pi RT}}{C_v P(\gamma+1)}K\left.\frac{\partial T}{\partial y}\right|_w \tag{9.23}$$

Using equation (9.14), we can write

$$C_{\text{jump}} = \frac{2-\sigma_T}{\sigma_T}\frac{1}{\gamma+1}\frac{K\sqrt{2\pi RT}}{C_v P} \tag{9.24}$$

Using the expression of viscosity μ, molecular speed (v_m), and gas law,

$$\mu = \frac{1}{2}\rho u_m\lambda, u_m = 2\sqrt{\frac{2RT}{\pi}}, \rho = \frac{P}{RT}$$

we have

$$C_{\text{jump}} = \frac{2-\sigma_T}{\sigma_T}\frac{2}{\gamma+1}\frac{K}{C_v\mu}\lambda = \frac{2-\sigma_T}{\sigma_T}\frac{2\gamma}{\gamma+1}\frac{K}{c_p\mu}\lambda \tag{9.25}$$

Hence, the temperature jump boundary condition, equation (9.14), can be written as

$$T_s - T_w = \frac{2-\sigma_T}{\sigma_T}\frac{2\gamma}{\gamma+1}\frac{\lambda}{Pr}\frac{\partial T}{\partial y} \tag{9.26}$$

9.7 Couette Flow with Viscous Dissipation

In a microjournal bearing arrangement, a shaft rotates with an angular velocity inside a housing. There is an air gap between the shaft and the housing. The movement of the shaft leads to Couette-like flow in the air gap. The microscale nature of the flow leads to viscous dissipation. There is a rise in temperature due to viscous dissipation. The present section reports an analogous example of Couette flow for demonstrating the temperature rise in similar situations.

Figure 9.5 shows a Couette flow with two infinitely large parallel plates separated by distance H. The upper plate moves with surface velocity, u_s. The lower plate is insulated. The upper plate exchanges heat with the ambient by convection. The heat transfer coefficient along the exterior surface of the moving plate is h_o and the ambient temperature is T_∞. Our objective is to determine the Nusselt number distribution of the upper plate.

Using the first-order slip flow boundary condition, the velocity profile can be derived as

$$\frac{u}{u_s} = \frac{1}{1 + 2\beta_v Kn}\left(\frac{y}{H} + \beta_v Kn\right) \tag{9.27}$$

Surface temperature is unknown in this example. The temperature jump condition at the upper plate can be written as

$$T_s = T(x, H) + \frac{2\gamma}{1+\gamma}\beta_T \frac{\lambda}{Pr}\frac{\partial T}{\partial y}\bigg|_{(x,H)} \tag{9.28}$$

The energy equation for incompressible two-dimensional, constant conductivity flow is given by

$$\rho c_p\left(\frac{\partial T}{\partial t} + u\frac{\partial T}{\partial x} + v\frac{\partial T}{\partial y}\right) = K\left(\frac{\partial^2 T}{\partial x^2} + \frac{\partial^2 T}{\partial y^2}\right) + \mu\phi$$

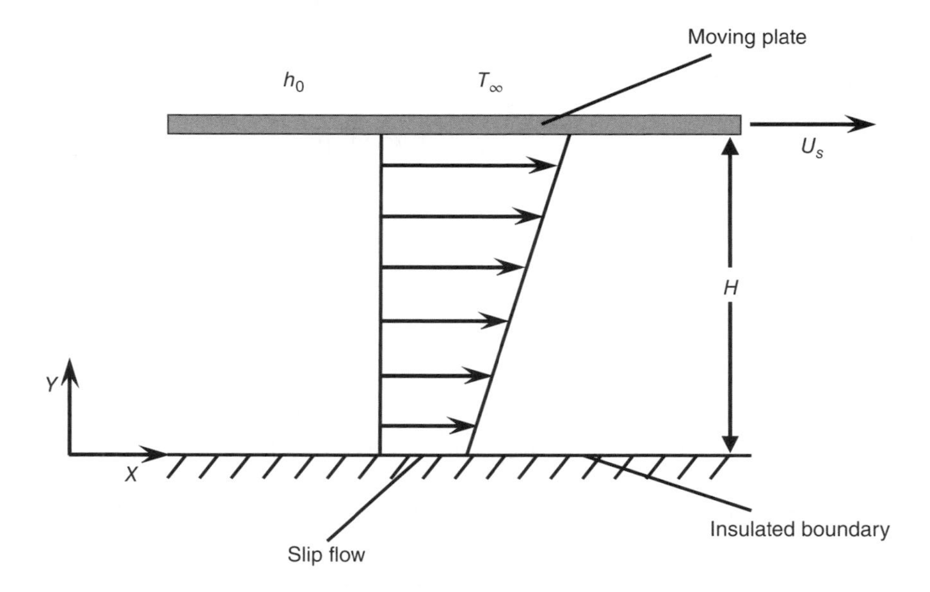

Figure 9.5 Schematic of Couette flow between two large parallel plates

where

$$\phi = 2\left[\left(\frac{\partial u}{\partial x}\right)^2 + \left(\frac{\partial v}{\partial y}\right)^2 + \left(\frac{\partial w}{\partial z}\right)^2\right]$$
$$+\left[\left(\frac{\partial u}{\partial y} + \frac{\partial v}{\partial x}\right)^2 + \left(\frac{\partial v}{\partial z} + \frac{\partial w}{\partial y}\right)^2 + \left(\frac{\partial w}{\partial x} + \frac{\partial u}{\partial z}\right)^2\right]$$
$$-\frac{2}{3}\left(\frac{\partial u}{\partial x} + \frac{\partial v}{\partial y} + \frac{\partial w}{\partial z}\right)^2$$

Noting that all derivatives with respect to x is negligible and $v = \frac{\partial}{\partial t} = 0$, the governing equation simplifies to

$$\frac{d^2T}{dy^2} = -\frac{\mu}{K}\left(\frac{du}{dy}\right)^2 \tag{9.29}$$

The boundary conditions are

$$\left.\frac{dT}{dy}\right|_{y=0} = 0 \text{ (adiabatic surface condition)} \tag{9.30}$$

$$-K\left.\frac{dT}{dy}\right|_{y=H} = h_o(T_s - T_\infty) \text{ (convective boundary condition)} \tag{9.31}$$

Substituting temperature jump condition (equation (9.28)), we have

$$-K\left.\frac{dT}{dy}\right|_{y=H} = h_o\left[T(x,H) + \frac{2\gamma}{1+\gamma}\frac{\lambda}{Pr}\beta_T\left.\frac{\partial T}{\partial y}\right|_{x,H} - T_\infty\right]$$

Substituting the velocity expression (equation (9.27)) in governing equation (9.29), we have

$$\frac{d^2T}{dy^2} = -\frac{\mu}{K}\left[\frac{u_s}{H\left(1+2\beta_v Kn\right)}\right]^2 = \psi$$

Integrating, we write

$$T = \frac{\psi y^2}{2} + Cy + D$$

where C and D are integration constants. Application of the first boundary condition (equation (9.30)) gives

$$C = 0$$

Application of the second boundary condition (equation (9.31)) gives

$$D = \frac{HK\psi}{h_o} + \frac{H^2\psi}{2} + \frac{2\gamma}{\gamma+1}\frac{Kn}{Pr}\beta_T H^2\psi + T_\infty$$

Thus, the temperature distribution is

$$T = -\frac{\psi}{2}y^2 + \frac{HK\psi}{h_o} + \frac{H^2\psi}{2} + \frac{2\gamma}{\gamma+1}\frac{Kn}{Pr}\beta_T H^2\psi + T_\infty \tag{9.32}$$

The Nusselt number for parallel plate geometry based on the equivalent diameter $D_h = 2H$ is defined as

$$Nu = \frac{2Hh}{K}$$

The heat transfer coefficient h for channel flow is defined as

$$h = \frac{-K\left.\frac{dT}{dy}\right|_{y=H}}{T_m - T_s}$$

Thus, the Nusselt number is given by

$$Nu = \frac{-2H\left.\frac{dT}{dy}\right|_{y=H}}{T_m - T_s} \tag{9.33}$$

Differentiating equation (9.32), we have

$$\left.\frac{dT}{dy}\right|_{y=H} = -H\psi \tag{9.34}$$

Using equations (9.32) and (9.28) gives

$$T_s = \frac{KH\psi}{h_o} + T_\infty \tag{9.35}$$

The mean temperature, T_m, is defined as

$$mC_pT_m = w\int_0^H \rho C_p uTdy \tag{9.36}$$

The mass flow rate, m, for the channel of width W is given by

$$m = W\int_0^H \rho udy$$

Substituting the u-velocity expression, we have

$$m = \rho W\int_0^H \frac{u_s}{1+2\beta_v Kn}\left(\frac{y}{H} + \beta_v Kn\right)dy = \rho WH\frac{u_s}{2}$$

Substituting m in equation (9.36), we have

$$T_m = \frac{2}{u_s H}\int_0^H uTdy \tag{9.37}$$

Substituting equations (9.27) and (9.32) in equation (9.37), we have

$$T_m = \frac{2}{H(1+2\beta_v Kn)}\int_0^H \left(\frac{y}{H} + \beta_v Kn\right)\left(-\frac{\psi}{2}y^2 + D\right)dy$$

Evaluating the integral, we have

$$T_m = \frac{2}{1+2\beta_v Kn}\left[-\frac{1}{18}H^2\psi - \frac{1}{6}\beta_v KnH^2\psi\left(\frac{1}{2} + \beta_v Kn\right)D\right] \tag{9.38}$$

Substituting D, we have

$$T_m = \frac{2}{1+2\beta_v Kn}\left[\frac{1}{4}H^2\psi + \frac{2}{3}\beta_v KnH^2\psi\right] + \frac{KH\psi}{h_o}$$
$$+\frac{2\gamma}{\gamma+1}\frac{Kn}{Pr}\beta_T H^2\psi + T_\infty$$

Substituting equations (9.34), (9.35), and (9.38) into equation (9.33), we have

$$Nu = \frac{2H^2\psi}{\frac{1}{1+2\beta_v Kn}\left[\frac{1}{4}H^2\psi + \frac{2}{3}\beta_v KnH^2\psi\right] + \frac{KH\psi}{h_o} + \frac{2\gamma}{\gamma+1}\frac{Kn}{Pr}\beta_T H^2\psi + T_\infty - \frac{KH\psi}{h_o} - T_\infty}$$

On simplification, we have the Nusselt number expression as

$$Nu = \frac{8(1+2\beta_v Kn)}{1+\frac{8}{3}\beta_v Kn + \frac{8\gamma}{\gamma+1}(1+2\beta_v Kn)\frac{\beta_v\beta_T Kn}{Pr}}$$

We can observe the following remarks from the above expression of the Nusselt number:

1. The Nusselt number for macrochannel flow can be obtained by setting $Kn = 0$, which is equal to 8.0.
2. The second term in the denominator of the above equation represents the effect of rarefaction (the Knudsen number), and the third term represents the temperature jump. Both these effects act to reduce the Nusselt number.
3. The change in heat transfer coefficient, h_o, does not affect the Nusselt number.
4. The Nusselt number is independent of the Reynolds number, similar to that of macrochannel flow.
5. The temperature of the bottom and top surface increases with an increase in viscous dissipation without effecting the Nusselt number.
6. When dissipation is neglected ($\psi = 0$), equation (9.32) gives $T = T_\infty$. Hence, the temperature is uniform and no heat transfer takes place. The above equation for the Nusselt number is no more applicable in that situation.

9.8 Isothermal Parallel Plate Channel Flow without Viscous Heating

For a fully developed flow inside the parallel plate channel (Figure 9.6), the N–S equation reduces to

$$\mu\frac{d^2u}{dy} = \frac{dp}{dx} \tag{9.39}$$

The boundary conditions are

$$\frac{du}{dy} = 0 \quad \text{at} \quad y = 0, [A]$$

$$u = -\frac{2-\sigma_v}{\sigma_v}\lambda\left.\frac{\partial u}{\partial y}\right|_{y=h} [B]$$

(first-order slip condition)

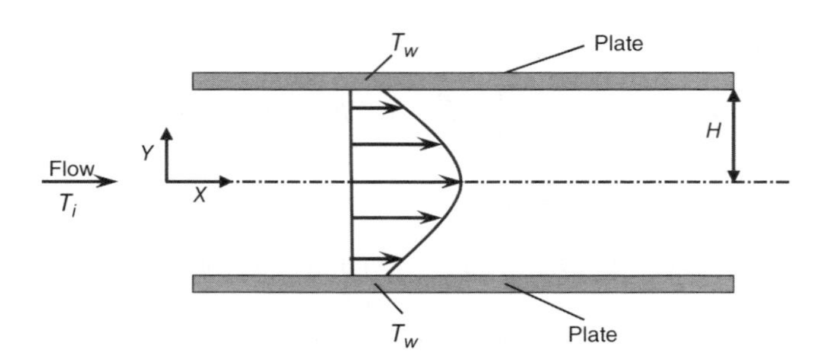

Figure 9.6 The schematic of the parallel plate channel flow

The solution of the governing equation (9.39) with boundary conditions A and B gives

$$u(y) = -\frac{H^2 \frac{dp}{dx}}{\mu}\left(1 + 4Kn\left(\frac{2-\sigma_v}{\sigma_v}\right) - \left(\frac{y}{H}\right)^2\right)$$

$$\text{where} \qquad kn = \frac{\lambda}{2H}$$

Let us assume:

1. The entrance temperature of fluid is uniform.
2. The channel surface temperature is uniform.
3. The flow is thermally developing.
4. The flow is hydrodynamically forced laminar flow.
5. Viscous dissipation is negligible.

Energy equation in 2-D is written as

$$u\frac{\partial T}{\partial x} + v\frac{\partial T}{\partial y} = \alpha\left(\frac{\partial^2 T}{\partial x^2} + \frac{\partial^2 T}{\partial y^2}\right) + \frac{1}{\rho c}\phi \tag{9.40}$$

For fully developed flow, $v = 0$. Assuming the axial conduction to be negligible, we have $\frac{\partial^2 T}{\partial x^2} = 0$. The energy dissipation (ϕ) is assumed negligible. Hence, the energy equation reduces to

$$u\frac{\partial T}{\partial x} = \alpha\frac{\partial^2 T}{\partial y^2} \tag{9.41}$$

The boundary conditions are

$$T = T_i \quad \text{at} \quad x = 0 \quad \text{(inlet temperature)}$$

$$\frac{\partial T}{\partial y} = 0 \quad \text{at} \quad y = 0 \quad \text{(symmetry condition)}$$

The temperature jump boundary condition is

$$T_s - T_w = -\frac{2-\sigma_T}{\sigma_T}\frac{2\gamma}{\gamma+1}\frac{\lambda}{Pr}\left(\frac{\partial T}{\partial y}\right)_{y=H}$$

Let us assume:

$$\beta_T = \frac{2-\sigma_T}{\sigma_T}\frac{2\gamma}{\gamma+1}\frac{1}{Pr}$$

Hence, the above temperature jump boundary condition can be written as

$$T_s - T_w = -\beta_T 2hKn\left(\frac{\partial T}{\partial y}\right)_{y=H}$$

Let the nondimensionalization variables be defined as

$$\theta = \frac{T-T_w}{T_i-T_w},\ Y = \frac{y}{H}$$

Hence, B.C. at $y = H$ is expressed as

$$\frac{T_s-T_w}{T_i-T_w} = -\beta_T 2Kn\left(\frac{\partial \theta}{\partial Y}\right)_{Y=1}$$

$$\theta(x,1) = -\beta_T 2Kn\left.\frac{\partial \theta}{\partial Y}\right|_{Y=1}$$

The above equation is the dimensionless representation of temperature jump boundary condition. Other boundary conditions are

$$\theta(0,Y) = 1 \quad \text{(inlet B. C.)}$$

$$\left.\frac{\partial \theta}{\partial Y}\right|_{Y=0} = 0 \quad \text{(symmetry B.C.)}$$

The above problem is known as Graetz problem and can be solved as an eigenvalue problem. A similar problem is also seen for thermally developing heat transfer. The solution can be obtained for local Nusselt number ($Nu(x)$).

The limiting Nusselt number expression is given as

$$Nu(\infty) = 2\left(1 + 4Kn\beta_v - \frac{1}{3}\right)m(1)^2$$

where $\beta_v = \dfrac{2-\sigma_v}{\sigma_v}$ and $m(1)$ is the eigenvalue, which is a function of β_T.

If $Kn = 0$ and $\beta_T = 0$, we have a classical flow situation. For this case, $m(1) = 1.681$ and $Nu(\infty) = 3.77$. In classical heat transfer text, it is equal to 3.77. For $Kn\beta_v = 0.1$ and $\dfrac{\beta_T}{\beta_v} = 10$ (a microchannel flow situation), the eigenvalue $m(1) = 0.6206$. For this case, the limiting Nusselt number is

$$Nu(\infty) = 2\left(1 + 0.4 - \frac{1}{3}\right)(0.6206)^2 = 2(1.066)(0.6206)^2 \simeq 0.7688$$

The above results indicate that with an increase in temperature jump, $Nu(\infty)$ decreases. It may be remembered that $Kn \times \beta_v$ controls the velocity slip and $\frac{\beta_T}{\beta_v}$ controls the temperature jump. The limiting Nusselt number ($Nu(\infty)$) results as a function of $Kn \times \beta_v$ at different values of β_T/β_v ratio has been shown in Figure 9.7. The heat transfer, that is, the Nusselt number, reduces with an increase in β_T/β_v, that is, the temperature jump. The heat transfer also increases with an increase in velocity slip. Hence, the velocity slip and temperature jump have opposite influences on the heat transfer of a microchannel flow.

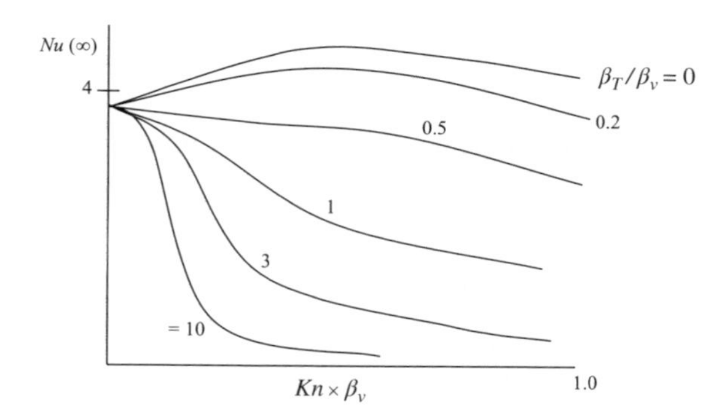

Figure 9.7 The limiting Nusselt number as a function of velocity slip and temperature jump of a microchannel

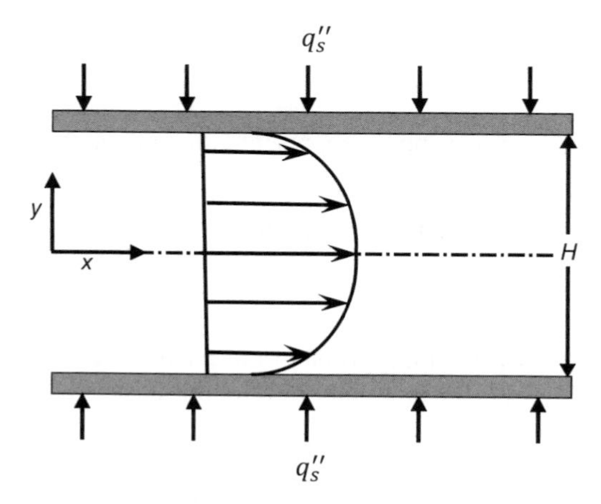

Figure 9.8 Schematic for heat transfer in fully developed parallel plate flow

9.9 Large Parallel Plate Flow without Viscous Heating: Uniform Surface Flux

Heat transfer in microsystems under pressure-driven flow conditions is considered here. Figure 9.8 shows two large infinite parallel plates, separated by distance, H, with uniform surface heat flux q_s''. This configuration can be used to model flow and heat transfer in rectangular channels with large aspect ratio. Velocity and temperature are assumed to be fully developed. Inlet and outlet pressures are P_i and P_o, respectively.

The velocity profile with slip flow boundary condition for this can be derived from N–S equation as

$$u = -\frac{H^2}{8\mu}\frac{dp}{dx}\left[1 + 4\frac{2-\sigma_v}{\sigma_v}Kn - 4\frac{y^2}{H^2}\right]$$

$$= -\frac{H^2}{8\mu}\frac{dp}{dx}\left[1 + 4\beta_v Kn - 4\frac{y^2}{H^2}\right] \tag{9.42}$$

where $\beta_v = \frac{2-\sigma_v}{\sigma_v}$.

The Nusselt number for a parallel plate channel based on the equivalent diameter $D_h = 2H$ is defined as

$$Nu = \frac{2Hh}{K}$$

The heat transfer coefficient is defined as

$$h = -\frac{q_s''}{T_s - T_m} = \frac{-K \left.\frac{\partial T}{\partial y}\right|_{\left(x,\frac{H}{2}\right)}}{T_m(x) - T_s(x)}$$

Hence,

$$Nu = \frac{2Hq_s''}{K(T_s - T_m)} \tag{9.43}$$

where

T_m = fluid mean temperature

T_s = plate temperature

It may be noted that the heat transfer coefficient in microchannels is defined in terms of surface temperature rather than fluid temperature at the surface. The surface temperature can be written as

$$\begin{aligned} T_s &= T\left(x, \frac{H}{2}\right) + \frac{2-\sigma_T}{\sigma_T}\frac{2\gamma}{1+\gamma}\frac{\lambda}{Pr}\left.\frac{\partial T}{\partial y}\right|_{\left(x,\frac{H}{2}\right)} \\ &= T\left(x, \frac{H}{2}\right) + \beta_T\frac{2\gamma}{1+\gamma}\frac{\lambda}{Pr}\left.\frac{\partial T}{\partial y}\right|_{\left(x,\frac{H}{2}\right)} \end{aligned} \tag{9.44}$$

where $\beta_T = \frac{2-\sigma_T}{\sigma_T}$.

Assuming invariant density and specific heat with respect to y, the mean temperature is defined as

$$T_m = \frac{\int_0^{H/2} \rho C_p u T dy}{\int_0^{H/2} \rho C_p u dy} = \frac{\int_0^{H/2} u T dy}{\int_0^{H/2} u dy} \tag{9.45}$$

The velocity distribution derived before is based on the isothermal flow assumption. However, the temperature is not uniform for this flow. We assume that the effect of temperature variation on the velocity distribution is negligible. Assuming negligible dissipation ($\phi = 0$), negligible axial conduction $\left(\frac{\partial^2 T}{\partial x^2} \ll \frac{\partial^2 T}{\partial y^2}\right)$, and nearly parallel flow ($v = 0$), the governing equation is simplified as

$$u\frac{\partial T}{\partial x} = \alpha\frac{\partial^2 T}{\partial y^2} \tag{9.46}$$

The boundary conditions are

$$\left.\frac{\partial T}{\partial y}\right|_{(x,0)} = 0 \tag{9.47}$$

$$K\left.\frac{\partial T}{\partial y}\right|_{\left(x,\frac{H}{2}\right)} = q_s'' \tag{9.48}$$

Let us define the dimensionless temperature as

$$\theta = \frac{T\left(x, \frac{H}{2}\right) - T(x, y)}{T\left(x, \frac{H}{2}\right) - T_m(x)}$$

For a fully developed condition, θ is independent of x, that is, $\theta = \theta(y)$. Hence,

$$\frac{\partial \theta}{\partial x} = 0$$

or

$$\frac{\partial}{\partial x}\left(\frac{T\left(x, \frac{H}{2}\right) - T(x, y)}{T\left(x, \frac{H}{2}\right) - T_m(x)}\right) = 0$$

Expanding and using the definition of θ, we have

$$\left.\frac{dT}{dx}\right|_{\left(x, \frac{H}{2}\right)} - \frac{\partial T}{\partial dx} - \phi(y)\left[\left.\frac{dT}{dx}\right|_{\left(x, \frac{H}{2}\right)} - \frac{dT_m}{dx}\right] = 0 \tag{9.49}$$

From the definition of θ, we can write

$$T(x, y) = T\left(x, \frac{H}{2}\right) - \left[T\left(x, \frac{H}{2}\right) - T_m(x)\right]\theta$$

Differentiating the above equation and evaluating the derivative at $y = \frac{H}{2}$, we have

$$\left.\frac{\partial T}{\partial y}\right|_{\left(x, \frac{H}{2}\right)} = -\left[T\left(x, \frac{H}{2}\right) - T_m(x)\right]\left.\frac{d\theta}{dy}\right|_{\frac{H}{2}}$$

Substituting into the expression for h, we have

$$h = -\frac{K\left[T\left(x, \frac{H}{2}\right) - T_m(x)\right]}{T_s(x) - T_m(x)}\left.\frac{d\theta}{dy}\right|_{\frac{H}{2}}$$

We can write another expression of h from Newton's law of cooling as

$$h = \frac{q_s''}{T_s(x) - T_m(x)}$$

Equating the above two equations, we write

$$T\left(x, \frac{H}{2}\right) - T_m(x) = \frac{-q_s''}{K\left.\frac{\partial \theta}{\partial y}\right|_{\frac{H}{2}}} = \text{Constant}$$

Differentiating the above equation, we have

$$\left.\frac{\partial T}{\partial x}\right|_{\left(x, \frac{H}{2}\right)} - \frac{\partial T_m}{\partial x} = 0$$

Combining the above equation with equation (9.49), we have

$$\left.\frac{dT}{dx}\right|_{\left(x,\frac{H}{2}\right)} = \frac{dT_m}{dx} = \frac{\partial T}{\partial x} \tag{9.50}$$

This equation shows that $\frac{\partial T}{\partial x}$ can be replaced with $\frac{dT_m}{dx}$. Let us derive the expression for $\frac{dT_m}{dx}$.

Application of conservation of energy to differential element in Figure 9.9 gives

$$2q_s''Wdx + mC_pT_m = mC_p\left[T_m + \frac{dT_m}{dx}dx\right]$$

where W is the width of the plate. Simplification of the above equation gives

$$\frac{dT_m}{dx} = \frac{2q_s''W}{mC_p} = \text{Constant} \tag{9.51}$$

We know that the mass flow rate in terms of the mean velocity u_m can be written as

$$m = WH\rho u_m$$

Substituting m, we have

$$\frac{dT_m}{dx} = \frac{2q_s''}{\rho C_p u_m H} = \text{Constant}$$

Substituting the above equation in equation (9.50), we have

$$\left.\frac{dT}{dx}\right|_{\left(x,\frac{H}{2}\right)} = \frac{dT_m}{dx} = \frac{\partial T}{\partial x} = \frac{2q_s''}{\rho C_p u_m H}$$

The above equation indicates that $T(x, y)$, $T_m(x)$, and $T_s(x)$ vary linearly with axial distance x.

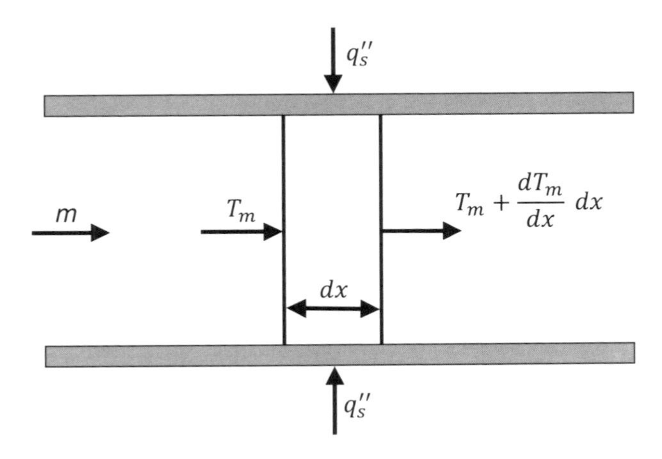

Figure 9.9 Differential element for application of conservation of energy equation

Substituting the above equation to the governing equation (equation (9.46)), we have

$$\frac{\partial^2 T}{\partial y^2} = \frac{2q_s''}{KH}\frac{u}{u_m} \tag{9.52}$$

The mean velocity can be expressed as

$$u_m = \frac{2}{H}\int_0^{H/2} u dy$$

Substituting equation (9.42), we have

$$\begin{aligned} u_m &= -\frac{H^2}{4\mu}\frac{dP}{dx}\int_0^{H/2}\left[1 + 4\beta_v Kn - 4\frac{y^2}{H^2}\right] dy \\ &= -\frac{H^2}{12\mu}\frac{dP}{dx}[1 + 6\beta_v Kn] \end{aligned}$$

Combining above equation with equation (9.42), we have

$$\frac{u}{u_m} = \frac{6}{1 + 6\beta_v Kn}\left[\frac{1}{4} + \beta_v Kn - \frac{y^2}{H^2}\right]$$

Substituting the above equation into equation (9.52), we have

$$\frac{\partial^2 T}{\partial y^2} = \frac{12q_s''}{(1 + 6\beta_v Kn)KH}\left[\frac{1}{4} + \beta_v Kn - \frac{y^2}{H^2}\right]$$

Integrating twice, we have

$$T(x, y) = \frac{12q_s''}{(1 + 6\beta_v Kn)KH}\left[\frac{1}{2}\left(\frac{1}{4} + \beta_v Kn\right)y^2 - \frac{y^4}{12H^2}\right] + f(x)y + g(x) + \cdots \tag{9.53}$$

where $f(x)$ and $g(x)$ are constants of integration. Using the symmetry boundary condition (equation (9.47)), we have

$$f(x) = 0$$

The second boundary condition (equation (9.48)) is automatically satisfied in the above equation. Therefore, $g(x)$ cannot be determined using this boundary condition. We can determine $g(x)$ by evaluating the mean temperature T_m using two methods.

In the first method, we integrate equation (9.51) between the inlet of the channel ($x = 0$) and an arbitrary location (x) as

$$\int_{T_{mi}}^{T_m} dT_m = \frac{2q_s''}{\rho C_p u_m H}\int_0^x dx$$

where $T_m(x = 0) = T_{mi}$ is the mean inlet temperature. Evaluating the above integral, we have

$$T_m(x) = \frac{2q_s''}{\rho C_p u_m H}x + T_{mi} \tag{9.54}$$

In the second method, T_m is evaluated using its definition (equation (9.45)). Substituting equations (9.42) and (9.53) in equation (9.45), we have

$$T_m(x) = -\frac{H^2}{8\mu}\frac{dP}{dx}\int_0^{H/2}\left[1 + 4\beta_v Kn - 4\frac{y^2}{H^2}\right]\left\{\frac{12q_s''}{(1 + 6\beta_v Kn)\, KH}\right\}$$
$$\times \frac{\left[\frac{1}{2}\left(\frac{1}{4} + \beta_v Kn\right)y^2 - \frac{y^4}{12H^2}\right] + g(x)dy}{-\frac{H^2}{8\mu}\frac{dP}{dx}\int_0^{H/2}\left[1 + 4\beta_v Kn - 4\frac{y^2}{H^2}\right]dy}$$

Evaluating the integrals, we have

$$T_m(x) = \frac{3q_s'' H}{K(1 + 6\beta_v Kn)^2}\left[\left(\beta_v Kn\right)^2 + \frac{13}{40}\beta_v Kn + \frac{13}{560}\right] + g(x) \quad (9.55)$$

Equating equations (9.54) and (9.55), we have

$$g(x) = T_{\mathrm{mi}} + \frac{2q_s''}{\rho C_p u_m H}x - \frac{3q_s'' H}{K(1 + 6\beta_v Kn)^2}\left[\left(\beta_v Kn\right)^2 + \frac{13}{40}\beta_v Kn + \frac{13}{560}\right]$$

Surface temperature $T_s\left(x, \frac{H}{2}\right)$ can be written by substituting equation (9.53) into equation (9.44).

$$T_s(x) = \frac{3q_s'' H}{K(1 + 6\beta_v Kn)}\left[\frac{1}{2}\beta_v Kn + \frac{5}{48}\right] + \frac{2\gamma}{\gamma + 1}\frac{q_s'' H}{KPr}\beta_T Kn + g(x) \quad (9.56)$$

Substituting equations (9.55) and (9.56) in the Nusselt number expression (equation (9.43)), we have

$$Nu = \frac{2}{\frac{3}{(1+6\beta_v Kn)}\left\{\frac{1}{2}\beta_v Kn + \frac{5}{48} - \frac{1}{1+6\beta_v Kn}\left[\left(\beta_v Kn\right)^2 + \frac{13}{40}\beta_v Kn + \frac{13}{560}\right]\right\} + \frac{2\gamma}{\gamma + 1}\frac{1}{Pr}\beta_T Kn}$$

Assuming $\sigma_v = \sigma_T = 1$, we have $\beta_v = \beta_T = 1$. Hence, the above expression for the Nusselt number can be written for this case as

$$Nu = \frac{2}{\frac{3}{(1+6Kn)}\left\{\frac{1}{2}Kn + \frac{5}{48} - \frac{1}{1+6Kn}\left[(Kn)^2 + \frac{13}{40}Kn + \frac{13}{560}\right]\right\} + \frac{2\gamma}{\gamma + 1}\frac{1}{Pr}Kn}$$

Table 9.2 Nusselt number variation as a function of the Knudsen number for air ($\gamma = 1.4$ and $Pr = 0.7$)

Kn	0.0	0.02	0.04	0.06	0.08
Nu	8.235	7.5	6.8	6.2	5.7

Table 9.2 shows the Nusselt number variation as a function of Knudsen number of air ($Pr = 0.7$ and $\gamma = 1.4$).

Note:

1. The Nusselt number for microchannel depends on the fluid as indicated by Pr and γ unlike macrochannels.
2. The no-slip Nusselt number for macrochannel flow is determined by setting $Kn = 0$. Substituting in the Nusselt number expression, we have

$$Nu = \frac{140}{17} = 8.235$$

   This is in agreement with that derived using no-slip boundary condition.
3. The effect of temperature jump on the Nusselt number is represented by the last term in the denominator.
4. The Knudsen number is a function of local pressure. Pressure varies along the microchannel. Therefore, it follows that the Nusselt number varies with distance, x. This behavior is contrary to that of a macrochannel, where the Nusselt number is a constant.

9.10 Fully Developed Flow in Microtubes: Uniform Surface Flux

Figure 9.10 shows the schematic of a microtube with fully developed convection heat transfer in uniform surface heat flux condition. The tube has radius r_0, surface heat flux q_s'', inlet

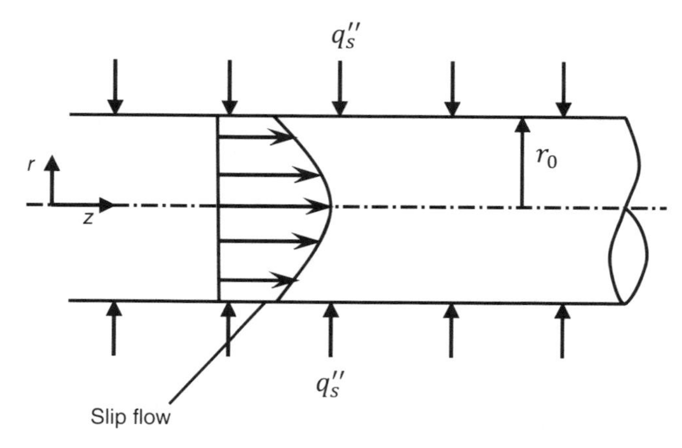

Figure 9.10 Schematic of the microtube flow with uniform surface heat condition and slip flow velocity profile

pressure P_i, and outlet pressure P_o. The incompressible fully developed Poiseuille flow in macrochannels has the following characteristics: (1) parallel streamlines, (2) zero lateral velocity component ($v = 0$), (3) invariant axial velocity with axial distance $\left(\frac{\partial u}{\partial x} = 0\right)$, and (4) linear axial pressure ($\frac{dp}{dx}$ = Constant).

However, in microchannels, compressibility and rarefaction change this flow pattern, and none of the above conditions hold good. Density changes in microchannel gaseous flows are appreciable, and the flow can no longer be assumed incompressible. The other effect is due to rarefaction. There is an increase in mean free path λ due to a decrease in pressure in microchannels. Thus, the Knudsen number increases along a microchannel. As a consequence, the axial velocity varies with axial distance, lateral velocity component does not vanish, streamlines are not parallel, and pressure gradient is not constant.

For the analysis, we make the following assumptions: (1) steady state; (2) laminar flow; (3) one dimensional (no variation with axial distance x and normal distance z); (4) slip flow regime ($0.001 < Kn < 0.1$); (5) ideal gas; (6) constant viscosity, conductivity, and specific heats; (7) negligible lateral variation of density and pressure; (8) negligible gravity; and (9) negligible inertia forces.

The isothermal flow assumption eliminates temperature as a variable in the momentum equation. Thus, density can be expressed in terms of pressure using the ideal gas law. Negligible inertial forces assumption is justified for microchannel flow due to low Reynolds number. The dominant viscous force is $\mu \frac{1}{r}\frac{\partial}{\partial r}\left(r\frac{\partial v_z}{\partial r}\right)$ as this term is of the order of r_o^{-2} and other viscous terms are the order of L^{-2} based on scale analysis. Based on these assumptions, the simplified N–S equation in cylindrical coordinates can be written as

$$\frac{1}{r}\frac{\partial}{\partial r}\left(r\frac{\partial v_z}{\partial r}\right) = \frac{1}{\mu}\frac{\partial p}{\partial z} \tag{9.57}$$

The boundary conditions are

$$\left.\frac{\partial v_z}{\partial r}\right|_{(o,z)} = 0 \quad \text{(symmetry condition)}$$

$$v_z(r_o, z) = -\lambda\beta_v \left.\frac{\partial v_z}{\partial r}\right|_{(r_o,z)}$$

Integration of equation (9.57) and multiplication of the above boundary conditions give

$$v_z = -\frac{r_o^2}{4\mu}\frac{dp}{dz}\left[1 + 4\beta_v Kn - \frac{r^2}{r_o^2}\right] \tag{9.58}$$

where the Knudsen number $Kn = \frac{\lambda}{2r_o}$.

The mean velocity for the flow can be calculated using

$$\begin{aligned} v_{zm} &= \frac{1}{\pi r_o^2}\int_0^{r_o} 2\pi r v_z dr \\ &= -\frac{r_o^2}{8\mu}\frac{dp}{dz}(1 + 8\beta_v Kn) \end{aligned} \tag{9.59}$$

Combining equations (9.58) and (9.59), we can write

$$\frac{v_m}{v_{zm}} = 2\frac{1 + 4\beta_v Kn - \left(\frac{r}{r_o}\right)^2}{1 + 8\beta_v Kn} \tag{9.60}$$

The heat transfer coefficient h for the uniform surface flux q_s'' is

$$h = \frac{q_s''}{T_s - T_m} \tag{9.61}$$

The Nusselt number is defined as

$$Nu = \frac{2r_o h}{K} = \frac{2r_o q_s''}{k(T_s - T_m)} \tag{9.62}$$

The surface temperature can be determined from temperature jump boundary condition as

$$T_s = T(r_o, z) + \frac{2\gamma}{1+\gamma}\frac{\lambda}{Pr}\beta_T \left.\frac{\partial T}{\partial r}\right|_{(r_o,z)} \tag{9.63}$$

The simplified energy equation in cylindrical coordinate is given as

$$\rho C_p v_z \frac{\partial T}{\partial z} = \frac{K}{r}\frac{\partial}{\partial r}\left(r\frac{\partial T}{\partial r}\right) \tag{9.64}$$

The boundary conditions are

$$\left.\frac{\partial T}{\partial r}\right|_{(0,z)} = 0 \tag{9.65}$$

$$K\left.\frac{\partial T}{\partial r}\right|_{(r_o,z)} = q_s'' \tag{9.66}$$

Let us define the nondimensional temperature profile as

$$\psi = \frac{T(r_o, z) - T(r, z)}{T(r_o, z) - T_m(z)} \tag{9.67}$$

where the mean temperature for the tube flow is given by

$$T_m = \frac{\int_0^{r_o} v_z T r dr}{\int_0^{r_o} v_z r dr} \tag{9.68}$$

For a fully developed case, ψ is assumed independent of z, that is, $\psi = \psi(r)$. Thus,

$$\frac{\partial \psi}{\partial z} = 0$$

or

$$\frac{\partial}{\partial z}\left(\frac{T\left(r_o, z\right) - T(r, z)}{T(r_o, z) - T_m(z)}\right) = 0$$

Expanding the above expression and using the definition of ψ, we have

$$\left.\frac{dT}{dz}\right|_{(r_o,z)} - \frac{\partial T}{\partial z} - \psi(r)\left[\left.\frac{dT}{dz}\right|_{(r_o,z)} - \frac{dT_m}{dz}\right] = 0 \tag{9.69}$$

We can rewrite equation (9.67) as

$$T(r,z) = T(r_o,z) - [T(r_o,z) - T_m(z)]$$

Differentiating with respect to r and evaluating the derivative at $r = r_o$, we have

$$\left.\frac{\partial T}{\partial r}\right|_{(r_o,z)} = -[T(r_o,z) - T_m(z)]\left.\frac{d\psi}{dr}\right|_{r_o} \tag{9.70}$$

The heat transfer coefficient is given by

$$h = \frac{-K\left.\frac{\partial T}{\partial r}\right|_{(r_o,z)}}{T_m(z) - T_s(z)}$$

Substituting equation (9.70) in the above expression, we have

$$h = \frac{-K[T(r_o,z) - T_m(z)]}{T_s(z) - T_m(z)}\left.\frac{d\psi}{dr}\right|_{r_o}$$

Equating the above expression with equation (9.61) and rearranging, we have

$$T(r_o,z) - T_m(z) = \frac{-q_s''}{K\left.\frac{d\psi}{dr}\right|_{r_o}} = \text{Constant}$$

Differentiating, we have

$$\left.\frac{\partial T}{\partial z}\right|_{(r_o,z)} - \left.\frac{\partial T_m}{\partial z}\right|_{z} = o$$

Combining with equation (9.69), we have

$$\left.\frac{dT}{dz}\right|_{(r_o,z)} = \left.\frac{dT_m}{dz}\right|_{z} = \frac{\partial T}{\partial z} \tag{9.71}$$

The application of conservation of energy to the element in Figure 9.11 gives

$$2\pi r_o q_s'' dz + mc_p T_m = mc_p\left[T_m + \frac{dT_m}{dz}dz\right]$$

On simplification, we have

$$\frac{dT_m}{dz} = \frac{2\pi r_o q_s''}{mc_p} \tag{9.72}$$

The mass flow rate can be expressed as

$$m = \rho\pi r_o^2 v_{\text{zm}} \tag{9.73}$$

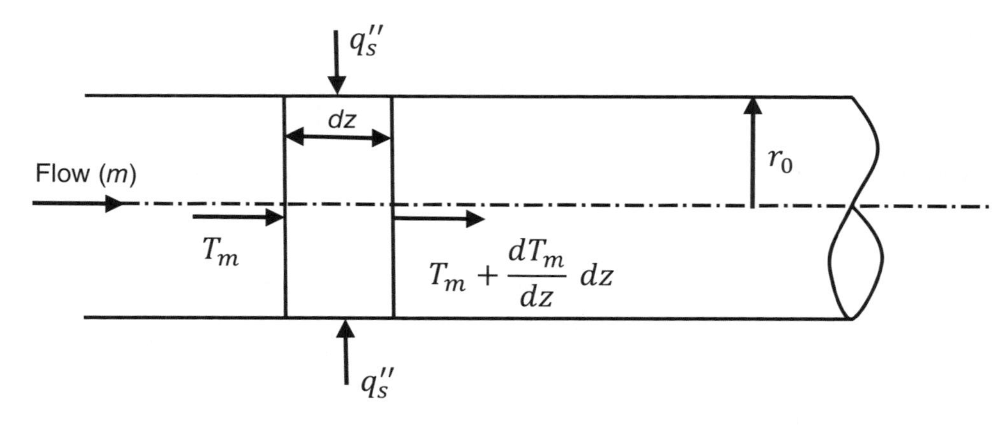

Figure 9.11 Schematic for the application of energy equation to a small element dz

where v_{zm} is the mean axial velocity. Combining equations (9.72) and (9.73), we get

$$\frac{dT_m}{dz} = \frac{2q_s''}{\rho c_p r_o v_{zm}} \tag{9.74}$$

Substituting equation (9.74) into equation (9.71), we get

$$\left.\frac{dT}{dz}\right|_{(r_o,z)} = \left.\frac{dT_m}{dz}\right|_z = \frac{\partial T}{\partial z} = \frac{2q_s''}{\rho c_p r_o v_{zm}} \tag{9.75}$$

The above expression shows that $T(r,z)$, $T_m(z)$, and $T_s(z)$ vary linearly with axial distance z. Substituting equation (9.75) into equation (9.64), we have

$$\frac{\partial}{\partial r}\left(r\frac{\partial T}{\partial r}\right) = \frac{2q_s''}{Kr_o}\frac{v_z}{v_{zm}} r \tag{9.76}$$

Substituting equation (9.60) in equation (9.76), we get

$$\frac{\partial}{\partial r}\left(r\frac{\partial T}{\partial r}\right) = \frac{4}{1+8\beta_v Kn}\frac{q_s''}{Kr_o}\left[1 + 4\beta_v Kn - \frac{r^2}{r_o^2}\right] r$$

Integrating twice, we get

$$T(r,z) = \frac{q_s''}{(1+8\beta_v Kn)Kr_o}\left[\left(1+4\beta_v Kn\right) r^2 - \frac{1}{4}\frac{r^4}{r_o^2}\right] + f(z)y + g(z)$$

Application of boundary condition (equation (9.65)) gives

$$f(z) = 0$$

Hence, the solution $T(r,z)$ becomes

$$T(r,z) = \frac{q_s''}{(1+8\beta_v Kn)Kr_o}\left[\left(1+4\beta_v Kn\right) r^2 - \frac{1}{4}\frac{r^4}{r_o^2}\right] + g(z) \tag{9.77}$$

The boundary condition is automatically satisfied by the above equation. For the determination of $g(z)$, the mean temperature T_m is evaluated using two methods. In the first method, equation (9.72) is integrated between the inlet of the tube and an arbitrary location z:

$$\int_{T_{\text{mi}}}^{T_m} dT_m = \frac{2q_s''}{\rho c_p r_o v_{\text{zm}}} \int_0^z dz$$

where

$$T_m(0) = T_{\text{mi}}$$

Evaluating the integrals, we get

$$T_m = \frac{2q_s''}{\rho c_p r_o v_{\text{zm}}} z + T_{\text{mi}} \tag{9.78}$$

In the second method, the definition in equation (9.68) is used for the evaluation of T_m. Substituting equation (9.58) into equation (9.77) and simplifying, we get

$$T_m = \frac{\int_0^{r_o} \left[1 + 4\beta_v Kn - \frac{r^2}{r_o^2}\right] \left\{ \frac{4q_s''}{(1+8\beta_v Kn)Kr_o} \left[\left(\frac{1}{4} + \beta_v Kn\right) r^2 - \frac{r^4}{16r_o^2}\right] + g(z) \right\} rdr}{\int_0^{r_o} \left[1 + 4\beta_v Kn - \frac{r^2}{r_o^2}\right] rdr}$$

Evaluating the integrals, we get

$$T_m = \frac{q_s'' r_o}{K(1 + 8\beta_v Kn)^2} \left[16\beta_v^2 Kn^2 + \frac{14}{3}\beta_v Kn + \frac{7}{24}\right] + g(z) \tag{9.79}$$

Equating equations (9.78) and (9.79), we get

$$g(z) = T_{\text{mi}} + \frac{2q_s'' z}{\rho c_p r_o v_{\text{zm}}} - \frac{q_s'' r_o}{K(1 + 8\beta_v Kn)^2} \left[16\beta_v^2 Kn^2 + \frac{14}{3}\beta_v Kn + \frac{7}{24}\right]$$

Using equations (9.63) and (9.77), and boundary condition (9.66), we get the surface temperature as

$$T_s(r_o, z) = \frac{4q_s'' r_o}{K(1 + 8\beta_v Kn)} \left[Kn\beta_v + \frac{3}{16}\right] + \frac{4\gamma}{\gamma + 1} \frac{q_s'' r_o}{KPr} \beta_T Kn + g(z) \tag{9.80}$$

Substituting equations (9.79) and (9.80) into equation (9.62), we get

$$Nu = \frac{2}{\frac{4}{(1+8\beta_v Kn)} \left(\beta_v Kn + \frac{3}{16}\right) - \frac{1}{(1+8\beta_v Kn)^2} \left[16\beta_v^2 Kn^2 + \frac{14}{3}\beta_v Kn + \frac{7}{24}\right] + \frac{4\gamma}{\gamma+1} \frac{1}{Pr} \beta_T Kn} \tag{9.81}$$

The no-slip Nusselt number can be obtained by setting $Kn = 0$ in equation (9.81) as

$$Nu_o = \frac{48}{11} = 4.364$$

Using air with $\gamma = 1.4$, $Pr = 0.7$, and $\sigma_v = \sigma_T = 1$ in equation (9.81), we see that the Nusselt number decreases with an increase in Kn. This indicates that the Nusselt number decreases due to rarefaction and compressibility effect.

9.11 Convection in Isothermal Circular Tube with Viscous Heating

The x-momentum equation in cylindrical coordinate is

$$u\frac{\partial u}{\partial x}+v\frac{\partial u}{\partial r}=-\frac{1}{\rho}\frac{\partial p}{\partial x}+\nu\frac{1}{r}\frac{\partial}{\partial r}\left(r\frac{\partial u}{\partial r}\right)+\frac{\partial^2 u}{\partial x^2} \tag{9.82}$$

Using the fully developed condition, $\partial u/\partial x=0$ and $v=0$ the governing equation simplifies to

$$\equiv\frac{1}{r}\frac{d}{dr}\left(r\frac{du}{dr}\right)=\frac{1}{\mu}\frac{dp}{dx} \tag{9.83}$$

Boundary conditions at $r=R$; $u=u_s$, at $r=0$; $u=$ finite where u_s is the slip velocity.
Energy equation in cylindrical coordinate is

$$u\frac{\partial T}{\partial x}+v\frac{\partial T}{\partial r}=\frac{\alpha}{r}\frac{\partial}{\partial r}\left(r\frac{\partial T}{\partial r}\right)+\alpha\frac{\partial^2 T}{\partial x^2}+\frac{\phi}{\rho C_p} \tag{9.84}$$

$$\phi=2\mu\left\{\left(\frac{\partial u}{\partial r}\right)^2+\left(\frac{v}{r}\right)^2+\left(\frac{\partial u}{\partial x}\right)^2\right\} \tag{9.85}$$

Using the fully developed assumption, $v=0$ and $\frac{\partial u}{\partial x}=0$, and negligible axial conduction assumption $\left(\frac{\partial^2 T}{\partial x^2}=0\right)$, the energy equation simplifies to

$$u\frac{\partial T}{\partial x}=\frac{\alpha}{r}\frac{\partial}{\partial r}\left(r\frac{\partial T}{\partial r}\right)+\frac{r}{C_p}\left(\frac{\partial u}{\partial r}\right)^2 \tag{9.86}$$

For uniform wall temperature case, the boundary conditions are
at $x=0$; $T=T_i$ (inlet condition) (A)
at $r=R$; $T=T_s$ (temperature jump condition) (B)
at $r=0$; $\frac{dT}{dr}=0$ (symmetry condition) (C)
The solution of the momentum equation (9.83) gives

$$u=u_m\frac{2\left\{1+4Kn\left(\frac{2-\sigma_\theta}{\sigma_v}\right)-\left(\frac{r}{R}\right)^2\right\}}{1+8Kn\frac{2-\sigma_\theta}{\sigma_v}}$$

For uniform wall heat flux, the boundary conditions are

at $x=0$; $T=T_i$ (A)
at $r=R$; $q=q''=K\left.\frac{\partial T}{\partial r}\right|_{r=R}$ (B)
at $r=0$; $\frac{\partial T}{\partial r}=0$ (C)

Let us look at the uniform wall heat flux condition. Let us use the nondimensional variables as

$$\zeta=\frac{x}{L},\eta=\frac{r}{R},u^*=\frac{u}{u_m},Gz=\frac{RePrD}{L},\theta=T_i\frac{T-}{q''R/K},Br=\mu u_m^2\Big|q''D,Kn=\frac{\lambda}{D}$$

Here, D is the diameter of the tube, Re is the Reynolds number, and Pr is the Prandtl number. The energy equation in dimensionless form becomes

$$\frac{Gz\left(1-\eta^2+4kn\frac{2-6v}{6}\right)}{2\left(1+8Kn\frac{2-\sigma_v}{\sigma_v}\right)}\frac{\partial\theta}{\partial\zeta}=\frac{1}{\eta}\frac{\partial}{\partial n}\left(\eta\frac{\partial\theta}{\partial\eta}\right)+\frac{32Br\eta^2}{\left(1+8Kn\frac{2-\sigma_v}{\sigma_v}\right)} \tag{9.87}$$

The boundary conditions are

at $\zeta=0, \theta=0$ (inlet condition) (A)

$\left.\frac{\partial\theta}{\partial\eta}\right|_{\eta=1}=1$ (wall condition) (B)

$\left.\frac{\partial\theta}{\partial\eta}\right|_{\eta=0}=0$ (center condition) (C)

The problem can be solved as an eigenvalue problem to obtain the temperature distribution.

Nusselt Number Determination

Heat transfer coefficient $h_x=\frac{q''}{T_w-T_b}=\frac{K\left.\frac{dT}{dr}\right|_{r=R}}{T_w-T_b}$

In dimensionless form, we have

$$h_x=\frac{\frac{K}{R}\left.\frac{\partial\theta}{\partial\eta}\right|_{\eta=1}}{\frac{T_w-T_b}{q''R/K}}$$

The expression for the Nusselt number is

$$Nu_x=\frac{h_xD}{K}=\frac{2}{\frac{T_w-T_b}{q''R/K}}$$

$$\frac{T_w-T_b}{q''R/K}=\frac{T_w-T_s}{q''R/K}-\frac{T_i-T_s}{q''R/K}-\frac{T_b-T_i}{q''R/K}$$

Here $\frac{T_i-T_s}{q''R/K}=-\theta_s$ and

$$\frac{T_b-T_i}{q''R/K}=\theta_b$$

Temperature jump boundary condition:

$$T_w-T_s=\frac{2-\sigma_T}{\sigma_T}\frac{2\gamma}{\gamma+1}\frac{\lambda}{Pr}\left.\frac{\partial T}{\partial r}\right|_{r=R}$$

$$=\frac{2-\sigma_T}{\sigma_T}\frac{2\gamma}{\gamma+1}\frac{\lambda}{Pr}\frac{q''R}{K}\frac{1}{R}\left.\frac{\partial\theta}{\partial\eta}\right|_{\eta=1}$$

$$\frac{T_w-T_s}{q''R/K}=\frac{2-\sigma_T}{\sigma_T}\frac{4\gamma}{\gamma+1}\frac{Kn}{Pr}\left.\frac{\partial\theta}{\partial\eta}\right|_{\eta=1}\quad \text{Since}\quad Kn=\frac{\lambda}{2R}$$

Substituting, the expression for the Nusselt number is

$$Nu_x=\frac{2}{\frac{2-\sigma_T}{\sigma_T}\frac{4\gamma}{\gamma+1}\frac{Kn}{Pr}+\theta_s-\theta_b}$$

Jeong and Jeong (2006b) for circular microchannel derived the fully developed Nusselt number including the effect of viscous heating for constant heat flux case as

$$Nu = \frac{48C_2^2}{C_2^2(6C_2^2 + 4C_2 + 1 + 48C_2^2C_3) + 8Br(2C_2^2 + 3C_2 + 1)}$$

where

$$C_2 = 1 + 8Kn\left(\frac{2-\sigma_v}{\sigma_v}\right)$$

$$C_3 = \frac{2-\sigma_T}{\sigma_T}\frac{2\gamma}{\gamma+1}\frac{Kn}{Pr}$$

For a heated surface ($Br > 0$), an increase in viscous heating leads to a decrease of the Nusselt number. For a cooled surface ($Br < 0$), the opposite trend is observed.

Jeong and Jeong (2006b) also derived the analytical solution of fully developed Nusselt number for constant wall temperature case with viscous heating ($Br \neq 0$).

Figure 9.12 shows the effect of viscous heating on heat transfer of a microtube. When $Br > 0$, fluid is being heated by heat flux.

The Nusselt number decreases with an increase in "Br" where $Br = \frac{T-T_i}{q''R/K}$.

Figure 9.13 demonstrates the effect of the Knudsen number and the Brinkman number on the fully developed Nusselt number of a microtube. *Note*: Positive "Br" means heat is transferred to the fluid and negative "Br" means heat is transferred from the fluid. Hence, the effect of "Br" is opposite for cooling and heating. Uniform wall temperature case: Figure 9.14 shows the Nusselt number distribution as a function of axial location and the Brinkman number for uniform wall temperature case. An increase in "Br" leads to an increase in "Nu." This behavior is opposite to that of the uniform wall flux case (Figure 9.13). This is because $Br > 0$ means cooling of the fluid for uniform wall heat flux case compared to the uniform temperature case.

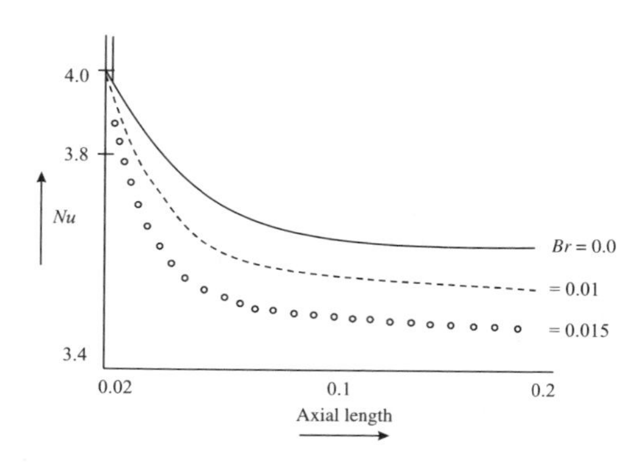

Figure 9.12 Effect of viscous heating on heat transfer at the microtube entrance for uniform heat flux at the wall ($Kn = 0.04$, $Pr = 0.7$)

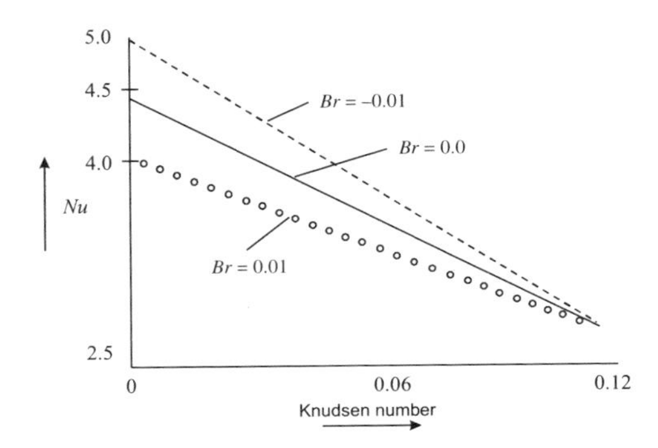

Figure 9.13 Variation of the fully developed Nu as a function of Kn, for uniform heat flux at the microtube wall

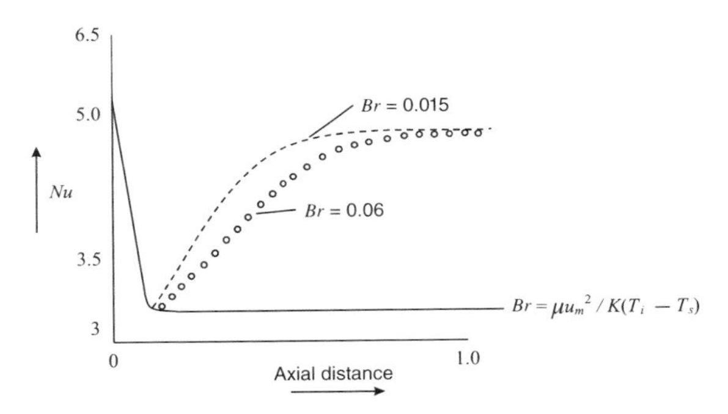

Figure 9.14 Effect of viscous heating for microtube with uniform wall temperature case ($Kn = 0.04$, $Pr = 0.7$)

9.12 Flow Boiling Heat Transfer in Mini-/Microchannels

Several applications such as lasers, X-ray medical systems, avionics, and computer data centers require dissipation of a large amount of heats from small areas. Two-phase cooling schemes are one of the actively pursued high heat flux thermal management systems. It has many unique attributes, that is, high heat dissipation-to-volume ratio, small coolant inventory, relative ease of fabrication, and compactness. Various experimental results have been reported to understand the heat transfer phenomena and develop appropriate correlations for accounting various features unique to microchannel heat sinks.

9.12.1 Minichannel versus Microchannel

A channel may behave as a microchannel for certain fluids and operating conditions and as a macrochannel for certain other conditions. Channel size can also be the distinguishing factor between macro- and microchannel flows. The boundary between macro- and microchannel flows is related to the ratio of bubble size to channel diameter for two-phase flows. The case with larger ratio between bubble size and channel size behaves more like a microchannel flow.

Table 9.3 Fluid properties and hydraulic diameter corresponding to transition from macro- to microchannel flow at 1 bar

Fluids	T_{sat} [C]	h_{fg} [kJ/kg]	ρ_f [kg/m^3]	ρ_g [kg/m^3]	σ [mN/m]	$D_{trans}(co = 0.5)$ [mm]
Water	99.6	2258	959	0.59	59.0	5.0
R134a	−26.4	217.2	1378	5.19	15.5	2.14
HFE 7100	59.6	111.7	1373	9.58	15.7	2.17

Kew and Cornwell (1997) used this rationale to define a confinement number for determining the boundary between microchannel and macrochannel flows as

$$co = \left[\frac{\sigma}{g\left(\rho_f - \rho_g\right) D_h^2} \right]^{1/2}$$

Here, σ is surface tension, ρ is density, D_h is hydraulic diameter of microchannel, subscript "f" corresponds to saturated liquid, and subscript "g" corresponds to saturated vapor. Confinement number is based on the ratio of surface tension force to buoyancy force. It is a good representation of pool boiling, that is, in a confined channel. The macrochannel assumption falls apart when the channel becomes too confining at $co > 0.5$.

Table 9.3 shows the hydraulic diameter data corresponding to $co = 0.5$ for several fluids. The Table 9.3 shows that water with high surface tension produces a microchannel flow in larger channels. Dielectric coolants HPF7100 and refrigerant R134a have lower hydraulic diameters corresponding to the transition from macro- to microchannel flow.

The confinement number criterion presented above is not a good representation of flow boiling situation. Bubble size in flow boiling is dominated by liquid drag rather than buoyancy. Lee and Mudawar (2009) proposed an alternative measure of the boundary between micro- and macrochannel flow boiling. The proposed criterion incorporates the influence of liquid drag on bubble size. Equating the drag force on the bubble with the surface tension force that holds the bubble on the wall gives

$$C_D \left(\frac{\pi D_b^2}{4} \right) \left(\frac{1}{2} \rho_f v^2 \right) \sim \pi D_b \sigma \tag{9.88}$$

where D_b is bubble departure diameter, C_D is drag coefficient, and v is liquid velocity. A channel tends to confine the flow when the diameter determined from the above equation approaches the diameter of the channel. Based on this observation, the criterion for transition from macro- to microchannel flow can be determined using the relation

$$D_{trans} \leq D_b$$

D_{trans} is equal to the diameter of the channel for circular channel. Two-phase microchannel flow applications of practical interest are characterized by modest Reynolds number, generally greater than 50. The drag coefficient for these conditions can be expressed as Liao (2002)

$$C_D = \frac{24}{Re_{trans}} \left(1 + \frac{3}{160} Re_{trans} \right) \tag{9.89}$$

Table 9.4 Fluid properties and hydraulic diameter corresponding to transition from macro- to microchannel flow for water and HFE 7100 at 1 bar

Fluids	μ_f [kg/m-s]	G [kg/m^2-s]	D_{trans} [mm]
Water	2.83×10^{-4}	500	3.99
		1000	0.99
		2000	0.243
HFE 7100	3.57×10^{-4}	500	1.49
		1000	1.49
		2000	0.0863

Combining equations (9.88) and (9.89) and substituting $G = \rho_f v$, we have the criterion for transition from macro- to microchannel flow as

$$D_{\text{trans}} = \frac{160}{9}\frac{(\sigma\rho_f - 3\mu_f G)}{\sigma^2}$$

Table 9.4 shows a decrease of transitional diameter with an increase in flow velocity. Hence, smaller channels behave as macrochannels with an increase in mass velocity.

9.12.1.1 Subcooled and Saturated Boiling

Subcooled boiling occurs at the surface with a bulk liquid temperature below the saturated temperature. Boiling commences when the applied heat flux is capable of superheating liquid adjacent to the surface for high subcooling. Subcooling is highest in the inlet region of the microchannel. Bubbles near the inlet region of the microchannel quickly recondense at the wall. Bulk liquid temperature increases along the length of the microchannel. Thus, bubbles are able to grow larger and detach from the surface, which mix into the bulk flow and undergo partial or full condensation. The magnitude of void fraction greatly reduces in subcooled boiling compared to saturated boiling due to condensation. Subcooled boiling is dominated by a bubbly flow pattern alone compared to drastic flow pattern transitions in a saturated flow (bubbly, churn, slug, annular).

Figure 9.15 illustrates different mechanisms for subcooled and saturated flow boiling. A large increase in void fraction triggers a succession of flow regime changes for saturated flow boiling, and the flow culminates in the high void fraction annular flow pattern. Cooling is sustained by evaporation of a thin liquid film along the surface in annular regime. The occurrence of critical heat flux (CHF) is associated with a sudden, large reduction in the heat transfer coefficient, due to the loss of liquid contact with the solid surface. CHF takes place due to dryout of the liquid film as the surface is exposed directly to the vapor. Dryout occurs primarily due to large length-to-diameter ratios and low mass velocities. Bubbly flow persists over much of the channel length for highly subcooled flow boiling. CHF for this case ensures when bubbles near the wall coalesce into a localized vapor blanket, which causes a sharp reduction in the local heat transfer coefficient. This form of CHF is termed as departure from nucleate boiling (DNB).

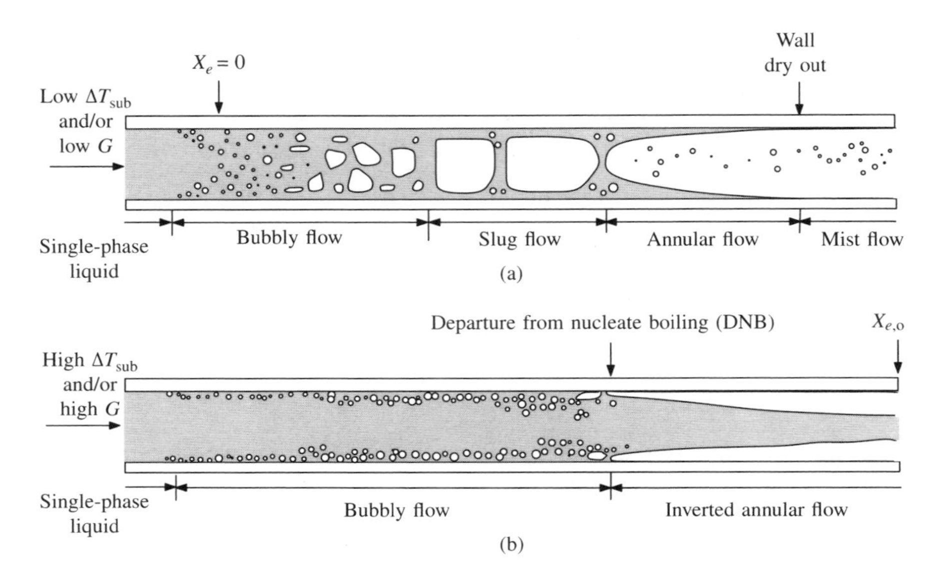

Figure 9.15 Mechanisms of flow boiling in uniformly heated channel: (a) saturated flow boiling and (b) subcooled flow boiling

DNB occurs with high inlet subcooling, high mass velocity, and small length-to-diameter ratio. The magnitude of CHF for DNB is much higher than with dryout. However, the ensuing surface temperature rise is more catastrophic. Subcooled boiling offers a thermal management solution for high heat flux defense electronics applications. Here, the dielectric coolant is precooled by a secondary refrigeration cooling system before entering a microchannel heat sink to which the electronic device is attached. One of the key benefits of this system is that it helps maintain a relatively low device temperature when dissipating very high heat fluxes. The other benefit is an increase in CHF for a given flow rate. Most of the subcooled boiling studies have been carried out with respect to electronic cooling applications.

9.12.2 Nucleate and Convective Boiling

Two types of heat transfer regimes are associated with saturated inlet conditions and terminating with dryout, that is, nucleate boiling and convective boiling. Figure 9.16 shows schematic describing the two heat transfer regimes. Bubbly and slug flow regimes occupy a significant portion of the channel length in nucleate boiling, and the heat transfer coefficient decreases due to gradual suppression of nucleate boiling.

In contrast, annular flow spans a significant fraction of the channel length in case of convective boiling. Gradual evaporation and thinning of the annular liquid film cause the heat transfer coefficient to increase along the channel length in case of convective boiling. The annular film becomes vanishingly thin with sufficiently high wall heat flux or sufficiently long channel for both the heat transfer regimes. Uneven evaporation causes initial dry patches to form, where the heat transfer coefficient begins to decrease appreciably. This phase is known as dryout incipience or onset of dryout or partial dryout. Dryout completion occurs eventually at a location farther downstream where the film is fully evaporated.

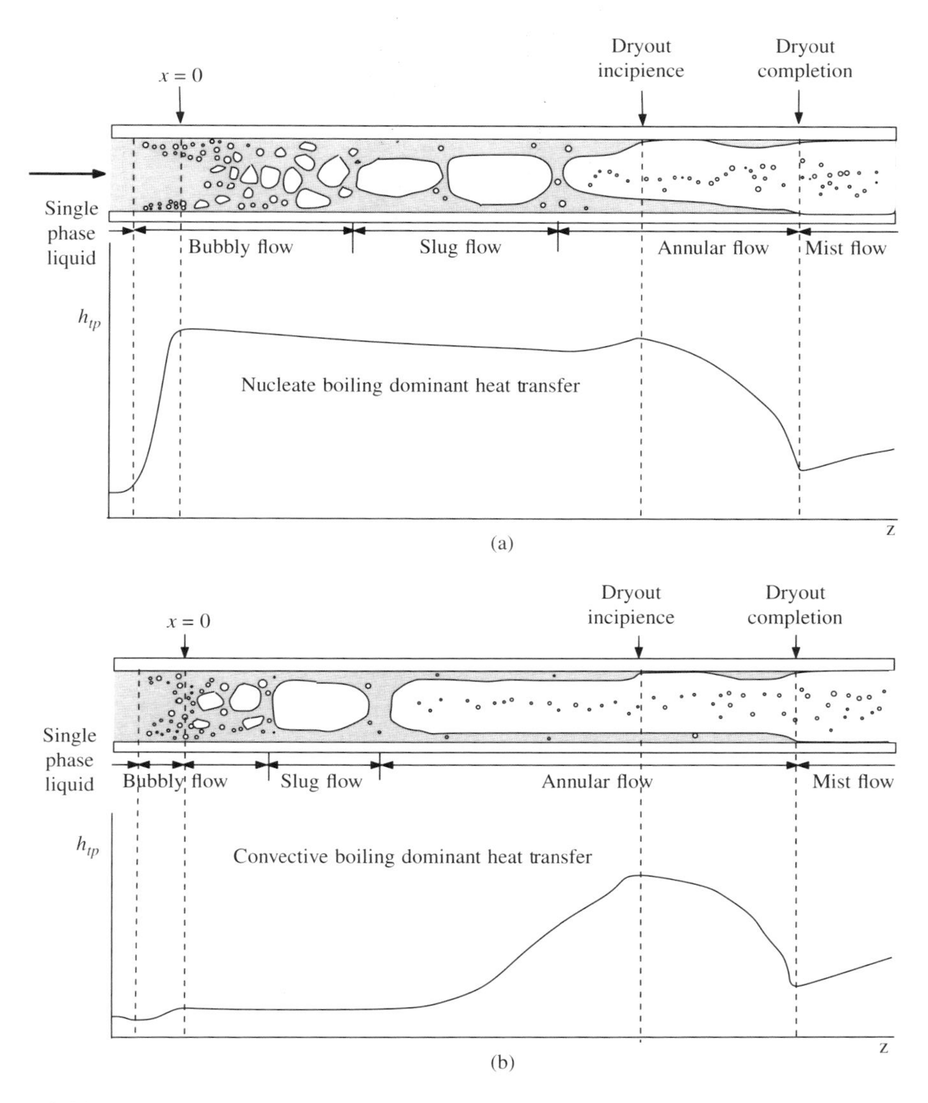

Figure 9.16 Schematic of boiling flow regimes: (a) nucleate boiling dominant heat transfer and (b) convective boiling dominant heat transfer

Dryout incipience can be identified from a shift in the slope of the measured boiling curve with increasing heat flux, where wall temperature starts to increase steeply following a small heat flux increment. Distinction between dryout incipience and dryout completion cannot be clearly pointed out, which is greatly influenced by working fluid. Boiling curve is the common method of studying the boiling performance in conventional pool boiling or flow boiling situations. The standard boiling curve is a plot of heat flux from the heater surface to the fluid as a function of temperature access of the heater above the saturation temperature of the fluid. The boiling curve can be generated through experimentation with the heater temperature excess or the heat input as the dependent variable. Figure 9.17 shows boiling curve of water and refrigerant R134a in rectangular microchannel. Water shows a narrow dryout region, high CHF value, and fast wall temperature excursion possibly because of the high latent heat. R134a refrigerant shows relatively broad dryout region, low CHF value, and slow temperature excursion due to relatively low latent heat.

9.12.3 *Dryout Incipience Quality*

The knowledge about the transition point or location of dryout incipience is essential as the dryout exhibits substantial reduction in the heat transfer coefficient. Various parameters, that is, working fluid, heat flux, mass velocity, channel diameter, and saturation pressure, affect the incipience quality for dryout. Kim and Mudawar (2013a) proposed a correlation for dryout incipience quality of saturated boiling mini-/microchannel flow as

$$x_{\mathrm{di}} = 1.4We_{fo}^{0.03}P_R^{0.08} - 15.0\left(Bo\frac{P_H}{P_F}\right)Ca^{0.35}\left(\frac{\rho_g}{\rho_f}\right)^{0.06}$$

where

We_{fo} = weber number = $\frac{G^2D_h}{\rho_f\sigma}$
P_R = reduced pressure = $\frac{P}{P_{\mathrm{crit}}}$
Bo = boiling number = $\frac{q''_H}{Gh_{\mathrm{fg}}}$
Ca = capillary number = $\frac{\mu_f G}{\rho_f\sigma} = \frac{We_{fo}}{Re_{fo}}$
G = mass velocity $(kg/m^2 - s)$
h_{fg} = latent heat of vaporization
P = pressure
P_{crit} = critical pressure
P_F = wetted perimeter of channel
P_H = heated perimeter of channel
q''_H = heat flux based on heated perimeter of channel
Re_{fo} = liquid only Reynolds number
σ = surface tension
ρ = density
D_h = hydraulic diameter

The wetted perimeter to heated perimeter ratio is used for considering the one sidewall heating effect.

9.12.4 *Saturated Flow Boiling Heat Transfer Correlation*

The dryout incipience correlation has been presented in the previous section. The generalized correlation for predryout two-phase heat transfer coefficient associated with saturated flow boiling in mini-/microchannels is discussed in the present section. Kim and Mudawar

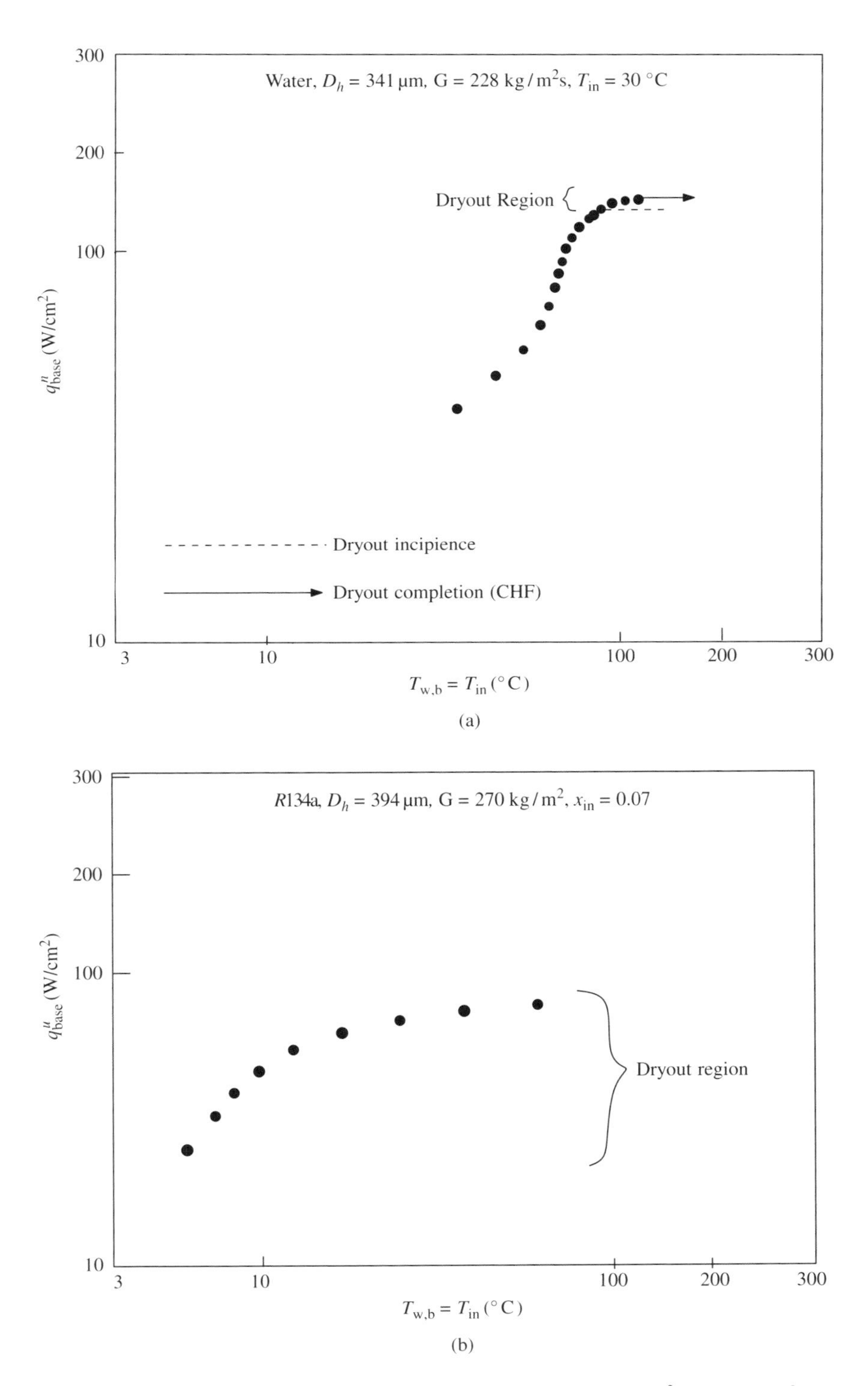

Figure 9.17 Boiling curve for (a) water for D_h = 341 μm, G = 228 kg/m²-s, T_{in} = 30 °C (Qu and Mudawar, 2004) and (b) R134a for D_h = 349 μm, G = 270 kg/m²-s, x_{in} = 0.07

(2013b) proposed the correlation for predryout saturated flow boiling heat transfer in mini-/microchannels as

$$h_{\text{tp}} = (h_{\text{nb}}^2 + h_{\text{cb}}^2)^{0.5}$$

$$h_{\text{nb}} = \left[2345\left(Bo\frac{P_H}{P_F}\right)^{0.7} P_R(1-x)^{-0.51}\right] \times \left(0.023Re_f^{0.8}Pr_f^{0.4}\frac{K_f}{D_h}\right)$$

$$h_{\text{cb}} = \left[5.2\left(Bo\frac{P_H}{P_F}\right)^{0.8} We_{fo}^{-0.54} + 3.5\left(\frac{1}{x_{++}}\right)^{0.94}\left(\frac{\rho_g}{\rho_f}\right)^{0.25}\right] \times \left(0.023Re_f^{0.8}Pr_f^{0.4}\frac{K_f}{D_h}\right)$$

where

Re_f = superficial liquid Reynolds number = $G(1-x)\frac{D_h}{\mu_f}$
x_{++} = Lockhart–Martinelli parameter based on turbulent liquid–turbulent vapor flows
x = thermometer equilibrium quality
h_{nb} = nucleate boiling dominant heat transfer coefficient
h_{cb} = convective boiling dominant heat transfer coefficient

9.12.5 SubCooled Flow Boiling Heat Transfer Correlation

Lee and Mudawar (2009) proposed the subcooled boiling CHF correlation based on the inlet condition as

$$Bo_{c,mc} = \frac{q''_{P,c}}{Gh_{\text{fg}}} = 0.0332We_{D_{\text{eq}}}^{-0.235}\left(\frac{\rho_f}{\rho_g}\right)^{-0.681} \frac{\left[1-0.684\left(\frac{\rho_f}{\rho_g}\right)^{0.832} x^*_{e,in}\right]}{1+0.0908We_{D_{\text{eq}}}^{-0.235}\left(\frac{\rho_f}{\rho_g}\right)^{0.151}\left(\frac{L}{D_{\text{eq}}}\right)} \times We_{D_{\text{eq}}}^{0.121}$$

where Bo is the boiling number, subscript "c" corresponds to CHF and subscript "mc" corresponds to microchannel, D_{eq} is the equivalent hydraulic diameter for uniformly heated channel, $x^*_{e,in}$ is the pseudo-inlet thermodynamic equilibrium quality with properties based on outlet pressure = $\frac{(h_{\text{in}}-h_{\text{fo}})}{h_{\text{fg},o}}$ where subscript "f" corresponds to test section outlet, and "L" is the length of the microchannel.

9.13 Condensation Heat Transfer in Mini-/Microchannel

Cutting-edge technologies such as high-performance computers, high-power lasers, power electronics, and avionics depend on dissipation of large amounts of heat from small surface areas. The fast increase in power is increasing pushing the existing single-phase cooling

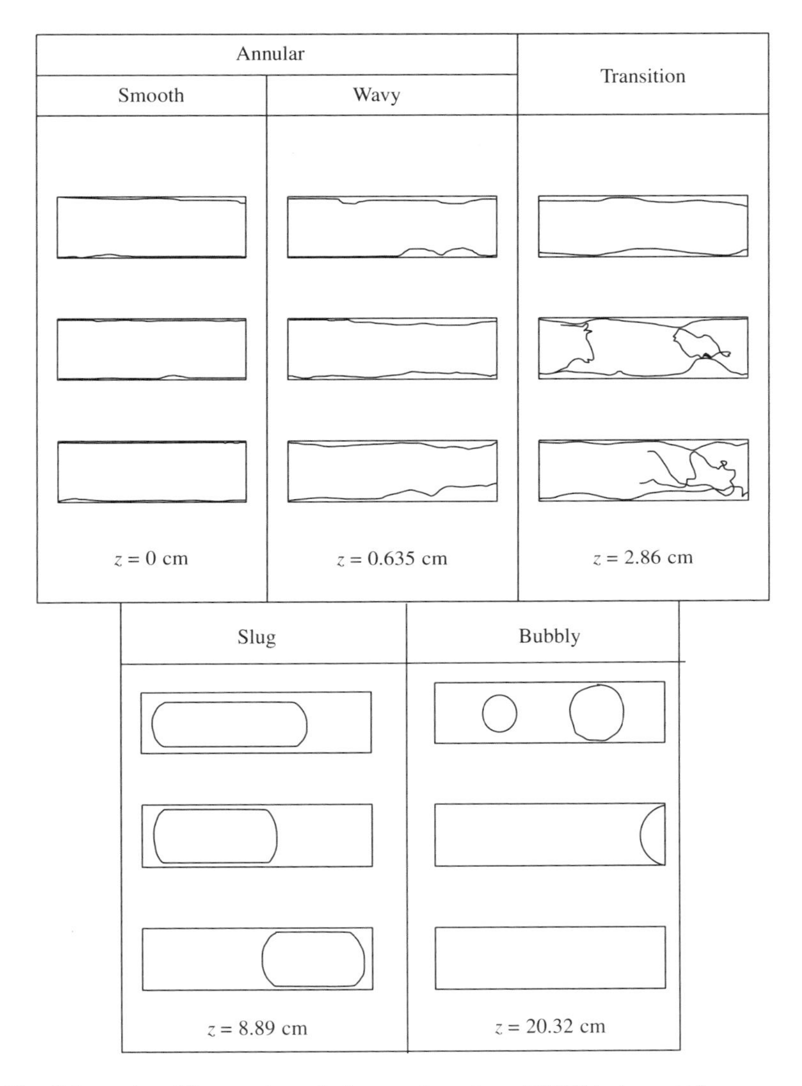

Figure 9.18 Schematics of flow regimes during condensation of FC72 in 1 mm × 1 mm square channel at mass flow rate of 6 g/s (Kim and Mudawar, 2012)

systems obsolete due to high coolant flow rate requirement with large pressure drop. Therefore, focus has shifted in the recent years in favor of two-phase cooling schemes. The two-phase cooling systems offer superior cooling performance due to high boiling and condensation heat transfer coefficients. Many present two-phase cooling systems employ fairy standard air-cooled condensers to reject heat to the ambient. Presently, there is a growing need for miniature condensers, which can reject the heat by condensing a primary coolant in a compact primary cooling loop. The heat from the primary coolant is subsequently transferred to a secondary liquid coolant and transported to a remote heat exchanger, which is ultimately rejected to ambient air or seawater for marine applications. Integration of miniature condensers containing parallel microchannels needs to be realized for implementation in the primary cooling loop.

9.13.1 Condensation Flow Regimes

Several flow patterns are observed during condensation inside a microchannel. Figure 9.18 shows representative flow regimes in a microchannel at different locations from the microchannel inlet. Five distinct flow regimes as identified are discussed below.

Smooth annular flow regime: This regime is characterized by a very thin and fairly smooth liquid film flowing along the channel wall. There is a simultaneous vapor flow in the core clear of any liquid droplets. This regime occurs in the inlet region of the microchannel.

Wavy-annular regime: This regime features a liquid film, which is thicker than that of smooth annular regime with discernible inter facial waves.

Transition regime: This regime is characterized by bridging of liquid ligaments across the vapor core.

Slug regime: This regime is characterized by elongated cylindrical bubbles, with length several times larger than the width of the channel.

Bubbly regime: This regime features spherical bubbles with a diameter approaching that of the microchannel width.

It may be noted that no droplets are entrained in the vapor core for the smooth annular and wavy-annular flow, contrary to the annular flow associated with flow boiling in microchannels.

Figure 9.18 shows the flow regime boundaries for FC72 condensation flow regimes as a function of mass velocity and quality. The smooth annular regime corresponds to high quality values. The smooth annular regime extends further downstream toward lower quality values with an increase in mass velocity. The use of dimensional plots to characterize flow regimes has a fundamental weakness, that is, the dimensional variables do not uniquely govern all flow regimes. Therefore, there has been attempt to slow the flow regimes as a function of dimensionless numbers. Soliman (1986) derived the expression of the modified Weber number assuming that the inertia of the vapor phase is the dominant destructive force acting on the liquid film and surface tension and liquid viscous forces are the stabilizing forces. Balancing the destructive and stabilizing forces, they derived the modified Weber number as

$$We^* = 2.45\frac{Re_g^{0.64}}{Su_g^{0.3}(1+0.09X_{++}^{0.039})^{0.4}}$$

for $Re_f < 1250$ and

$$We^* = 0.85\frac{Re_g^{0.79}X_{++}^{0.157}}{Su_g^{0.3}\left(1+0.09X_{++}^{0.039}\right)^{0.4}}\left[\left(\frac{\mu_g}{\mu_f}\right)^2\left(\frac{\nu_g}{\nu_f}\right)\right]^{0.084}$$

where Re_g is the Reynolds number at saturated vapor condition, Su_g is the Suratman number, v_g is the specific volume of saturated vapor, v_f is the specific volume of saturated liquid, and X_{++} is the Martinelli parameter.

$$Su_g = \frac{\rho_g \sigma D}{\mu_g^2} X_{++} = \left(\frac{\mu_f}{\mu_g}\right)^{0.1} \left(\frac{1-x}{x}\right)^{0.9} \left(\frac{v_f}{v_g}\right)^{0.5}$$

Soliman proposed that the flow is always laminar for $We^* < 20$. Kim and Mudawar (2012) proposed the following criteria for boundaries between flow regimes: smooth-annular to wavy-annular; $We^* = 90X_{++}^{0.5}$, wavy-annular to transition; $We^* = 24X_{++}^{0.41}$, and transition to slug; $We^* = 7X_{++}^{0.2}$.

9.13.2 Condensation Heat Transfer Correlation

Kim and Mudawar (2013) proposed a new correlation for local annular flow (smooth annular, wavy annular, and transition to slug) corresponding to $We^* < 7X_{++}^{0.2}$ as

$$\frac{h_{\text{ann}} D_h}{K_f} = 0.048 Re_f^{0.69} Pr_f^{0.34} \frac{\phi_g}{X_{++}}$$

where Re_f = Reynolds number for saturated liquid = $\frac{G(1-x)D_h}{\mu_f}$, Pr_f = Prandtl number for saturated liquid = $\frac{\mu_f c_{p,f}}{K_f}$, X_{++} = Lockhart–Martinelli parameter based on turbulent liquid and turbulent vapor (t–t) = $\left(\frac{\mu_f}{\mu_g}\right)^{0.1}\left(\frac{1-x}{x}\right)^{0.9}\left(\frac{\rho_g}{\rho_f}\right)^{0.5}$, ϕ_g = two-phase pressure drop multiplier based on gas flow = $1 + CX + X^2$, $X^2 = \frac{\left(\frac{dp}{dz}\right)_f}{\left(\frac{dp}{dz}\right)_g}$, $\left(\frac{dp}{dz}\right)_f = -\frac{2f_f v_f G^2 (1-x)^2}{D_h}$, $\left(\frac{dp}{dz}\right)_g = -\frac{2f_f v_f G^2 x^2}{D_h}$, $f_K = 16Re_K^{-1}$ for $Re_K < 2000$,
$f_K = 0.079Re_K^{-0.25}$ for $2000 \le Re_K \le 20,000$,
$f_K = 0.046Re_K^{-0.2}$ for $Re_k \ge 20,000$
Re_g = Reynolds number for saturated vapor = $\frac{GxD_h}{\mu_g}$ The constant "c" can be determined based on the following conditions: turbulent(liquid)–turbulent(gas): $C = 0.39Re_{\text{fo}}^{0.03} Su_{\text{go}}^{0.1} \left(\frac{\rho_f}{\rho_g}\right)^{0.35}$ for $Re_f \ge 2000, Re_g \ge 2000$

turbulent(liquid)–laminar(gas): $C = 8.7 \times 10^{-4} Re_{\text{fo}}^{0.17} Su_{\text{go}}^{0.5} \left(\frac{\rho_f}{\rho_g}\right)^{0.35}$ for $Re_f \ge 2000, Re_g <$ 2000

laminar(liquid)–turbulent(gas): $C = 0.0015Re_{\text{fo}}^{0.59} Su_{\text{go}}^{0.19} \left(\frac{\rho_f}{\rho_g}\right)^{0.36}$ for $Re_f < 2000, Re_g \ge 2000$

laminar(liquid)–laminar(gas): $C = 3.5 \times 10^{-5} Re_{\text{fo}}^{0.44} Su_{\text{go}}^{0.5} \left(\frac{\rho_f}{\rho_g}\right)^{0.48}$ for $Re_f < 2000, Re_g < 2000$

where Re_{fo} = Reynolds number for liquid only = $\frac{GD_h}{\mu_f}$, Su_{go} = saturation number for vapor only = $\frac{\rho_g \sigma D_h}{\mu_g^2}$. Similarly, the heat transfer correlation for the slug and bubbly flow regime is given by

$$\frac{h_{\text{non-ann}} D_h}{K_f} = \left[\left(0.048 Re_f^{0.69} Pr_f^{0.34} \frac{\phi_g}{X_{++}}\right)^2 + \left(3.2 \times 10^{-7} Re_f^{-0.38} Su_{\text{go}}^{1.39}\right)^2\right]^{0.5}$$

Problems

9.1 For an infinite parallel plate geometry with separation distance between plates equal to H, inlet pressure P_i, $\sigma_v = \sigma_T = 1.0$, outlet pressure P_o, length L and Knudsen number at the outlet Kn_o, derive that the pressure distribution in stream wise direction x for compressible flow is given as

$$\frac{p(x)}{p_o} = -6Kn_o + \sqrt{\left[6Kn_o + \frac{p_i}{p_o}\right]^2 + \left[\left(1 - \frac{p_i^2}{p_o^2}\right) + 12Kn_o\left(1 - \frac{p_i}{p_o}\right)\right]\frac{x}{L}}$$

9.2 A microchannel heat exchanger uses rectangular channel of H = 1.26 μm, W = 90 μm, and L = 10 mm at uniform surface heat flux to remove heat from a device. Inlet and outlet pressure respectively are P_i = 210 kPa and P_o = 105 kPa. Air at T_i = 20 °C is used as cooling fluid. Assuming steady state fully developed conditions, determine the Nusselt number at $\frac{x}{L} = 0, 0.4$, and 0.8. Assume slip flow condition with $\sigma_v = \sigma_T = 1.0$.

9.3 Repeat problem 2.0 with $\sigma_v = 0.5$ and 0.75. Comment on the effect of velocity slip on the Nusselt number variation in a microchannel.

9.4 Repeat problem 2.0 with $\sigma_T = 0.5$ and 0.75. Comment on the effect of temperature jump on the Nusselt number variation in a microchannel.

9.5 Consider a microchannel heat exchanger with height H = 1.26 μm, width, w = 90 μm, and length, L = 10 mm. Air at inlet temperature, T_i and inlet pressure 210 kPa enters the channel as coolant fluid. The accommodation coefficient, $\sigma_v = \sigma_T = 1.0$. The surface temperature is maintained at T_s = 22 °C. Assuming negligible viscous dissipation, calculate the Nusselt number at $\frac{x}{L} = 0.2, 0.6$, and 1.0 for negligible axial conduction and with axial conduction case.

9.6 FC72 coolant flow condensation takes place in a microchannel of 1 mm × 1 mm cross-section with coolant flow rate $C(G) = 186\ \text{kg/m}^2\text{-s}$. Calculate the condensation heat transfer coefficient at quality (x) = 0.8,0.4, and 0.2. The quality of FC72 is close to unity at the inlet to the condensation channel. The FC72 is fully condensed at the exit of the condensation module. The thermophysical properties of FC72 at T_{sat} = 60 °C are provided in Table 9.5:

Use annular condensation correlation.

Table 9.5 The thermophysical properties of FC72 at $T_{sat} = 60$ °C.

h_{fg} [kJ/kg]	ρ_f [kg/m³]	ρ_g [kg/m³]	μ_f [kg/m-s]	K_f [W/m-K]	$C_{p,f}$ [kJ/kg-K]	σ [mN/m]
93.7	1583.4	14.9	$4.18{=}\times{=}10^{-4}$	0.0534	1.1072	8.0

9.7 FC72 with $P_R = 0.2$ and $P_{sat} = 3.7$ bar flows inside a rectangular channel with $D_h = 1$ mm and $q''_H = 5$ W/cm². Calculate the incipience dryout quality at mass velocity (a) 800 kg/m²-s, (b) 600 kg/m²-s, and (c) 400 kg/m²-s.

9.8 Repeat the above problem with water at $P_{sat} = 44.1$ bar as working fluid.

9.9 Water with $P_R = 0.2$ flows inside rectangular microchannel at mass velocity, $G = 600$ kg/m²-s. Calculate the incipience quality with (a) $D_h = 0.5$ mm, (b) $D_h = 2$ mm and (c) $D_h = 5$ mm for $q''_H = 300$ W/cm².

9.10 Calculate the dryout incipience quality of water inside a rectangular channel, with $D_h =$ 1 mm and $q''_H = 50$ W/cm², mass velocity, $G = 600$ kg/m²-s for different saturation pressures: (a) $P_R = 0.0045$, $P_{sat} = 132$ bar (b) $P_R = 0.2$, $P_{sat} = 44.1$, bar (c) $P_R = 0.6$, and $P_{sat} = 132$ bar.

9.11 FC72 flows in a rectangular channel of $D_h = 0.5$ mm, $P_{sat} = 1$ bar mass velocity, $G =$ 150 kg/m²-s with heat flux $q''_H = 10$ W/cm². Calculate the nucleate boiling, convective boiling, and two-phase boiling heat transfer coefficient at quality $x = 0.1$ and 0.2.

9.12 Repeat the heat transfer coefficient calculation in the above problem for water at $x = 0.1$, and 0.2, and 0.4 for the same condition.

9.13 Calculate the boiling heat transfer coefficient of CO_2 flow at $T_{sat} = -16.2$ °C, mass velocity inside micro-/minichannels with heat flux $q''_H = 5$ W/cm² for hydraulic diameter $D_h = 0.5$ mm, 1 mm, 2 mm and at quality $x = 0.1$ and comment on the effect of channel size on the boiling heat transfer coefficient.

9.14 Calculate the boiling heat transfer coefficient of CO_2 flow at $T_{sat} = -16.2$ °C, in a rectangular microchannel of $D_h = 0.5$ mm with heat flux $q''_H = 3$ W/cm² for mass velocity $G = 200$ kg/m²-s, 400 kg/m²-s, and 600 kg/m²-s at quality 0.2, 0.4, and 0.6 and comment on the effect of mass velocity on the boiling heat transfer coefficient.

9.15 Calculate the boiling heat transfer coefficient of CO_2 flow at mass velocity $G = 250$ kg/m²-s inside a rectangular microchannel with $D_h = 0.5$ mm, heat flux $q''_H = 5$ W/cm² at quality $x = 0.1$ and 0.5 for (a) $P_{sat} = 36.9$ bar, $T_{sat} = 2.2$ °C, $P_R = 0.5$ and (b) $P_{sat} = 1$ bar, $T_{sat} = -73.8$ °C, $P_R = 0.014$. Comment on the effect of saturation temperature on the effect of boiling heat transfer coefficient in microchannel.

9.16 Calculate the boiling heat transfer coefficient of CO_2 flow at mass velocity $G =$ 200 kg/m²-s, $T_{sat} = -16.2$ °C inside a rectangular microchannel with $D_h = 0.5$ mm at quality $x = 0.2, 0.4$, and 0.6 for heat flux $q''_H = 0.5\ \mathrm{W/cm^2}$, $1\ \mathrm{W/cm^2}$ and $5\ \mathrm{W/cm^2}$. Comment on the effect of heat flux on the boiling heat transfer coefficient.

9.17 Calculate the mass flow rate (g/s) at HFE 7100 inside a microchannel with $D_h = 0.5\ \mu\mathrm{m}$, $\frac{L}{D_h} = 50$ for achieving $q''_{\mathrm{eff}} = 1000\ \mathrm{W/cm^2}$ at inlet temperature $T_{\mathrm{in}} = 0\ °\mathrm{C}$ and $-20\ °\mathrm{C}$. Comment on the effect of sub-cooling on boiling heat transfer.

References

Jeong H and Jeong J 2006b Extended Graetz problem including axial conduction and viscous dissipation in micro tube. *J. Meas. Sci. Technol.*, **20**(1), pp. 158–166.

Kew PA and Cornwell K 1997 Correlations for the prediction PF boiling heat transfer in small diameter channels. *Appl. Thermal Eng.*, **17**(8-10), pp. 705–715.

Kim S and Mudawar I 2012 Flow condensation in parallel micro-channels-Part 2: heat transfer results and correlation technique. *International Journal of Heat and Mass Transfer*, **55**, pp. 984–994.

Kim S and Mudawar I 2013a Universal approach to predicting saturated flow boiling heat transfer in mini/micro-channels-Part I. Dryout incipience quality. *Int. J. Heat Mass Transfer*, **64**, pp. 1226–1238.

Kim S and Mudawar I 2013b Universal approach to predicting heat transfer coefficient for condensing mini/micro channel flow. *Int. J. Heat Mass Transfer*, **56**, pp. 238–250.

Lee J and Mudawar I 2009 Critical heat flux for sub-cooled flow boiling in micro-channel heat sinks. *Int. J. Heat Mass Transfer*, **52**, pp. 3341–3352.

Liao SJ 2002 An analytic approximation of the drag coefficient for the viscous flow past a sphere. *Int. J. Non Linear Mech.*, **37**, pp. 1–18.

Qu W and Mudawar I 2004 Measurement and correlation of critical heat flux in two phase micro channel heat sinks. *Int. J. Heat Mass Transfer*, **47**, pp. 2045–2059.

Soliman HM 1986 The mist-annular transition during condensation and its influence on the heat transfer mechanism. *Int. J. Multiphase Flow*, **12**, pp. 277–288.

Supplemental Reading

Bertman B and Sandiford DJ 1970 Second sound in solid helium. *Sci. Am.*, **222**(5), pp. 92–101.

Colin S 2012 Gas micro flows in the slip flow regime: a critical review on convective heat transfer. *ASME J. Heat Transfer*, **134**, pp. 020908(1-11).

Hadjiconstantinou NG and Simex O 2002 Constant-wall-temperature Nusselt number in micro and nano channels. *J. Heat Transfer*, **124**, pp. 356–364.

Jeong HE and Jeong JT 2006a Extended Graetz problem including stream wise conduction and viscous dissipation in micro channel. *Int. J. Heat Mass Transfer*, **49**, pp. 2151–2157.

Kim S, Kim J, and Mudawar I 2012 Flow condensation in parallel micro-channels-Part I: experimental results and assessment of pressure drop correlations, *Int. J. Heat Mass Transfer*, **55**, pp. 971–983.

Latif MJ 2009 *Heat Convection*. Springer Publication.

Lee J and Mudawar I 2005 Two phase flow in high heat flux micro channel heat sink for refrigeration cooling applications: Part II-heat transfer characteristics. *Int. J. Heat Mass Transfer*, **48**, pp. 941–955.

10

Microfabrication

10.1 Introduction

The technology of making microfluidic devices is going through a rapid change. It started with silicon micromachining, where a number of microfluidic devices with integrated sensors and actuators were made of silicon. The basic difference between microelectronics and microfluidics is the transport quantity. Microelectronics manipulates electrons in integrated circuits, while microfluidic devices transport macromolecules and fluids in larger channels. There is a size difference between the electrons and the macromolecules transported inside the microfluidic systems. Because of this size difference, fewer microfluidic devices can be placed on a silicon wafer in comparison to the electronic devices. The microfluidic devices based on silicon technology are too expensive because of the material cost, processing cost, and the yield rate. Thus, there is a philosophical change in microfluidic fabrication. The plastic technology has resulted in simple microfluidic devices with a passive system of microchannels. The actuating and sensing devices are not necessarily integrated into the microdevices. But these are used as replaceable elements in bench top. Batch fabrication of plastic devices is possible with replication and forming techniques. The other possibilities are the combination of plastic and silicon micromachining technologies, where part of the device is manufactured by one technology and the other with another technology. Another requirement of microfluidic applications is the use of highly corrosive chemicals. Thus, microfabrication in new materials such as stainless steel or ceramics is desired. Laser machining and electrodischarge machining are some of the alternative fabrication techniques for these purposes.

The range of microfabrication techniques available today is quite large. Among all technologies, silicon microfabrication is the most matured with maximum know-how and therefore is a resource of knowledge for other technologies. Figure 10.1(a) compares the absolute size of different objects. Figure 10.1(b) shows the size ranges corresponding to different types of microtechnologies. Microtechnologies are based on etching, lithography, and deposition in the size range of 0.2–500 μm. These technologies using hard materials such as glass and silicon are therefore known as hard technologies. Plastic microtechnologies use elastomers such as polydimethylsiloxane (PDMS) or plastic materials like polymethylmethacrylate (PMMA) in the size range of 0.5–500 μm. Plastic technologies are attractive from the cost point of view,

Transport Phenomena in Microfluidic Systems, First Edition. Pradipta Kumar Panigrahi.

Companion Website: www.wiley.com/go/panigrahi/microfluidic

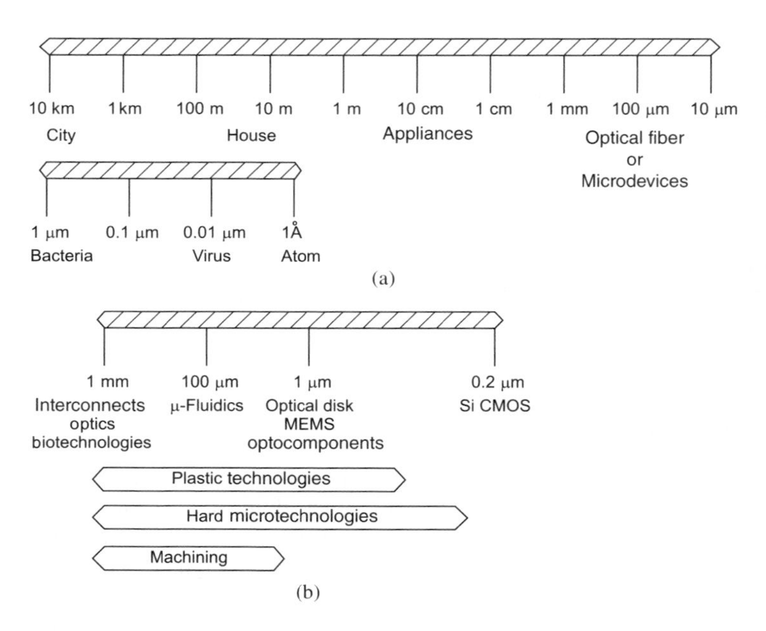

Figure 10.1 (a) Comparison of different scales for adopting general fabrication strategies and (b) different types of microtechnologies

that is, plastic are about 100 times less expensive than silicon. The price is a critical factor when a large number of disposable miniaturized elements are required. The other attractive feature of plastic technology (soft technology) is rapid prototyping, that is, a complete microfluidic circuit can be developed in a few hours time. The other advantages of soft technologies are surface effects, transparency, and diversity of materials. Hard microfabrication technologies are usually complicated and involved. However, the advantage of silicon is its resilience to aggressive media and therefore is more suitable to chemical analysis.

This chapter introduces different fabrication methodologies of microdevices. Different functional materials used in these devices are discussed first. Subsequently, different steps adopted for photolithography-based micromanufacturing are discussed next. This is followed by the plastic-based micromanufacturing process. Subsequently, the laser-based microfabrication is introduced. Finally, different bonding processes adopted during fabrication of microdevices are discussed.

10.2 Microfabrication Environment

Microfabrication needs to be carried out in an extremely clean environment known as clean room, because, in dusty environment, standard particles of micrometer size tend to absorb on the surface and change the nature of the microstructure being fabricated. A clean room is an environment, that is, regulated for temperature and humidity and is permanently traversed by flux of air. The particle introduced to the workplace by human beings and the chemical processes at work are filtered out.

The classification of a clean room is based on the number of particles of a size less than 4 μm contained in a volume of cubic inch. Typical clean room for MEMS microfabrication has a rating in the range of 1000–10,000. For microelectronics, the cleanliness requirement is higher, and thus class number of the room is as low as 10 or 1. For clean room work, it is necessary to wear specialized clothing to cover hair and to put on gloves and shoe covers. The human beings go through air curtain for flushing any dusts from their body before entering the room.

10.3 Functional Materials

This section presents various common materials used for microfabrication.

10.3.1 Monocrystalline Silicon

Monocrystalline silicon is basically of "face cubic center" structure (Figure 10.2). It has a total of 18 atoms, that is, 8 at vertices (shared between 8 unit cells), 6 at faces (shared between two neighborhood unit cells), and 4 completely inside the unit cell.

The single-crystalline wafers are classified according to the crystalline orientation of the surfaces based on the Miller indices. Miller indices are referenced with respect to the crystallographic axes of a crystal and therefore do not have to be oriented at right angles. However, they correspond to the x-, y-, and z-axis of a cubic lattice structure.

In Figure 10.3, a plane intersects x, y, and z coordinates at a, b, and c, respectively. The equation of the plane is

$$\frac{x}{a} + \frac{y}{b} + \frac{z}{c} = 1 \tag{10.1}$$

or

$$hx + ky + mz = 1 \tag{10.2}$$

where $h = \frac{1}{a}$, $k = \frac{1}{b}$, and $m = \frac{1}{c}$ are the Miller indices. Miller indices designate $\langle hkm \rangle$ as direction perpendicular to the (hkm) plane. In a cubic silicon crystal, $a = b = c = 1$. With reference to Figure 10.2, plane (001) is the top face, plane (010) is the right face, plane (100) is the front face, plane (110) is the diagonal face, and plane (111) is the inclined face.

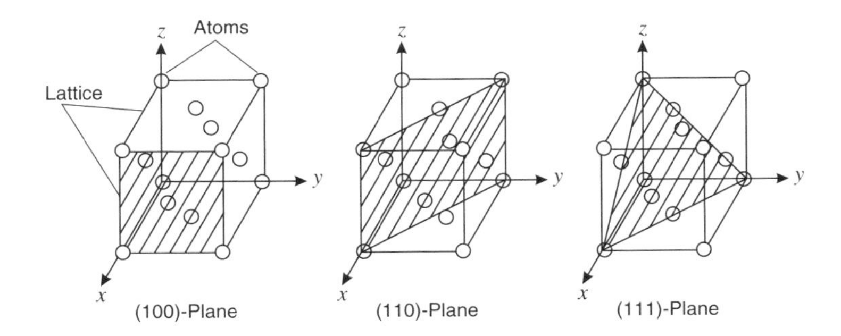

Figure 10.2 Different crystal planes in the cubic lattice of monocrystalline silicon

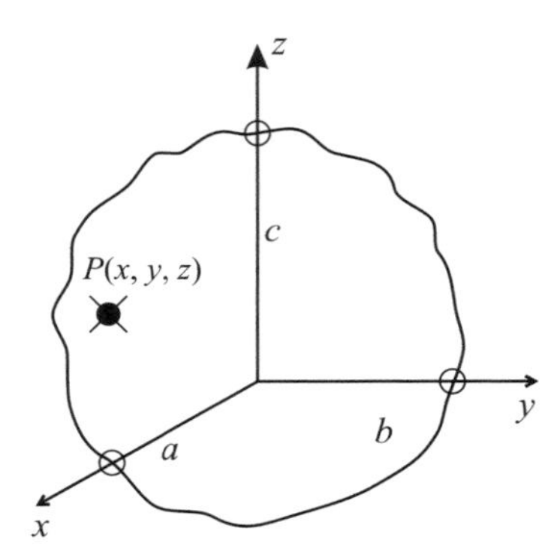

Figure 10.3 Schematic showing the definition of Miller index

The procedure to determine the Miller indices is the following:

1. Determine the points at which a given crystal plane intersects the three axis, say, $(a, 0, 0)$, $(0, b, 0)$, and $(0, 0, c)$. If the plane is parallel to an axes, it is said to intersect at infinity.
2. The Miller index for the face is $\left(\frac{1}{a}\,\frac{1}{b}\,\frac{1}{c}\right)$ where the three numbers are expressed as smallest integers, that is, common factors are removed.

Example: Let a plane intersects the crystallographic axes at (2, 0, 0), (0, 4, 0), and (0, 0, 4). Hence, the Miller indices are $\left(\frac{1}{2}\,\frac{1}{4}\,\frac{1}{4}\right) = (211)$, that is, this is a (211) plane.

Silicon is produced by a well-controlled Czochralski crystal growth process in a very clean environment, that is, class of 1 or 10. In this process, a small seed crystal is dipped into a highly purified silicon melt. This seed is slowly pulled while the crucible containing the melt is rotated. The silicon crystal grows along the selected orientation of the seed to the rod. A cylindrical crystal is obtained from which slices are cut. This is followed by the atomic polishing phase. The side of one cubic face is 5.43Å. The mechanical, electrical, and thermal properties of silicon have been presented in Table 10.1.

Table 10.1 Properties of monocrystalline silicon

Properties	Magnitude
Young's modulus	190 GPa
Density	2.33 g/cm^3
Thermal conductivity	2.33 W/cm-K
Thermal expansion	2.33×10^{-6} K^{-1}
Electrical permittivity	11.9 F/m
Electrical field breakdown	3×10^5 V/cm
Electrical resistivity	2.33×10^5 Ω cm
Thermal diffusivity	0.9 cm^2/s
Specific heat at constant pressure	0.7 J/g-K

Table 10.2 Mechanical properties of silicon by principal planes

Miller index for orientation	Young's modulus (GPa)	Shear modulus (GPa)
100	129.5	79.0
110	168.0	61.7
111	186.5	57.5

The characteristics of silicon depend on the principal planes (see Table 10.2). The (100) plane is the weakest plane and easiest to work with. The (110) plane offers the cleanest surface for microfabrication. The (111) plane is the strongest plane and toughest to work with.

Young's modulus of silicon is comparable in magnitude to that of conducting materials like stainless steel. Silicon is also a good heat conductor in comparison to metal and thus is favorable material for heating the miniaturized systems. Thermal expansion coefficient of silicon is weak, that is, about 2.33×10^{-6} K^{-1} at temperature of 300 K, a value comparable to glass. This value changes significantly with temperature and becomes negative at around 100 K.

Some ideal properties of silicon for its popularity in micromanufacturing are the following:

1. It is a semiconducting material having ideal electronic properties for integration with electronics.
2. It is an ideal structural material. It has about same Young's modulus as steel ($\sim 2 \times 10^5$ MPa).
3. The meeting point of silicon is about 1400 °C, which is about twice higher than that of aluminum. Thus, it is dimensionally stable even at elevated temperature.
4. Its thermal expansion coefficient is about 8 times smaller than that of steel and about 10 times smaller than that of aluminum.
5. Silicon has no mechanical hysteresis and therefore is an ideal candidate for sensors and actuators.
6. Silicon wafers are extremely flat for coatings and thin film deposition and therefore ideal for performing precise electromechanical functions.

Silicon wafers are available with one or both sides polished and marking of crystallographic orientation at different diameters and thicknesses. Currently, a 4-in. wafer is frequently used, and there is a trend toward increase in this size. The advantage of larger wafer is the possibility of constructing more microsystems in parallel. Typical thickness of wafer is 500 μm. Another wafer available in the market is the silicon on insulator (SOI) type. These wafers possess three layers, two made of silicon and one made from silicon dioxide. These wafers facilitate fabrication of devices employing sacrificial layers.

10.3.2 Polysilicon

Polysilicon is an aggregation of pure silicon crystals with random orientation deposited on the top of silicon substrate (see Figure 10.4). Polycrystalline silicon is deposited during

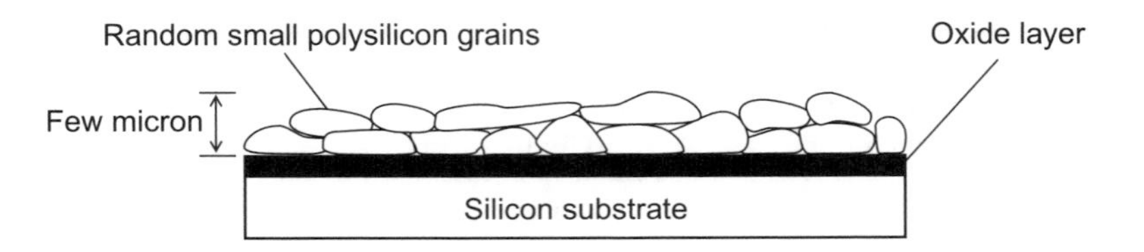

Figure 10.4 Schematic showing polycrystalline silicon

low-pressure chemical vapor deposition (LPCVD) process with silane atmosphere in the temperature range of 57 – 650 °C:

$$SiH_4 \xrightarrow{630\,^{\circ}C,\ 60\ Pa} Si + 2H_2 \uparrow$$

After annealing at 900–1000 °C for several minutes, crystallization and grain growth occur. They have columnar structure with a grain size between 0.03 and 0.3 μm. In microfluidics, polysilicon is used for making channel walls and sealing etched channel structures. Being randomly oriented, polysilicon is stronger than single silicon crystals.

10.3.3 Silicon Dioxide

Silicon dioxide (SiO_2) is a least expensive material with good thermal and electrical insulation properties. Its resistivity is $<10^{16}$ (Ω-cm) and thermal conductivity is about 0.014 (W/cm- °C). It is used as a low-cost material for masks in microfabrication process such as etching and deposition. It is also used as a sacrificial material in surface micromachining.

Silicon dioxide is grown using thermal oxidation procedure. Thermal oxidation can be categorized as both *dry oxidation* and *wet oxidation*. Silicon reacts with dry oxygen at high temperature (800 – 1200 °C) during dry oxidation process with the following reaction:

$$Si + O_2 \rightarrow SiO_2$$

In wet oxidation process, water vapor reacts with silicon at high temperature, and the corresponding reaction is

$$Si + 2H_2O \rightarrow SiO_2 + 2H_2 \uparrow$$

The growth rate during wet oxidation is higher than that of dry oxidation. However, the quality of SiO_2 from dry oxidation is superior. Figure 10.5 shows an oxidation furnace for dry and wet oxidation of silicon. Substrate sample is loaded inside the furnace, and the temperature of the furnace can be raised as per the requirement. Nitrogen gas is used for flushing the chamber from impurity. The dry oxygen or wet oxygen is supplied depending on the requirement.

The growth rate of thermal oxidation decreases with increase in the thickness of oxide layer, because it relies on the diffusion of oxygen. The chemical vapor deposition (CVD) process can make thicker silicon dioxide layer and does not require silicon substrate. Silane is a toxic extremely flammable chemical compound with chemical formula SiH_4. The silane CVD process using silane gas is based on the following reaction:

$$SiH_4 + O_2 \xrightarrow[1\ bar]{430\,^{\circ}C} SiO_2 + 2H_2 \uparrow$$

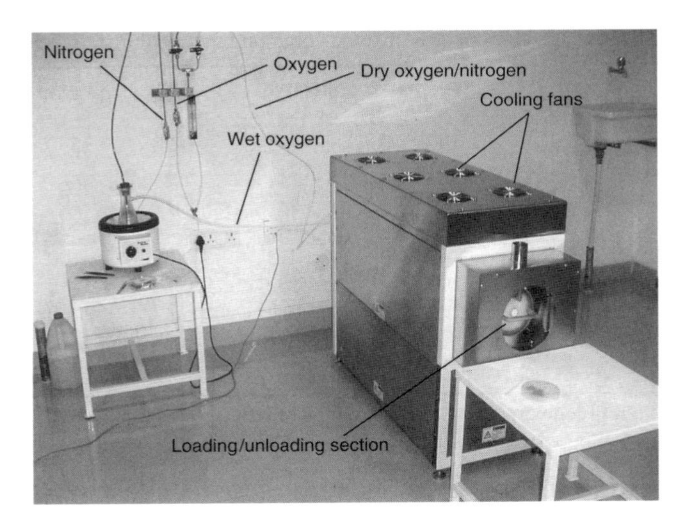

Figure 10.5 An image of an oxidation furnace. (Courtesy of Micro Fabrication Laboratory, ME Department, IIT Kanpur)

The plasma-enhanced CVD (PECVD) process uses the reaction:

$$SiH_4 + 4N_2O \xrightarrow[40\ \text{Pa}]{350\ ^\circ\text{C}} SiO_2 + \text{Gas}$$

Silicon dioxide has selectivity to many silicon etchants and therefore is a good mask material for self-aligned etching process. In combination with silicon nitride, multistep etching process of three-dimensional structure is possible. Silicon dioxide is also used to seal microchannels. The insulating property of silicon dioxide makes it a good coating layer of channels in microfluidics.

10.3.4 Silicon Nitride

Silicon nitride is created by CVD process of silane (SiH_4) and dichlorosilane (SiH_2Cl_2) in ammonia atmosphere at high temperature. The reactions for LPCVD is

$$3SiH_2Cl_2 + 4NH_3 \xrightarrow[30\ \text{Pa}]{750\ ^\circ\text{C}} Si_3N_4 + 6HCl + 6H_2$$

The reactions for PECVD is

$$3SiH_4 + 4NH_3 \xrightarrow[40\ \text{Pa,Plasma}]{700\ ^\circ\text{C}} Si_3N_4 + \text{Gas} \uparrow$$

Silicon nitride is a good insulator and acts as barrier against all kinds of diffusion to water and ions. Due to thermal insulation properties, heater structures are suspended on silicon nitride membrane or fixtures. Its ultrastrong resistance to oxidation and many etchants makes it a superior material for masks in deep etching. It is also used as high-strength electrical insulator.

Table 10.3 Mechanical properties of some common functional materials

Materials	T_m(°C)	E(GPa)	σ_y(GPa)	γ	ρ(kg/m^3)	H
Silicon	1415	160–200	—	0.22	2330	5–3
Polysilicon	1415	181–203	—	—	—	10–13
Silicon dioxide	1700	70–75	8.4	0.17	2200	15–18
Pyrex glass	—	64	—	0.2	2230	—
Silicon nitride	1800	210–380	14	0.25	3100	8
Silicon carbide	—	300–430	21	0.19	3210	24–27
Aluminum	661	70	0.2	0.33	2700	—
Platinum	1772	170	0.137–0.17	0.38	21440	—
Stainless steel	—	200	2.1	0.3	7900	6.5

T_m is the melting temperature, E is the Young's modulus, σ_y is the yield strength, γ is the Poisson's ratio, ρ is the density, and H is the Knoop hardness.

10.3.5 Metals

Metals are deposited using plasma vapor deposition, chemical vapor deposition, and electroplating. Aluminum is a common material for electrical connections. It is also used as a sacrificial layer for surface micromachining. Some metals are used for sensing and actuation process. Thermoelectrical properties of metals are used for temperature sensing. Permalloy, an iron–nickel alloy, is used for magnetic sensing and actuating. Metals like platinum and palladium have catalytic properties and are useful in chemical sensors and microreactors. Table 10.3 compares the properties of some functional materials used during microfabrication.

10.3.6 Polymers

Microfluidic devices are relatively large compared to other MEMS devices. Therefore, the substrate material price is one of the primary costs. A glass substrate, that is, borofloat glass, borosilicate glass, and photostructurable glass, costs 10–100 times more than a polymer substrate. The other advantage is the availability of wider range of polymers with different surface chemistries, which can be tailored depending on the applications.

Polymers consist of macromolecules having more than 1000 monomeric units. The cross-linking of monomers can be chemically triggered by pressure, temperature, and photons. The polymer is called *homopolymer* when one type of monomer is used and is called *copolymer* when more monomer units are polymerized to form it. Plastics are the polymers containing specific additives. The macromolecules in a polymeric material have different lengths. Therefore, there is no fixed melting temperature of polymers. *Glass transition temperature* and *decomposition temperature* are two characteristic temperatures. The polymer can keep its solid shape while losing its strength at glass transition temperature. At temperature above glass transition, it loses its solid shape. The polymeric material is soft and can be machined by molding above the glass transition temperature. The glass transition is the reversible transition of a material from a hard brittle state into a molten rubberlike state. Despite the massive change in physical properties of the material, the transition is not a phase transition of any kind. There may be 17 orders of change in the viscosity of the material without any pronounced change in material structure.

Table 10.4 Common properties of some polymers

Materials	ρ (kg/m^3)	K (W/m-K)	T_g (°C)	$\beta \times 10^{-6}$ (K^{-1})
Parylene C	1290	0.08	290	35
Parylene N	100	0.13	410	69
Polyamide 6(PA6)	1130	0.29	60	80
Polycarbonate (PC)	1200	0.21	150	65
Polymethylmethacrylate (PMMA)	1180–1190	0.186	106	70–90
Polystyrene (PS)	1050	0.18	80–100	70

ρ is the density, K is the thermal conductivity, T_g is the glass transition temperature, and β is the thermal expansion coefficient.

Many microfluidic applications use fluorescence for sensing. Many polymers are self-fluorescent at low excitation wavelengths affecting the sensitivity of the device. The other drawback is the poor chemical resistance to solvents and aging. In the case of electroosmosis, the surface properties play an important role. Stable and controllable electroosmotic flow requires high charge density on the surface. However, most polymers have low charge density due to lack of ionizable groups compared to the glass. Biocompatibility is another important advantage of polymers compared to glass and silicon. Many polymers are compatible to tissue and blood and are suitable for implantable microfluidic devices for drug delivery applications. Polymeric devices are ideal for cell handling, clinical diagnostics, DNA analysis, and polymerase chain reactions. In addition to substrate materials, polymers can be spin coated and vapor deposited on other substrates. It can be used as photoresist or passivation layer in traditional microelectronic applications. Some common properties of polymers have been presented in Table 10.4.

10.4 Surface Preparation

The most preliminary step for microfabrication is the surface preparation of the wafer. It involves exposure to DI water and different types of acid, that is, nitric acid, buffered hydrofluoric acid, and so on, for different time duration in normal or warm temperature environment. The surface preparation step can have various steps with different objectives: (1) general clean, (2) particle removal, (3) oxide removal, and (4) metal contaminant removal. During *general cleaning*, a mixture of sulfuric acid and hydrogen peroxide is used for 2–10 min during which heat is generated. This step can remove organic and inorganic contamination from the wafer. Removal of silicon *particles* from the wafer can be achieved by megasonic cleaning (at about 70 °C) in a 5:1:1 ratio mixture of DI water:ammonium hydroxide:hydrogen peroxide. *Removal of oxide layer* from the wafer surface can be achieved by 15–60 s dip in 1:20 HF:DI water solution megasonic clean at about 70 °C in a 6:1:1 ratio mixture of DI water:HCl:hydrogen peroxide for 2–10 min can remove ionic and *metal surface contamination*. It may be noted that strong rinsing in DI water is required in between each of the above cleaning steps. It is followed by priming using hexamethyldisilazane (HMDS) to form polar surface for enhancing adhesion of photoresist to the wafer surface. Figure 10.6

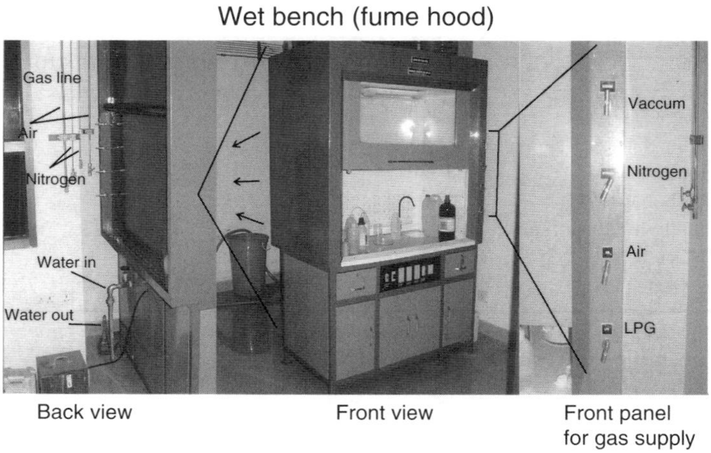

Figure 10.6 The image of a fume hood system for wet etching and surface treatment. (Courtesy of Micro Fabrication Laboratory, ME Department, IIT Kanpur)

shows the picture of wet bench for surface preparation. The wet bench is equipped with water lines, gas lines, and electrical supply lines for executing the surface preparation step. This is also used for wet etching, which is described in following section.

10.5 General Micromachining Procedure

The flow chart in Figure 10.7 shows the standard processes involved in a micromanufacturing of silicon-based materials. It consists of the basic functions, that is, cleaning the substrate, application of thin film using deposition techniques, lithography, etching to form required shape, removal of the mask material using chemical and plasma etching, and final characterization of the microstructure using microscope. These steps may be replaced for complex geometries and quality control objective. The lithography process involves cleaning, application of masking material, exposure of pattern, development, and curing of mask material. The final microstructure is then released from the substrate.

Overall, the micromanufacturing involves three basic steps, that is, *addition, multiplication, and subtraction* (see Figure 10.8). The *addition* process can involve liquid film coating on the substrate using electroplating or spray coating. The thin solid film can also be created by oxidation, doping in an atmospheric or vacuum chamber. The other possible method is the fusion bonding of a solid to the substrate. The *multiplication* step is a necessary step in most micro- and nanofluidic devices to create microscale structures. Multiplication of features can be achieved by a number of processes, that is, direct writing by using electron beam, ion beam, and atomic force microscopic techniques. Lithography and microstamping are other techniques for deposition of features. *Subtraction* step can be achieved by etching using chemical reaction in liquid or gas and plasma. The other techniques for metal removal are focused laser beams and mechanical means, that is, milling and water jets, and so on. Mechanical material removal processes have much higher material removal rates compared to other approaches.

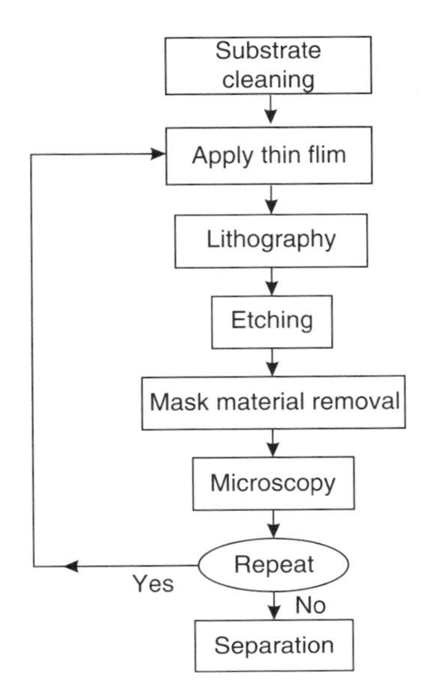

Figure 10.7 Various processing steps during standard micromachining procedure of silicon-based material

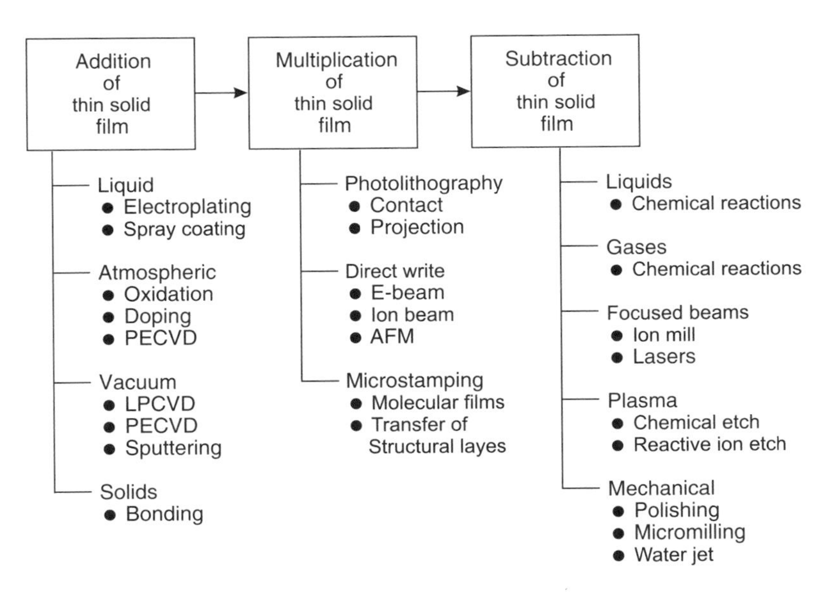

Figure 10.8 The decomposition of microfabrication procedure to three primary steps, that is, addition, multiplication, and subtraction of thin solid film

Combination of different addition, multiplication, and subtraction techniques can be used to produce features of different sizes, shapes, scales, and so on.

10.6 Photolithography

Photolithography is used for transferring the pattern of the devices during microfabrication. It can be divided into *optical lithography* (wavelength, 10–400 nm), *electron lithography* (wavelength, 2–15 pm), and *ion lithography* depending on the type of energy beam. Figure 10.9 shows the image of a photolithography system. A photosensitive layer called resist is used during photolithography, which transfers a desired pattern from a transparent mask to the material. The mask is generally a plate of quartz glass with chromium deposits forming a pattern. The small wavelength of the energy sources is related to the micromachining precision. It is not possible to fabricate a structure with superior geometric precision than that of the mask. Therefore, masks are made with superior precision, that is, electron lithography with precision of the order of submicrometer. If submicron precision is not required, other methods based on high-quality printout of transparency is used. This is most common for microfluidic applications due to low cost and faster processing. Photolithography consists of the following four basic steps: (1) photoresist deposition, (2) positioning process, (3) exposure process, and (4) development process.

10.6.1 Photoresist Deposition

The photosensitive polymer is deposited on a solid substrate of silicon or glass by using a spin coater. The spin coater consists of a disk that turns at high velocity (typically between 1000

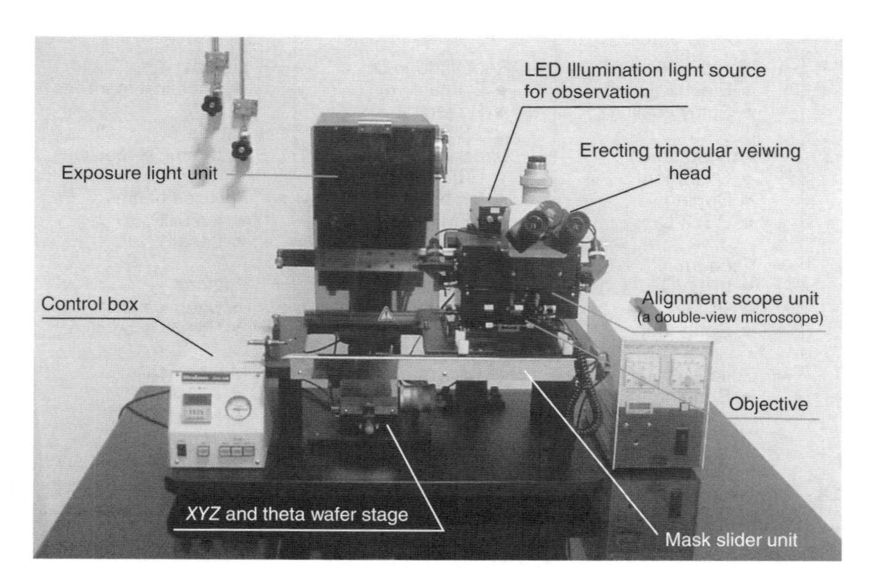

Figure 10.9 An image of a photolithography system. (Courtesy of Micro Fabrication Laboratory, ME Department, IIT Kanpur)

and 10,000 rpm), which allows the spread of a liquid droplet initially deposited at the center of the disk. Figure 10.10 shows the picture of a spin coater. Figure 10.11 shows the basic steps during spin coating process. Table 10.5 presents the film thickness of different SU-8 photoresist types at rotation speed of 1000 rpm. The film thickness increases with increase in viscosity of photoresist. The coating thickness can be estimated on the basis of balance between viscous and centrifugal force. For a cylindrical polar coordinate (r, θ, z) and rotation at angular velocity, ω, the balance between the viscous and centrifugal forces per unit volume can be written as

$$-\mu\frac{\partial^2 v}{\partial z^2} = \rho\omega^2 r \tag{10.3}$$

Figure 10.10 An image of a spin coater. (Courtesy of Micro Fabrication Laboratory, ME Department, IIT Kanpur)

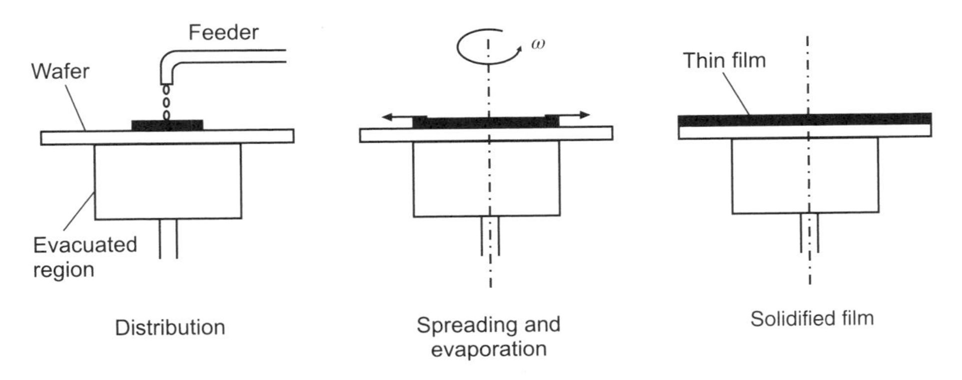

Figure 10.11 Schematic showing the spin coating process

Table 10.5 Film thickness at a spin speed of 1000 rpm for different SU-8 photoresists

Type	Kinematic viscosity (m^2/s)	Thickness (μm)
SU-8 2	4.3×10^{-1}	5
SU-8 5	29.3×10^{-5}	15
SU-8 20	105×10^{-5}	30
SU-8 25	252.5×10^{-5}	40
SU-8 50	1225×10^{-5}	100
SU-8 100	5150×10^{-5}	250

The above equation can be solved by using the boundary condition at the surface ($z = 0$) of the wafer and the free surface of the photoresist ($z = h$) as

$$v = 0 \quad \text{at} \quad z = 0 \tag{10.4}$$

$$\frac{\partial v}{\partial z} = 0 \quad \text{at} \quad z = h \tag{10.5}$$

Integrating the above equation and using the above boundary condition, one can get

$$v = \frac{1}{\mu}\left(-\frac{1}{2}\rho\omega^2 rz^2 + \rho\omega^2 rhz\right) \tag{10.6}$$

Using mass conservation and assuming the uniform initial height of the fluid layer, h_0, it can be derived that

$$h = \frac{h_0}{(1 + 4Kh_0^2 t)^{1/2}} \tag{10.7}$$

where $K = \frac{\rho\omega^2}{3\mu}$.

The above equation indicates that a thick layer thins out more rapidly than a thin one. This in turn indicates that a nonuniform layer should become increasingly more uniform as centrifugation continues. The following empirical relation approximately models the height of the thin film, h, for a constant evaporation assumption:

$$h = \kappa C\left(\frac{\mu}{\omega^2}\right)^{1/3} \tag{10.8}$$

where κ is a constant, C is the initial concentration of polymer in the solution, μ is the viscosity, and ω is the angular rotation velocity. The equilibrium thickness is achieved over long times, that is, a few minutes. For resist having non-Newtonian behavior, the equilibrium thickness would not exist, and the film thins out progressively with time. The surface to volume ratio increases during spreading favoring the evaporation of the resist concentration leading to activation of polymerization process. The film is no more a liquid structure at the end of the deposition process; rather, it is a glass-like substance with equilibrium thickness. The precision of the pattern, that is, uniformity, decreases with increase in thickness of the resist. Therefore, thin deposits are usually preferred. The range of film thickness achieved by spin coating is between 1 and 200 μm. For thicker film, higher viscosity and low spin speed are desired. However, these parameters affect the uniformity of the coat. Multiple coatings are preferred for a film thickness greater than 15 μm.

During evaporation, complete evaporation of the solvent does not take place, and about 15% of solvent remains. This remaining solvent contributes toward the appearance of cracks or fissures. For complete elimination of the solvent, the resin is heated slightly at about 70 °C for few minutes before the exposure. Note that thick resist layers can be achieved by either multiple spin coating or with viscous resist at a slower spinning speed.

10.6.2 Positioning

The positioning step involves the proper alignment of the mask and the substrate, which is coated with a resist. The distance between the mask and the substrate is adjusted during this step. Based on this adjustment, the photolithography is categorized as *contact printing*, *proximity printing*, and *projection printing*. Mask is brought close to the substrate for the first two techniques. In contact printing, the mask even touches the photoresist layer. In contact printing, certain amount of pressure (0.05 – 0.3 atm) is applied against the resist. The uniformity or resolution of the structure is affected by the nature of the contact. Therefore, mask used in contact printing should be thin and flexible to allow better contact over the whole wafer. The resolution of proximity printing is not as good as that of contact printing due to the diffraction of light (near-field diffraction known as Fresnel diffraction). The mask used in proximity printing has longer useful life. In contrast, contact printing requires regular disposal of mask after certain use. The resolution (L_r) depends on the wavelength, λ, and the separation distance, S, between the mask and the photoresist layer:

$$L_r = 1.5\sqrt{\lambda S} \tag{10.9}$$

Projection printing involves no contact between the mask and the wafer. It employs a large gap between the wafer and the mask such that Fresnel diffraction is no longer valid. Instead, Fraunhofer diffraction is valid. It has well-designed objective lens between the mask and the wafer. The resolution of projection printing system governed by Rayleigh's criterion is

$$L_r = \frac{\lambda}{2\text{NA}} \tag{10.10}$$

where NA is the numerical aperture of the imaging system. The numerical aperture is defined as $\text{NA} = n \sin \theta$, where n is the index of refraction of the recording medium and θ is the half angle subtended by the aperture of the recording lens. The typical NA values used are in the range of 0.16–0.40. Higher NA leads to better resolution. However, the depth of focus is inversely proportional to the square of NA. Thus, improving the resolution reduces the depth of focus. Poor depth of focus can result in some points of the wafer to be out of focus as no wafer is perfectly flat. Thus, a compromise is required. Contact printing and projection printing offer resolution of the order of 1 μm, and proximity printing has lower resolution of the order of few microns. The resolution depends on the wavelength of the light source, and for deeper channels, the separation distance from the mask is higher, leading to blur images.

10.6.3 Exposure

The substrate after deposition of the photosensitive polymer is placed in an aligner. This is subsequently exposed to a luminous flux produced by a light source crossing the mask. The luminous flux initiates physicochemical reactions in the polymer, which modify the solubility of the solvents. An ideal photoresist should have the following properties:

1. High photosensitivity
2. Large contrast between the solubilities of the exposed and unexposed parts
3. High resistance to certain chemical agents.

The photosensitive resist solution consists of the resist material, solvent to reduce the viscosity of solution for easy spreading on the spin coater, and additive to control the kinetics of the photoreactions. The resists are of two types, that is, positive and negative based on their nature of reactions with the incident light. The solubility of the light zones in a solvent is influenced by the nature of reactions:

- *Positive resist:* In positive resist, the light zones become soluble in a particular solvent, and other zones remain insoluble.
- *Negative resist:* Here, the light zones become insoluble in the solvent, and the other zones can be dissolved in the solvent.

For positive resist, the zone exposed to the radiation forms holes in the resist film during development stage, that is, when immersed in the solvent. The unexposed portion remains permanently polymerized. For positive resist, the light weakens or breaks the internal bonds of the resist inducing rearrangement of molecule to a more soluble form. In negative resist, the light induces the formation of covalent bonds between the primary and secondary chains. Common positive resists are PMMA and DQN (mixture of diazoquinone and phenoline novolak). These resists are soluble in basic solutions like KOH, ketones, and acetates. The examples of negative resist are KTFR (a polyisoprene elastomer produced by Kodak) and SU-8. SU-8 is a popular resist developed by IBM, which can be used as a mold and for the formation of structures with high aspect ratio.

The choice between the positive and negative resist for microfabrication is not trivial. Negative resist adheres better to the substrate than positive resist. However, the photosolubility of negative resist is weaker than that of positive resist. Positive photoresist does not swell during development. The resist must be sufficiently transparent to the illumination such that whole thickness of the deposited layer is illuminated and induces chemical reactions. During the exposure, the half-light effect and diffraction limit the precision of a pattern exposed to the resist. Figure 10.12 shows a light source illuminating a mask. The light dark zone around the mask reduces the spatial resolution of the pattern defined by the mask. The zone with half-light of the mask is subjected to intermediate intensity between darkness and light. The diffraction is another phenomenon affecting the precision in transfer of pattern from

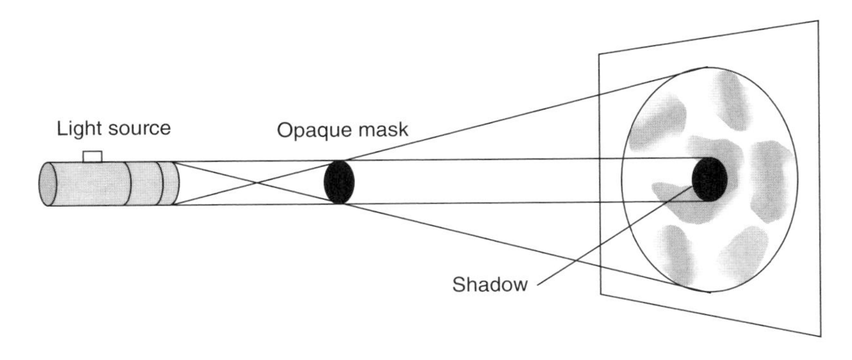

Figure 10.12 Schematic showing half-light effect, that is, zone with light and dark region

the mask. The size of the zone, δ, effected by the diffraction phenomena is estimated by the following formula:

$$\delta \approx 3\sqrt{\lambda t} \tag{10.11}$$

where λ is the wavelength of the monochromatic light source and t is the thickness of the resist film. This is the basis of preference for short wavelength of illumination source and limited thickness of the resist.

10.6.4 Development

Development step is carried out after the exposure to radiation. The substrate with resist is immersed in the solvent during development. The temperature of the solvent is carefully controlled as the development phase is physicochemical and has high temperature dependence. During development, either the exposed part (positive resist) or the unexposed part (negative resist) is eliminated. The final step is the polymerization of the resist by incubating the system above the glass transition temperature. Figure 10.13 shows the transfer of similar pattern (hole) by using both positive and negative resists. The type of mask used is different between the two technologies: one uses the clear field mask and the other uses the dark field mask. However, the final structure of the wafer is same.

10.7 Subtractive Techniques

Etching is a subtractive technique, where a certain portion of the substrate is attacked by chemical or ionic species. Its genesis can be related to decoration of armor during primitive years. The armor was covered with wax, and portion of it was cut out for exposure. The armor was subsequently dipped into a liquid chemical bath for a specific time duration. The wax was eliminated by heating after taking the armor out of the bath. An etched pattern of controllable depth was obtained. Large amount of data on etching velocities of different chemicals at different conditions and discovery of lithography has led to etching being an important step for microfabrication processes.

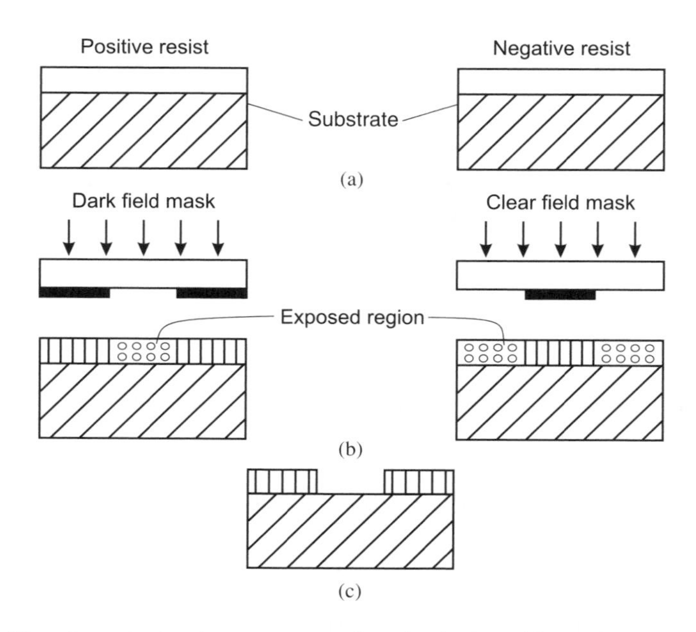

Figure 10.13 The schematic showing pattern transfer using both positive and negative resists: (a) coating of resist, (b) photolithography, and (c) development

10.7.1 Wet Etching

The etching of a solid material using chemical solution is known as wet etching. During this process, the substrate is either inside the chemical solution or the solution is sprayed on the substrate.

The etching process is also classified as *isotropic* and *anisotropic* etching (see Figure 10.14). It may be noted that the wafer after development shown in Figure 10.13 is used in Figure 10.14 for etching a hole inside the substrate. Etching takes place equally in all three directions for isotropic etching, and spherical cavities are formed. The example of isotropic etching is fluoric acid in glass and EDP in silicon. In anisotropic etching, chemical attack take place preferentially along one of the crystallographic planes. Anisotropic etching results in cavities with facets. Anisotropic etching is not possible in glass as it is an amorphous solid. KOH results in anisotropic etching of silicon.

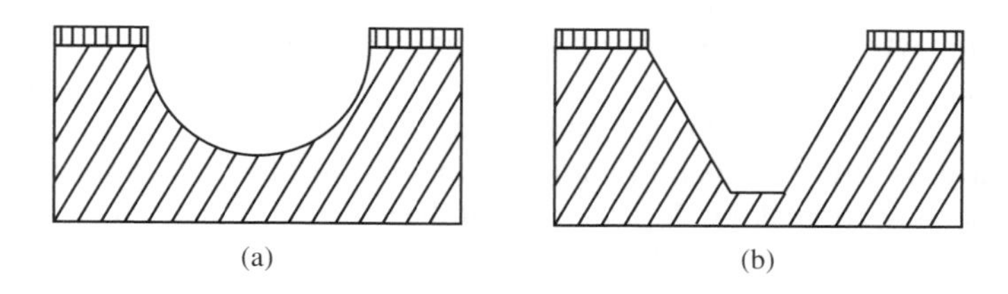

Figure 10.14 (a) Isotropic etching of Si with EDP and (b) anisotropic etching of Si with KOH

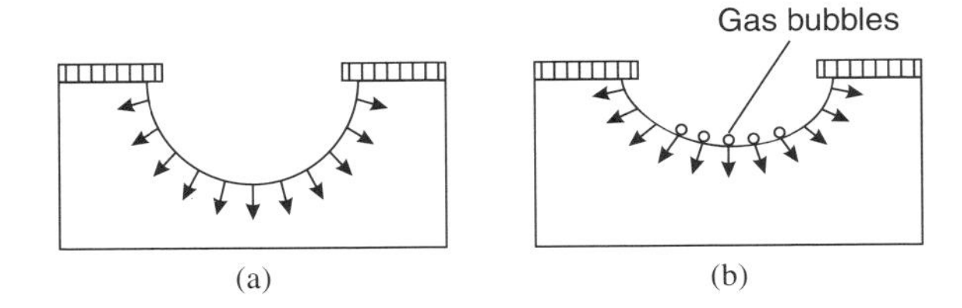

Figure 10.15 Etching profiles during wet etching due to influence of stirring: (a) well stirred and (b) not stirred

Silicon commonly uses a mixture of $HF/HNO_3/CH_3COOH$ for isotropic etching, and glass uses fluoric acid. The etching velocities are in the range of tens of micrometers per minute. During etching, the chemical attack takes place in four steps: (1) The reactants are transported toward the surface. (2) Chemical reaction takes place at the surface. (3) The cavity is formed. (4) The evacuation of the product is carried out. The transport of the reactant is controlled by mixing of the reactants inside the bath (see Figure 10.15); otherwise, the transport is diffusion dominated, and the velocity of diffusion font is slower than etching velocity. The chemical reaction between the reactants and substrate is a subtle process. For silicon, the following global reaction takes place:

$$Si + HNO_3 + 6HF \rightarrow H_2SiF_6 + HNO_2 + H_2O + H_2(gas)$$

During the reaction, the covalent bond of the crystal breaks, producing a fluoride compound of silicon (H_2SiF_6), which is soluble in fluoric acid. This leads to the formation of holes in the valence bond of silicon. For proper control of chemical reaction, the temperature of the bath must be maintained constant. Regulation of temperature up to few tenths of a degree is necessary to guarantee precision on the order of 10% of the depth. In an ideal situation, the mask is not attacked by the reactant. However, in practice, there is a chemical attack of the order of few percent of the original process. Therefore, it is proper to work with thick masks such that the masked part remains covered during the entire etching process. However, it must be remembered that lesser geometric precision is achieved with thicker mask during photolithographic process. Therefore, a compromise is essential. The main advantages of wet etching are (1) high selectivity, (2) high repeatability, and (3) controllability. Etchants for some selective functional materials have been presented in Table 10.6.

Table 10.6 Examples of wet etchants for some common materials

Material	Etchants	Selective to
Al	H_3PO_4, HNO_3, H_2O	SiO_2
Si	KOH	SiO_2
Si	HF, HNO_3, CH_3COOH	SiO_2
SiO_2	HF, HNO_3, H_2O	Si
SiO_2	NH_4, HF	Si

Anisotropic etching in silicon allows the walls with trapezoidal cross-section. The reactants used are alkanes of high concentration, that is, several moles per liter. Unlike isotropic etching, which uses acids, the medium used for etching is strongly basic. The etching velocity is faster along some planes and slower along other orientations.

10.7.2 Anisotropic KOH Etching

KOH is one of the most commonly used silicon etch chemistry for micromachining silicon wafers.

Table 10.7 shows that the KOH etch rate is strongly affected by the crystallography orientation of the silicon (anisotropic).

The (110) plane is the fastest etching surface. The (111) plane is an extremely slow etching plane that is tightly packed and is overall atomically flat.

Table 10.8 shows the etch rate dependence of KOH as a function of composition, temperature, and orientation. It may be observed that etch rate is a strong function of temperature. Significant fast etch rates at high temperature may not be ideal as it can influence the etched surface properties.

Figure 10.16 shows the etching behavior of Si as a function of temperature and concentration. It is observed that the size of the bubbles decreases significantly and the density of the bubble increases drastically as concentration is increased. Changing the temperature and concentration simultaneously shows a similar trend. The other factor is the rate of diffusion, which increases at high temperature and concentration. The other parameters influencing the bubble properties are density of the KOH solution, the surface tension, and solubility of hydrogen in the solution. The etch rate is proportional to the number of hydrogen bubbles. It has been observed that the density of bubbles increases nearly linearly and the solubility decreases with concentration. The best quality of the etched surface is usually seen at 30% KOH and 70 °C.

The nature of etching is responsible in deciding the precision of structures obtained during etching process. Figure 10.17 shows the anisotropic etching of (100) silicon by KOH solution.

Table 10.7 Etch rate of silicon as a function of crystallographic orientation in KOH at temperature of 70 °C for different KOH concentrations

Crystallographic orientation	Rates of different KOH concentration (μm/min)		
	40%	40%	50%
(100)	0.797	0.599	0.539
(110)	1.455	1.294	0.870
(111)	0.005	0.009	0.009
(210)	1.561	1.233	0.959
(211)	1.319	0.950	0.621
(221)	0.714	0.544	0.322

Table 10.8 Effect of composition and temperature on the etch rate of silicon by KOH

Etchants	Temperature (°C)	Direction (plane)	Etch rate (μm min^{-1})
30% KOH	20	(100)	0.024
70% H_2O	40	(100)	0.108
	60	(100)	0.41
	80	(100)	1.3
	100	(100)	3.8
	20	(110)	0.035
	40	(110)	0.16
	60	(110)	0.62
	80	(110)	2.0
	100	(110)	5.8
40% KOH	20	(100)	0.020
60% H_2O	40	(100)	0.088
	60	(100)	0.33
	80	(100)	1.1
	100	(100)	3.1

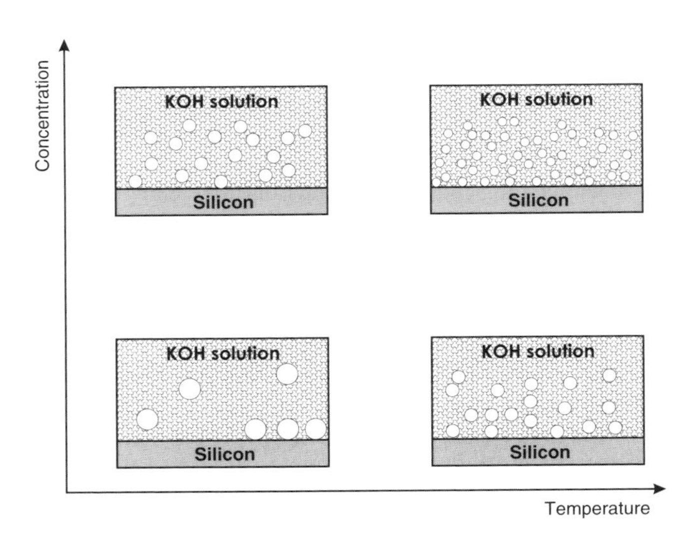

Figure 10.16 Bubble formation during Si etching in KOH solution as a function of temperature and concentration

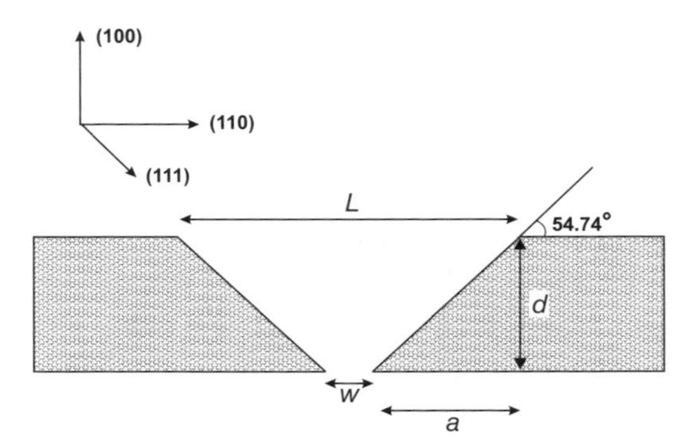

Figure 10.17 Formation of a V-shaped groove in (100) silicon by using a KOH solution

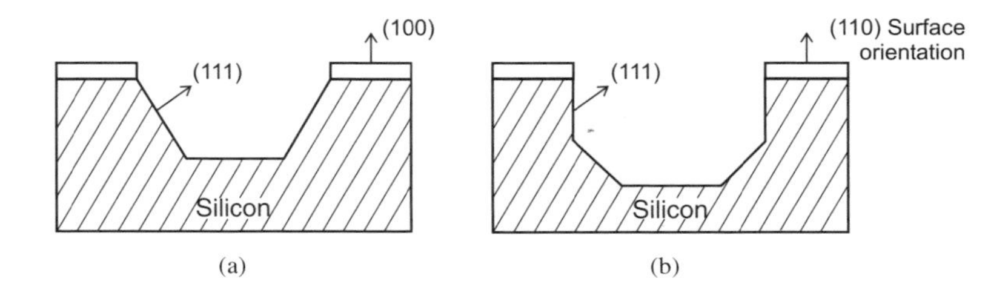

Figure 10.18 Schematic of different shapes resulting from anisotropic etching of Si as a function of crystallographic orientation

Here, the opening of the etching material depends on the etching angle. Thus, the size of exposed surface L needs to be accordingly decided. Here,

$$W = L - 2a = L - \frac{2d}{\tan 54.74\,°} = L - \frac{2d}{1.414}$$

where d is the etch depth and w is the opening size.

Note that the etch velocity is also dependent on the crystallographic orientation. Figure 10.18 shows the typical shape of the etched cavity as a function of the orientation of the Si wafer. It shows that the edge of the cavity for (110) Si wafer etching has vertical orientation compared to the (100) Si wafer because of higher etching velocity of the (110) orientation compared to that of the (100) orientation.

10.7.3 Dry Etching

Dry etching is the process in which the substrate is attacked by ionic species contained in a gaseous or plasma phase. The shapes obtained during dry etching can be isotropic or anisotropic depending on the etching conditions. The anisotropic behavior is controlled by the system, not by the crystalline structure. Etchant gases like Cl_2/CHF_3, Cl_2/He, CF_4, and so on can be used for etching of Si which are also selective to SiO_2. Etchant gases like

CF_4/H_2, C_2F_6, C_3F_8, and so on can be used to etch SiO_2, which are selective to Si. Similarly, different dry etchant gas combinations can be used for etching of different materials. Square channels in glass can be obtained with this technique. The dry etching can be classified as four main types depending on the nature of the plasma containing the ions: (1) physical etching, (2) chemical etching, (3) physicochemical etching, and (4) physicochemical etching with inhibitor.

10.7.3.1 Physical Etching

Physical etching uses beams of ions, electrons, or photons to bombard on the material surface. The ions are accelerated in an electric field and bombarded on the surface of a target. The ions knock out the atoms from the substrate surface due to the high kinetic energy, that is, the etching action takes place due to the physical action of the incident ion flux.

Figure 10.19(a) shows the mechanism of physical process involved in a physical dry etching process. Plasma is a low-pressure gas where a high energy field is used to drive ionization creating a large number of ions and free electrons. Figure 10.19(b) shows the schematic of plasma production for etching. The voltage between the anode and cathode controls the acceleration of the ions. For 10 eV energy setting, the ions are not sufficiently energetic to eject material from the substrate. The ejection of material is produced for setting between 10 eV and 500 eV. Ions penetrate the depth of target for energy range between 10 keV and 20 keV. This is known as ionic implantation useful for doping the semiconductors.

Physical dry etching is carried out in low pressures, and the ejected material must be transported out before undergoing redeposition. The target, that is, the substrate to be etched, is coupled with electrode. The plasma produced by the electrodes is directed to the target. The

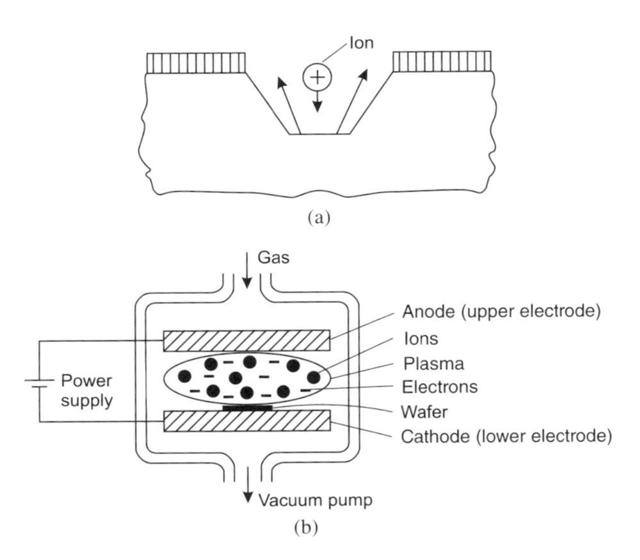

Figure 10.19 (a) The schematic of a physical dry etching process and (b) a setup to sustain plasma for etching

ions produced are sometimes directed toward the target by magnetic field, and this is known as *magnetron-enhanced ion etching*. Almost all materials can be etched by this process. The etching rate is between 0.6 and 18 μm/h. The drawbacks of this method are (1) slow etch rates, (2) low selectivity as ions attack all materials, and (3) reflected ions that can cause trench effects.

10.7.3.2 Chemical Etching

In chemical etching, etchant gases attack material surface by chemical reaction. It is isotropic similar to wet etching and exhibits relatively high selectivity. The etching process can be decomposed into six steps as shown in Figure 10.20. These steps are: (1) Generation of reactive species, (2) Transport/diffusion of reactive species toward the substrate, (3) Adsorption on the surface of the substrate, (4) Reaction with the substrate material and formation of volatile component, (5) Desorption of the volatile component, (6) Diffusion/transport in the gas contrast to physical etching, the reactive molecule movement is diffusive in the case of chemical etching (not ballistic). Chemical etching is carried out under pressure of the order of 10^{-1} to 1 Torr. The simplest chemical etching of Si is carried out by XeF_2, which contains strongly electronegative element, that is, fluoride ion. The reaction involved is

$$2XeF_2 + Si = 2Xe + SiF_4 \text{ (volatile)}$$

where xenon and SiF_4 are the volatile components which desorb.

10.7.3.3 Physicochemical Dry Etching (Reactive Ion Etching)

Physicochemical dry etching also known as reactive ion etching (RIE) is an important technique for micromachining. Here, the reactant gases are excited to ions. The ions are bombarded on the substrate activating the reaction. The basic configuration is that the object, that is, the target to be etched, is placed on the cathode of a cold plasma, which is made up of reactive species. The movement of the ions is ballistic at low pressure. The ions

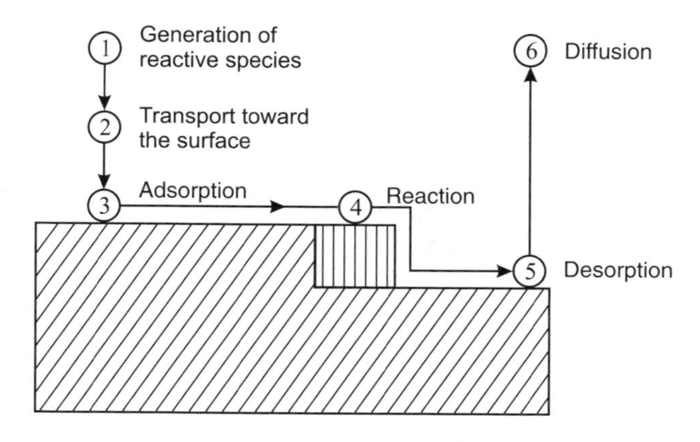

Figure 10.20 Schematic showing the physicochemical process during etching

are accelerated by the electric field which are localized near the cathode, and the target is bombarded by ions. This method can achieve relatively high aspect ratios. The etch rates range between that of physical etching and chemical etching.

10.7.4 Deep Reactive Ion Etching

RIE techniques with extremely high aspect ratios are called deep reactive ion etching (DRIE). In this technique, chemical vapor is deposited to protect the sidewalls. The etching cycle consists of two parts: etching and deposition. In the first step of etching of silicon, SF_6 is used for about 5–15 s during which 25–60 nm of silicon is etched. During the next deposition step, supply gas is switched to C_4F_8. A film of fluorocarbon polymer of about 10 nm thickness is deposited on the trench wall. In the next step, the polymer film is removed by ion bombardment, while the film in the sidewalls is intact. Thus, the etch front advances into the substrate at rates ranging from 1.5 to 4 μm/min.

10.8 Additive Techniques

Deposition plays an important role during all processes of microfabrication. Deposition techniques allow deposition of large variety of materials, that is, metals, insulators, semiconductors, polymers, proteins, and so on. Some of the common additive techniques are discussed in the following section.

10.8.1 Physical Vapor Deposition

During physical vapor deposition (PVD), the object/target is put in contact with a holding gas. Certain species in the gas adsorb, forming a layer that constitutes the deposit. PVD is primarily used for depositing of electrically conducting layers, that is, metals or silicates. It is categorized as two types: *thermal evaporation* and *sputtering*.

10.8.1.1 Thermal Evaporation

Here, the material to be deposited is placed in a container and positioned facing the target. The entire setup is then placed in a chamber maintained at low pressure. The system is taken to high temperature. The sublimation of the heated source material is initiated. The atom flux produced during sublimation step condenses and adsorbs on the target surface. Other molecules contained in the chamber also get deposited simultaneously. Therefore, it is necessary to work at low pressures of the order of 10^{-8} Torr to avoid unwanted parasite deposition. This method can be categorized as *vacuum thermal evaporation* (VTE) and *electron beam evaporation* (EBE) based on the heating method used. For example, in EBE, electron beam is focused on the source material, which locally melts. This method is suitable for metallic deposits and specialized coatings for scratch resistance, antireflection, and so on. Alloys also can be deposited by evaporation using two or more material sources. However, thermal evaporation method is not very stable compared to sputtering.

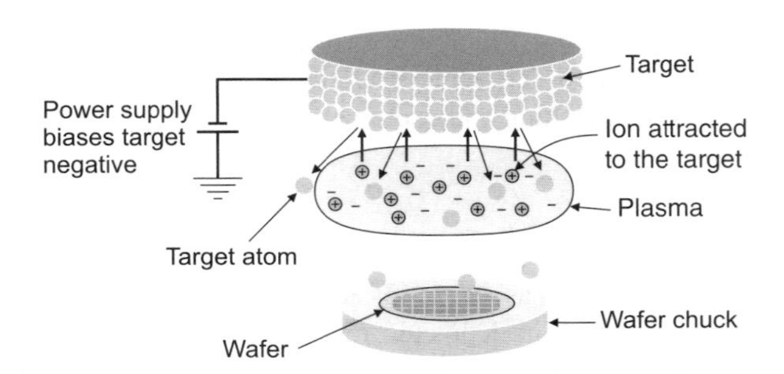

Figure 10.21 Schematic showing the schematic of sputtering procedure

10.8.1.2 Sputtering

Sputtering is similar to physical dry etching with the difference that the wafer material to be deposited is placed on the anode and not on the cathode (see Figure 10.21).

The container with the material destined for deposition is placed on the cathode. The cathode is exposed to an energetic flux such as argon gas, which in a strong electromagnetic field becomes plasma. The target atoms are knocked out by the ions, which deposit on the substrate as a thin film. There is good adhesion between the layer and substrate. Sputtering can deposit all kinds of materials such as alloys, insulators, and piezoelectric ceramics. Sputtering has higher deposition rates.

10.8.2 Chemical Vapor Deposition

In a CVD process, gaseous reactants are introduced into a reaction chamber, where reactions occur on heated substrate surface leading to deposition of the solid product. The reactions can take place in two different manners: (1) The reactions take place in the gas, and the products of reactions are adsorbed on to the surface of the target with homogeneous reaction. (2) The reaction takes place on the surface of the target with heterogeneous reaction. The adherence of the film is superior for heterogeneous reaction compared to that of the homogeneous reaction. Most CVD equipment are based on heterogeneous reaction. Depending on the conditions of reaction, CVD techniques are categorized as:

1. *Atmospheric pressure chemical vapor deposition (APCVD)* Here, the deposition takes place at atmospheric pressure.
2. *Low pressure chemical vapor deposition (LPCVD)* Here, deposition takes place at low pressure (1 Torr). This process is primarily used for polysilicon, which is a fundamental material for MEMS. The deposition speed for this process is about 1μm/h.
3. *Plasma enhanced chemical vapor deposition (PECVD)* This system uses plasma for systems in which the thermal activation is insufficient for heterogeneous chemical reaction. Plasma activates the chemical reaction on the target surface using ionic bombardment.

Figure 10.22 shows the image of a dual sputtering and PECVD system.

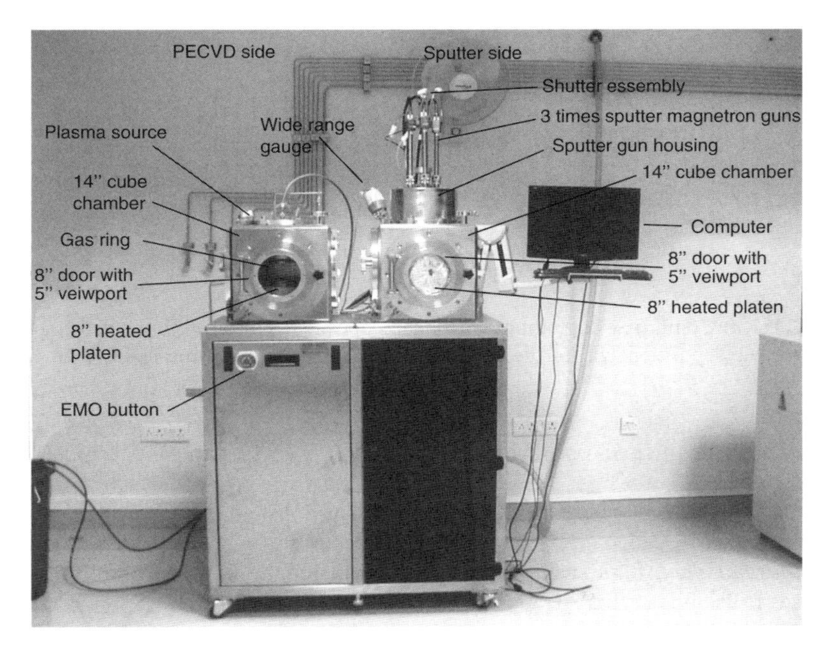

Figure 10.22 An image of a dual PECVD-sputtering system. (Courtesy of Micro Fabrication Laboratory, ME Department, IIT Kanpur)

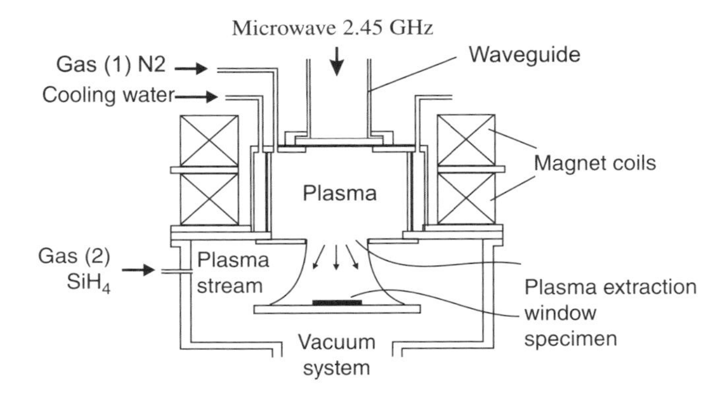

Figure 10.23 Schematic showing the principle of PECVD process

Presence of plasma in PECVD process changes the thermodynamics of surface reactions. It considerably lowers the temperature at which reactions are possible. For example, the deposition reaction of TiC is not possible below 1218 °K. However, the reaction is possible at 700 K in presence of the plasma. Figure 10.23 shows a silane-based PECVD process.

The operating temperature for APCVD and LPCVD is in the range of 500–800 °C. PECVD process has part of its energy in the plasma. Thus, lower substrate temperature on the order of 100–300 °C is used.

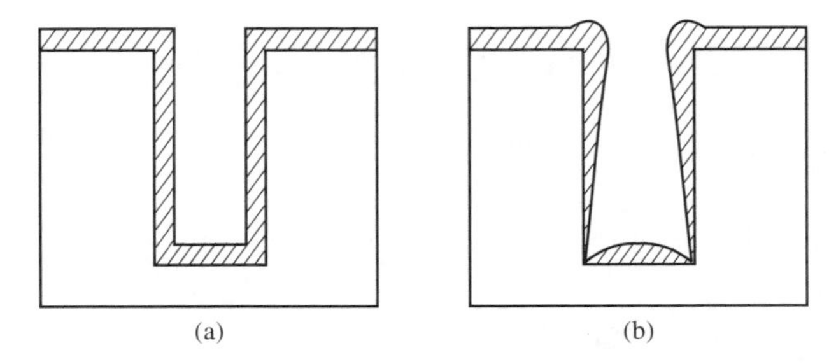

Figure 10.24 (a) Conformal deposition, where the deposited film takes form of the substrate, and (b) nonconformal deposition, where the deposited film has different forms/shape than those of the substrate

Depending on the thickness of layers from the CVD process, two cases arise:

1. *Conformal deposition*, where the deposited film has constant thickness at all points on the target
2. *Nonconformal deposition*, where the deposit film has nonuniform thickness, that is, it possesses rolls and crevices

Figure 10.24 shows the schematic of conformal and nonconformal deposition.

For obtaining conformal deposition, the incident particles must be sufficiently energetic to diffuse the target surface before forming chemical bonds. This is dependent on the initial energy of incident particles and the mean free path. Therefore, it is advantageous to work at low pressures, as the system has high mean free path at low pressures. Some of the chemical reactions during CVD have been presented in Table 10.9. The table shows different chemical reaction processes for depositing Si, SiO_2, and Si_3N_4.

Table 10.9 Some representative materials with their chemical reactions and CVD techniques

Material	Techniques	Chemical reactions
Silicon	Silane CVD	$SiH_4 \rightarrow Si + 2H_2 \uparrow$
Polysilicon	LPCVD	$SiH_4 \xrightarrow{630\ ^\circ C,\ 60\ Pa} Si + 2H_2 \uparrow$
Silicon	Silane oxide CVD	$SiH_4 \xrightarrow{630\ ^\circ C,\ 1\ bar} SiO_2 + 2H_2 \uparrow$
dioxide	PECVD	$SiH_4 + 4N_2O \xrightarrow{350\ ^\circ C,\ 40\ Pa,\ (plasma)} SiO_2 + Gas \uparrow$
	High-temperature oxide CVD	$SiH_2Cl_2 + 2N_2O \xrightarrow{900\ ^\circ C,\ 40\ Pa} SiO_2 + Gas \uparrow$
Silicon	LPCVD	$SiH_2Cl_2 + 4NH_3 \xrightarrow{450\ ^\circ C,\ 30\ Pa} Si_3N_4 + Gas \uparrow$
nitride	PECVD	$3Si + 4NH_3 \xrightarrow{300\ ^\circ C,\ plasma,\ 30\ Pa} Si_3N_4 + 6H_2 \uparrow$
	PECVD	$3SiH_4 + 4NH_3 \xrightarrow{700\ ^\circ C,\ plasma,\ 30\ Pa} Si_3N_4 + Gas \uparrow$

10.8.3 *Doping*

The behavior of silicon can be changed into a conductor by doping, that is, mixing a small amount of impurity into the silicon crystal. Ion implantation adds impurities to semiconductors. Addition of dopant atoms, that is, boron having three valence electrons with silicon having four valence electrons, creates positively charged carriers called holes. Silicon of this type is called p-type. Adding dopant atoms with phosphorus having five valence electrons creates negatively charged carriers called electrons. Silicon of this type is called n-type. A minute amount of either n-type or p-type doping turns a silicon from crystal form with good electrical insulation properties into a viable conductor – therefore the name semiconductor.

Ion implantation is also used to fabricate insulating layer such as silicon dioxide. At high temperatures, oxygen ion implantation creates oxide layer with depths ranging from 0.1 to 1 μm on the surface.

10.8.4 *Electrolytic Deposition*

During electrolytic deposition, a current is passed through a solution of metallic salt such as $CuSO_4$. Metallic ions migrate toward the anode, capturing electrons there. The molecules adsorb on the electrode. There is metallic deposit of copper on the anode. This process is also known as electroplating, where various metals can be deposited on the electrode.

10.9 Example of a Silicon Membrane Fabrication

The detailed fabrication procedure of silicon membrane with controlled thickness using the photolithography and etching procedure is discussed here. Figure 10.25 pictorially shows the stepwise procedure of fabrication. The first step is the deposition of silicon dioxide layer (SiO_2) on the silicon wafer. This is followed by the deposition of photosensitive resist. The photoresist is exposed under a mask during lithography, and the exposed portion of the resist dissolves after

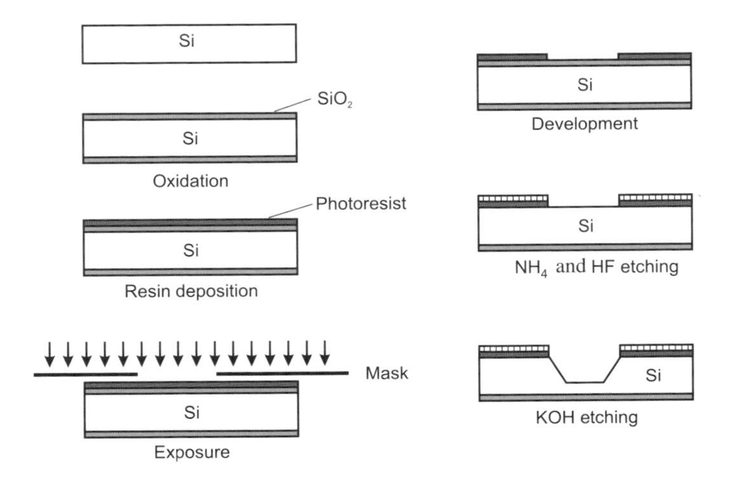

Figure 10.25 Photolithography and etching steps carried out for fabrication of a silicon membrane

development. The exposed portion of SiO_2 is dissolved by chemical etching (NH_4 and HF). The exposed silicon is subsequently etched by KOH to form the cavity. Note that SiO_2 often acts as the electrical insulation of the metallic structures, that is, silicon or metal substrate.

10.10 PDMS-Based Molding

PDMS-based molding technique plays a very important role in microfluidics. PDMS is a family of polymers containing silicon. Its formula is $(Si(CH_3)_2O)_n$, where n is the number of repeating monomer. Figure 10.26 shows the semistructural representation of PDMS.

Some representative properties of PDMS material have been presented in Table 10.10. Due to the special properties of PDMS, the following observations can be made on its role in microfluidics:

1. It allows visualization due to its transparency in the visible spectrum.
2. Its Young's modulus is a function of ratio between the PDMS and reticulating agent concentration. The value given in Table 10.10 is a higher end value.
3. The elastomeric quality of PDMS facilitates the watertightness of microfluidic connections. The elasticity of the material facilitates the fabrication of valves and pumps using membranes.
4. Untreated PDMS is hydrophobic in nature. It becomes hydrophilic after exposure to the oxygen plasma or after immersion in strong base. Oxidated PDMS adheres by itself to glass and silicon as long as the other surfaces are also exposed to oxygen plasma.
5. The permeability of PDMS to gas facilitates filling of the channel. However, its permeability to other nonpolar organic solvents makes it unsuitable for other applications.
6. The surface energy of PDMS is weak, which facilitates the process of peeling off the mold from the substrate.

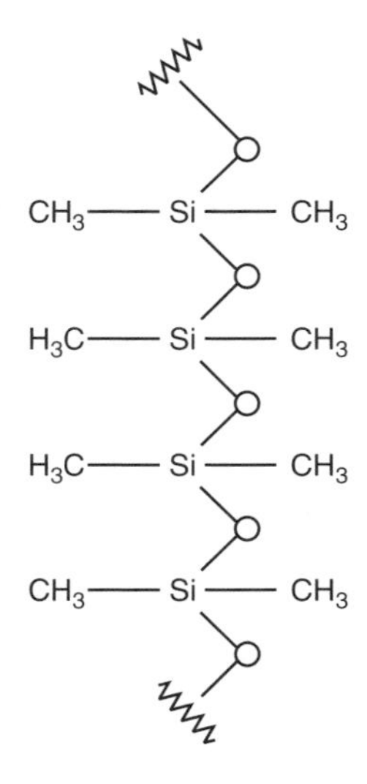

Figure 10.26 Semistructural formula of PDMS

Table 10.10 Representative properties of PDMS

Properties	Characteristic values
Density	$\simeq 0.9\ kg/m^3$
Thermal conductivity	$\simeq 0.2$ W/m-K
Young's modulus	$\simeq 750$ kPa (elastomeric)
Surface energy	$\simeq 20$ mN/m (low)
Optical	Transparent between 300 nm and 2200 nm
Electrical	Insulating, breakdown field 20 kV/cm
Permeability	Permeable to gas, nearly impermeable to water and polar organic solvents
Reactivity	Inert, oxidizable by a plasma
Toxicity	Nontoxic

7. A disadvantage of PDMS is the aging of reticulated PDMS with time, which is difficult to predict. The exposure of PDMS to water vapor also modifies the hydrophobicity of the surface.
8. It is not possible to evaporate the metallic electrodes on PDMS due to its adherence properties and the high temperature requirement of the deposition technique. However, PDMS structures can be placed on silicon wafer with evaporated electrodes. Therefore, the hybrid technology has large amount of potential.

During PDMS molding process, a mixture containing a catalyst and polymer is poured on a mold and heated. After reticulation, the structure is peeled off the mold, which contains the pattern of the mold in negative form. The schematic of molding process has been shown in Figure 10.27.

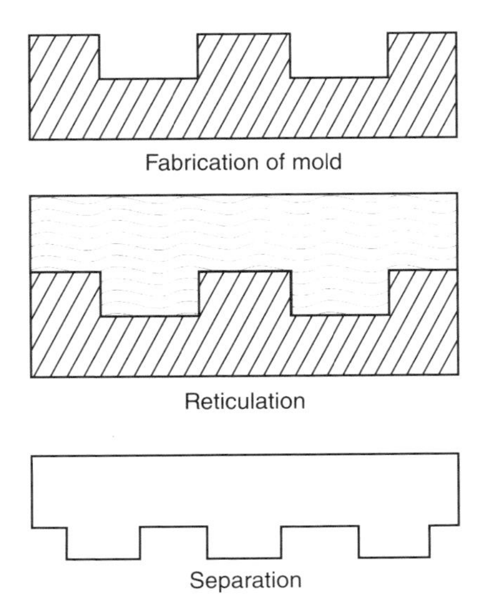

Figure 10.27 The schematic procedure showing the molding procedure of PDMS structure

The mold is made from a hard material separately using the microfabrication processes discussed earlier in a clean room. The mold material can be either silicon or reticulated polymer such as SU-8. The regular molding step can be carried out in ordinary environment with proper temperature and humidity control. After the fabrication of the mold, the mixture of PDMS and reticulating agent is poured onto the mold. The system is taken to a moderately elevated temperature of about 70 °C. During this phase, the PDMS polymerizes and reticulates and the mixture becomes solid. During the third step, the PDMS is peeled off and the negative structure of the mold structure is obtained. The precision of the structure obtained through the process is high, that is, it is possible to obtain structures with submicrometric precision. However, considering the elastomeric properties of the PDMS materials, the dimensions between 5 and 500 μm are suitably obtained.

The other methods used for fabrication of PDMS-based MEMS device are (1) casting and (2) injection molding. During casting, the mold is pressed into the heated deformable material. After cooling and separation, the negative structure of the mold is obtained. In microinjection, the heated PDMS material is injected in a liquid state into the mold. After separation, the negative structure of the mold is obtained.

10.10.1 Example of Microchannel Fabrication

Microchannels are the most important components of a microfluidic device. This section describes two common methods of microchannel fabrication.

10.10.2 Soft Lithography

Soft lithography using PDMS can be used for rapid prototyping of microfluidic devices. It consists of three major steps: (1) SU-8 master fabrication, (2) PDMS pouring and lift-off, and (3) surface treatment and bonding with glass (see Figure 10.28).

The first step involving the preparation of master is spin coating of silicon wafer with SU-8 negative resist. Subsequently, UV exposure is carried out with a clear field mask. This is followed by the development of SU-8 master. Glass posts are placed on SU-8 master to define the inlets and reservoirs. PDMS is mixed with prepolymers having weight ratio of the base and

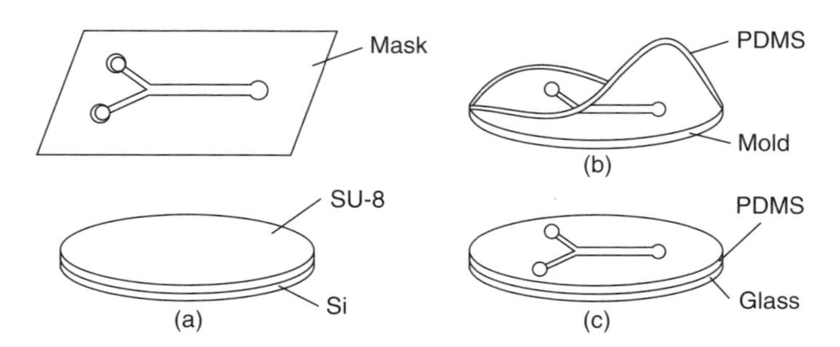

Figure 10.28 Fabrication of Y-channel using soft lithography: (a) SU-8 master fabrication, (b) PDMS pouring and lift-off, and (c) surface treatment by plasma and bonding by glass

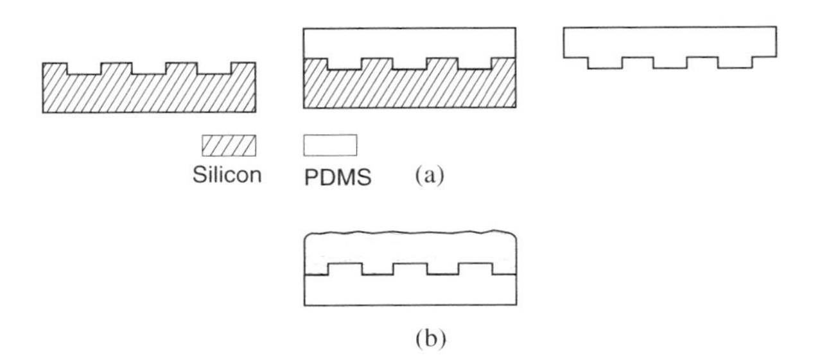

Figure 10.29 (a) PDMS master fabrication and (b) micromolding for microchannel fabrication

curing agent to be 10:1 or 5:1. The PDMS mixture is poured into the master and allowed to settle for few minutes to self-level. The whole set is cured at temperature range of 60-80 °C for several hours. This is followed by peel-off of the PDMS structure. The structured PDMS membrane is surface treated with low-temperature oxygen plasma. The surface-activated PDMS structure is brought into contact with clean glass or another piece of surface-activated PDMS for bonding. Watertight leakproof channel capable of withstanding 5 bar is produced when pressed against glass or another PDMS detachable fluidic device.

10.10.3 Replica Molding

The PDMS master is used as replica master in replica molding. In this process, the original master cannot be damaged or degraded as multiple copies are made. Figure 10.29 shows the schematic step in fabrication of PDMS replica master and replica molding. The silicon master can be prepared by conventional micromachining techniques. The prepolymer is coated on the silicon master. The curing is carried out at elevated temperature, and the PDMS device is subsequently peeled off. The PDMS part is subsequently used as replica master for the prepolymer. The final structure is obtained after curing and peel off.

10.11 Sealing

The microfluidic fabrication is carried out separately for different parts. These parts need to be subsequently sealed together in watertight manner. Sealing is not necessarily an easy procedure. For ensuring watertightness, adhesion must be applied on all surfaces brought into contact with one another. Defects in flatness and interstitial spaces may form due to improper sealing. These defects cause hydrodynamic fluxes and often render the system unusable. Some of the sealing techniques are discussed in the following sections.

10.11.1 Anodic Field-Assisted Bonding

Anodic bonding technique allows adhesion of glass onto a metal or semimetal such as silicon. In anodic bonding, glass wafer and silicon wafer are bonded together at elevated temperatures

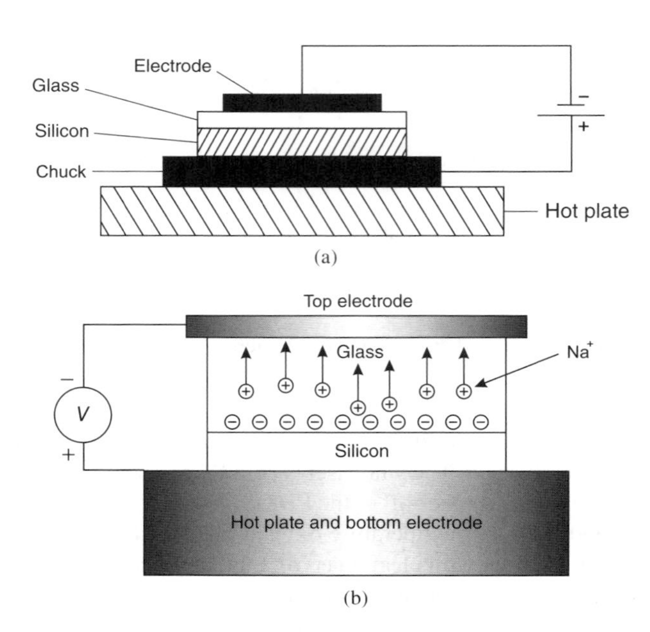

Figure 10.30 (a) Schematic showing the anodic bonding process and (b) principle of anodic bonding

of the order of 400 °C and high electric field with bonding voltage about 1 kV. Figure 10.30 shows a typical setup of anodic bonding. Here, the silicon structure is connected to the positive electrode that works as anode. Therefore, the name of this bonding technique is known as anodic bonding.

10.11.1.1 Principle

With reference to Figure 10.30(b), when external electric field is applied at elevated temperature, positive sodium ions Na^+ in the glass migrate toward the negative pole. This creates a space charge depletion region adjacent to the glass–silicon interface. The voltage drop over this depletion layer creates a large electric field that pulls the wafer into intimate contact. The elevated temperature also enables covalent bonds to form between silicon and oxygen atoms at the surface of the two materials, resulting in an oxide layer. In the absence of electric field, the system should be brought to about 1000 °C for bonding by simple fusion. One important factor for success of this bonding process is minimization of thermal deformation between glass and silicon. This means that the glass should be selected such that it has similar thermal expansion coefficient as that of silicon. Commonly used glasses are Pyrex corning 7740, Corning 7750, Schott 8329, and Schott 8330. The conditions of cleanliness at the surface also influence the success of the sealing.

A special type of glass named Pyrex 7740 is most commonly used in anodic bonding because of large sodium cation content and matched thermal expansion coefficient ($\simeq 32.5 \times 10^{-7}$cm/°C) with that of silicon. The approximate Composition of Pyrex 7740 is

SiO_2, 80.6%; B_2O_3, 13%; Na_2O, 4.0%; Al_2O_3, 2.3%; and other traces, 0.1%. The sodium cation plays a key role in the process. It migrates under applied external electric field, leading to electric potential drop at the bonding interface. The large potential drop and the relatively small gap between the two substrates induce an extremely large electrostatic force. This electrostatic force overcomes the effects of microscale surface roughness of the substrates and puts them into intimate contact. At elevated temperature, oxidation occurs at the bonding surface. A hermetic bond is realized after keeping the bonding conditions (voltage/temperature) for a specific time. It has been mostly observed that bonding strength increases with increase in bonding temperature and voltage.

10.11.2 Direct Bonding

Bonding between two substrates without an intermediate layer is known as direct bonding. When direct bonding of same material takes place, it is advantageous due to lack of thermal stresses because of the same thermal expansion coefficient. Three possible cases of direct bonding are as follows:

1. For silicon–silicon bonding, the reaction between hydroxyl (OH) groups at the surface of the oxide layers (deposited or native) is utilized. First, the silicon wafers are immersed inside the H_2O_2/H_2SO_4 mixture, boiling nitric acid, or dilute H_2SO_4, leading to hydration. The bonding process is accomplished subsequently at elevated temperatures between 300 and 1000 °C.
2. For bonding between two soda-lime glass slides, the glass wafers are cleaned in an ultrasonic bath in the beginning. It is subsequently immersed in a solution of [$5H_2O$:$1NH_3$ (25%):$1H_2O_2$ (20%)] or [$6H_2O$:1HCl (37%):$1H_2O_2$ (20%)] for 10 min. The moisture is removed next by annealing at 130 °C. Finally, the two wafers are thermally bonded together at 600 °C for 6–8 h.
3. For polymer direct bonding, that is, PDMS, the surface is treated with oxygen plasmas and sealed at room temperature. PDMS surface is oxidized and forms a good seal with silicon and glass surface when in contact with each other.

10.11.3 Indirect Bonding

Indirect bonding takes place with intermediate layer. The intermediate layer can be epoxies, photoresists, or other polymers depending on the type of substrate materials. A number of epoxies, UV-curable epoxies, and photoresists can be used for adhesive bonding. The advantage of polymers is the low process temperature requirement. The advantage of this bonding is that it can be used for all kinds of materials.

10.12 Laser Microfabrication Techniques

Laser microfabrication technique is also commonly present in most of the microfluidic laboratories. This technique is growing very fast recently and is contributing toward

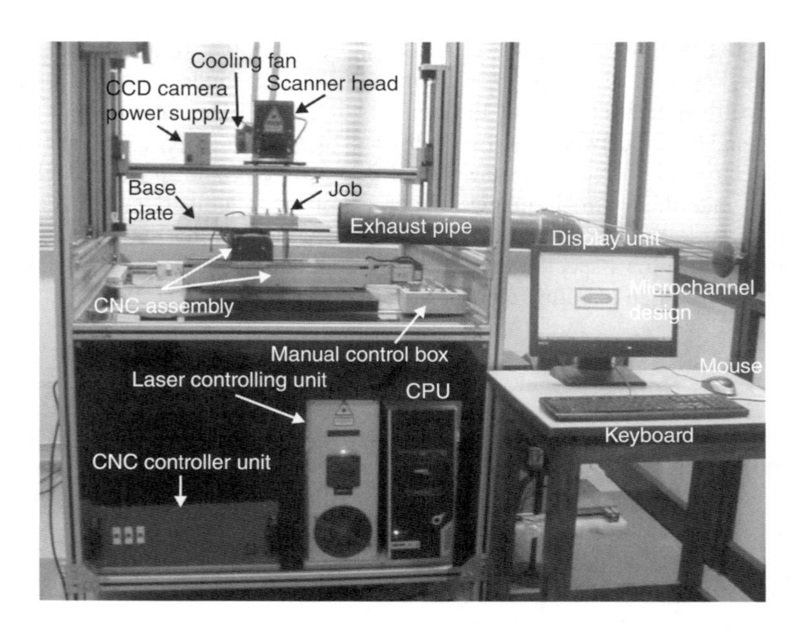

Figure 10.31 An image of a laser micromachining center (CHE/ME Department, IIT Kanpur)

simplification of the overall microfabrication process. Suitable design of a laser-based microfabrication system can lead to attaining of high material removal rate, high resolution, and superior surface quality.

Figure 10.31 shows the picture of a laser micromachining center. In laser microfabrication, a highly directional light beam, that is, laser, is focused close to diffraction-limited spot size, which is incident on the matter to be machined. The availability of pulse laser at higher repetition rates adds to the capability of laser-based microfabrication. This technique avoids a lengthy and wasteful multistep approach based on lithography.

There are two modes of laser micromachining: *direct writing* and *using a mask*. In the direct writing mode, the laser beam is focused on the substrate surface, and the pattern is scanned using precision x–y stage or galvano scanner consisting of f-theta lens and scanning mirrors. Here, the smallest structure depends on the accuracy of the scanning system, that is, of the order of 20 μm. In the mask mode, the mask determines the shape of the structure, that is, the maximum structure size can be brought down to twice the wavelength of the laser. Laser micromachining is suitable for fabrication of microchannels and fluidic access holes.

10.12.1 Minimum Spot Size

The minimum spot size of a laser beam determines the size of the structure that can be micromachined. Figure 10.32 shows the schematic of a laser beam being focused by a

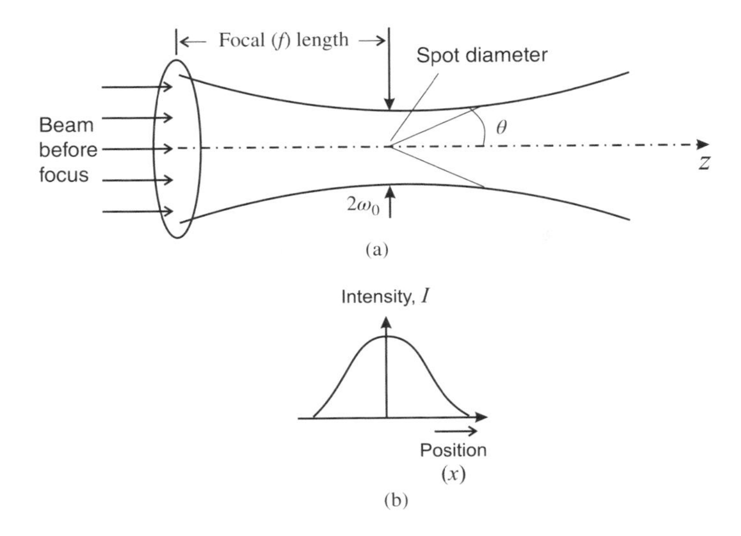

Figure 10.32 (a) Propagation of Gaussian laser beam and (b) its intensity profile

converging lens. Here, θ is the half divergence angle, ω_0 is the half beam waist size, and λ is the wave length.

For an ideal beam (see Figure 10.32),

$$\theta\omega_0 = \frac{\lambda}{\pi} \tag{10.12}$$

For a real beam, the above equation is modified as

$$\theta\omega_0 = M^2\frac{\lambda}{\pi} \tag{10.13}$$

where M^2 is a beam quality number which is equal to one for a perfect beam. The beam quality number is defined as the square ratio of the multimode beam diameter to the diffraction-limited beam diameter $\left(m^2 = \left(\frac{D_m}{D_0}\right)^2\right)$.

The minimum spot size is governed by the equation

$$\theta_f = \frac{D}{2f} \tag{10.14}$$

where f is the focal distance and D is the lens diameter.

Using the above two equations, the minimum spot size δ can be written as

$$\delta = 2\omega_0 = 2\frac{M^2\lambda}{\pi\theta_f} = \frac{4M^2\lambda f}{\pi D} \tag{10.15}$$

Hence, for micromachining small spot size, we should have short wavelength, short focal length, and $M^2 = 1$. The laser type influences the beam quality. The He–Ne ($\lambda = 0.63$ μm), Nd:Yag ($\lambda = 1.06$ μm), and CO_2 laser (10.6 μm) have beam quality equal to 0.98, 10, and 1.5, respectively.

10.12.2 Physical Mechanism

The mechanism that lasers use to remove material is called ablation. Ablation is achieved by melting and vaporizing the material, which is subsequently ejected from the vicinity of the surface (see Figure 10.33). Ablation of metals is caused by the absorption of laser energy in a three-step process: (1) absorption of photon by electrons, (2) energy transferred to the lattice, and (3) heat transferred to the lattice. Ablation at the end of the pulse is due to a superheated layer, which continues to evaporate till the surface is above the boiling temperature. The vaporization rate depends on the power density of the laser beam. The high vaporization rate can cause shock wave of speed equal to 3 km/s. The expulsion of the material takes place due to the generation of high pressure. The size of the heat-affected zone (HAZ) is given by

$$\mathrm{HAZ} \sim (\alpha t)^{1/2} \tag{10.16}$$

where α is the thermal diffusion coefficient and t is the pulse duration.

The quality of fabrication depends on the laser–matter interaction, which is related to the laser parameters, that is, *laser power*, *laser frequency*, *pulsing frequency*, and *pulse duration*. The absorptivity of the laser energy inside the matter depends on the frequency of laser. Hence, the frequency of laser is decided such that maximal portion of incident laser photon is absorbed by the material being machined. The complete description of laser–matter interaction requires the solution of Maxwell equation for the laser field coupled with the matter. Hence, the total problem is interconnected by simpler problems involving *absorption of laser light*, *ionization*, *energy transfer from photon to electrons and ions*, *heat conduction*, and *hydrodynamic*

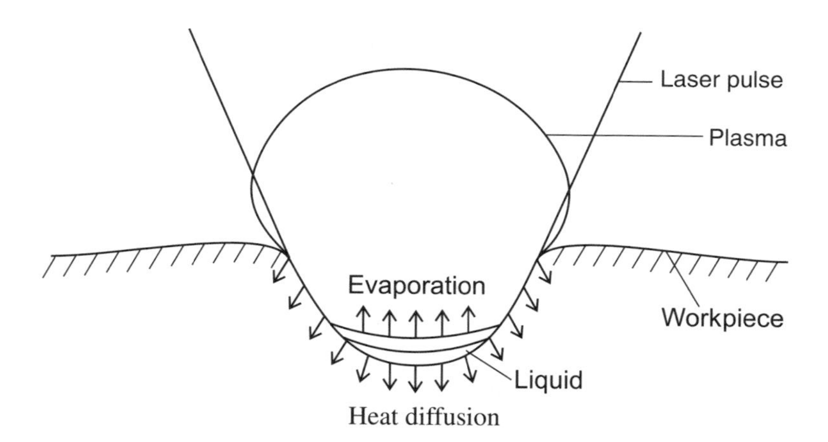

Figure 10.33 Laser pulse interaction with the workpiece material

expansion. Thus, it is not easy to specify the required processing conditions. It is found by experimental trial and error and software based on self-learning algorithm. The laser–matter interaction is drastically different at lower intensity (below ionization threshold) from that at high intensity. At high intensity, that is, at above the ionization threshold, the interaction takes place in laser–plasma interaction mode. With further increase in energy density, a strong shock wave is generated in the interaction region, which propagates into the surrounding cold material. Shock wave propagation leads to compression of the solid material at the wave front and decompression behind it, leading to the formation of void inside the material. In some solids, chemical decomposition may take place at relatively low beam density. The decomposed matter can get released in the gas phase and then expands, producing a bubble inside the material. The other factor of importance is the thermal diffusion coefficient (α) which is of the order of $10^{-7} m^2/s$ for dielectric and $10^{-4} m^2/s$ for metal. Thus, for micron-sized region of dielectric, it will cool in a time of about 10 μs ($t \sim l^2/\alpha \sim 10^{-5}$s) compared to 10^{-8} s for metals. Therefore, multiple pulses in a dielectric will accumulate energy if the period between the pulses is shorter than the cooling time. Hence, if a single pulse energy is too low to produce any modification of the material, a change can be induced using high-pulse repetition rate, because of this accumulation phenomenon. Thus, repetition rate is another means for controlling the size of structure produced by the laser. In practical terms, generation of high intensity using low-pulse energy requires that energy be focused to the smallest possible volume, whose dimensions are of the order of laser wavelength, λ.

The laser pulses of very short duration also eliminate heat flow to surrounding materials. Hence, short-pulse lasers achieve clean and accurate structures. The following lasers are commonly used for laser micromachining: (1) excimer lasers ($\lambda = 351, 308, 248$, and 193 nm) (2) Nd:Yag laser ($\lambda = 1067, 533, 355$, and 266 nm), and (3) CO_2 lasers (λ = 10.6 μm).

Rapid heating can damage the material to a greater depth. The shorter-pulse duration decreases the HAZ; therefore, femtosecond laser pulses machine with minimal heat generation. The material at the top of the surface is removed by evaporation, and the sidewall material is removed by plasma.

The ejected material can be deposited near the melt region due to freezing and is known as the "recast" layer. Redeposition may occur in the path of the track, which is yet to be machined. In this case, more material is to be removed.

In addition to the laser-based technique discussed here, the following techniques are also available for general microfabrication: (1) focused ion beam micromachining, (2) ultrasonic micromachining, (3) microelectrodischarge machining, (4) powder blasting, (5) microgrinding, and (6) micromachining using water droplets.

Problems

10.1 You have to purchase a laser for the microfabrication center of your microengineering laboratory. Two lasers, that is, Nd:Yag (λ = 1.06 μm) and CO_2 (λ = 10.6 μm) having beam quality parameter M^2 equal to 5.0 and 1.5, respectively, are available to you. Assume that the material to be machined has same laser energy absorption characteristic for both the lasers. If the cost, power, and pulse characteristics for both the lasers are same, which laser will you purchase? Justify your answer.

10.2 A silicon wafer (4 cm × 4 cm) of 1 mm thickness is first coated with a photoresist. After subsequent exposure by the lithography apparatus followed by development process, 2 mm diameter region of the silicon wafer is exposed to outside. This exposed region of silicon is subsequently etched by KOH etchant. If the etching angle during anisotropic etching by KOH solution is equal to 54°, calculate the minimum diameter of the etched hole after completion of the etching process.

10.3 1 mm^3 of volume of SU-8 5 photoresist is spin coated at 10,000 rpm. The equilibrium coating thickness is equal to 15 μm. Estimate the speed of rotation for obtaining 5 μm thickness of the photoresist.

10.4 You would like to fabricate some microchannel-based waveguide in borosilicate glass. A 800 nm wavelength laser is available for the microfabrication process. The size HAZ needs to be minimized for best quality. The size of the HAZ is 1 μm for pulse duration of the 100 ps. Calculate the approximate size of the HAZ for pulse duration of 200 fs.

10.5 Explain the stepwise procedure for fabrication of a T-shaped microchannel using PDMS material for a microfluidic application.

Supplemental Reading

Emslie AG, Bonner FT, and Peck LG 1958 Flow of a viscous liquid on a rotating disk. *J. Appl. Phys.*, **29**(5), pp. 858–862.

Kew PA and Cornwell K 1997 Correlations for the prediction PF boiling heat transfer in small diameter channels. *Appl. Thermal Eng.*, **17**(8-10), pp. 705–715.

Kim S and Mudawar I 2012 Flow condensation in parallel micro-channels-Part 2: heat transfer results and correlation technique. *Int. J. Heat Mass Transfer*, **55**, pp. 984–994.

Kim S and Mudawar I 2013a Universal approach to predicting heat transfer coefficient for condensing mini/micro channel flow. *Int. J. Heat Mass Transfer*, **56**, pp. 238–250.

Kim S and Mudawar I 2013b Universal approach to predicting saturated flow boiling heat transfer in mini/micro-channels-Part I. Dryout incipience quality. *Int. J. Heat Mass Transfer*, **64**, pp. 1226–1238.

Kim S, Kim J, and Mudawar I 2012 Flow condensation in parallel micro-channels-Part I: experimental results and assessment of pressure drop correlations. *Int. J. Heat Mass Transfer*, **55**, pp. 971–983.

Lee J and Mudawar I 2005 Two phase flow in high heat flux micro channel heat sink for refrigeration cooling applications: Part II-heat transfer characteristics. *Int. J. Heat Mass Transfer*, **48**, pp. 941–955.

Lee J and Mudawar I 2009 Critical heat flux for sub-cooled flow boiling in micro-channel heat sinks. *Int. J. Heat Mass Transfer*, **52**, pp. 3341–3352.

Liao SJ 2002 An analytic approximation of the drag coefficient for the viscous flow past a sphere. *Int. J. Non Linear Mech.*, **37**, pp. 1–18.

Nguyen NT and Wereley ST 2006 *Fundamentals and Applications of Microfluidics*. Artech House, Inc.

Qu W and Mudawar I 2004 Measurement and correlation of critical heat flux in two phase micro channel heat sinks. *Int. J. Heat Mass Transfer*, **47**, pp. 2045–2059.

Sato K, Shikida M, Matsushima Y, Yamashiro T, Asaumi K, Iriye Y, and Yamamoto M 1998 Characterization of orientation-dependent etching properties of single-crystal silicon: effects of KOH concentration. *Sens. Actuators, A*, **64**, pp. 87–93.

Schwartz LW and Roy RV 2004 Theoretical and numerical results for spin coating of viscous liquids. *Phys. Fluids*, **16**(3), pp. 569–584.

Soliman HM 1986 The mist-annular transition during condensation and its influence on the heat transfer mechanism. *Int. J. Multiphase Flow*, **12**, pp. 277–288.

Tabeling P 2005 *Introduction to Microfluidics*. Oxford University Press.

Yu H, Zhou G, and Chau FS 2008 Yield improvement for anodic bonding with suspending structure. *Sens. Actuators, A*, **143**, pp. 462–468.

Yun M 2000 Investigation of KOH anisotropic etching for the fabrication of sharp tips in silicon-on-insulator (SOI) Material. *J. Korean Phys. Soc.*, **37**(5), pp. 605–610.

11

Microscale Measurements

11.1 Introduction

Different flow physics are encountered in microdevices compared to macrodevices. Therefore, the performance of microdevices cannot be analyzed using similar techniques as macrodevices. The measurement procedure being adopted for macrosystems cannot be directly used in microdevices. Therefore, it is essential to have microscale measurements tools for analysis, design, and optimization of microdevices. Velocity and temperature are the primary performance parameters of microfluidic systems. This chapter reports some salient experimental techniques for measurement of velocity and temperature of microfluidic systems.

11.2 Microscale Velocity Measurement

The recent surge of microfluidic devices has created a need for diagnostic tools with spatial resolutions on the order of several microns. These diagnostic techniques can be used as engineering tools to measure the flow performance of microfluidic devices and to understand the physics of transport processes at microscale.

The μ-PIV has been used for flow studies in microfluidic devices by many researchers. Meinhart *et al.* (1999) studied pressure-driven flow in a microchannel. Kim *et al.* (2001) studied electroosmotically driven flows in various microchannel configurations to examine their feasibility for micropumping and microvalve applications. Shinohara *et al.* (2004) studied two-phase transient flow in microfluidic devices by using high-speed μ-PIV. The laminar to turbulent flow transition in microtubes was studied by Sharp and Adrian (2004). Nguyen *et al.* (2005) used μ-PIV for visualization and quantitative measurement of magnetic microflows. Liu *et al.* (2005) used infrared μ-PIV in silicone-based microdevices and demonstrated the effectiveness of this technique by measuring the velocity profiles for laminar flows. Lima *et al.* (2006) used confocal micro-PIV for measurements of three-dimensional profiles of cell suspension flow in a square microchannel. Considerable efforts to describe theoretically various aspects of μ-PIV for flow measurements have been made. Special interrogation techniques and filtering schemes have been developed in order to improve the quality of μ-PIV measurements

Transport Phenomena in Microfluidic Systems, First Edition. Pradipta Kumar Panigrahi.

Companion Website: www.wiley.com/go/panigrahi/microfluidic

(see e.g., Santiago *et al.*, 1998; Wereley and Meinhart, 2001; Wereley *et al.*, 2002, and Meinhart *et al.*, 2000b), and modifications have been attempted in order to improve the resolution of μ-PIV systems (see e.g., Shinohara *et al.*, 2004).

Three major factors distinguish μ-PIV from its macroscopic counterpart: (1) the seeding particles are small compared to the wavelength of illuminating light, requiring application of fluorescence imaging in μ-PIV; (2) due to small particle size, Brownian motion is a source of random error in the measurement of the particle displacement between images, especially for slow flows; and (3) flow illumination differs considerably from the conventional PIV. In PIV, a thin sheet of laser light is generated in order to illuminate a single plane within the fluid flow, thus defining the measurement plane of the PIV system. However, in μ-PIV, laser sheet generation is not feasible, and so the entire volume of the flow is illuminated.

We discuss below the fundamentals of regular PIV technique followed by μ-PIV due to many similarity between the two techniques.

11.3 PIV Fundamentals

In macroscopic system, the particle image velocimetry is a nonintrusive technique for measuring the spatial distribution of velocity in a single plane inside the flow. The measurement is carried out indirectly via the displacement of moving particle groups within a certain time interval. For this purpose, the flow is seeded homogeneously with appropriate tracer particles. The concentration of the particle must be well adjusted with regard to the finest flow structures. It is assumed that the particles are small enough to move with the local flow velocity. The flow field is illuminated twice within a short time interval by a laser sheet. The duration of the illumination light pulse must be short enough such that the motion of the particle is *frozen* during the pulse exposure in order to avoid blurring of the image. The light from each light pulse scattered by the tracer particles is recorded by a CCD sensor on separate frames. The time delay between the illumination pulses must be long enough for proper determination of the displacement between the images of the tracer particles with sufficient resolution. It should be short enough to avoid particles with an out-of-plane velocity component leaving the light sheet between subsequent illuminations. One image pair identifies the path a particle has traveled. Knowing the time delay between the two pulses, velocity can be calculated. The time interval between the two pulses has to be adjusted according to the mean flow velocity and the magnification of the camera lens. The particle displacement Δx must be small relative to the finest flow scale to be resolved. After finding the displacement of each interrogation spot, the displacement is divided by Δt, and the magnification factor M of the image system to calculate the first-order approximation of the velocity field is

$$u \simeq \frac{\Delta x}{\Delta t} = \frac{\Delta X}{M\Delta t}, \quad v \simeq \frac{\Delta y}{\Delta t} = \frac{\Delta y}{M\Delta t} \tag{11.1}$$

Cross Correlation: Displacement of particles in PIV image pairs is calculated based on correlation approach in contrast to the particle tracking algorithm, where particle path is followed. Here, the average motion of a small group of particles contained in the interrogation spot is calculated by spatial autocorrelation or cross correlation. Autocorrelation is performed when images for both laser pulses are recorded on the same sensor, while in cross correlation, each pulse is collected into separate frames. Cross-correlation calculation can be carried out faster in the frequency domain due to availability of FFT algorithm. There is directional ambiguity in autocorrelation technique. Hence, in case of reverse flow, this technique is not suitable. The drawback is eliminated using the cross-correlation technique. Cross correlation allows us to use a small interrogation area compared to autocorrelation and leads to a reduction of the random error due to spatial velocity gradients. The criteria involving depth of field of recording optics and laser light sheet thickness is that the generally, depth of the field should not be smaller than the thickness of the light sheet in order to avoid imaging of out-of-focus particles.

11.3.1 Implementation Issues

There are various aspects that should be taken into consideration during PIV measurements. Starting from image capturing to ultimate conversion of image information into velocity vectors require proper validation. The tracer particles should follow the flow faithfully without much velocity lag and should be homogeneously distributed in the flow field. They should be small enough to follow the fluid movement and large enough to be visible. A set of six nondimensional parameters that are most significant for optimization of PIV measurements were identified by Keane and Adrian (1990). These are the data validation criterion, the particle image density, the relative in-plane image displacement, the relative out-of-plane displacement, a velocity gradient parameter, and the ratio of mean image diameter to the interrogation spot diameter. Two terms, which are frequently used in PIV measurements, are source density

$$N_s = \frac{C\Delta z_0}{M_0^2}\frac{\pi}{4}d_\tau^2 \tag{11.2}$$

and the image density

$$N_I = \frac{C\Delta z_0}{M_0^2}D_I^2 \tag{11.3}$$

Here, C is the tracer particle concentration [m^{-3}], Δz_0, light sheet thickness [m], M_0, image magnification, d_τ, particle image diameter [m], and D_I is the interrogation spot diameter [m]. The source density represents the type of image that is recorded. Source density represents whether the particle images are overlapping ($N_s > 1$) or can be recognized individually ($N_s < 1$). In case of particles overlapping ($N_s > 1$), no individual particles can be recognized anymore. By using a coherent light source, speckle pattern is generated due to interference of the coherent light. The displacement of speckle patterns can provide velocity information similar to that of PIV. This is known as laser specie velocimetry. The image density represents the mean number of particle images in an interrogation region. It should be larger than 10–15 for a good PIV measurement.

The optimal pulse separation between two images is influenced by a number of parameters. Two main types of error affect the choice of pulse separation (Δt), that is, random error and acceleration error. Random error arises from noise during recording of images and subsequent interrogation of the particle images. Acceleration error arises from approximation of Lagrangian motion of tracer particle to local Eulerian velocity based on small particle displacement:

$$\frac{dV}{dt} = \frac{\partial V}{\partial t} + \vec{V}\cdot\vec{\nabla}V \tag{11.4}$$

These two kinds of error contradict the selection of pulse separation. The random error contribution can be reduced by increasing Δt but acceleration error increases as Δt increases. At some intermediate value of Δt, the total error should be a minimum. The optimum separation can be derived by considering these two errors as (Boillot and Prasad, 1996)

$$\Delta t_{\text{opt}} = \sqrt{\frac{2cd_{\tau}}{M|dv/dt|}} \tag{11.5}$$

In order to evaluate the correlation correctly with a single pass, the correlation function must be small with respect to the displacement correlation peak. This is accomplished when the following requirements are fulfilled:

1. There should be at least 7–10 particle image pairs in each interrogation spot.
2. The in-plane displacement should be limited to 1/4 of the size of the interrogation region.
3. The out-of-plane displacement should be limited to 1/4 of the thickness of the light sheet.
4. The displacement difference over the interrogation volume should be less than 3–4% of the size of the interrogation region.
5. The optimum particle diameter should be about 2 pixels.

11.3.2 Recording of the Particle Images

Besides particle dynamics, the registration, storage, and readout of the individual particle images are other key elements in PIV. This is because the accuracy of the technique strongly depends upon the precision with which the image displacement can be related to particle locations and their respective particle displacements. The continuous intensity distribution of the particle image is transformed into a discrete signal of limited bandwidth. When the discretization of the image signal matches with the minimum sampling rate, the frequency contents of the original signal, that is, the particle location, can be reconstructed without any losses. According to the well-known Nyquist criterion of signal processing, the bandwidth-limited signal can be perfectly reconstructed from its discrete samples when the sampling rate of the signal is at least twice the signal bandwidth.

11.3.3 Evaluation of Image Pairs

Cross-correlation technique is used to extract displacement information from the two single exposed gray level patterns acquired at time, t and $t + \delta t$. A two-dimensional gray level sample $I(x, y)$ is extracted from the source image at time, t. It is cross correlated with the corresponding sample $I'(x, y)$ from the second image at time, $t + \delta t$ (see Figure 11.1). In general, the cross-correlation function is given by

$$R(x, y) = \sum_{i=-k}^{k} \sum_{j=-l}^{l} I(i, j) I'(i + x, j + y) \tag{11.6}$$

Here, I and I' are intensity values of the image pair, and (x, y) is the image shift. For shift values, when the particle images align with each other, a high correlation value is obtained

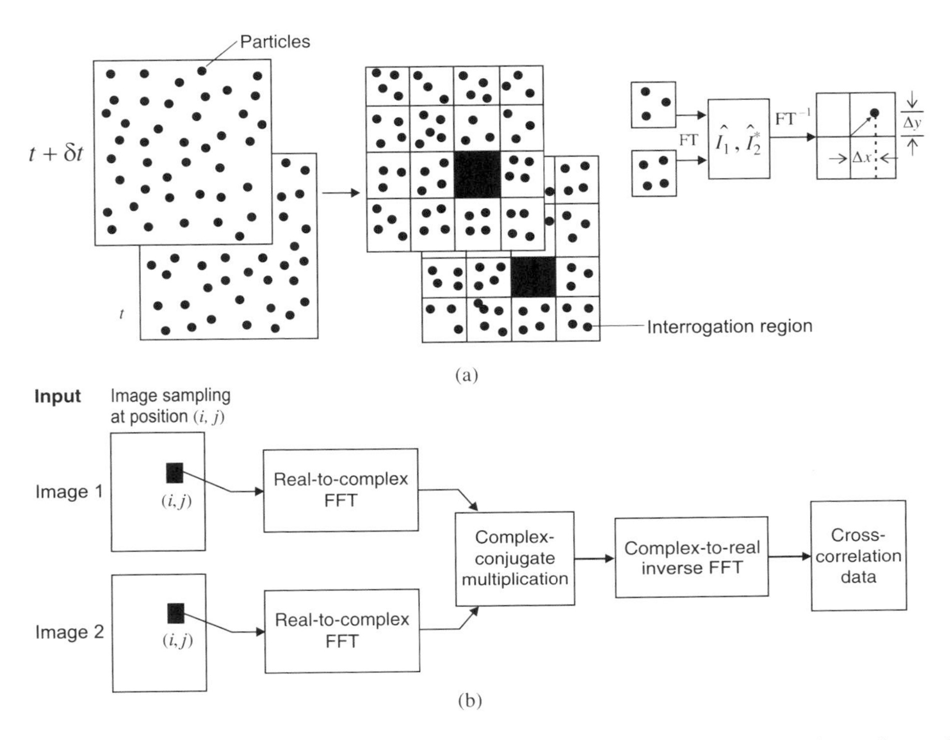

Figure 11.1 (a) PIV image pair and interrogation zones and (b) the computation scheme for cross-correlation analysis

at that position. The cross-correlation function produces a signal peak when the image aligns with each other, since the sum of the product of the pixel intensities will be larger than elsewhere. The highest value of cross-correlation value is used as a direct measure of particle displacement.

In general, calculation of the correlation function is done in the Fourier space. Each subregion is transformed into the Fourier space via a Fourier transformation. The subregions are spatially shifted to obtain good correlation between the image pairs.

The cross correlation of two functions is equivalent to a complex conjugate multiplication of their Fourier transforms, that is,

$$R \leftrightarrow \hat{I}\hat{I}' \tag{11.7}$$

where $\hat{I}$ and $\hat{I}'$ are the Fourier transforms of the image intensities I and I', respectively.

Once the correlation is obtained, the Fourier space information are converted back into the physical space. The displacement that yields a maximum in the correlation function over the interrogation area is regarded as the particle displacement. Actually, it is not the particle displacement, which is computed but the displacement of the interrogation area. The displacement vector is of first order, that is, the average shift of the particles is geometrically linear within the interrogation window. The size of the interrogation should be sufficiently small such that the second-order effect, that is, displacement gradients, can be neglected.

11.3.4 Peak Detection and Displacement Estimation

One of the important steps in the evaluation of PIV images is to measure the position of correlation peak accurately to subpixel accuracy. To increase the accuracy in determining the location of the displacement peak from ±0.5 pixel to subpixel accuracy, an analytical function is fitted to the highest correlation peak by using the adjacent correlation values. Various methods of estimating the location of the correlation peak have been proposed. Some of these are peak centroid fit, Gaussian peak fit, and the parabolic peak fit. Of the three, the Gaussian fit is most frequently used to estimate the shape of the signal around its peak assuming ideal imaging conditions as

$$f(x) = C_0 \exp\left[-\frac{(x_0 - x)^2}{k}\right] \tag{11.8}$$

where x_0 is the exact location of the maximum peak and C_0 and k are the coefficients. Using this expression for the main and the adjacent correlation values and based on the fact that the first derivative of this expression at x_0 must be zero, the peak position can be estimated with subpixel accuracy. When the particle image size is small, the displacement tends to bias toward integer values. The assumed peak shape does not match the actual shape of the peak and the three-point Gaussian estimator cannot represent the true shape of the correlation function. This is called the *peak-locking* effect. In actual displacement data, the presence of the peak-locking effect can be detected from histogram plot.

11.3.5 Data Validation

The evaluation of raw images from particle image velocimetry measurements may contain a number of *spurious* vectors. These vectors deviate unphysically in magnitude and direction from the nearby physically meaningful vectors. They originate from those interrogation spots that contains insufficient number of particle images or whose signal-to-noise ratio (SNR) is very low. In postprocessing process, the first step is to identify these spurious vectors and subsequently discard them to form the valid data set. The detection of either a valid or spurious displacement depends on the number and spatial distribution of particle image pairs inside the interrogation spot. In practice, there should be at least four particle image pairs to obtain an unambiguous measurement of the displacement (Westerweel, 1997). An average of 10 particle images per interrogation spot with an average in-plane displacement of $\frac{1}{4}D_I$ will give a probability of 95% of finding at least four particle image pairs. Here, D_I is the size of interrogation spot. The valid data yield can be improved by increasing the seeding density. But by increasing the seeding density, we increase the influence of the seeding on the flow.

There are various way to detect spurious vector in a velocity field (see Figure 11.2). Three mainly used tests are the global mean test, local mean test, and local median test. The global mean and the local mean are both linear estimators of valid vector. The local median test is a nonlinear estimator that is often used in outliers identification. The outliers in turn are identified by the median of the sample data. Out of the above three, Westerweel (1997) has shown that the local median test has the highest efficiency. In these techniques, the value at a grid point is compared with the neighboring grid points; if it exceeds a certain threshold, the value is discarded.

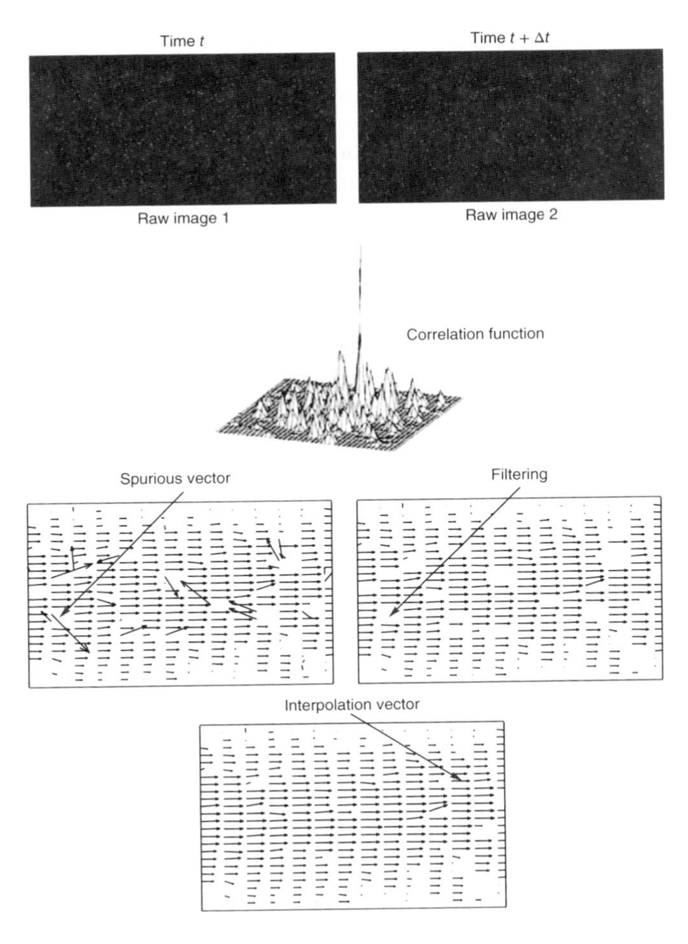

Figure 11.2 Processing of PIV images

11.3.6 Dynamic Velocity Range

Dynamic velocity range (DVR) is related to the fundamental velocity resolution and hence the accuracy of PIV measurement. DVR specifies the range of velocity over which measurements can be made. It is the ratio of the maximum velocity to the minimum resolvable velocity or equivalently the root mean square (rms) error in the velocity measurement, that is,

$$\mathrm{DVR} = \frac{U_{\mathrm{max}}}{\sigma_u} = \frac{U_{\mathrm{max}}}{\sigma_{\Delta x} M_0^{-1} \Delta t^{-1}} \tag{11.9}$$

where M_0 is the image magnification and Δt is the maximum time interval used for the experiments. The rms error of the displacement field on the pixel plane ($= \sigma_{\Delta x}$) generally lies between

1% and 10% and can be expressed as

$$\sigma_{\Delta x} = 0.1(d_e^2 + d_r^2)^{1/2} \quad (11.10)$$

where d_r represents the resolution of the recording medium that is taken to be equivalent to the pixel size and d_e is the diameter of the particle image prior to being recorded at the pixel plane. Assuming that the particle image is diffraction limited and its image intensity is Gaussian, the diameter of the diffracted image of the particle is expressed as

$$d_e^2 = M_0^2 d_p^2 + \left[2.44(1 + M_0)f^{\#}\lambda\right]^2 \quad (11.11)$$

where d_p is the seeding particle diameter, $f^{\#}$ is the F-number of the imaging lens, and λ is the laser wavelength. For an experiment with image magnification, $M_0 = 0.04$, pulse interval, $\Delta t = 70$ μs, d_r the pixel size (= 6.7 μm), $f^{\#} = 1.4$, particle diameter, $d_p \approx 2$ μm, the dynamic velocity range is equal to 30.

11.3.7 Optimum Pulse Separation Time

The pulse separation time between successive light pulses is one of the important parameters during PIV measurements. For a velocity magnitude in x-direction u, pulse separation time Δt, the average displacement, d_x, we have:

$$u = \frac{d_x}{\Delta t} + \frac{\varepsilon_x}{\Delta t} \quad (11.12)$$

where ε_x is the error in the evaluation of displacement, d_x.

For small time interval, the displacement d_x varies linearly with time, Δt. Thus, the first term of the above equation will remain unchanged with reduction in time Δt.

The error ε_x will not reduce below a certain limit due to the reduction of the pulse separation time as the uncertainty in determination of the particle image positions is unaffected to the selection of pulse interval time. Hence, the second term of the above equation increases rapidly with a decrease in pulse separation:

$$\lim_{\Delta t \to 0} . \frac{\varepsilon_x}{\Delta t} = \infty \quad (11.13)$$

This indicates that the measurement error decreases by an increase in separation time. However, for high values of Δt, the measurement error ε_x increases due to loss of pairs because of the particle displacement. For large value of Δt, number particles illuminated twice reduces resulting in loss of correlation between image pairs. The choice of Δt on the quality of PIV data has been illustrated using Figure 11.3. The curve g illustrates the effect of residual error, and the curve f represents the influence of loss of pairs. The product of function, f and g, represents the quality function, Q_{PIV}. It may be noted that function f is chosen arbitrarily as it is difficult to define the quality of measurement.

Hence, there is an optimum pulse separation time, Δt_{opt}, at which the combined error because of residual error magnitude and loss of image pairs is minimum. This is found iteratively by slowly increasing the pulse separation time until the number of outliers within the vector map increases. The normalized strength of displacement correlation can also be optimized with respect to pulse separation time.

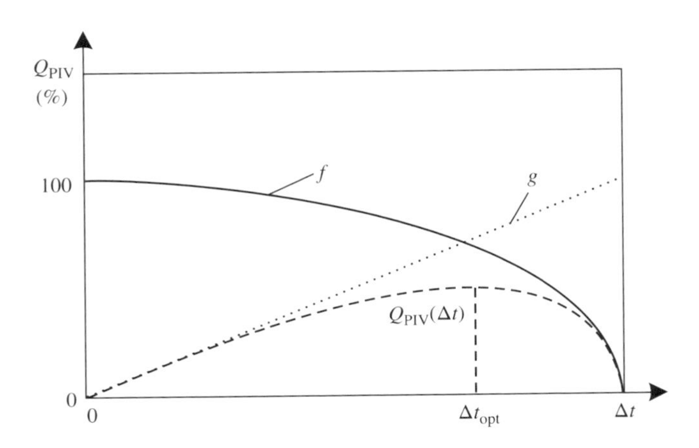

Figure 11.3 Schematic representation of PIV data quality as a function of pulse delay time

11.3.8 Image Preprocessing

Image intensity variation strongly influences the correlation signal magnitude. Brighter particle images have stronger influence on the correlation value compared to weaker particle images. Nonuniform illumination of particle image intensity can be produced due to light sheet nonuniformities, pulse-to-pulse variation in intensity, irregular particle shape, and out-of-plane motion. Therefore, filtering operation is carried out before cross-correlation calculation to bring particle image intensity variation to smaller level.

Background subtraction from the PIV recordings can reduce the effects of laser flare and other stationary image features. Background image can be recorded in the absence of seeding. The average background image can also be obtained by computation of average intensity images from sufficiently large number of raw PIV recordings. High-pass filtering can also be used for removing low-frequency background variations leaving the particle images unaffected. Here, the filter kernel width should be larger than the diameter of the particle images. High-pass filtering followed by thresholding or image binarization can be used to have particle with same intensity. However, it may be noted that binarization may result in an increase of measurement uncertainty. The application of low-pass filtering can be used to remove high-frequency noise from camera noise, pixel anomalies, and so on. Low-pass filtering can lead to broadening of correlation peaks, thus allowing better performance of subpixel peak-fitting algorithm. In case of undersampling of the images, that is, ($d_{\tau} < 2$), it reduces the peak-locking effect. However, low-pass filter can also contribute to an increase in measurement uncertainty. It must be noted that the implementation of the above filtering scheme may increase the measurement uncertainties. This has to be balanced against the increase in data yield. It is recommended to apply contrast enhancement filter selectively in areas of low data.

11.3.9 Advanced PIV Interrogation Schemes

Many advanced PIV processing algorithms have been developed in the last few years to improve data yield and get higher accuracy. The overview of these schemes is provided in the following sections.

11.3.9.1 Multiple Pass Interrogation

In multiple pass interrogation scheme, a window offset equal to the local integer displacement from the previous interrogation step is implemented in the subsequent interrogation step. This increases the fraction of matched particle images resulting in an increase of correlation peak or SNR. It is also advised to filter out the outliers and replace them by interpolating valid neighbors after each interrogation pass and before application of the window offset.

11.3.9.2 Grid Refining

Multiple pass interrogation scheme described in the previous section can also be further improved by continuously refining the sampling grid size. This procedure can allow utilization of interrogation window sizes smaller than the particle displacement. This procedure is useful in PIV recordings with high image density and high dynamic range. In those cases, standard evaluation schemes cannot use small interrogation windows without losing correlation signal due to larger displacements.

11.3.9.3 Image Deformation Scheme

The usual assumption during PIV processing is that the motion of particle images is uniform within the interrogation window. However, many flows have large velocity gradients. The velocity differences inside the interrogation window can contribute toward border correlation peaks or multiple peaks. Thus, the presence of velocity gradient contributes toward larger uncertainty in measurements. This problem can be circumvented by iteratively deforming the windows. It is very useful in shear flows and turbulent flows in general. Figure 11.4 shows the window deformation of the second exposure image before calculation of the correlation.

11.3.9.4 Subpixel Estimation

It should be noted that the intensity values are discretized, and therefore, the correlation values exist only for integral shifts. A variety of methods are used to estimate the location of

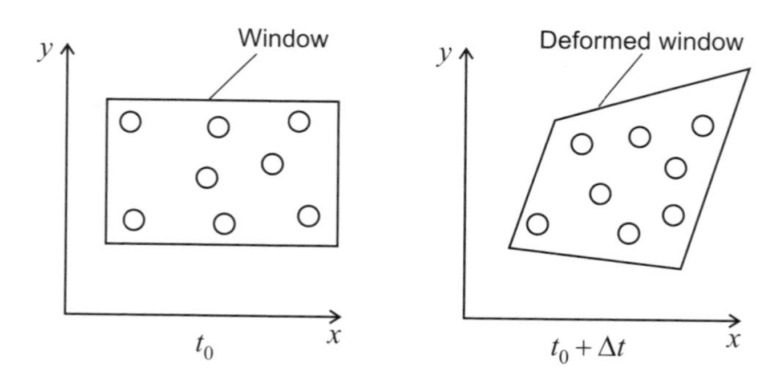

Figure 11.4 Schematic showing the deformation of the interrogation window in the second exposure image

the correlation peak. The coordinates (i, i) corresponding to the maximum correlation values are first identified. The neighboring correlation values are subsequently curve-fitted centroid, parabolic peak fit, Gaussian peak fit, or Whittaker reconstruction.

11.3.10 Accuracy of PIV Measurements

The total error in PIV measurements can be decomposed into bias error and random error as

$$\varepsilon_{\text{tot}} = \varepsilon_{\text{bias}} + \varepsilon_{\text{rms}} \tag{11.14}$$

Bias error, $\varepsilon_{\text{bias}}$, corresponds to over or under estimation of the velocity vector. This may be due to error in calibration. The rms error, ε_{rms}, corresponds to the randomness of experimental parameter during measurements. One approach in the evaluation of rms error during PIV measurements is PIV recording of the static (quiescent) flow. This approach permits the study on experimental parameters such as particle image diameter and background noise.

11.4 Micro-PIV System

A typical μ-PIV system consists of a CCD camera, a microscope (inverted or upright) equipped with fluorescent filter cube, and a light source for flow illumination (laser or microstrobe), along with appropriate optics. A schematic diagram of a μ-PIV system is shown in Figure 11.5.

The laser beam enters the microscope through appropriate beam attenuation optics and fiber-optics cable. The light is transmitted through the microscope objective, and the flow field of interest is illuminated. The fluorescence signal from tracer particles inside the

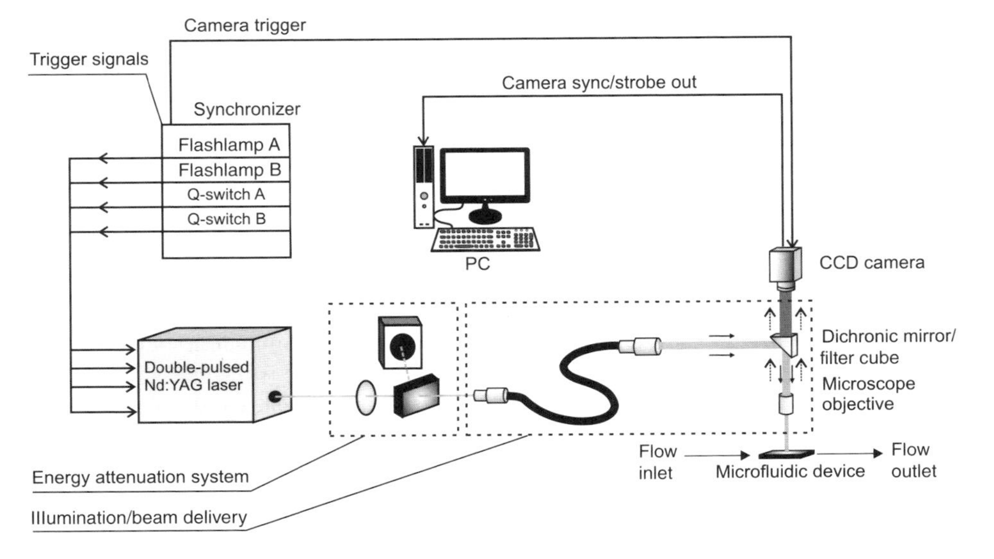

Figure 11.5 A schematic illustration of μ-PIV system setup

microfluidic device is collected by the microscope objective through the dichromatic mirror. The dichromatic mirror filters out reflected light (noise) and only transmits the fluorescence signal from the tracer particles to the image recording system. Subsequently, the signal from the tracer particles is recorded onto a CCD chip, which is then transferred to the PC for analysis. A synchronizer is used for synchronization between the light source and camera.

11.4.1 Volume Illumination

In PIV, a thin sheet of laser light is generated in order to illuminate the measurement plane of the fluid flow. The particles within the measurement plane scatter light, which is captured by the camera. But in μ-PIV, illumination of a single plane within the fluid flow is not feasible because of the following reasons:

- *Optical access*: Many microfluidic devices are manufactured by some treatment (e.g., etching, micromachining, and so on) of a silicon wafer, PDMS, and PMMA and sealed by bonding a glass cover on top of the device. Thus, optical access is limited to one direction, making the use of laser sheet impossible.
- *Applicability*: Application of a laser sheet in microscale devices requires special optical guides, waveguides, inside the microfluidic device. Because of increased complexity and higher cost during fabrication, the use of waveguides is not a desirable approach (Meinhart *et al.*, 2000a).
- *Dimension*: The dimensions of channels in MEMS devices are typically in the range of tens or few hundreds of microns, requiring that the measurement plane thickness should be only a few micrometers. Obtaining such a thin light sheet is difficult, while its alignment with the imaging optics focal plane is nearly impossible for practical purposes (Meinhart *et al.*, 2000a).

Figure 11.6 demonstrates the difference between the illumination of μ-PIV and macro-PIV system. Therefore, in μ-PIV, the entire flow field is illuminated, and the measurement plane is determined by the focal characteristics of the microscope objective. An idea of field of view for various microscope objectives (*Model: Leica DMI 5000M*) can be found in the Table 11.1.

11.4.2 Fluorescence

Due to the mode of illumination and accompanying noise issues, elastic scattering (i.e., light scattering by pure reflection) is not suitable for μ-PIV. Instead, the seeding particles must emit light at a wavelength λ_{emit} different from the wavelength of the illumination source λ_{laser}. This is accomplished by means of fluorescence imaging. Fluorescence is the emission of photons by an atom or molecule following a temporary excited electronic state caused by absorption of photons of a certain wavelength from an external radiation source. A molecule capable of fluorescence is called fluorochrome or fluorescent dye. When a fluorochrome absorbs a photon, an electron is excited from its ground state E_0 to a higher state (excited state 1) corresponding to the energy of the absorbed photon (see Figure 11.7). Due to the fluorochrome's interactions

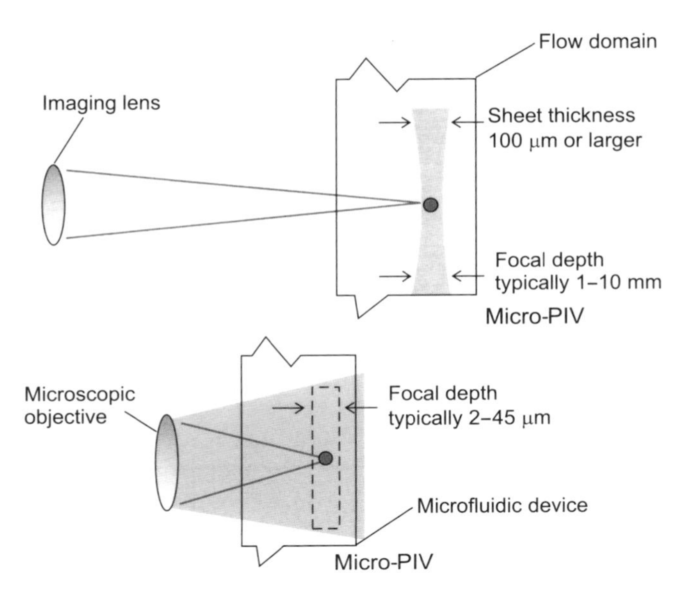

Figure 11.6 Difference between light sheet and volume illumination

Table 11.1 Field of view for various microscope objectives (Model: Leica DMI 5000M) used in μ-PIV applications

Mag/NA	WD (mm)	Coverglass thickness (mm)	Field of view (h × w) (μm^2)
X5/0.12	14	—	1651 × 1321
X10/0.22	7.8	—	826 × 660
X20/0.40	3.2–1.9	0–2	413 × 330
X40/0.55	3.3–1.9	0–2	206 × 165
X63/0.70	2.6–1.8	0.1–1.3	131 × 105

with its molecular environment during the excited state, there is some loss of energy resulting in a relaxed excited state with energy E_2 (excited state 2) from which fluorescence emission originates; after a short time period (typically in the order of 10^{-9}–10^{-12} s), the excited electron collapses back to its ground state, emitting a photon with energy corresponding to the difference between the electron's excited and ground states, that is, $E_2 - E_0$. The energy of a photon is given by $E = \frac{hc}{\lambda}$ where h is Planck's constant, c is the speed of light, and λ is the wavelength. As the energy of the emitted photon is lower than that of absorbed photon, that is, $E_2 < E_1$, the wavelength of the fluorescent emission λ_{emit} is longer than that of the absorbed radiation λ_{abs}. Table 11.2 shows some of the dye with their absorbing wavelength and emitting wavelength.

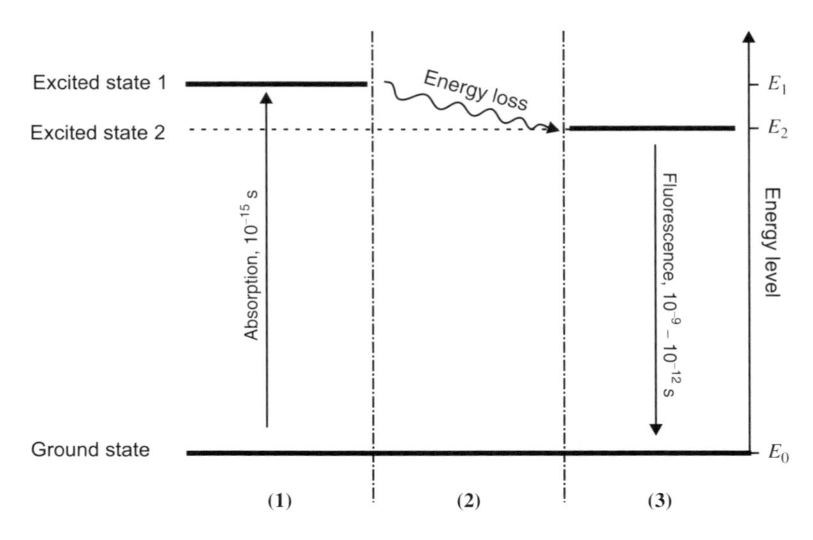

Figure 11.7 Principle of fluorescence: (1) a fluorochrome is excited by a photon to an energy level higher than the ground state (excited state 1); (2) the excited electron loses energy due to interactions with the environment, thereby falling to a relaxed excited state (excited state 2); and (3) the electron collapses back to its ground state emitting a photon with energy corresponding to the difference between excited state 2 and ground state

Table 11.2 Absorbing and emitting wavelength frequency of different fluorescent media

Dye	Absorbing wavelength (nm)	Emitting wavelength (nm)
Acridine orange (with DNA)	502	525
Acridine orange (with RNA)	460	650
Carboxyfluorescein	492	517
Cyanine	550	570
DAPI (4,6 diamidino-2-phenylindole)	358	461
Dylight fluor	493–770	518–794
Fluorescein	494	521
Hilyte fluor	497–753	525–778
Phycoerythrin	566	575±10
Quinine	355	450
Rhodamine 6G	530	555–585
Rhodamine 6B	520	570
SYBR Green I	498	522
Texas red	595–605	615

11.4.3 Seeding Particles

The seed particle size must be small enough to faithfully follow the flow without affecting the flow field and clogging the device. At the same time, particles must be large enough to be adequately imaged with proper signal strength and without any effects of Brownian motion. Making accurate μ-PIV measurements with spatial resolutions on the order of several microns requires that the diameter of flow-tracing particles, d_p, may be on the order of 100–300 nm diameter. For visible light with a wavelength of $\lambda = 532$ nm used for illuminating the particles, the particle diameter will be smaller than the wavelength of the light, that is, $d_p < \lambda$. In this regime, it is difficult to image particles using normal elastic scattering techniques. However, inelastic scattering techniques such as epifluorescence can be used to image submicron particles. Fluorescently labeled polystyrene particles with diameters of 100–300 nm and a specific gravity of $\rho_g = 1.055$ are suitable for many liquid flow-related microfluidic applications.

When submicron particles are used to trace slow flows, one must consider errors due to particle diffusion resulting from Brownian motion. A first-order estimate of this error relative to the displacement in the x-direction is given by Santiago *et al.* (1998):

$$\varepsilon_B = \frac{\langle s^2 \rangle^{1/2}}{\Delta x} = \frac{1}{u}\sqrt{\frac{2D}{\Delta t}} \tag{11.15}$$

where s^2 is the random mean square particle displacement associated with Brownian motion, D is the Brownian diffusion coefficient, u is the characteristic velocity, and Δt is the time interval between pulses. In practice, errors due to Brownian motion put a lower limit on the size of particle that can be used to achieve the desired velocity measurement accuracy.

Due to volume illumination during μ-PIV measurements, all the particles in the illuminated fluid volume emit light. Particles outside the focal plane of the imaging optics create a background noise and thus reduce the SNR of the particle images. Two prime parameters influence SNR, namely, particle concentration and test section depth. Higher SNR can be obtained by either reducing the channel depth or the particle concentration. The dimensions of channels are predefined, and hence, the channel depth cannot be altered. The particle concentration is the only adjustable parameter. It is apparent that particle concentration should be chosen with great care in order to obtain particle images of adequate quality for correlation analysis. Lowering the particle concentration is, however, not purely advantageous as fewer particles within the measurement plane require larger interrogation areas in order to obtain good correlation signal leading to poor spatial resolution. In order to avoid loss of spatial resolution, special interrogation or image processing techniques are required for μ-PIV analysis.

11.4.4 Particles Dynamics

It is required that there is proper matching between the particle density and fluid density for proper matching of the fluid velocity with the particle velocity. In case of mismatch between the two, there will be a gravitationally induced velocity V_g. Using Stokes law for drag and balancing the drag force with the gravity force in the low Reynolds number regime, one can drive the gravitational induced velocity as

$$V_g = \frac{d_p^2(\rho_p - \rho)g}{18\mu} \tag{11.16}$$

where d_p is the diameter of particles, ρ_p is the density of tracer particle, ρ is the velocity of the fluid, and μ is the viscosity.

In similar line, for an accelerating fluid with acceleration, a the velocity lag between the particle fluid can be written as

$$V_{\text{lag}} = V_p - V = \frac{d_p^2(\rho_p - \rho)a}{18g} \tag{11.17}$$

A particle with density, ρ_p, much greater than the fluid density, ρ, the velocity V_p corresponding to a steep increase in the velocity of fluid V will show an exponential response as

$$V_p(t) = V\left[1 - \exp\left(-\frac{t}{\tau_r}\right)\right] \tag{11.18}$$

where the relaxation time τ_r is given by

$$\tau_r = d_p^2 \frac{\rho_p}{18\mu} \tag{11.19}$$

For PIV measurements in liquid flow, it is not difficult to identify particles with matching densities. Therefore, the velocity lag (V_{lag}) and gravity-induced velocity (V_g) issues are not so severe. For a typical μ-PIV measurement in water using 300 nm diameter polystyrene latex spheres, the particle response time, τ_r, can be calculated to be equal to 10^{-9} s. This response time is much smaller than the time scale requirement of any realistic flow field. For the case of slip flow in gases, it is possible to use the corrections of Stokes drag relation to quantity particle dynamics (see Beskok *et al.*, 1996).

11.4.5 Brownian Motion

Brownian motion is a random thermal motion of a particle inside a fluid medium. The collision between the fluid molecules and suspended microparticles are responsible for the Brownian motion. The Brownian motion consists of high frequencies and is not possible to be resolved easily. Average particle displacement after many velocity fluctuations is used as a measure of Brownian motion. The mean square diffusion distance, σ_x^2, is proportional to $D\Delta t$, where D is diffusion coefficient of the particle given by Einstein relation as

$$D = \frac{KT}{3\pi\mu d_p} \tag{11.20}$$

where K is Boltzmann's constant, T is the absolute temperature of the fluid, and μ is the dynamic viscosity of the fluid.

For a particle with velocity u in x-direction, the travel distance is $\Delta x = u\Delta t$ during the pulse interval of Δt. The relative error ε_x due to the Brownian motion during this time interval can be expressed as the ratio of the rms of Brownian displacement to the average motion, that is,

$$\varepsilon_x = \frac{\sqrt{2\sigma_x^2\Delta t}}{\Delta x} = \frac{\sqrt{2D\Delta t}}{u\Delta t} = \frac{1}{u}\sqrt{\frac{2D}{\Delta t}} \tag{11.21}$$

The above expression indicates that the error due to Brownian motion increases as the time of measurement decreases. Larger time intervals produce flow displacements proportional to Δt while the Brownian motion goes as $\Delta t^{1/2}$. The above expression also indicates that influence of Brownian motion reduces with faster flows. However, it should also be noted that for higher value of u, Δt generally decreases. The other point to be noted that longer Δt can decrease the accuracy of PIV measurements as the velocity value is calculated based on first-order approximation of velocity. The use of second-order technique, that is, central difference interrogation, can allow for longer Δt without increasing the error.

The other effect of Brownian motion is its influence on the depth of correlation. Depth of correlation defines the depth over which particles significantly contribute to the correlation function. Olsen and Adrian (2000) derived an expression of correlation depth as

$$Z_{\text{corr}} = \left[\frac{1-\sqrt{\varepsilon}}{\sqrt{\varepsilon}} \left(f^{\neq 2} d_p^2 + \frac{5.95(M+1)^2 \lambda^2 f^{\# 4}}{M^2} + 8\beta^2 D \Delta t f^{\neq 2} \right) \right]^{1/2} \tag{11.22}$$

The last term in the right-hand side of the above expression represents the contribution of Brownian motion Z_{corr}. The depth of correlation increases due to the presence of the Brownian motion.

11.4.6 Microscope Recording and Imaging

The microscope is an important component of μ-PIV. A microscope consists of an optical system mounted on a stable mechanical base allowing magnified observation of small objects (specimen). In most cases, microscope has some built-in light source for illumination and a rear aperture permitting external light source for fluorescence observation. The microscope objective is the main magnifying unit of the microscope, and its optical properties are of great importance for image quality. Dichromatic mirrors, filters, and so on, for fluorescence observation are mounted inside the housing in the light path. The oculars for visual observation and a camera mount are usually located on top of the microscope body. The illumination mode when the specimen is illuminated by means of built-in illumination source (lamp) is termed as transillumination. This mode requires that the specimen is transparent in order to be imaged. The light emerging from the lamp is collected by a set of lenses (a collector) and then concentrated on the specimen via another set of lenses (a condenser). When an external light source is applied, light is reflected via a dichromatic mirror and through the objective lens thereby illuminating the specimen. The objective then functions both as a condenser for the illuminating light and as an objective lens for magnification and imaging of the specimen. This mode of illumination is termed as epi-illumination. Epi-illumination and epifluorescence can be combined and is the mode of illumination commonly applied in μ-PIV.

11.4.7 Resolution and Depth of Field

The resolution of any optical system is limited by diffraction (Inoue and Spring, 1997), occurring when waves of light, upon passing through lens system, converge and interfere with each other. For μ-PIV, the spatial resolution is limited by the effective diameter of particle images when projected back into the flow field. For magnifications much larger than unity, the diameter of the diffraction-limited point spread function, in the image plane, is given by

$$d_s = 2.44M\frac{\lambda}{2\text{NA}} \tag{11.23}$$

where M is the total magnification of the microscope and NA is the numerical aperture of the lens (Born and Wolf, 1997). Considering an objective with $M = 60$ and NA = 1.4 and assuming the wavelength of the emitted light to be $\lambda = 600$ nm, the lateral resolution calculated from the above equation is equal to 15.7 μm in the image plane, and when projected back into the object plane, this corresponds to a limit of resolution of 261 nm. Another important parameter of a microscope objective is the depth of field, which is defined as (Inoue and Spring, 1997)

$$\delta_z = \frac{n\lambda}{(\text{NA})^2} + \frac{ne}{(\text{NA})\text{M}} \tag{11.24}$$

where δ_z is the depth of field, n is the refractive index of the imaging medium, λ is the wavelength of the imaged light in vacuum, and e is the smallest distance that can be resolved by the image detector placed in the microscope's image plane. In case of a CCD camera, e equals the pixel-to-pixel spacing on the CCD chip. As can be seen from the above equation, the focal depth or depth of field δ_z is the sum of focal depth due to diffraction (first term) and that due to geometrical effects (second term). The depth of field δ_z gives an estimate of the distance a microscope slide may be moved in vertical direction while still maintaining focus on an infinitely thin specimen. Overall, an increase in magnification leads to a decrease in the depth of fluid. With reference to Table 11.1, it can be seen that for 63× magnification objective for $\lambda = 600$ nm and $e = 6.3$ μm, the depth of field is equal to 1.362 μm. Similarly for 5× objective, the depth of field is 52.1 μm. Thus, δ_z is a measure of vertical resolution of a microscope. This resolution is, however, not equal to the vertical resolution, or measurement depth, of a μ-PIV system, as particles slightly out of focus can be sufficiently bright for contributing to the correlation function applied for PIV interrogation analysis.

11.4.8 Measurement Depth

The measurement depth of a μ-PIV system does, in general, differ from the microscope's depth of field, as discussed above. Hence, the expression for depth of field should not be directly applied to determine the measurement depth for μ-PIV systems. The measurement depth is defined as twice the distance from the center of the object plane to a location such that the imaged particle is sufficiently unfocused, and thus has low enough intensity, and it does not contribute significantly to the velocity measurement. This occurs when the intensity of the imaged particle drops below 10% of the maximum intensity of a focused particle. The total measurement depth of a μ-PIV system can be obtained by adding the effects of diffraction and those of geometrical optics. Further, the finite size of the particles must be taken into consideration during the calculation of the total measurement depth of the μ-PIV system.

11.4.9 Particle Visibility

Particle visibility is an important issue in μ-PIV measurements due to the use of volume illumination. Fluorescence technique is able to filter out the background light scattered from the test section surfaces. However, the background light fluoresced from unfocused tracer particles cannot be easily removed because it occurs at same wavelength. The influence of unfocused particles can be lowered by selected proper experimental parameters. A parameter termed as visibility, that is, ratio of the focussed particle intensity to the average background intensity, provides good guidance on selection of the experimental conditions. Olsen and Adrian (2000) have derived the expression for particle visibility. The visibility of an infocus particle can be derived as

$$V = \frac{4M^2\beta^2(S_0 - a)(s_0 - a + L)}{\pi CLS_0^2(M^2d_p^2 + 1.49(M+1)^2\lambda^2/\mathrm{NA}^2)} \tag{11.25}$$

where C is the particle concentration, L is the test section length, S_0 is the object distance, a is the measurement plane location from one edge of the measurement volume, β is the cutoff level to define the edge of the particle, M is the magnification, d_p is the particle diameter, λ is the wavelength, and NA is the numerical aperture.

According to this relation, visibility can be increased by decreasing particle concentration or by decreasing the test section length. For a fixed particle concentration, visibility increases with a decrease in particle diameter and an increase in NA of the recording lens. The visibility is higher for thinner test section allowing higher particle concentrations and thus adequate spatial resolution. Visibility depends weakly on magnification and object distance.

11.4.10 Data Interrogation in μ-PIV

In standard PIV, the particle images are subdivided into a fine mesh (interrogation area), where each individual region is analyzed by means of correlation schemes and thus a velocity vector field is calculated from a single image pair. However, in μ-PIV, because of low SNR, it often requires either (1) large interrogation window enclosing enough particles to obtain good correlation, thus seriously limiting the spatial resolution of the system, or (2) interrogation windows of size such that acceptable level of resolution is maintained, at the cost of number of particles in each individual window. The later approach does generally generate an erroneous vector field due to lack of correlation and high noise levels in the images. In order to overcome this limitation, special interrogation techniques or image processing prior to interrogation is commonly applied in μ-PIV. Some of these schemes are described later.

11.4.10.1 Image Overlapping

Image overlapping is an image processing method without any special interrogation schemes, thus permitting the use of standard PIV interrogation software. In this method, a series of images or image pairs, sampled at statistically independent times, are added (or overlapped) to form a single, well-seeded particle image (see Figure 11.8). The maximum numbers of recordings that may be overlapped depend on the number of particles in the individual images. Adding too many recordings will lead to interference in terms of overlapping of particles from different images, without any further improvement to the resolution. In the upper limit, excessive overlapping will result in an image containing no distinguishable individual particles. This procedure can only be adopted for laminar flow field, where no difference in the velocity vector is expected between the first recording and the second recording.

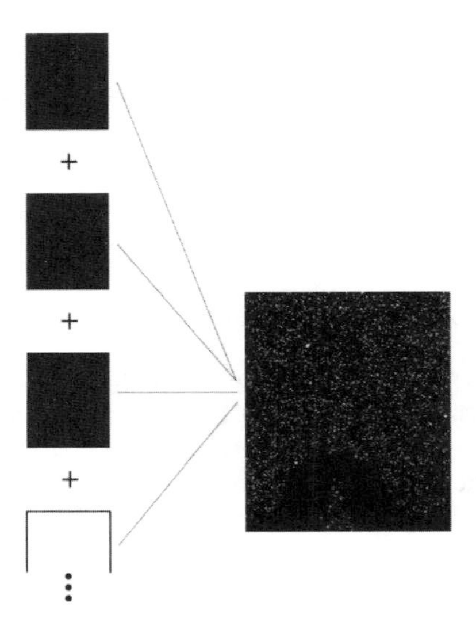

Figure 11.8 Image overlapping. Left: individual μ-PIV recordings, right: result of overlapping of 16 images (background noise subtracted)

11.4.10.2 Average Correlation

The cross-correlation function at a certain interrogation region is represented by

$$\Phi(x, y) = \sum_{j=1}^{q} \sum_{i=1}^{p} f(i,j).g(i + x, j + y) \tag{11.26}$$

where $f(i,j)$ and $g(i,j)$ are the gray level distributions of the first and second exposures, respectively, and $\Phi(x, y)$ is the value of correlation function for a displacement vector of $\delta_s = (x, y)$. For a single, well-seeded image pair, the correlation function Φ has a distinct maximum peak corresponding to the particle displacement of δ_s in the interrogation window. There may be a number of substantially lower subpeaks, resulting from noise and mismatch of particle images, which are randomly scattered in the correlation plane. In μ-PIV, however, when the number of particles contained within the interrogation window is insufficient, or the noise levels become too high, the peak representing the true particle displacement may become smaller than some of the subpeaks. This results in the generation of an erroneous velocity vector. For laminar, steady flows, the velocity field is independent of time and so the position of the correlation peak corresponding to true particle displacement does not change for PIV recordings sampled at different times. The random subpeaks, however, appear at varying positions and with different intensities in the different image pairs. By acquiring a large number of image pairs, calculating their individual correlation functions, and then ensemble averaging the correlation functions for corresponding interrogation windows in the individual PIV recordings, the true displacement peak remains in the same position, while the randomly scattered subpeaks average to zero. The average (or ensemble) correlation function may be written as

$$\Phi_{\text{avg}}(x, y) = \frac{1}{N} \sum_{k=1}^{N} \Phi_k(x, y) \tag{11.27}$$

where $\Phi_k(x, y)$ is the kth PIV recording pair, and N is the total number of recordings. In contrast to image overlapping, there is no limitation on the number of recordings used for the calculation of average correlation. For the same reason, average correlation is not restricted to low image density PIV recordings. Both image overlapping and average correlation are based on the assumption that the flow field under consideration is steady and laminar. However, transient phenomenons commonly encountered in flows such as two-phase flows, electrokinetic flows, mixing, and so on cannot use these techniques for flow field calculations. The measurement of instantaneous velocity field in such cases requires better hardware and setting and processing techniques.

11.5 Temperature Measurement

In this section, we discuss various techniques used for measurement of thermophysical properties at microscale.

11.5.1 3ω Technique

The 3ω technique utilizes a microfabricated metal line deposited on the specimen to act as a heater/thermometer (see Figure 11.9). The metal film acts simultaneously as a resistance heater and resistance thermometer detector (RTD) by passing an alternating current (AC) signal through it. An AC voltage signal is used to excite the heater at a frequency ω. This periodic heating generates oscillations in the electrical resistance of the metal line at a frequency of 2ω. In turn, this leads to a third harmonic (3ω) in the voltage signal. The frequency dependence of the oscillation amplitude and phase can be analyzed to obtain the thermal conductivity of the specimen.

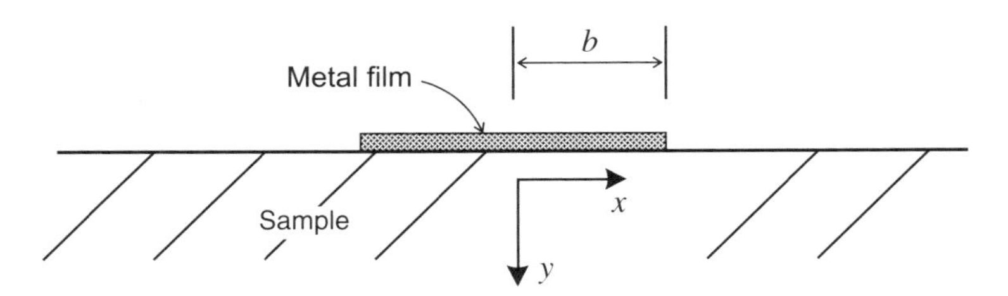

Figure 11.9 Sample geometry (metal film heater mounted on the sample) for the implementation of the 3ω method

For small temperature changes, the resistance of the filament varies with temperature as

$$R = R_0(1 + \beta\Delta T)$$

where β is the temperature coefficient of resistance and R_0 and R are the resistances at temperatures T_0 and $T_0 + \Delta T$, respectively. Since β for most metals is on the order of 10^{-3}K^{-1}, a Wheatstone bridge is used to measure the minute changes in resistance by passing a direct current (DC) signal through RTD. The steady-state temperature measurements of low thermal conductivity specimens are plagued with long equilibration times. Therefore, it is preferable to conduct a transient measurement using an AC excitation signal. The power dissipated by the heater/RTD due to Joule heating is defined as

$$P = I_h^2 R_h$$

where I_h and R_h are the heater current and resistance, respectively. The AC passing through the heater is given as

$$I_h(t) = I_{h,0}\cos\omega t$$

where $I_{h,0}$ is the peak amplitude of the nominal heater current at a frequency ω. Let us assume that the change in resistance is negligible compared to the amplitude of the current. Thus, the instantaneous power can be written as

$$P(t) = \frac{1}{2} I_{h,0}^2 R_{h,0}(1 + \cos(2\omega t))$$

Here, $R_{h,0}$ is the nominal heater resistance. The above equation for instantaneous power can be written as combination of two parts, a constant component independent of time (P_{DC}) and an oscillating component (P_{AC}):

$$P_{\text{DC}} = \frac{1}{2} I_{h,0}^2 R_{h,0}$$

$$P_{\text{AC}} = \frac{1}{2} I_{h,0}^2 R_{h,0}\cos(2\omega t)$$

The average power dissipated by heater can be written as

$$\bar{P} = \frac{1}{\tau}\int_0^{\tau} P(t)dt = \frac{1}{2} I_{h,0}^2 R_{h,0}\left(\frac{\omega}{2\pi}\int_0^{2\pi/\omega}(1 + \cos(2\omega t))dt\right) = \frac{1}{2} I_{h,0}^2 R_{h,0} = P_{\text{DC}}$$

The average power dissipated by the heater is also called the rms power, which is half of the power dissipated by a DC current of the same amplitude. It may be noted that the oscillating component of the instantaneous power does not dissipate any average power over one cycle. This observation is important when discussing the frequency response of the 3ω voltage.

The rms power can also be defined as

$$P_{\text{rms}} = I_{h,\text{rms}}^2 R_{h,0}$$

where the rms current $I_{h,\text{rms}}$ is defined as

$$I_{h,\text{rms}} = \sqrt{\frac{1}{\tau}\int_0^{\tau} I_h^2(t)dt} = I_{h,0}\sqrt{\frac{\omega}{2\pi}\int_0^{2\pi/\omega}\cos^2(\omega t)dt} = \frac{I_{h,0}}{\sqrt{2}}$$

Let us assume that the heater circuit is stable and all transients decay over time. The temperature oscillation of the metal filament produces the harmonic variation of heater resistance, which can be written as

$$R_h(t) = R_{h,0}(1 + \beta_h \Delta T_{DC} + \beta_h |\Delta T_{AC}| \cos(2\omega t + \phi))$$

$$V_h(t) = I_{h,0} R_{h,0} \begin{bmatrix} (1 + \beta_h \Delta T_{DC}) \cos(\omega t) + \frac{1}{2}\beta_h |\Delta T_{AC}| \cos(\omega t + \phi) \\ + \frac{1}{2}\beta_h |\Delta T_{AC}| \cos(3\omega t + \phi) \end{bmatrix}$$

The voltage component at 3ω in the above equation results from the multiplication of the oscillating current with the periodic portion of the heater resistance at 2ω. One can infer the in-phase and out-of-phase components of the temperature oscillations by measuring the voltage signal at the 3ω frequency. We can write the 3ω voltage as

$$V_{h,3\omega} = \frac{1}{2} I_{h,0} R_{h,0} \beta_h \Delta T_{AC} = V_{h,3\omega,x} + V_{h,3\omega,y}$$

We can also write the temperature fluctuation as

$$\Delta T_{AC} = \Delta T_{AC,x} + i\Delta T_{AC,y}$$

where

$$\Delta T_{AC,x} = |\Delta T_{AC}| \cos(\phi)$$
$$\Delta T_{AC,y} = |\Delta T_{AC}| \sin(\phi)$$

The magnitude and phase of the temperature oscillation vary with the excitation frequency due to the finite thermal diffusion time of the specimen. The thermal diffusion time is written as

$$\tau_D = \frac{L^2}{\alpha}$$

where L is the specimen size and α is the thermal diffusivity. In the limit of infinite thermal diffusivity (i.e., infinite thermal conductivity), heat propagates with infinite velocity such that the temperature is constant throughout the specimen. This leads to undamped temperature oscillations with zero phase lag. Conversely, zero thermal diffusivity (i.e., infinite volumetric heat capacity) results in no heat propagation, zero oscillation amplitude, and large phase lag. It may be noted that the dissipation of the rms power is independent of the frequency response of the specimen since it is constant with time. Next, the heat equation can be solved to determine the values of ΔT_{DC}, ΔT_{AC}, and φ for a given input power in order to construct a mathematical model for a finite width line heater deposited on the surface of the specimen. Cahill (1990) showed the temperature oscillation for the heater arrangement (Figure 11.9) in the limit $|qr| \ll 1$ as

$$\Delta T = \frac{P}{\pi l k}\left(\frac{1}{2}\ln\frac{\alpha}{r^2} + \ln 2 - 0.5772 - \frac{1}{2}\ln(2\omega) - \frac{i\pi}{4}\right)$$

where P/l is the amplitude of power per unit length, α is the thermal diffusivity, $r = x^2 + y^2$, and k is the thermal conductivity. The magnitude of complex quantity $1/q$ is defined as

$$\frac{1}{q} = \left(\frac{\alpha}{i2\omega}\right)^{1/2}$$

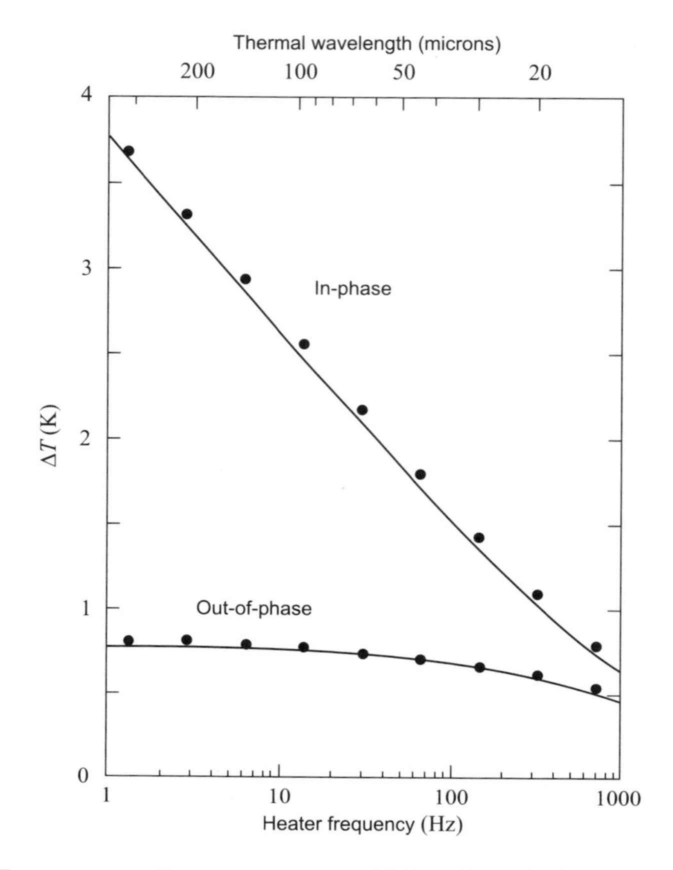

Figure 11.10 Temperature oscillation measurement of SiO_2 at $T = 300$ K measured by the 3ω method

The above quantity is called as thermal penetration depth or the wavelength of the diffusive thermal wave. Either the real or the imaginary part of the temperature oscillations ΔT can be used to determine the thermal conductivity. The imaginary part (out-of-phase oscillations) gives the thermal conductivity directly. The slope of the real part (in-phase oscillations) versus ln ω also gives the thermal conductivity.

Figure 11.10 shows the measured value of in-phase and out-of-phase temperature oscillation of SiO_2 at $T = 300$ K using 3ω method. The measured thermal conductivity from the slope of the curve is equal to 1.35×10^{-2} W/cm-K. The thermal conductivity can be obtained directly without any fitting parameters from either the slope of the in-phase magnitude or the magnitude of the out-of-phase temperature oscillations.

11.5.1.1 Implementation of 3ω Method

Figure 11.11 shows the schematic of a typical 3ω measurement apparatus. The function generator generates a low-distortion AC signal, which is fed through the microfabricated metal

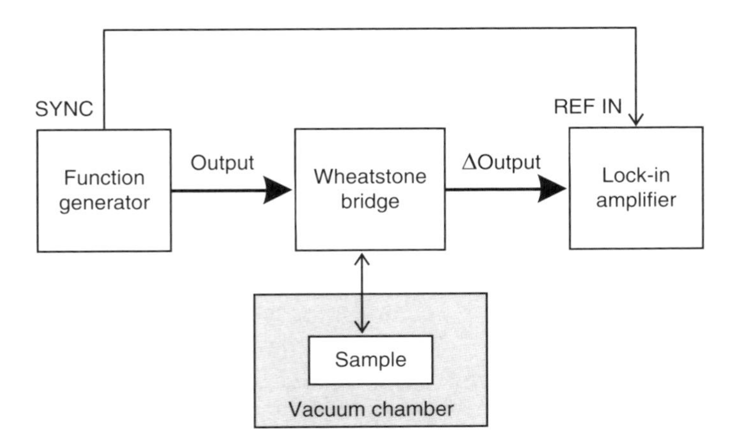

Figure 11.11 Schematic of a typical 3ω measurement apparatus

film mounted on the specimen surface. The heater is connected to Wheatstone bridge for detection of weak 3ω voltage. The lock-in amplifier is used to measure the output from the bridge. The primary objective is to accurately measure the in-phase and out-of-phase components of the 3ω voltage. Figure 11.12 shows the common-mode cancellation approach of the Wheatstone bridge. The third harmonic of the Wheatstone bridge output ($W_{3\omega}$) is related to the 3ω voltage ($V_{h,3\omega}$) as

$$W_{3\omega} = \frac{R_1}{R_{h,0} + R_1} V_{h,3\omega}$$

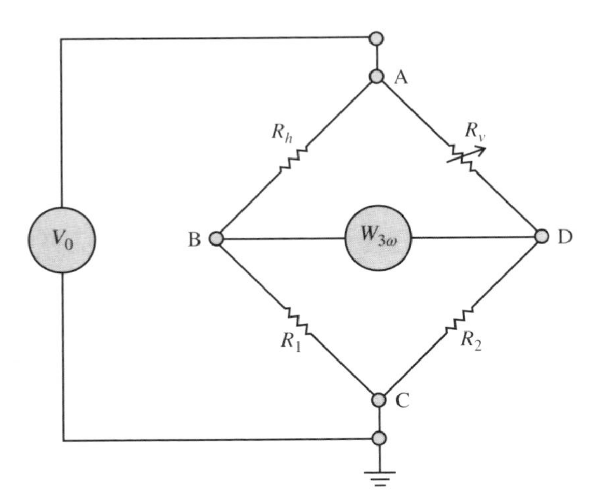

Figure 11.12 Schematic diagram of the Wheatstone bridge common-mode cancellation method. R_h is the heater, R_v is the variable resistor, V_0 is the fundamental signal from the function generation, and $W_{3\omega}$ is the differential output voltage sent to the lock-in amplifier

where $R_{h,0}$ is the heater at room temperature and R_1 is the in-series resistor as shown in Wheatstone bridge circuit. The 3ω voltage signal is measured using a lock-in amplifier, which can accurately measure the amplitude and phase of very small voltage or current signals.

The most important aspect of any 3ω measurement apparatus is a precise measurement of the minute third harmonic signal across the specimen heater. One must also ensure that the output from the heater itself is an accurate indicator of the temperature fluctuations in the specimen. There should be negligible thermal resistance between the heater and the specimen as per assumption used in derivation of the analytical model. This assumption requires fabrication of heaters having intimate thermal contact with the specimen surface.

11.5.2 Scanning Thermal Microscope Based on AFM

Atomic force microscope is traditionally used for obtaining surface topography by controlling the contact force between tip and sample. In AFM, a very fine tip is brought close to the surface. The tip is located at the end of a cantilever with low stiffness. The resultant force exerted between the tip and surface causes a deflection of the cantilever, which can be measured with an optical deflectometer (laser and photodiodes). The image of the forces (essentially topographic) is obtained by scanning the tip above the surface. The AFM technique can be modified for thermal measurement, depending on the operating principle of the probe. Figure 11.13 shows the schematic of a scanning thermal microscope based on AFM. The cantilever used in this technique is flexible, which limits damage to the probe tip during contact. There is a servomechanism controlling the distance or force between the sample and the probe.

11.5.2.1 Heat Probe

Various types of heat probe are used in the AFM-based thermal microscope: (a) thermocouple tip and (b) thermoresistive tip. The thermoelectric voltage is directly proportional to the temperature when using a thermocouple tip. A sample thermocouple tip is shown in

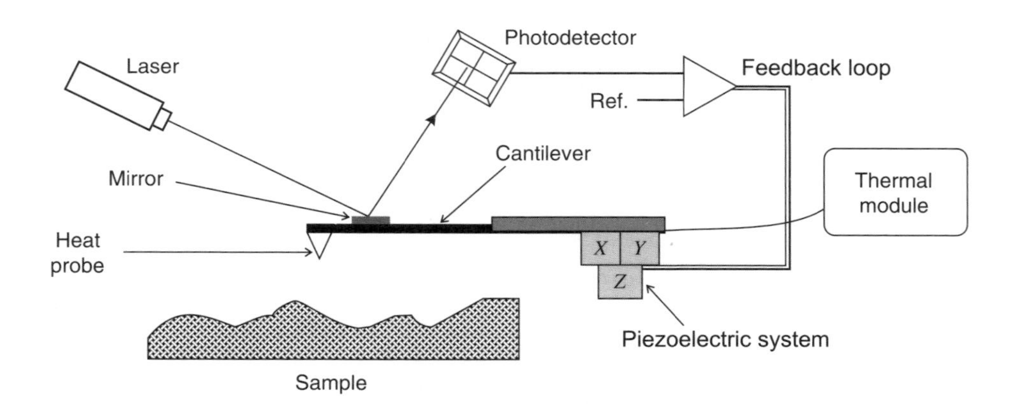

Figure 11.13 Schematic of scanning thermal microscope based on AFM

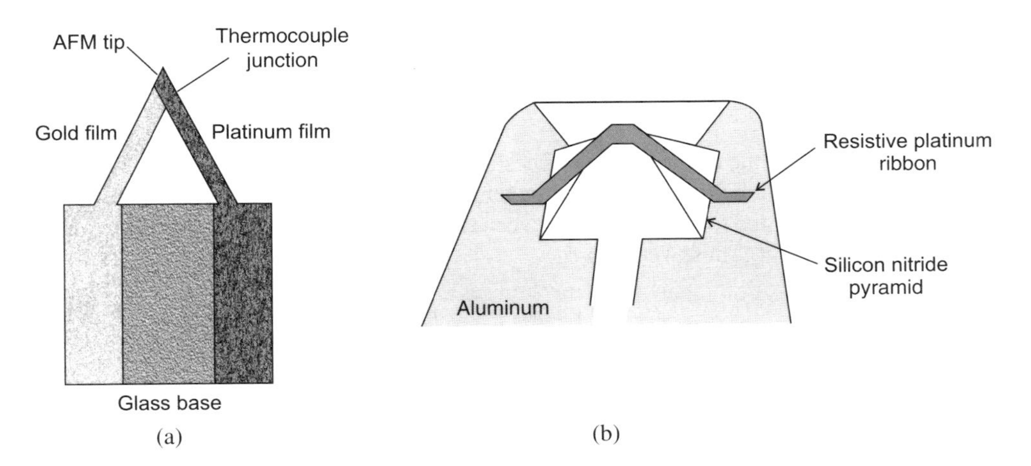

Figure 11.14 Schematic of a (a) thermocouple tip and (b) passive thermoresistive probe used with a AFM tip for scanning thermal microscopy

Figure 11.14(a). Here, gold and platinum film is deposited on different arms of a V-shaped cantilever. Temperature sensitivity of about 0.1 K can be attained using these types of thermocouple probe. Figure 11.14(b) shows the schematic of a thermoresistive probe. Here, aluminum film is first coated on the cantilever to form two electrodes. Subsequently, the aluminum film is joined by resistive platinum ribbon. The temperature sensitivity of this probe is about 1°.

11.5.2.2 Measurement with the Thermoresistive Probe

Figure 11.15 shows the schematic of the electric circuit used for temperature measurement using thermoresistive probe. The thermal probe constitutes one leg of the Wheatstone bridge.

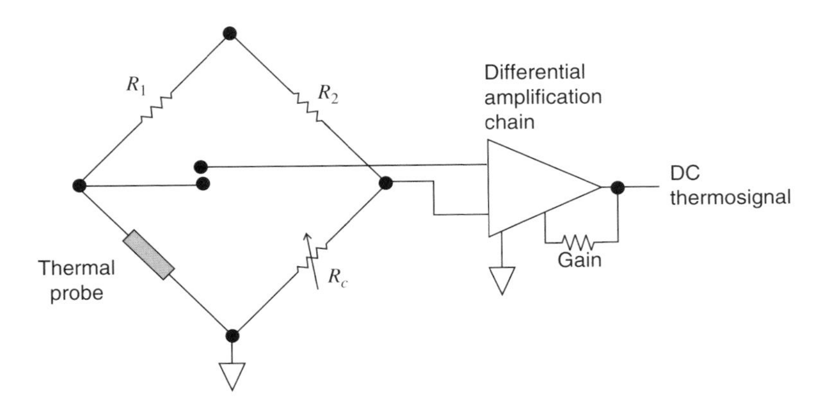

Figure 11.15 Electrical circuit for measurement by thermoresistive probe

The variable resistance, R_C, of the bridge at equilibrium can be written as

$$R_C = \frac{R_2}{R_1}(R_P + R_{\text{wire}})$$

where R_P and R_{wire} are the electrical resistances of the tip and the wires connecting the probe to the circuit of the thermal unit. The average temperature of the probe can be set using the variable resistance R_C due to the linearity between electrical resistance and temperature of the thermoresistive element. Thus, we can write

$$R_C = \frac{R_2}{R_1}\{R_{P0}[1 + \alpha_P(T_{\text{op}} - T_{P0})] + R_{\text{wire}}\}$$

Here, R_{P0} is the electrical resistance of the tip at reference temperature T_{P0} and α_P is the temperature coefficient of electrical resistivity.

The thermal probe can be used in either passive mode or active mode as described later.

Passive Mode

The thermal probe is used as a resistive thermometer in passive mode. A small constant current (1–2 mA) is applied such that Joule heating of the tip can be assumed negligible. The temperature change at the sample surface can cause a temperature variation (ΔT_P) in the tip during scan. This temperature variation can cause a change in electrical resistance of the probe given by

$$\Delta R_P = R_{P0}\alpha_P\Delta T_P$$

The change in electrical resistance leads to change in the voltage output from the circuit. The variation of this voltage is an indication of local heating at the sample surface.

Active Mode

A large current passes through the resistive element of the probe in active mode. The tip is heated by Joule effect and used as a source to excite the sample. There are two possible active measurement modes: (a) constant current mode and (b) constant temperature mode.

The current passing through the probe is held constant in *constant current mode*, and the voltage output from the amplification unit is recorded. The evolution of the sensor temperature due to variation in heat flux is monitored from the voltage signal.

The Wheatstone bridge is equipped with a feedback loop for *constant temperature measurement*. The heat loss from the heated tip to its surroundings (including sample) is compensated electrically by servo controlling the equilibrium voltage of the bridge.

11.5.2.3 Thermal Signal Analysis

The analysis of thermal image obtained from scanning thermal microscopy is not always trivial. The thermal signal is a function of the electrical power dissipated (P) in the tip, which

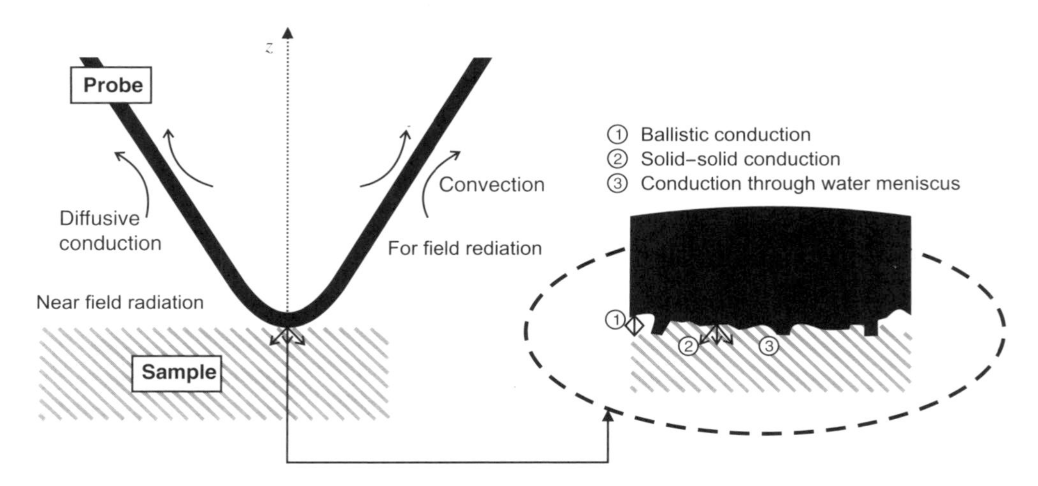

Figure 11.16 Different mode of heat transfer between the tip and the sample of scanning thermal microscope

depends on the forms of heat transfer to and from the tip. At equilibrium during DC operation of the scanning thermal microscope, we have

$$P = Q_W + Q_{\text{conv}} + Q_S$$

Here, Q_W is the heat rate transferred to the Wollaston support of the filament, which acts as a heat sink for the tip. The flux transferred to it is very large, and it causes a temperature gradient along the thermoresistive filament. The Q_{conv} is the heat rate exchanged by convection with the surrounding fluid (air). Its contribution is about 10% of the total power, P. Q_S is the heat rate exchanged with the sample while the sensor is displaced. Its variations lead to the contrast of the thermal image. Variations in the power dissipated in the tip, and hence in the thermal signal, will be larger when the variations in this flux are greater.

Different modes of heat transfer between tip and sample have been shown in Figure 11.16. The heat transfer can be classified to two types: (a) short-range heat transfer between two solids and (b) contact heat transfer between two solids.

The short-range heat transfer has two parts: (a) radiative transfer and (b) thermal conduction through air. Radiation mode of heat transfer depends on the temperature and size of the hot spot on the sample. The tip for which the distance from the sample is less than few microns is exposed to near-field radiative heat transfer. There are three transfer regimes via air depending on the distance between the relevant region of the tip and the sample surface. Diffusive transfer acts on areas of the tip situated at distances greater than 100 times the mean free path of the air molecules. For distance below this and up to the mean free path of air, heat transfer occurs in the slip regime. The region of the tip situated at a distance less than mean free path from the sample undergoes ballistic exchange (Knudsen regime) with the surface.

Contact heat transfer between two solids can be by two mechanisms: (a) heat transfer via water film and (b) direct solid–solid thermal conduction. The probe temperature can affect the thickness of the water meniscus formed between the tip and sample when contact is made in ambient air with nonzero relative humidity. The heat transfer through this water film falls off as the tip is heated. The thickness of the adsorbed water layer depends on the hydrophilic properties of the sample. It is recommended when studying a given sample in constant temperature mode to determine the tip temperature at which the water contribution becomes negligible. Therefore, it is also recommended to work at a higher tip temperature. Transfer by direct solid–solid conduction is revealed for good heat conductors by a sudden jump in the heat flux exchanged between the thermal probe and the sample when they come into contact. This heat flux also increases as the tip gradually penetrates into the material.

11.5.2.4 Precautions During Scanning Thermal Microscopy Measurements

Certain numbers of precautions are necessary in order to obtain good measurements. The following points should be taken into account:

1. The environment of the setup, for example, ambient temperature, relative humidity of the air, must be perfectly checked and controlled to guarantee stability.
2. The contact and displacement of the tip relative to the sample must occur in optimal conditions. Special attention must be given to suitable adjustment of the tip–sample force (which must be small enough to avoid damage to either tip or sample), the gain of the feedback system controlling the force, and the scan rate.
3. The response time of the system must be determined and optimized in order to optimize the speed of the sensor and the spatial interval for data acquisition so as to obtain the most accurate measurement possible.

One must estimate the experimental error and the influence of the environment and the sensor on the measurement while reporting the data.

11.5.3 Transient Thermoreflectance Technique

Determination of thermophysical properties is essential for thermal management of micro- and optoelectronics systems as device sizes are reducing day-by-day. Modulated thermoreflectance technique is one of the techniques for thermophysical property measurement of micro-/nanosystems. This technique examines the propagation of thermal waves generated by laser pulse.

A schematic showing laser pulse and sample interaction has been shown in Figure 11.17. Pulsed laser with short pulse duration acts as source of energy. A given volume on the sample surface heats up to a temperature level above ambient during each pulse due to the laser light energy absorbed by the sample. The heating area can be adjusted using the laser aperture and the optics of the system. The depth of the volumetric heating is determined by the optical penetration depth, which is a function of laser wavelength and surface material properties. The heating level inside the light penetration depth (δ_λ) obeys an exponential decay law. The influence of a pulsed laser irradiation on a given material depends both on the optical properties

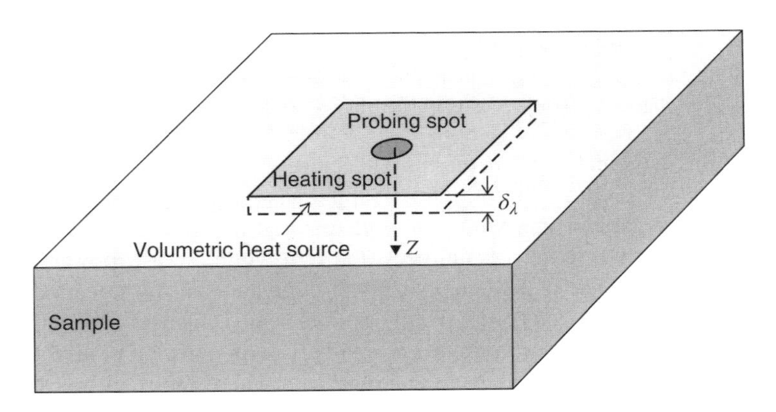

Figure 11.17 Schematic showing the heating and probing spot of a laser interacting on a sample

of that material as well as on the wavelength and pulse duration of the laser itself. Therefore, both the wavelength and pulse duration are important parameters in transient thermoreflectance technique.

After completion of each laser pulse, the sample begins to cool down to the initial ambient temperature. A probing CW laser light is reflected from the sample surface at the heating spot center during this process, which is collected by a detector to provide instantaneous reflectivity.

Multilayered sample of various sizes is usually used in various devices. There is no effect of size on the thermal conductivity for large sample. However, the thermal conductivity is dependent on the size of the smallest dimension for thin film sample (one of the size is in submicron range). Figure 11.18 shows the schematic of the interaction of the laser pulse with two multilayered samples. Here, h is the thickness of a layer and δH is the penetration depth of

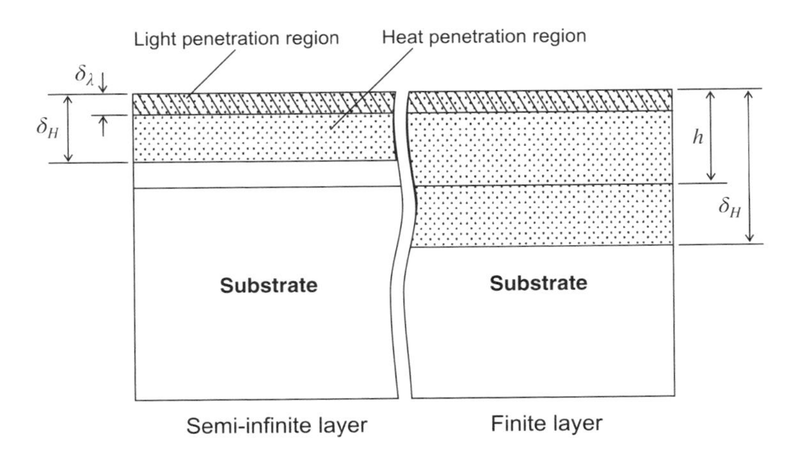

Figure 11.18 Schematic showing the interaction of laser pulse with layered material

the laser pulse energy into the sample. If the penetration depth during a particular cycle only involves partially a particular layer of the sample, then this layer can be considered *semi-infinite* layer. If the heat energy penetrates entirely through a layer, the layer should be classified as a finite layer.

The schematic of a typical transient thermoreflectance setup is shown in Figure 11.19 as proposed by Stevens *et al.* (2005). The 76 MHz Ti:Sapphire laser generates pulse with pulse width of 200 fs which are separated into two beams with an intensity ratio of 9:1 by a nonpolarizing beam splitter. The intense "pump" beam heats the film while the low-power "probe" beam monitors the reflectivity. The pump beam passes through an acousto-optic modulator, which creates a pulse train at a frequency of 1 MHz. The pump beam is focused at an incident angle of about 30°. The probe beam passes through a dovetail prism mounted on a variable delay stage, which increases the optical path length of the probe beam. This establishes time delay between the pump and probe pulses. A half-wave plate rotates the probe beam's polarization parallel to the plane of incidence. The silicon photodetector monitors the probe beam's reflection on the sample. A polarizer is placed ahead of the detector for only the probe light to pass and not the pump. A lock-in amplifier set at a frequency of 1 MHz monitors the photodiode response. The probe pulse is delayed in time (in picoseconds) for recording the transient temperature profile of the sample being cooled after the pump pulse. The reflectivity of the material is assumed to be a linear function of sample temperature. Thus, the change in temperature of the sample modulates the probe beam.

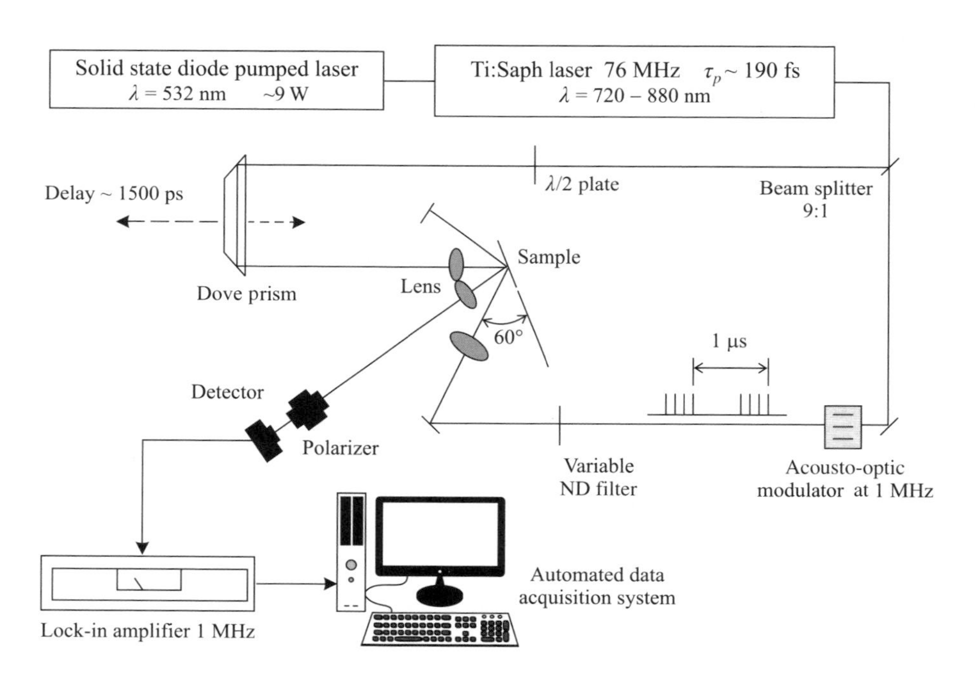

Figure 11.19 Schematic of a typical transient thermoreflectance setup

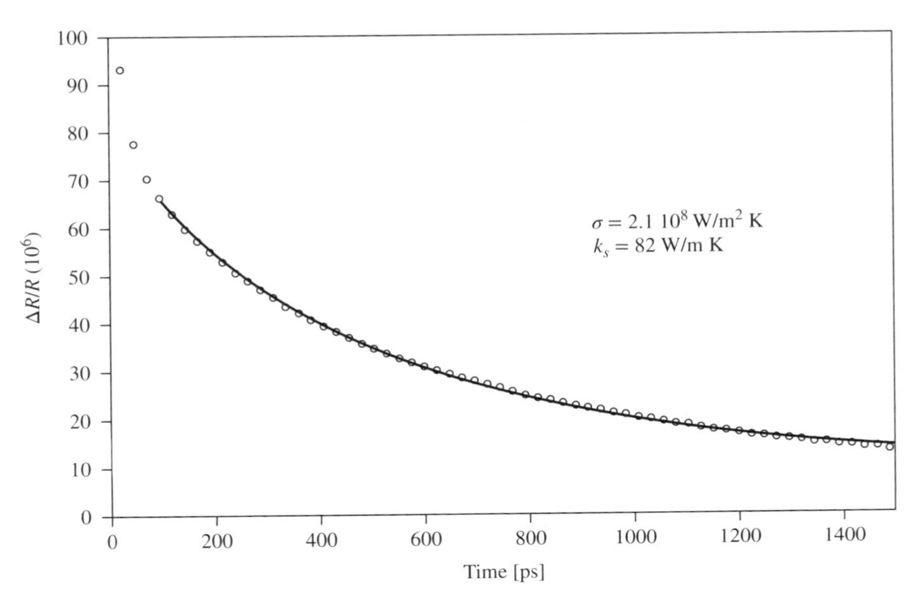

Figure 11.20 Experimental results (symbols) from thermoreflectance measurement of 30 nm Cr film on Si along with the model fit (solid line)

A sample thermal response from transient thermoreflectance measurement of 30 nm Al film on a sapphire substrate is shown in Figure 11.20. The experimental data can be compared with the best-fit equation from modeling the thermal transport in the sample. The unknown parameters, the thermal conductivity (k) of the sample and the thermal boundary conductance (α), can be obtained from the best-fit parameters.

11.5.3.1 Mathematical Model

Different mathematical models can be adopted to compare the model behavior with the experimental data obtained from the thermoreflectance technique. The thermoreflectance technique is applicable for a wide range of time scales from femtosecond to microsecond. In femtosecond regime, the heat absorption process involves two stages. Here, the heat absorption process involves two stages: photon–electron interactions in the first hundreds of femtoseconds and electron–phonon interaction subsequently. Therefore, Fourier conduction mechanism cannot be assumed to be valid in this range. Here, we assume the range where the Fourier conduction is applicable. The range of heating pulse widths is of the order of tens of picoseconds and above.

Let us consider a thin film of size d mounted on a substrate. The temperature above ambient is denoted as θ, C is the thermal capacitance, and k is the thermal conductivity. We assume the heat transfer within the metal film and the substrate heated by short-pulsed laser in time scale of about 100 ps is governed by the time-dependent heat conduction diffusion equation.

The governing equations are

$$C_f \frac{\partial \theta_f}{\partial t} = k_f \frac{\partial^2 \theta_f}{\partial x^2}$$

$$C_s \frac{\partial \theta_s}{\partial t} = k_s \frac{\partial^2 \theta_s}{\partial x^2}$$

Here, subscript f is for film and subscript s is for the substrate. The initial condition for the film and the substrate, respectively, are

$$\theta_f(x) = \frac{F(1-R)}{C_f \delta} e^{-x/\delta}$$

$$\theta_s(x) = 0$$

Here, F is the fluence, R is the reflectance, and δ is the energy deposition depth. The boundary condition at the interface ($x = d$) can be written as

$$-k_f \frac{\partial \theta_f}{\partial x}(x = d) = \sigma(\theta_f - \theta_s)$$

$$-k_s \frac{\partial \theta_s}{\partial x}(x = d) = \sigma(\theta_f - \theta_s)$$

Here, σ is the thermal boundary conductance. The Crank–Nicholson method can be used to numerically obtain the temperature response of the sample. The temperature response from the model corresponding to σ and k value which best fits the experimental data is considered as the appropriate thermophysical property of the sample.

The time constant of the film can be approximated using

$$\tau \approx \frac{d^2}{\alpha}$$

where α is the thermal diffusivity of the metal film. The time constant of the interface is given by

$$\tau_i = \frac{C_f d}{\sigma}$$

The time constant of the film should be significantly smaller than the time constant of the interface to resolve the thermal boundary conductance.

11.5.4 *Microlaser-Induced Fluorescence Thermometry*

Laser-induced fluorescence is traditionally used in analytical chemistry to identify and measure the concentration of various chemical species. The fluorescence intensity (I_f) in LIF is a linear function of concentration (C), which can be expressed mathematically as

$$I_f = \beta_c i_0 b \varepsilon \varphi C$$

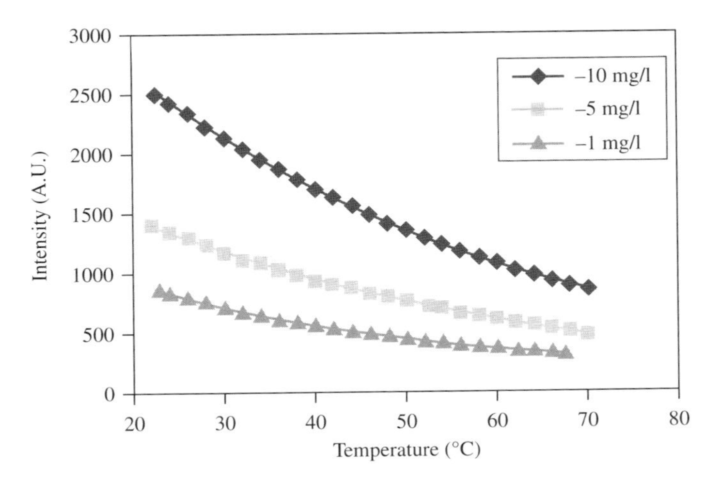

Figure 11.21 Fluorescence intensity versus temperature of Rhodamine B dye at different dye concentrations

where β_c, i_0, b, ε, and φ are the collection efficiency, incident irradiation, absorption path length, molar absorptivity, and quantum efficiency. If the fluorescence intensity is a function of temperature, the LIF technique can be used for temperature measurement. Figure 11.21 shows a typical fluorescence intensity plot as a function of temperature and concentration of Rhodamine B dye at 542 nm excitation wavelength (Chamarthy *et al.*, 2010). At a fixed value of dye concentration, using the knowledge of fluorescence intensity at a reference temperature, one can measure the temperature distribution from the measured fluorescence intensity at various locations. A μ-PIV apparatus with appropriate selection of temperature-sensitive dye and optical filter can be used for temperature measurement. The laser can excite the dye molecule inside the microchannel, and the fluorescence intensity is collected by the CCD camera through an optical filter. The optical filter should transmit the emission intensity from the dye and block the excitation intensity. Nonuniform illumination, fluctuation of light, and nonuniform dye concentration influence the accuracy of the measurement. Normalized intensity using a reference image of the temperature-sensitive dye can be used to account for the nonuniform laser illumination. The normalized intensity (I_N) can be expressed as

$$I_N = \frac{I_f}{I_0}$$

where I_0 is the LIF intensity at the reference temperature. The calibration curve of IN versus temperature can be used for measurement of local temperature. Various dyes, that is, Rose Bengal (temperature range 25–60 °C), Rhodamine 700 (temperature range 34–43 °C), Rhodamine 800 (temperature range 25–55 °C), Rhodamine 6G (temperature range 29–54 °C), Nile red (temperature range 33–46 °C), and Nile blue A (temperature range 23–71 °C) can be used for temperature measurement. The laser wavelength for these dyes can be 532 nm, and the cutoff wavelength for the optical filter is 560 nm.

References

Beskok A, Karniadakis GE, and Trimmer W 1996 Rarefaction and Compressibility. *J. Fluids Eng.*, **118**, pp. 448–456.

Boillot A and Prasad AK 1996 Optimization procedure for pulse separation in cross-correlation PIV. *Exp. Fluids*, **21**, pp. 87–93.

Born M and Wolf E 1997 *Principle of Optics: Electromagnetic Theory of Propagation Interference and Diffraction of Light*. Cambridge University Press.

Cahill DG 1990 Thermal conductivity measurement from 30 to 750 K: the 3ω method. *Rev. Sci. Instrum.*, **61**, pp. 802.

Chamarthy P, Garimella SV, and Wereley ST 2010 Measurement of the temperature on uniformity in a micro channel heat sink using micro scale laser induced fluorescence. *Int. J. Heat Mass Transfer*, **53**, pp. 3275–3283.

Inoue S and Spring KR 1997 *Video Microscopy: The Fundamentals*, 2nd Edition. Plenum Press, New York.

Keane RD and Adrian RJ 1990 Optimization of particle image velocimeters. *Meas. Sci. Technol.*, **2**, pp. 1202–1215.

Kim MJ, Kim HJ, and Kihm KD 2001 Micro-scale PIV for electroosmotic flow measurement. Proceedings of PSFVIP-3, March 18–21, 2001, Maui, Hawaii, USA.

Lima R, Wada S, Tsubota K, and Yamaguchi T 2006 Confocal micro PIV measurements of three dimensional profiles of cell suspension flow in a square microchannel. *Meas. Sci. Technol.*, **17**, pp. 797–808.

Liu D, Garimella SV, and Wereley ST 2005 Infrared micro-particle image velocimetry in silicon based microdevices. *Exp. Fluids*, **38**, pp. 385–392.

Meinhart CD, Wereley ST, and Santiago JG 1999 PIV measurements of a microchannel flow. *Exp. Fluids*, **27**, pp. 414–419.

Meinhart CD, Wereley ST, and Gray MHB 2000a Volume illumination for two-dimensional particle image velocimetry. *Meas. Sci. Technol.*, **11**, pp. 809–814.

Meinhart CD, Wereley ST, and Santiago JG 2000b A PIV algorithm for estimating time-averaged velocity fields. *J. Fluids Eng.*, **122**, pp. 285–289.

Nguyen NT, Wu ZG, Huang XY, and Wen CY 2005 The application of μ-PIV technique in the study of magnetic flows in a microchannel. *J. Magn. Magn. Mater.*, **289**, pp. 396–398.

Olsen MG and Adrian RJ 2000 Brownian motion and correlation in particle image velocimetry. *Opt. Laser Technol.*, **32**, pp. 621–627.

Santiago JG, Wereley ST, Meinhart CD, Beebe DJ, and Adrian RJ 1998 A particle image velocimetry system for microfluidics. *Exp. Fluids*, **25**, pp. 316–319.

Sharp KV and Adrian RJ 2004 Transition from laminar to turbulent flow in liquid filled microtubes. *Exp. Fluids*, **36**, pp. 741–747.

Shinohara K, Sugii Y, Aota A, Hibara A, Tokeshi M, Kitamori T, and Okamoto K 2004 High-speed micro-PIV measurement of transient flow in microfluidic devices. *Meas. Sci. Technol.*, **15**, pp. 1965–1970.

Stevens RJ, Smith AN, and Norris PM 2005 Measurement of thermal boundary conductance of a series of metal-dielectric interfaces by the transient thermoreflectance technique. *J. Heat Transfer*, **127**, pp. 315–322.

Wereley ST, Gui L, and Meinhart CD 2002 Advanced algorithms for microscale particle image velocimetry. *AIAA J.*, **40**, pp. 1047–1055.

Wereley ST and Meinhart CD 2001 Adaptive second-order accurate particle image velocimetry. *Exp. Fluids*, **31**, pp. 258–268.

Westerweel J 1997 Fundamentals of digital particle image velocimetry. *Meas. Sci. Technol.*, **8**, pp. 1374–1392.

Supplemental Reading

Adrian RJ 1997 Dynamic ranges of velocity and spatial resolution of particle image velocimetry. *Meas. Sci. Technol.*, **8**, pp. 1393–1398.

Cahill DG, Goodson K, and Majumdar A 2002 Thermometry and thermal transport in micro/nanoscale solid-state devices and structures. *J. Heat Transfer*, **124**, pp. 223–241.

Komarov PL and Raa PE 2004 Performance analysis of the transient thermo-reflectance method for measuring the thermal conductivity of single layer materials. *Int. J. Heat Mass Transfer*, **47**, pp. 3233–3244.

Saffmann PG 1965 The lift on a small sphere in a slow hear flow. *J. Fluid Mech.*, **22**, pp. 385–400.

12

Microscale Sensors and Actuators

12.1 Introduction

Flow control systems are critical components of most of the energy systems involving fluid flow and heat transfer. These systems are essential for performance optimization of both macroscale and microscale devices. Micropumps, microvalves, microshear stress sensors, and microflow sensors are integral components of flow control systems. Capillary micropump, MHD micropump, thermocapillary micropump, and electrokinetic micropump have been presented in earlier chapters. The present chapter reports various microactuators and shear stress sensors for flow control systems. More details on microvalves and microflow sensors can be found in other references (Nguyen and Wereley, 2006).

12.2 Flow Control

Pressure, velocity, and temperature information are used to characterize a flow field. A flow field is broadly classified as laminar or turbulent. In turbulent flows, these flow variables are random functions of space and time. The coherent and random flow structures/vortices and their mutual interactions influence the overall behavior of a turbulent flow field. The location of the wall surface, Reynolds number, Mach number, buoyancy force due to temperature, concentration gradients, and so on influence the turbulent flow field characteristics. The nature of flow field also depends on the convective or absolute instability mechanism leading to turbulent flow.

The flow control is a process by which the flow field is manipulated to obtain the required behavior in comparison to the natural uncontrolled case. The flow control is broadly classified as (1) *passive control*, where no auxiliary power source is required, and (2) *active control*, where there is expenditure of energy. In passive control, parameters like geometry, compliance, temperature, porosity, and so on are varied. Boundary oscillation, acoustic waves, blowing, suction, and so on are used for active control. The active control schemes use actuators for manipulating the flow behavior. The size of these actuators depends on the nature of the flow field. When the Reynolds number is increased, the required size of the actuator is reduced.

Transport Phenomena in Microfluidic Systems, First Edition. Pradipta Kumar Panigrahi.

Companion Website: www.wiley.com/go/panigrahi/microfluidic

The availability of MEMS fabrication technique has contributed toward small-scale actuator development and enhanced the capability of flow control systems.

12.2.1 Applications of Flow Control

The turbulent flow control finds application in various industrial applications: (1) drag reduction, (2) lift enhancement, (3) mixing and/or heat transfer enhancement, (4) flow-induced noise suppression, and so on. Reduction in drag of aircraft and underwater body leads to reduced fuel cost and overall size. Enhanced heat transfer is desired for heat exchangers and electronic packaging, while enhanced mixing improves the performance of combustors. The reduction in boundary layer noise due to pressure fluctuations helps in the operation of underwater sonar. To achieve these end results, the flow control techniques aim to control either the transition (delay or advance) or separation (suppression or enhancement) in a wall-bounded flow. For an example, when the boundary layer is turbulent on an aircraft wing, the resistance to separation is enhanced and more lift can be attained by operating at higher angle of attack. On the other hand, the skin friction drag for a laminar boundary layer is smaller than that of turbulent boundary layer, and therefore, the delay in transition leads to lower skin friction drag and flow-induced noise.

12.2.2 Flow Control Implementation Strategy

The turbulent flow control can be implemented through different control schemes (Gad-el-Hak, 2000). They are (1) predetermined open-loop control, (2) reactive feed-forward open-loop control, and (3) reactive feedback closed-loop control. The physical arrangement of sensors and actuators depends on the nature of control schemes and flow characteristics. The criteria for selection of appropriate actuators have been discussed in the next section. It should be noted that the actuators require parasitic power supply for its operation. This fact should be taken into consideration for overall performance analysis of actuators.

Turbulent flow control schemes assume the significant role played by coherent large-scale structures or vortices. These structures are responsible for more than 50% turbulence productions. The growth of large coherent structures leads to mixing enhancement. Similarly, the annihilation of large-scale structures to small-scale structures leads to lower drag. These coherent structures are generated or amplified due to the flow instability and get convected from the upstream to the downstream locations. The boundary conditions imposed by the actuators located in between the upstream and downstream locations modify the approaching turbulent structures. One possible turbulent flow control implementation scheme is shown in Figure 12.1. The physical system on which the flow control scheme is implemented can be subdivided into three subsystems: (1) flow sensors, which characterize the upstream coherent structures; (2) actuators, which modify the wall boundary conditions experienced by the approaching coherent structures; and (3) control sensors, which establish the coherent structures evolution between upstream and downstream locations and the effect of actuations.

The control sensors are used for verification of control objectives. When the objective is to reduce drag, the control signal is the shear stress. Similarly, for reduction of pressure fluctuations, the control sensors provide the rms value of pressure fluctuations. For implementation of the control scheme, the relationship/transfer function between the upstream sensors and

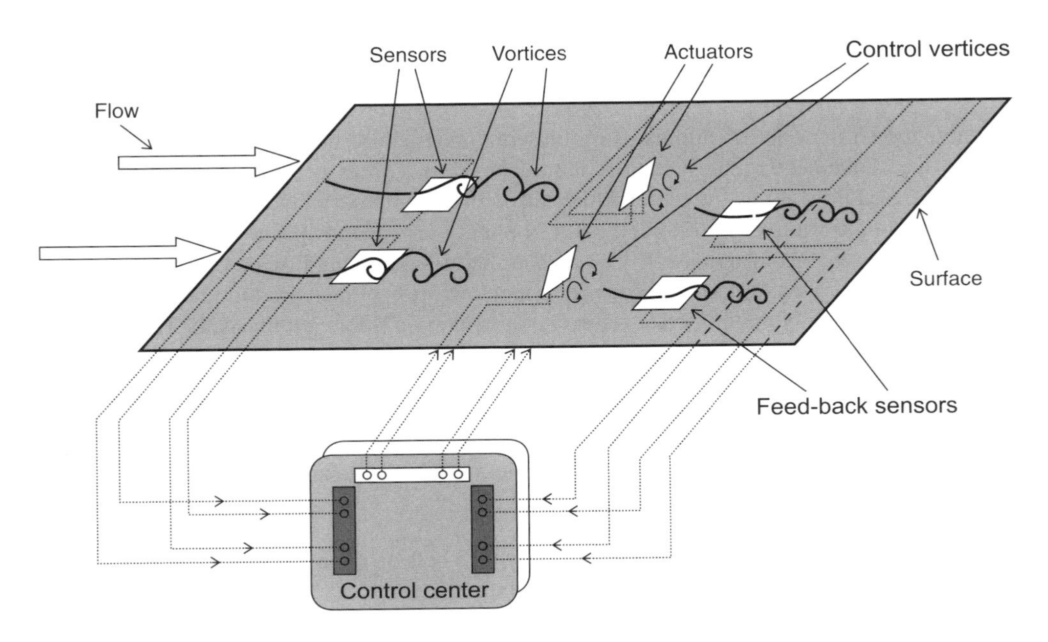

Figure 12.1 A schematic of flow control arrangement

control sensors is established. Similarly, the transfer function between the actuator properties and the control sensors is obtained. These two transfer functions are used by the control center to obtain the actuator state for optimal control.

12.2.3 Actuator Requirements for Flow Control

Due to chaotic nature, the turbulent flow is high dimensional in nature. Therefore, the implementation of turbulent flow control schemes for practical applications is challenging. However, the form and structures of turbulent flow can be diagnosed, and control of turbulent flow is realizable by suitable actuation. It is difficult to specify the requirement of an ideal actuator for turbulent flow control. The ideal characteristics of an actuator for turbulent flow control can be described based on the following guidelines:

1. Turbulent flow is quite sensitive to the surface roughness. The actuator should not introduce additional roughness to the surface of the device. It should not have parasitic effect on the flow when the actuator is switched off.
2. The frequency to be introduced by the actuators should match with the spatial and temporal frequencies present in the turbulent flow. The high Reynolds number flow has spatial and temporal frequency of the order of microns and kilohertz, respectively. Therefore, MEMS-based actuators are ideal for turbulent flow control due to their smaller physical dimension and lower inertia.
3. Strength is the other requirement of actuator for maintaining the integrity against the load from the fluid forces.

4. During implementation of distributed control with multiple actuators for optimal turbulent flow control, the behavior of the actuators should be uniform for all actuators.

The earlier flow control studies using blowing and suction have shown that the velocity due to the movement of the actuator should be of the order of friction velocity u_τ. For blowing-based flow control, the flow blows through the boundary layer when the jet velocity is higher and the control is not effective for boundary layer control. The other observation is that when the size of the actuator is higher than the turbulent length scale ($l = \nu/u_\tau$, where ν is the kinematic viscosity), the surface roughness effect becomes predominant. The size of the actuator is usually about 10 times the characteristic turbulent length scale. The operating frequency of the actuator is set based on two criteria:

1. The actuator frequency is matched with the instability frequency of the flow, and this approach is traditionally successful for separation control.
2. The actuator frequency is set higher than the unstable frequency of the flow, and the frequency content from the actuator is damped.

The actuator acts like a constant flux source, which acts on the main flow and is effective in turbulent boundary layer control for drag reduction.

12.3 Actuator Classification

The actuators can be broadly classified based on their principle of operation as (1) thermal actuators, (2) electrohydrodynamic actuators, (3) magnetohydrodynamic actuators, (4) momentum injection actuators, and (5) moving surface actuators.

Thermal actuators are surface-mounted electrical resistors, which inject heat into the near-wall flow. The heating of the working fluid influences the instability of the flow in different ways depending on whether the working field is gas or liquid. The injected heat reduces the fluid density when the working fluid is gas leading to generation of buoyant force. When the working fluid is liquid, there is a decrease in fluid viscosity due to increase in temperature influencing the generation of Tollmien–Schlichting waves. The drawback of these actuators is the low frequency response due to the parasitic heat loss to the solid on which it is mounted and high power consumption.

For an electrically conducting fluid (water or ionized gas), the electrical field can induce motions due to the movement of charged ions or the body forces can be generated due to the interactions between magnetic and electric fields. These options are used in *electrohydrodynamic* and *magnetohydrodynamic* actuators. The drawbacks of these actuators are the high-voltage requirement and significant loss due to the thermal effect.

The *momentum injection actuators* act like a source of momentum to the flow. One popular approach in this category is blowing. The other popular actuator in this category, which does not require any fluid supply, is the synthetic jet. The moving surface actuators provide motion of the wall beneath the flow. These actuators inject momentum to the flow by modification of the surface boundary condition. They introduce localized boundary layer disturbances. Flap actuator and balloon actuator belong to the moving surface actuators. The fabrication procedure of microsynthetic jet, microballoon, and microflap actuators is established in literature. The practical applications of these actuators have been demonstrated. The following sections discuss the details of these three actuators, that is, microsynthetic jet, microballoon, and microflap. Other special MEMS-based actuators are electrokinetic actuator, microbubble actuator, microvalve actuator, and micropolymer actuator.

12.3.1 Microsynthetic Jet Actuator

The schematic of synthetic jet actuator is shown in Figure 12.2. It consists of a cavity open to the flow through a slit or opening. The bottom wall of the cavity is driven by a moving surface, which can be either a piston or a vibrating membrane. The vibration of the membrane can be achieved using electrostatic actuation, piezoceramic material, and differential thermal expansion of the electrode material with silicon layer underneath. When the membrane or piston moves inward of the cavity, fluid is pushed out of the cavity, and when the membrane or piston moves back, the fluid is sucked into the cavity. The nature of flow through the slit opening depends on (1) the frequency of membrane oscillation, (2) the amplitude of the membrane displacement, (3) the size of the orifice, and (4) the geometry of the orifice. When the movement of the membrane or piston is small, the flow through the slit is reversible during ingestion and expulsion phase. Above an amplitude threshold, the flow through the slit generates acoustic or pressure wave similar to a loud speaker. With subsequent increase in the membrane or piston displacement above a threshold, the outflow separates at the exit corner and a vortex ring moving away from the orifice is formed (see Figure 12.2). The jet flow from the orifice draws fluid from the sides contrary to the uniform inward flow during the ingestion phase. Hence, the jet provides a net point source of momentum to the flow. The synthetic jet has zero net

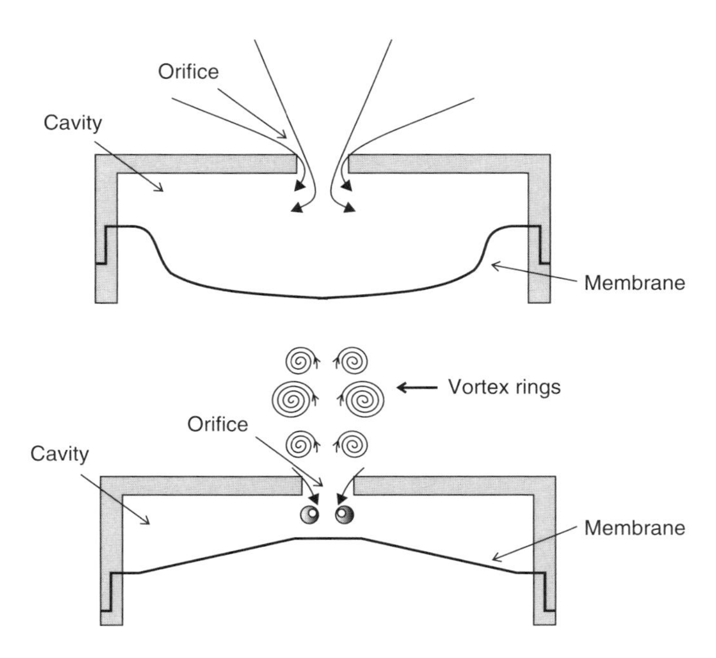

Figure 12.2 A schematic of synthetic jet actuator

mass flux contrary to a normal jet. It imparts a net momentum to the flow without adding any net mass to the system. There is no need of reservoir or piping system for the operation of a synthetic jet. Temporal frequency of the order of 4 kHz can be achieved for a microsynthetic jet. The spatial resolution of the microsynthetic jet depends on the orifice diameter which is of the order of 500 μm. However, the combination of the oscillating membrane size and the substrate increases the overall dimension of the microsynthetic jet, limiting the capability of distributed control using multiple actuators.

12.3.1.1 Fabrication

A possible procedure for synthetic jet actuator fabrication has been shown in Figure 12.3. The first step in the fabrication process is the wet etching of the silicon wafer using KOH to form a cavity. The cavity is filled with electroplated nickel in the second step. A polyimide membrane is deposited over the silicon wafer using spin coating in the third step. The back side of the silicon substrate is etched using KOH to obtain the orifice hole. The orifice hole extends till the filled nickel inside the cavity and the nickel material are also etched away. The piezoceramic material is deposited on top of the membrane for actuation. In the case of electrostatic-based actuation, aluminum electrode is deposited on the membrane. The disadvantage of the KOH-based wet etching is the limited possible control over the size and shape of the orifice. To have superior control on the dimensional and shape accuracy, dry reactive etching is preferable.

12.3.1.2 Characteristics

Different nondimensional numbers and their critical values characterize the operation of a microsynthetic jet. Reynolds number and Stokes number are used to describe the viscous and

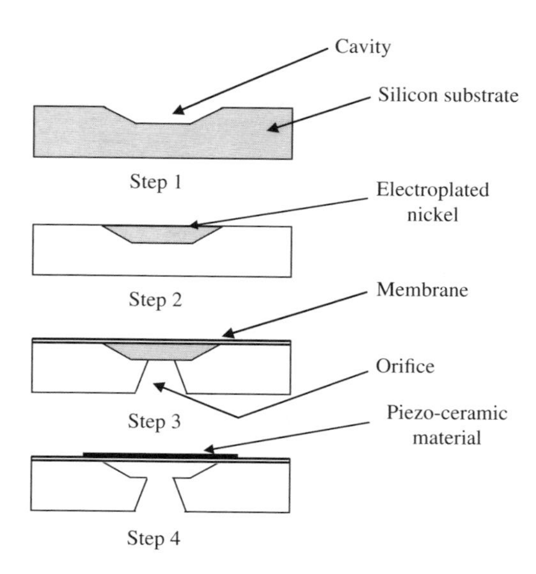

Figure 12.3 A possible fabrication procedure of synthetic jet actuator

unsteady effects, respectively. The Reynolds number (*Re*) is defined based on the maximum jet velocity, U, at the exit of the orifice and orifice diameter (d) as

$$Re = \frac{Ud}{\nu}$$

where υ is the kinematic viscosity of the fluid. When the jet Reynolds number is less than a critical value of about 50, the flow is reversible. For this case, the flow is identical during suction and blowing phase. When the Reynolds number exceeds the critical Reynolds number, the jet separates from the orifice edge. The Stokes number (*Stk*) is defined as

$$Stk = \sqrt{\frac{\omega d^2}{\nu}} = \sqrt{\frac{d^2}{\delta_\upsilon^2}}$$

where ω is the oscillation frequency and δ_υ is the thickness of the unsteady boundary layer. Stokes number is the ratio between the orifice diameter and the thickness of the boundary layer at the exit of the orifice. When *Stk* is large, the viscous effect is not significant. When *Stk* is small, the viscous effect dominates and the exit velocity from the orifice reduces. This phenomenon is known as *viscous choking*. The increase in orifice size minimizes the viscous effect; but there is a simultaneous reduction in the exit velocity due to large flow area. Hence, a compromise solution is essential for the design of the microsynthetic jet. The ratio between the square of the Stokes number and Reynolds number is known as Strouhal number (*St*):

$$St = \frac{\omega d}{U}$$

The high value of Strouhal number indicates that the number of membrane cycle is higher than the frequency of fluid element passage through the orifice. The low Strouhal number value indicates similar actuation cycle of the actuator and the fluid element. Low Strouhal number indicates higher strength of the jet with low viscous losses. The *Stroke ratio* which is defined as the ratio of the expelled stroke (L) to the diameter of the orifice (d) is also used to characterize the vortex ring formation of a synthetic jet. At higher stroke ratio (L/d), the vortex ring formed is disconnected from the trailing jet, and at lower stroke ratio, only single vortex ring is formed. The transition between these two cases is known as *formation number* and is usually within the range of 3.6–4.5.

The Stroke ratio $L/d = U/(\omega d)$ is equal to the inverse of the Strouhal number. However, the limitation of stroke ratio definition is the assumption of slug-velocity profile. Holman *et al.* (2005) discussed the formation criteria of synthetic jet based on both numerical simulation and experiment. They observed that vortex ring needs to withstand the suction velocity during ingestion stroke, which is a function of the vortex strength. From dimensional reasoning, they proposed the following criterion for jet formation:

$$\frac{Re}{Stk^2} = \frac{1}{St} > K$$

The above equation indicates that vortex escapes when the Strouhal number is less than a critical value. The constant K is a function of orifice geometry, the aspect ratio of the orifice, and the shape of the exit edge, that is, sharp or rounded. The constant (K) is approximately equal to 1.0 and 0.16 for two-dimensional (2-D) and axisymmetric synthetic jet, respectively.

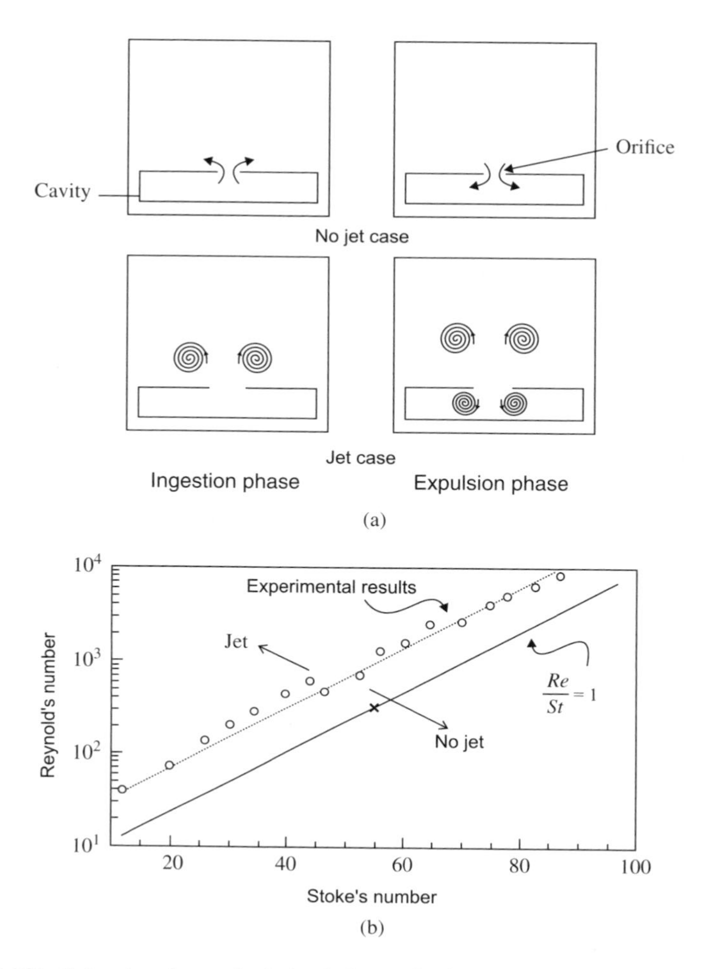

Figure 12.4 (a) Vorticity plots for synthetic jet during no-jet formation (top) and jet formation (bottom) cases. (b) Reynolds number ($U_o d/\lambda$) versus Stokes number ($\sqrt{\omega d^2/\lambda}$) of synthetic jet for no-jet and jet formation cases

Figure 12.4(a) shows the vorticity distribution during the ingestion and expulsion phase of a synthetic jet. The no-jet case corresponds to the K value less than 1.0, and the jet formation case corresponds to K value greater than 1.0 for a 2-D synthetic jet. For the jet formation case, the separated vortex ring is not dragged into the cavity during the ingestion stroke leading to a net momentum transport to the flow. Figure 12.4(b) compares the Reynolds number versus Stokes number relationship between the jet and no-jet formation case from both experimental and theoretical data (Holman *et al.*, 2005). The small discrepancy between the two is attributed to the orifice geometry or curvature effect.

Lee *et al.* (2003) presented the effectiveness of microsynthetic jet for modification of turbulent boundary layer under adverse gradient. They showed significant mixing enhancement of the boundary layer when the forcing frequency is closer to the natural instability frequency. The boundary layer modification is more sensitive to forcing frequency compared to the forcing amplitude. The near-wall mean velocity and turbulence intensity are significantly enhanced due to the operation of the synthetic jet. The synthetic jet needs to have sufficient velocity output and strong longitudinal vortices for effective flow control.

12.3.2 Microballoon Actuator

The schematic of a microballoon actuator is shown in Figure 12.5(a). This consists of silicone rubber with manifold underneath for pressurized air actuation. The actuation takes place by the deformed shape of the silicone rubber, which is a function of the actuator pressure. The compression and decompression are achieved by a solenoid valve with external compressed air supply. Each microballoon consists of about 120 μm thick silicon membrane over two/three holes on a silicon chip. Figure 12.5(b) shows the shape and deflection of a balloon actuator at two different pressure levels. The advantage of these actuators is that they can support very large force (>100 mN) and actuation length (>1 mm). A microballoon can be operated for about 11,000 cycles of inflation and deflation and is robust against harsh environment. The disadvantage of microballoon actuators is the low step response. The other disadvantage is the plastic deformation of the membrane due to stretching at high load.

12.3.2.1 Fabrication

A possible fabrication procedure of the microballoon actuator has been shown in Figure 12.6. In the first step of the fabrication, silicon nitride is deposited using low-pressure chemical

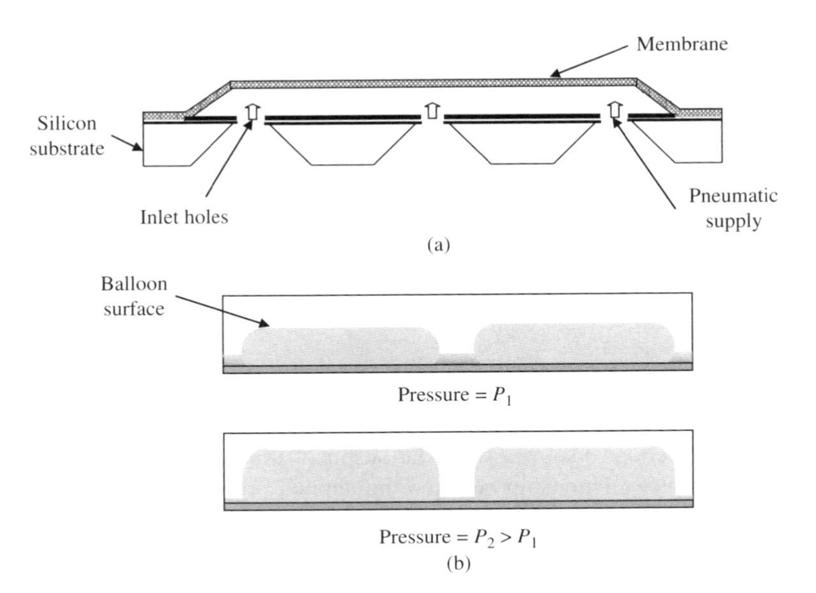

Figure 12.5 (a) Schematic of a balloon actuator and (b) the picture of a deformed balloon actuator at two different pressure levels

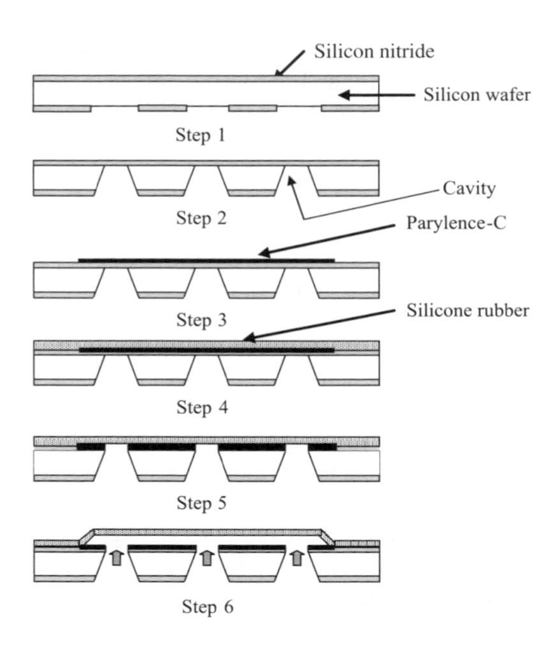

Figure 12.6 Fabrication procedure of microballoon actuator

vapor deposition (LPCVD) technique at both sides of the silicon wafer. The back side of the wafer is patterned using reactive ion etching. In the second step, the cavities are etched using KOH-based wet etching procedure. In the third step, parylene C is deposited on the front side and patterned using oxygen plasma. Parylene C is used here to act as an intermediate layer between the silicone rubber and the sacrificial materials during the fabrication process. This is required as the sacrificial materials as oxide of silicon, photoresist, and etchants are incompatible with silicone membrane. Before removing the photoresist, the front side of the wafer is roughened using SF_6/O_2 plasma. The photoresist is removed, and silicone rubber is spin coated on the front side of the wafer in the fourth step. In the fifth step, silcone nitride and parylene C are removed from the back side of the wafer using SF_6/O_2 plasma, followed by O_2 plasma etching. In the last step, pressurized compressed air is sent through the cavity to release the membrane.

12.3.2.2 Characteristics

The deflection of a microballoon actuator depends on the actuation air pressure. Figure 12.7 shows the typical deflection versus pressure relationship of a silicone-based microballoon actuator. It shows the direct relationship between the applied pressure and the deflection. The plastic deformation of the balloon and the effect due to cyclic variation are used to decide the operating range of the microballoon. Tung *et al.* (2004) investigated the compressible flow separation control of a three-dimensional (3-D) wing section using microballoon actuators. A glass plenum was used for supply of compressed air to the balloons. An initial pressure was applied for overcoming the stiction of the diaphragm to the substrate. The balloon deflection

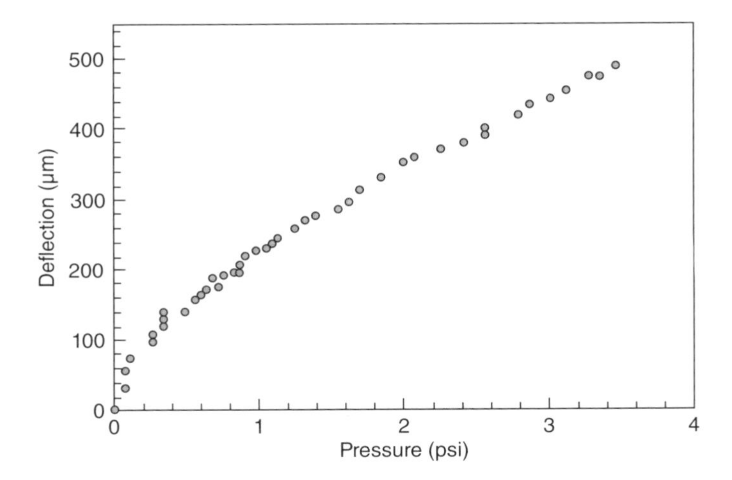

Figure 12.7 Typical deflection versus actuation pressure plot of a balloon actuator

of about 1.2 mm was obtained during the testing experiment. The actuation pressure was varied between 4.5 and 7.5 psig depending on the Mach number of the flow. The microballoon actuator was shown to be successful in control of separation characteristics. A large pressure recovery was observed during the activation of the microballoon actuator. However, the shear stress sensors showed reduction in lift to drag ratio due to the microballoon actuation indicating the necessity of proper optimization of control technique. The microballoon was shown to be robust demonstrating its ability to be operated in high-speed flow control of real environment.

12.3.3 Microflap Actuator

The schematic of flap actuator is shown in Figure 12.8. The actuator surface is supported at one end as cantilever support, and the actuation takes place due to the external applied force at the other end. The actuation principle can be based on microcoil, piezoelectric, electrostatic, and electromagnetic principle. The disadvantage of microcoil, piezoelectric, and electrostatic actuation is the large voltage requirement. Thus, the devices heat up, leading to increase in temperature of the substrate with parasitic heat loss to the fluid. Magnetic actuation can achieve large force (order of hundreds micronewton) and large displacement (tens to hundreds of micrometer). The actuator consists of suspended magnetic piece supported by cantilever beams. The benefit of magnetic actuation is that magnetic field can be applied externally without any required connections for voltage supply.

Silicon, silicon nitride, silicon elastomer, polymer, and so on have been used as flap/membrane materials in literature. One mode of operation is the application of current through the coil over electromagnet consisting of thin film coil with current passing through it. The other mode of operation is using a flap fabricated from a permanent magnet (permalloy) and actuated by an external magnetic field. During the rest mode, no magnetic field is applied, and the membrane is flushed with the wall. When external magnetic field is applied by using either a permanent magnet or electromagnet, the flaps on the membrane surface are magnetized, and

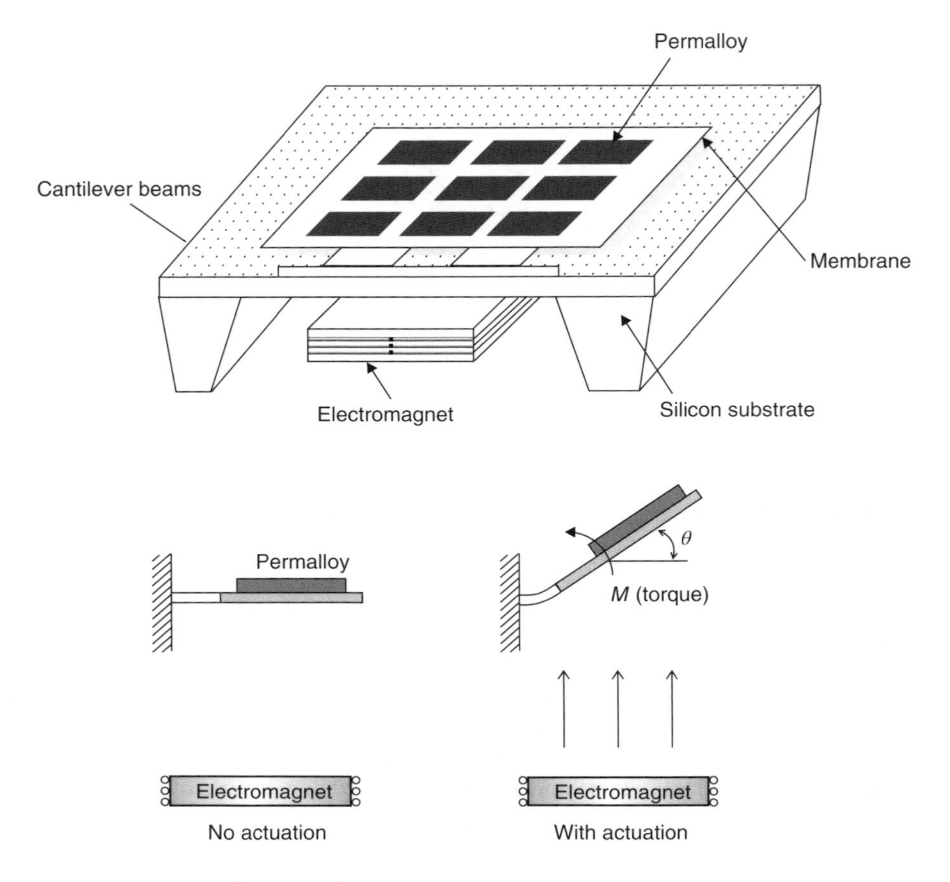

Figure 12.8 Schematic of a magnetic flap actuator

a torque is generated due to the interaction between the internal and external magnetic field (see Figure 12.8, bottom). The application of torque on the cantilever leads to deflection of the membrane/flap. The strength of the external magnetic field controls the deflection of the flap membrane. The frequency of the voltage supply to the electromagnet can be controlled for dynamic actuation.

12.3.3.1 Fabrication

The fabrication procedure of a magnetic flap actuator is shown in Figure 12.9. The details on the fabrication procedure can be found in Liu *et al.* (1999). In step 1, phosphosilicate glass sacrificial thin film is deposited on the top of the silicon substrate. This thin film is patterned on which the actuator will be deposited. The mesas help in limiting the possible undercut during the etching process. The photoresist layer is removed, and the wafer is annealed in nitrogen at 1000 °C for 1 h. Subsequently, a layer of LPCVD polycrystalline silicon is coated on which thin phosphosilicate glass layer is deposited for doping purpose. Annealing at 950 °C

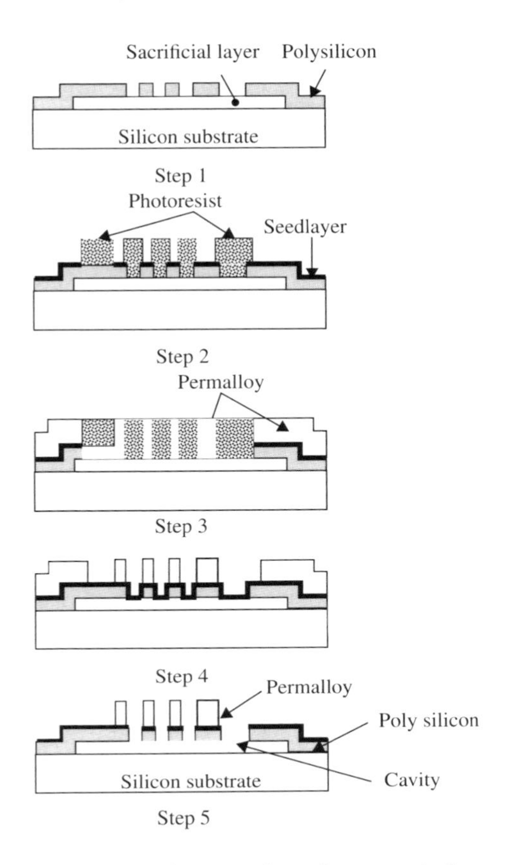

Figure 12.9 Fabrication procedure of a magnetic flap actuator

in nitrogen for 1 h is carried out next for stress relieving and doping. Conductive seed layer (200Å thick chromium and 1800Å thick copper) is thermally evaporated in step 2. The permalloy (80% nickel and 20% iron) piece is electroplated in step 3. During electroplating, the wafer is affixed to the cathode, and a pure Ni piece acts as anode. An external magnet is applied with the field lines being parallel to the wafer substrate. This establishes the easy axis of the permalloy parallel to the length of the cantilever beams. In step 4, the photoresist material is removed. The exposed seed layer material is then etched away. For etching copper, etchant (water:acetic acid:hydrogen peroxide) is used. For chromium etching, etchant (water:HCl = 10:1) is used. In step 5, HF etching is used for about 20 min to release the actuators. The sacrificial layer is removed by etching, and there is little effect on polysilicon. The permalloy sustains no structural and chemical damage.

12.3.3.2 Characteristics

Liu *et al.* (1999) investigated the effect of microflap actuator on a delta wing airfoil. The flap was 400 μm long and 100 μm wide. The calibration of the flap actuator was carried out in

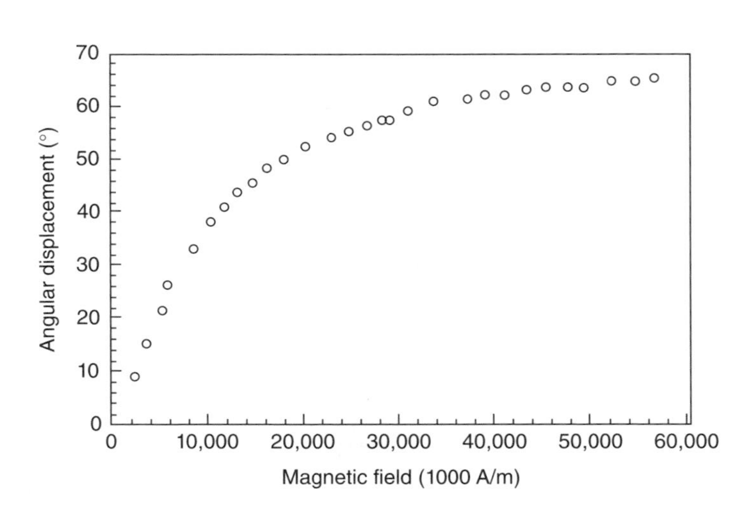

Figure 12.10 Angular displacement versus magnetic field strength of flap actuator

still air. An electromagnet was used to provide uniform external magnetic field. The magnetic field intensity was measured using Gauss meter. The angular displacement with respect to the magnetic field intensity is shown in Figure 12.10. Large angular deflections (about 60 °) and vertical deflections (about 1–2 mm) were demonstrated. It shows the asymptotic variation of flap displacement angle with respect to the magnetic field intensity. Arrays of flaps positioned near the leading edge of delta wing showed generation of rolling moment by flap actuation. The effect of actuator location on the rolling moment of the delta wing was investigated from the force balance. Clear effect on rolling moment as a function of the actuator location was observed. The flow visualization images showed the vortical structures generated by the actuator.

12.4 Shear Stress Sensors

The fluid flow past a surface or boundary leads to surface forces acting on it. These surface forces depend on the rate at which fluid is strained by the velocity field. Stress tensor with nine components is used to describe the surface forces on a fluid element. The tangential component of the surface forces with respect to the boundary is known as shear stress. The nature or origin of shear stress depends on the nature of flow, that is, laminar or turbulent. The stress components for a laminar flow are functions of the viscosity of the fluid and are known as viscous stresses. The turbulent flow has additional contribution known as Reynolds stresses due to velocity fluctuation, that is, the stresses of a laminar flow are increased by additional stresses known as apparent or Reynolds stresses. Hence, the total shear stresses for a turbulent flow are the sum of viscous stresses and apparent stresses. In a turbulent flow, the apparent stresses may outweigh the viscous components.

The wall shear stress is an essential quantity of interest in a wall-bounded flow. The time-averaged value of wall shear stress can be used to determine the skin friction drag acting on the body by the fluid flow. The time-resolved behavior of surface shear stress indicates the unsteady flow structures responsible for individual momentum transfer events and turbulence activities. The instantaneous shear stress values at distributed locations on the surface can be used for feedback control of the turbulence events inside a boundary layer. Shear stress measurement also helps in the assessment and control of power consumption rate and therefore is an important quantity of interest in various industrial applications.

The importance of shear stress measurement is even more crucial for small-scale devices, that is, MEMS applications, due to their higher surface to volume ratio. There have been many efforts in literature for successful shear stress measurements. The success of these efforts primarily relies on the complexity of the flow, the nature of solid boundaries, and the limitations of the measurement techniques. The other drawback of shear stress measurements is its smaller magnitude, that is, the estimated value of shear stress of a typical car moving at 100 km/h is about 1 Pa. Hence, highly sensitive shear stress measuring devices are required for successful measurements of surface shear stress, and it is essential to have proper understanding of the various noise sources that can effect shear stress measurements.

Shear stress sensors provide qualitative information about the flow field, that is, separation, transition, reattachment, and so on, and proper calibration of shear stress sensor is required for quantitative measurement/information. Here, we concentrate on the principle of operation, noise sources, and calibration of various shear stress sensors. The spatial and temporal resolution requirements for turbulent flow applications and benefit of MEMS sensors are also discussed.

12.4.1 Sensor Requirements for Turbulent Flow Control

Turbulence is characterized by the existence of numerous length and time scales. The time scale is a measure of the fact that at a given point in a turbulent flow field, distinct patterns get repeated regularly in time known as time scale. Similarly, distinct patterns may be repeated in space at a given instant, and the sizes of the patterns are known as length scales. Different length scales and time scales have been defined for characterization of turbulent flow field. The smallest eddy of turbulent flow is expressed by Kolmogorov length (l) and time scale (τ), which is uniquely determined by dissipation (ε) and kinematic viscosity (ν):

$$l = \left(\frac{\nu^3}{\varepsilon}\right)^{1/4} \tag{12.1}$$

$$\tau = \left(\frac{\nu}{\varepsilon}\right)^{1/2} \tag{12.2}$$

Using the inviscid assumption for the calculation of the dissipation rate, the nondimensional length scale and time scale, respectively, reduce to

$$\frac{l}{l_S} = (Re)^{-(3/4)} \tag{12.3}$$

$$\frac{\tau}{l_S} u = (Re)^{-(1/2)} \tag{12.4}$$

Here, l_S is the characteristic length of the inertial sublayer, u is the characteristic velocity of the eddy, and Re is the Reynolds number (ul_S/ν). Equations (12.3) and (12.4) indicate that with the increase in the Reynolds number, the length and time scale of eddies reduce. To spatially resolve the small eddies, sensors that are of same size as the Kolmogorov length scale for that particular flow are needed. Hence, smaller sensors are required as the Reynolds number is increased. For a flat plate boundary layer with momentum thickness Reynolds number equal to 4000, the Kolmogorov length scale is of the order of 50 μm. Large-dimensional sensors integrate the fluctuation due to small eddies over its spatial extension. Therefore, when measuring fluctuating quantities, small eddies are counted as mean flow, and the fluctuating energy is lost. This leads to lower estimated value of turbulence intensity. The other important issue is the spacing between the sensors, which is primarily determined by the spacing between the coherent structures present in the flow. For the turbulent wall boundary layer, the spacing between the sensors needs to be less than the distance between the low-speed streaks, that is, the near-wall region coherent structures.

12.4.2 Benefits of MEMS-Based Sensors

MEMS-based shear stress sensors offer many advantages over traditional shear stress sensors as follows:

1. The MEMS-based shear stress sensors can meet the spatial and temporal requirement of turbulent flow due to their small size.
2. The MEMS-based sensors have low thermal inertia leading to higher dynamic response.
3. The MEMS-based sensors are less intrusive compared to the traditional sensors due to their small size and therefore do not contribute any parasitic effects.
4. The energy consumption is low for MEMS-based sensors.
5. MEMS-based sensors are ideal for distributed measurements and control as large number of sensors can be fabricated on the same chip in a cost-effective manner.

12.5 Classification of Shear Stress Sensors

Various shear stress measurement techniques have been proposed in literature. Some principal measurement techniques are Stanton tube, Preston tube, electrochemical technique, velocity measurements, thermal method, floating element sensors, sublayer fence, oil-film interferometry, and shear stress-sensitive liquid crystal.

Stanton tube is a rectangular-shaped pitot tube located very close to the boundary wall, and the mean velocity from this pitot tube is directly related to the shear stress. Preston tube is similar to the concept of Stanton tube using a pitot static tube close to the surface, and the difference between the stagnation pressure at the center of the tube and the static pressure is related to the shear stress. The electrochemical or mass transfer probe is flush mounted with the wall, and the concentration at the wall element is maintained constant. The measurement of mass transfer rate between the fluid and the wall element is used for determination of the wall shear stress. One of the limitations of the mass transfer probe is that at very high flow rates, the mass transfer rate becomes large and it may not be possible to maintain the wall concentration constant. The detailed discussion on above three techniques can be found in Hanratty and Campbell (1996). These shear stress measurement techniques are not ideal MEMS-based technique.

Due to numerous benefits of MEMS-based shear stress sensors, the following shear stress measurement techniques having great promise for future MEMS applications have been discussed in the following sections: (1) velocity measurements, (2) thermal sensors, (3) floating element sensors, (4) sublayer fence, (5) oil-film interferometry, and (6) shear stress-sensitive liquid crystal.

12.5.1 Shear Stress from Velocity Measurements

Velocity measurements at close to the wall or the velocity profile measurement in the near-wall region can be used to determine the wall shear stress. Clauser proposed an approach for shear stress measurement of turbulent flow (Hanratty and Campbell, 1996). Here, the mean velocity measurements away from the wall are used with assumption that the mean velocity ($\bar{U}$) varies with the logarithmic distance from the wall (y), that is,

$$\frac{d\bar{U}}{dy} = \frac{u^*}{\kappa y} \tag{12.5}$$

where $u^* = (\tau_w/\rho)^{1/2}$ is the friction velocity and κ is the von Karman constant. The slope of a semilogarithmic plot between $\bar{U}$ and ln y gives u^*/κ and therefore τ_w. The difficulty in this approach is the validity of the logarithmic region assumption and the value of the von Karman constant, which is approximated to be equal to 0.40. For flows with large pressure gradient, well-defined logarithmic layer does not exist. If the wall is flat and in equilibrium situation, Clauser proposed a relationship as

$$\frac{\bar{U}}{u^*} = f\left(\frac{yu^*}{V}, \Delta\right) \tag{12.6}$$

where $\Delta = |dP/dx|(V/u^*\tau_w)$ and dP/dx is the pressure gradient. Comparison of measured variation of mean velocity with distance from the wall gives the wall shear stress τ_w.

Spalding formula represents a good fit to the velocity profile in a turbulent boundary layer, that is, from the wall to the end of the log region, thus including both the linear sublayer and the logarithmic region. The Spalding formula is given by

$$y^+ = \frac{yu^*}{V} = u^+ + \frac{1}{E}\left[e^{-\kappa u^+} - 1 - \kappa u^+ - \frac{(\kappa u^+)^2}{2} - \frac{(\kappa u^+)^3}{6} - \frac{(\kappa u^+)^4}{24}\right] \tag{12.7}$$

where $u^+ = u/u^*$ and E is a constant equal to 8.6. The iterative procedure is carried out for curve fitting the experimental velocity profile to the Spalding formula, and the shear stress is determined from the best fit. The Spalding formula represents the mean velocity profile in a turbulent boundary layer at all pressure gradients and till about $y^+ = 200$.

Various optical techniques, that is, laser Doppler velocimetry (LDV), particle image velocimetry (PIV), and holographic PIV, have matured as successful nonintrusive velocity measurement techniques for large-scale applications. Panigrahi *et al.* (2005) have obtained the shear stress from PIV measurements by assuming the validity of law of the wall for turbulent flow. In recent years, the μ-PIV has matured as a successful velocity measurement technique for MEMS applications (Nguyen and Wereley, 2006). The PIV techniques provide instantaneous velocity field information. The spatial and temporal resolution of PIV can be controlled with proper selection of lighting source, that is, laser; imaging system, that is, camera and frame grabber; and optical arrangements/components. The continuous development of these hardware in the near future is expected to result in very high spatial and temporal resolution velocity measurements. Therefore, the μ-PIV and μ-holographic PIV techniques are expected to develop as high-resolution indirect instantaneous shear stress measurement techniques in the near future.

12.5.2 *Thermal Shear Stress Sensors*

The operational principle of thermal sensor is based on the relationship between the heat transfer and the sensor exposed to the flowing fluid and the shear stress. For the heat transfer to take place, the sensor element temperature must differ from the temperature of the flowing medium, that is, the sensor is raised to a temperature above the medium temperature. The thermal sensor forms one of the resistances of the Wheatstone bridge circuit (see Figure 12.11). The resistance of the thermal sensor is given by

$$R = R_0[1 + \alpha(T - T_0)] \tag{12.8}$$

where R is the sensor resistance at temperature T, R_0 is the sensor resistance at reference temperature (T_0), T is the temperature of the sensor, and α is the temperature coefficient of resistance. The heat transfer rate depends on the nature of fluid flow past the sensor, leading to drop in temperature of the sensor. The drop in sensor temperature affects the sensor resistance leading to imbalance of the Wheatstone bridge circuit. The relationship between the bridge circuit imbalance and the heat transfer can be obtained from two approaches, that is, (1) *deflection method* and (2) *null method*. In the deflection method, the change in bridge balance due to the input signal is measured. In the null method, an electronic feedback is used to keep the

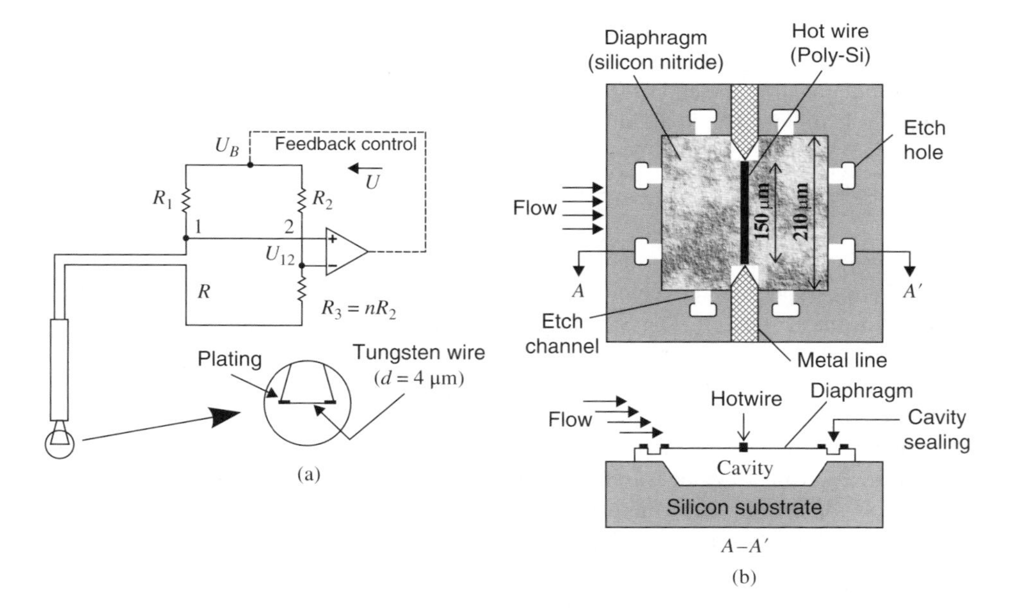

Figure 12.11 (a) The schematic of a hot-wire sensor connected to the Wheatstone bridge circuit and (b) the schematic of a flush-mounted thermal shear stress sensor

bridge balanced. The null method is usually preferred due to better frequency response. For the electronic feedback, there are three possible modes of operation:

1. *Constant current (CC) mode*: Here, the current through the sensor is kept constant, and the change in input current needed to maintain its constant value is related to the flow signal, that is, shear stress.
2. *Constant voltage (CV) mode*: Here, the voltage across the sensor is kept constant, and the voltage or current change needed to keep the voltage constant is related to the shear stress.
3. *Constant temperature (CT) mode*: Here, the temperature of the sensor is kept constant, and the voltage or current change needed to keep the temperature constant represents the wall shear stress.

The thermal sensors for shear stress measurements can have two principal modes of variation: (1) elevated hot-wire probe and (2) surface-mounted hot-wire probe. The principles of operation of these two types of thermal shear stress sensors are discussed in the following sections.

12.5.2.1 Elevated Hot-Wire Probe

In elevated hot-wire approach, the hot wire is mounted at small distance away from the wall. The velocity increases linearly with the wall distance in the near-wall region of both laminar and turbulent flow. For turbulent flow, this assumption is valid for the instantaneous velocity profile till the wall normal location $y^+ \left(\frac{yu^*}{\nu}\right) < 5$. The linear relationship between the shear stress (τ_w) and the velocity u at distance y from the wall is given as

$$\tau_w = \frac{\rho \nu u}{y} \tag{12.9}$$

The true velocity in the near-wall region determined from the hot wire calibrated in the free stream region is affected due to the heat conduction in the near-wall region. Therefore, the velocity obtained from the hot wire needs to be corrected based on the empirical relationship. The substitution of this corrected velocity in equation (12.9) provides the wall shear stress. The sensitivity of this approach depends on the wall distance, that is, there is an optimal wall distance at which the hot-wire sensitivity is optimal. In addition to that, the hot wire should be located inside the inertial sublayer to satisfy the basic assumption essential for the shear stress measurement.

12.5.2.2 Surface-Mounted Hot-Wire Sensor

The sensor element is either positioned on the top of a substrate or on a thin diaphragm above a vacuum cavity. As the thermal shear stress sensor is flush mounted along the surface of the wall, this technique is considered as nonintrusive with negligible disturbance to the mean flow. The ohmic heating (Q_{ohmic}) due to the current flow through the sensor element transfers heat to both fluid (Q_{fluid}) and the substrate ($Q_{\text{substrate}}$), that is,

$$Q_{\text{ohmic}} = Q_{\text{fluid}} + Q_{\text{substrate}} \tag{12.10}$$

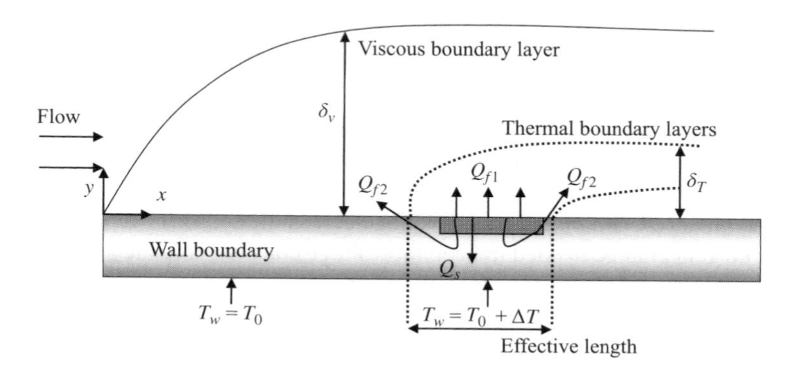

Figure 12.12 The flow and heat transfer mechanism of a flush-mounted thermal shear stress sensor

Figure 12.12 shows the heat transfer mechanism from a surface-mounted thermal sensor. The total heat transfer to the fluid from the thermal sensor (Q_{ohmic}) has two components, that is, the heat transfer to the fluid (Q_{fluid}) and the heat lost to the substrate ($Q_{\text{substrate}}$). The heat transfer to the fluid has two parts, that is, direct heat transfer from the sensor element (Q_{f1}) and indirect heat transfer from the substrate heated by the conduction of heat from the sensor to the substrate (Q_{f2}). The heat transferred to the fluid via the substrate effects the temperature distribution near the sensor. This affects the net heat transfer rate from the sensor element and limits the performance of thermal shear stress measurement. The effective length of the thermal sensor is higher than the size of the sensor element, thus limiting the spatial resolution of shear stress measurement. Therefore, effective thermal isolation between the sensor element and substrate is an important issue for optimum performance, fabrication, and packaging of thermal shear stress sensors. For thermal isolation, the resistor of the sensor sits on the top of a diaphragm above a vacuum cavity (see Figure 12.12). The presence of vacuum cavity and thin diaphragm reduces the convective and conductive heat transfer to the substrate. Better insulation improves the thermal sensitivity of the sensor, that is, higher temperature rise ($T - T_0$) of the thermal sensor is achieved for a particular power input (P).

In experimental aerodynamics, surface hot-wire probe has proved to be the most successful standard measurement technique to determine the laminar to turbulent flow transition, local separation, and shear stress fluctuations. The flush-mounted thermal shear stress sensor is one of the most successful techniques for shear stress measurement and is available in various forms, that is, sensor skin and so on (Xu *et al.*, 2003), due to rapid development of MEMS manufacturing processes. The quantitative determination of shear stress depends on the proper calibration with a reference method. The calibration issues have been described in a later section.

12.5.3 Floating Element Shear Stress Sensors

Direct measurement of tangential force exerted by the fluid on the sensor element is possible by floating element shear stress sensor. Therefore, floating element-based shear stress sensor is classified as direct measurement device. Schematic of a floating element shear stress sensor

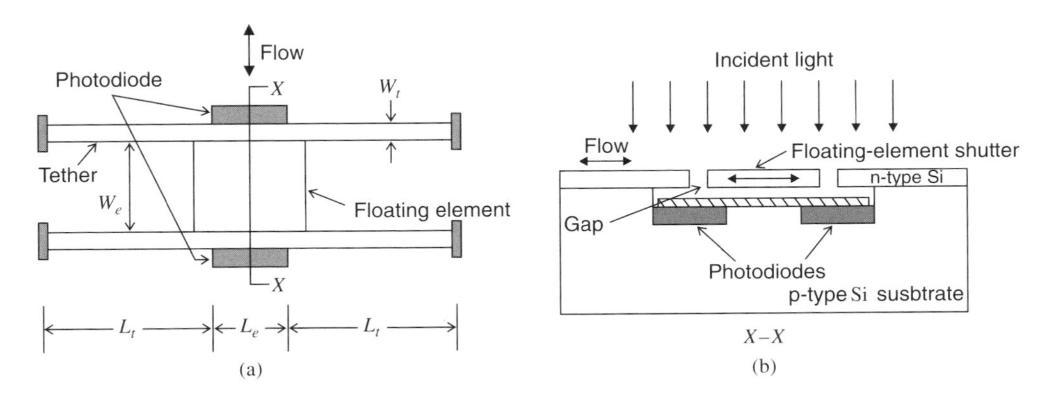

Figure 12.13 The schematic of a floating element shear stress sensor: (a) top view and (b) cross-sectional view

has been shown in Figure 12.13. The floating element is connected to the substrate by four tether elements. The movement of the floating element due to the surface force, that is, shear stress acting on it, is sensed by suitable transducer and is related to the shear stress. The movement of the floating element can be inferred from using the capacitive sensing, piezoresistive, and optical detection schemes. The axial compression and expansion of the tethers due to the imposed shear stress cause a change in piezoresistive effect for piezoresistive sensing element. The change in resistance is a measure of the shear stress magnitude and direction. In capacitive scheme, the sensor element displacement leads to change in the capacitance value. In photodiode-based scheme, photodiodes are integrated below the floating element at its leading and trailing edge. The coherent light source illuminates the sensor element from above. The differential photocurrent generated due to the movement of floating element is related to its displacement (see Figure 12.13).

No assumption regarding the flow field is required for the floating shear stress sensor. The displacement (δ) of the floating element sensor as a function of shear stress (τ_w) can be derived from Euler–Bernoulli beam theory:

$$\delta = \tau_w \frac{L_e W_e}{4Et} \left(\frac{L_t}{W_t} \right)^3 \left(1 + 2 \frac{2L_t W_t}{L_e W_e} \right) \tag{12.11}$$

where L_t is the tether length, W_t is the tether width, E is the elastic modulus of the tether, t is the floating element thickness, L_e is the length of the floating element, W_e is the width of the floating element, and g is the recessed gap of the floating element (see Figure 12.13). Equation (12.10) indicates a linear relationship between the displacement of the floating element and the shear stress. The above linear relationship has been verified from the static calibration results of floating element sensors. The flow under the floating element and the pressure gradient acting on its top introduce error to the measurement of skin friction. This relationship between the effective shear stress (τ_{eff}) and the actual shear stress (τ_w) is given by Naughton and Sheplak (2002)

$$\tau_{\text{eff}} = \left(1 + \frac{g}{h} + \frac{2t}{h} \right) \tau_w \tag{12.12}$$

where h is the channel height of the wind tunnel used for calibration. The second and third terms of the above equation relate to the effect of gap height and pressure gradient on the measurement error, respectively. Equation (12.12) indicates the benefit of MEMS-based shear stress sensor as both gap height and the floating element thickness can be maintained at very small value (~1 μm), leading to considerable reduction in errors. In addition, the MEMS-based sensor element results in high spatial resolution and higher frequency bandwidth due to lower effective size and mass, respectively. Due to the rapid advancement of MEMS-based fabrication technique, it is also possible to place pairs of sensor in close proximity at right angles to each other for detection of both streamwise and spanwise shear stress.

The capacitive and piezoresistive sensing-based floating element sensors may show sensitivity to electromagnetic interference due to the impedance of the electronic elements used for detection. The size of the floating element, the misalignment and gaps around the floating element, pressure gradient, sensitivity to acceleration, vibration, and thermal expansion are some of the general drawbacks of the floating element sensors.

12.5.4 MEMS Skin Friction Fence

The generic skin friction fence and its principle have been explained by Hanratty and Campbell (1996). Recently, Schober *et al.* (2004) developed a MEMS-based skin friction fence (see Figure 12.14(a)). The surface fence is a fence protruding from the wall. There is a slot below the surface fence with support at both ends. Two pairs of piezoresistors are located at both ends of the fence. These piezoresistors are connected to the Wheatstone bridge circuit. The voltage output of the Wheatstone bridge circuit is related to the strain acting on the fence. The strain due to the deflection of the fence is directly related to the pressure difference across the fence. Hence, the voltage output is indirectly related to the shear stress acting on the fence. The surface fence should remain inside the viscous sublayer ($y^+ < 5$). However, there is no requirement on validity of the law of the wall for turbulent boundary layer. The pressure difference

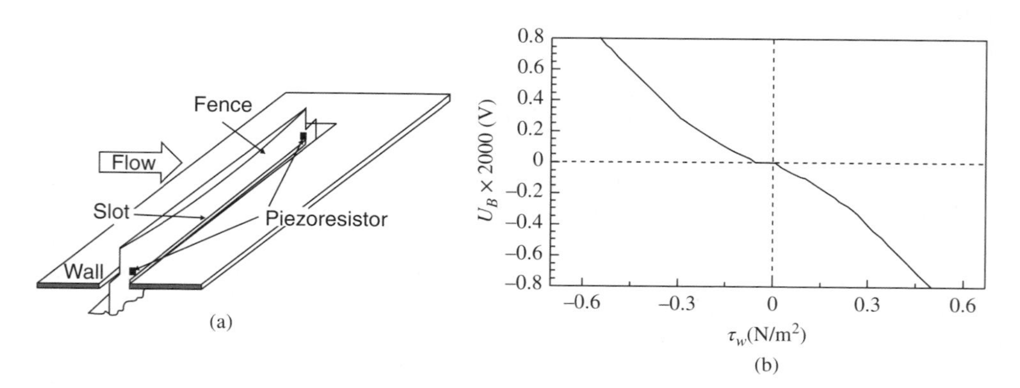

Figure 12.14 The schematic of (a) MEMS skin friction fence and (b) its calibration curve between the bridge voltage (U_B) and the shear stress (τ_w)

between the front and back of the fence is related to the shear stress, that is,

$$\Delta P = A\tau_w + B\tau_w^2 \tag{12.13}$$

where A and B are constants. From equation (12.13), the measured voltage (V_{br}) from the Wheatstone bridge circuit can be related to the shear stress as

$$V_{br} = A_{br}\tau_w + B_{br}\tau_w^2 \tag{12.14}$$

where A_{br} and B_{br} are the constants to be obtained from the calibration. Figure 12.14(b) shows the calibration curve between the bridge voltage and the shear stress of a microfence inside a zero-pressure-gradient turbulent boundary layer. It may be observed that the sensitivity of the microfence is higher at higher shear stress.

12.5.5 Optical Shear Stress Sensors

The optical-based shear stress sensor can be classified into two categories based on the measurement principle, that is, oil-film interferometry and shear-sensitive liquid crystal. These techniques are discussed in the following sections.

12.5.5.1 Oil-Film Interferometry

Oil-film technique is a direct method for skin friction measurement based on the movement of interference fringes of a thin film. In this method, an oil film is applied to the smooth solid wall, which spreads out by the flow into a thin layer whose thickness is of the order of 1 μm. The thickness of the oil film is dependent on the shear force acting upon it. The interferometry is used to measure the thickness of the oil film. The monochromatic light of wavelength λ from a light source is incident on the film at an angle θ (Figure 12.15(a)). The light partially gets reflected from the oil/air interface. The remainder of the light passes through the film

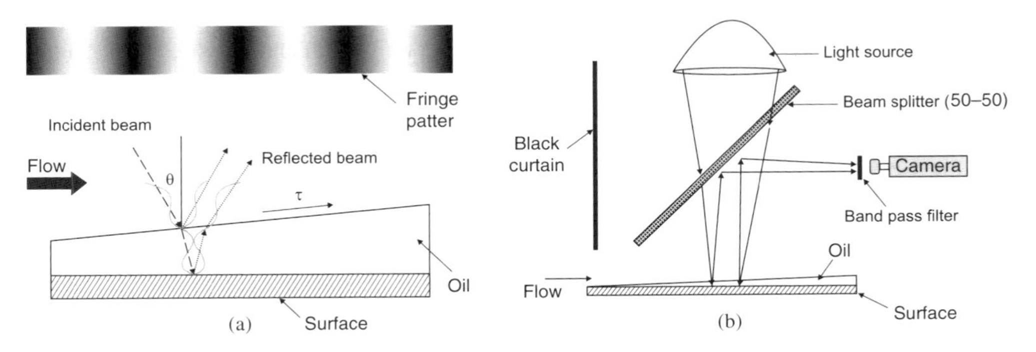

Figure 12.15 The schematic showing (a) the fringe formation for oil height measurement and (b) an experimental setup using oil-film interferometry

and reflects from the solid surface and travels back through the oil film. These two beams interfere constructively and destructively with each other in the image plane depending on the phase difference between the two, leading to formation of bright and dark bands known as interference fringes.

The local height of the oil film can be obtained from the fringe patterns. The height of the oil film at kth black fringe is given by

$$h_k = h_0 + k\Delta h; k = 0, 1, 2, \ldots \tag{12.15}$$

where h_0 is the height of the zeroth black fringe at the film edge, that is, at $k = 0$, and Δh is the difference in height between two consecutive fringes given by

$$\Delta h = \frac{\lambda}{2(n^2 - \sin^2\alpha)^{1/2}} \tag{12.16}$$

Here, n is the refractive index of oil and λ is the wavelength of light. A general oil-film interferometry experimental setup has been shown in Figure 12.15(b). Here, the light from preferably a He–Ne laser (wavelength = 632.8 nm) or sodium lamp (wavelength = 589 nm) is incident on the thin oil film through a beam splitter. A camera collects the reflected light from the oil film and a portion of the incident light from the beam splitter against a black background. The motion of the interference fringes is collected by imaging through the camera.

For oil-film interferometry, it is assumed that the oil film is so thin that it does not influence the flow above it and is driven by the skin friction distribution of the flow. Using a control volume analysis of the thin oil film with its height h in wall normal (y) direction as a function of streamwise (x) and spanwise (z) coordinate and assuming the shear stress contribution to be dominant compared to the pressure gradient and surface tension force, the governing equation for the thin-oil-film flow is

$$\frac{\partial h}{\partial t} + \frac{\partial}{\partial x}\left(\frac{\tau_{w,x}h^2}{2\mu}\right) + \frac{\partial}{\partial z}\left(\frac{\tau_{w,z}h^2}{2\mu}\right) = 0 \tag{12.17}$$

Using the distribution of oil-film height from interferometry measurements and the integration of equation (12.17), the skin friction can be determined. The benefit of oil-film interferometry is that no calibration is required for the measurement of shear stress and the basic analytical expression is used. It can be used for 3-D flow situation and provides both magnitude and direction of shear stress. The disadvantage is that it does not possess any temporal resolution and therefore cannot be used for fluctuating skin friction measurement.

12.5.5.2 Liquid Crystal Sensors

Liquid crystal is a phase of matter that exists between the liquid and solid phase. It exhibits optical properties similar to the solid crystalline material. The molecular arrangement of liquid crystal is a function of either temperature or shear stress. When the molecular arrangement is sensitive to temperature, the liquid crystal coating can be used for temperature measurement, and when the molecular arrangement is sensitive to shear stress, the liquid crystal can be used for shear stress measurement.

Shear-sensitive liquid crystal coatings applied to a solid surface consist of helical aggregates of long planar molecules arranged in layers parallel to the coated surface. This molecular

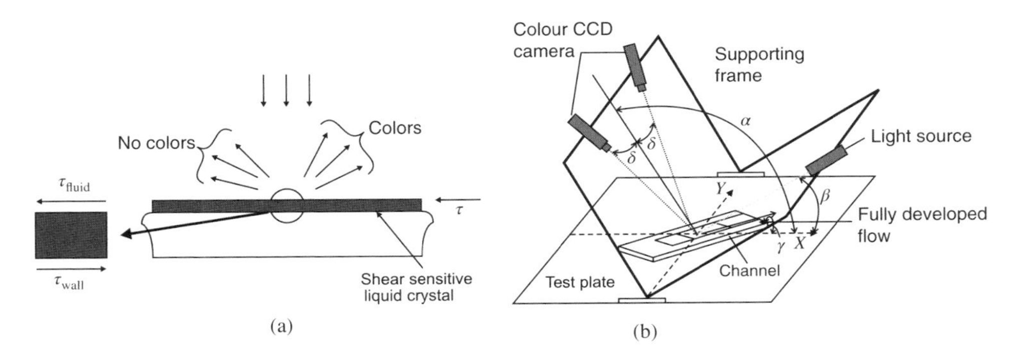

Figure 12.16 The schematic showing (a) the principle of shear-sensitive liquid crystal and (b) the experimental setup for shear stress measurement using shear-sensitive liquid crystal

arrangement is affected by the application of shear stress on the thin film of liquid crystal. Each layer of molecules is rotated relative to the layer above and below it due to the application of shear stress. The incident light on the coating surface selectively scatters at a wavelength proportional to the pitch of the helix. The local pitch of the helical structure is altered due to application of shear stress relative to the no-shear state. The liquid crystal selectively scatters the incident white light as a 3-D spectrum in color space depending on the molecular arrangement. The color change observed at a particular direction by the imaging device is a function of both shear stress vector magnitude and direction (see Figure 12.16(a)). The shear stress vectors with components directed away from the observer exhibit color change responses of the liquid crystal depending on the shear magnitude and direction. At any surface point, maximum color change occurs when the shear stress vector is aligned and directed away from the observer. The location with shear stress directed to the observer exhibits no color change response. The above behavior of the shear-sensitive liquid crystal is used for determination of the shear stress magnitude and direction.

A sample experimental setup for the application of shear-sensitive liquid crystal has been shown in Figure 12.16(b) similar to the one proposed by Fujisawa *et al.* (2003). The off-axis angle of the camera and the light source with respect to the horizontal are α and β, respectively. Two cameras are situated on opposite sides of the axis with an off-axis angle δ. As the color images are acquired at an oblique angle to the test surface, it is important to convert the images at oblique angle to the normal observation angle by using geometrical transformation technique. The two color cameras are synchronized with the lighting source. The test section can be rotated leading to variation of the shear stress angle (λ) for calibration. There are different color formats, that is, RGB and HSI, for quantitative representation of color. The hue has been observed to be the best indicator for expressing the change in color. It can be calculated from

$$H = \left[\pi/2 - \tan^{-1}\left(\frac{2R - G - B}{\sqrt{3(G - B)}}\right) + \pi\right] /2\pi; G > B \tag{12.18}$$

The color change of the liquid crystal depends on the illumination and observation angle with respect to the liquid crystal coating surface. Therefore, the calibration of the liquid crystal

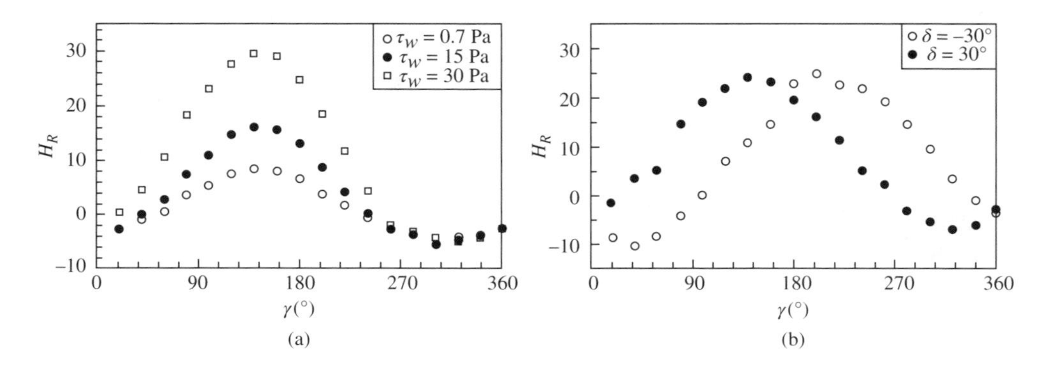

Figure 12.17 (a) The relative hue (H_R) versus the shear stress angle (γ) for different shear stress magnitude (τ_w) and (b) the relative hue (H_R) versus the shear stress angle (γ) for two cameras (left, $\delta = -30°$; right, $\delta = 30°$) at a particular shear stress magnitude

is carried out by the relative hue ($H_R = H_\tau - H_{\text{no}\tau}$) where H_τ is the hue value of the liquid crystal with applied shear stress and $H_{\text{no}\tau}$ is the hue value of the liquid crystal without applied shear stress. The hue value for a particular shear stress magnitude and direction depends on the camera and lighting angle. Therefore, two separate calibration curves for two cameras in stereoarrangement can be used to determine the two unknowns, that is, the shear stress magnitude and direction. Figure 12.17(a) shows the calibration curve between the relative hue (H_R) and the shear stress angle (λ) at different shear stress amplitudes (τ_w) for a particular camera and lighting angle. Figure 12.17(b) shows the relative hue (H_R) versus the shear stress angle (λ) for two different cameras (left and right).

For quantitative determination of the shear stress, the two cameras and lighting source are set at optimum off-axis (λ), observation (α), and illumination (β) angles, respectively. The calibration curve between the relative hue of both the cameras (left and right), that is, $H_{R\text{-left}}$ (τ_w, λ) and $H_{R\text{-right}}$ (τ_w, λ), with respect to the shear stress magnitude (τ_w) and direction (λ) is established during the calibration process. The determination of shear stress magnitude and direction for an actual measurement is carried out by an iterative process. Initially, the shear stress magnitude (τ_w) and direction (λ) are assumed. The corresponding hue values for both left and right cameras are determined from the calibration curve. The measured hue values for both left and right cameras ($H_{R\text{-left-}M}(\tau, \lambda)$ and $H_{R\text{-right-}M}$ are compared with the calibrated hue values. The iteration is carried out, that is, the shear stress magnitude (τ_w) and direction (λ) are changed till the error (E) between the calibrated hue and actual hue value for both left and right cameras is minimized, which is written as

$$\begin{aligned} E = {} & (H_{\text{R-left}}(\tau,\gamma) - H_{\text{R-left-M}}(\tau,\gamma))^2 \\ & + (H_{\text{R-right}}(\tau,\gamma) - H_{\text{R-right-M}}(\tau,\gamma))^2 \end{aligned} \quad (12.19)$$

12.6 Calibration of Shear Stress Sensors

The calibration is one of the important issues for quantitative determination of the shear stress vector from measurement. The suitability of the shear stress sensor for a particular application is based on the complexity of the calibration procedure. The calibration of thermal-based shear stress sensor and the calibration setup required have been described in the following. Similar setup can be used for calibration of all types of shear stress sensors. The calibration process is categorized as static and dynamic.

12.6.1 Static Calibration

The heat transfer rate from the flush-mounted shear stress sensor depends on the near-wall flow, that is, the magnitude of the velocity gradient. For a laminar 2-D thermal boundary layer developing over the heated sensor with an approaching linear velocity profile (Figure 12.12) and negligible free convection effect, the heat loss from the thermal element can be derived from the thermal boundary layer equation as

$$\frac{q}{\Delta T} = \frac{I^2 R}{\Delta T} = 0.807 \frac{C_p^{1/3} k^{2/3}}{L^{1/3} \mu^{1/3}} (\rho \tau_w)^{1/3} \tag{12.20}$$

where ρ is the density, k is the thermal conductivity, C_p is the specific heat, L is the length of heating element, T is the temperature, μ is the viscosity of the fluid, and τ_w is the shear stress. Equation (12.20) for the ohmic heating of the thermal sensor can be written as

$$\frac{I^2 R}{\Delta T} = A(\rho \tau_w)^{1/3} + B \tag{12.21}$$

where I is the heating current and R is the resistance of the sensor. The term B represents the heat loss from the substrate and depending on the design can be more than the heat loss to the fluid, that is, the first term of the above equation. The term A is a weak function of temperature. The length of the sensor calculated from experimentally determined value of A is more than the actual sensor length due to the heat transfer to the fluid from both substrate and the sensor (see Figure 12.12). The forced convection heat transfer assumption is limited by the effective length of the sensor. It is also difficult to satisfy the thermal boundary layer to be within the linear velocity profile. Hence, for a turbulent flow, the above equation is modified as

$$\frac{I^2 R}{\Delta T} = A_T (\rho \bar{\tau}_w)^{1/n} + B_T \tag{12.22}$$

Here, the constants A_T, B_T, and n should be determined empirically from calibration procedure. If the resistance of the heating element is held constant and the voltage (E) or current (I) is measured, then the above equation can be rearranged as

$$\tau_w = (A_{T,\text{New}} E^2 + B_{T,\text{New}})^n \tag{12.23}$$

Equation (12.23) between shear stress and the voltage can also be represented as a polynomial function:

$$\tau_w = C_0 + C_1 E + C_2 E^2 + \cdots + C_n E^n \tag{12.24}$$

One possible experimental arrangement for shear stress calibration is rotational rig consisting of a rotational disk above a bottom stationary disk (see Figure 12.18(a)). The gap

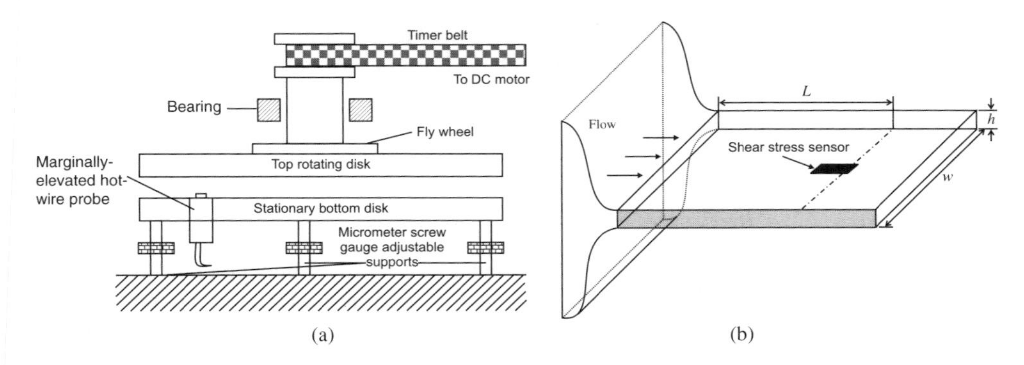

Figure 12.18 Schematic of static calibration apparatus: (a) using disk and (b) long, high aspect ratio smooth channel flow

size (δ) between the disks is very small (0.3–0.45 mm) to maintain a small Reynolds number ($Re_s = \omega\delta^2/\upsilon \gg 5$) where ω is the rotating disk angular velocity. For small Reynolds number, the tangential shear stress can be derived as

$$\tau = \mu(\omega r/\delta)\left[1.0 + (1/1050)Re_s^2 + O(Re_s^4)\right] \tag{12.25}$$

During calibration, the shear stress sensor is mounted on the stationary plate at various radial positions with different gap size and rotational speed to achieve wide range of shear stress for static calibration.

Another possible setup for the static calibration uses a wind tunnel (Figure 12.18(b)). The wind tunnel test section should be sufficiently long to establish fully developed flow. The width over height ratio of the test section also needs to be high, that is, greater than 30 for the flow to be 2-D. The shear stress (τ_w) for this configuration is related to the local pressure gradient as

$$\frac{dP}{dx} = -\frac{\tau_w}{h} \tag{12.26}$$

Here, P is the local pressure, x is the streamwise coordinate, and h is the half height of the channel. The pressure gradient can be measured by mounting pressure taps along the channel before and after the shear stress sensor location.

Figure 12.19(a) shows one representative calibration curve between the Wheatstone bridge output (E) and the shear stress (τ_w) of the shear stress sensor operating in CT mode. The same calibration curve has been replotted using the heating power ($P = E^2/R$) divided by the temperature difference (ΔT) with respect to shear stress to one-third (0.33) power. A linear variation as expected from equation (12.21) is clearly observed. The laminar value of exponent (0.33) may change due to the restriction on the experimental condition during calibration, that is, wall presence, mounting location, and size of the sensor element. Hence, the exponent value around 0.4 contrary to expected laminar value equal to 0.33 has been observed by various investigations.

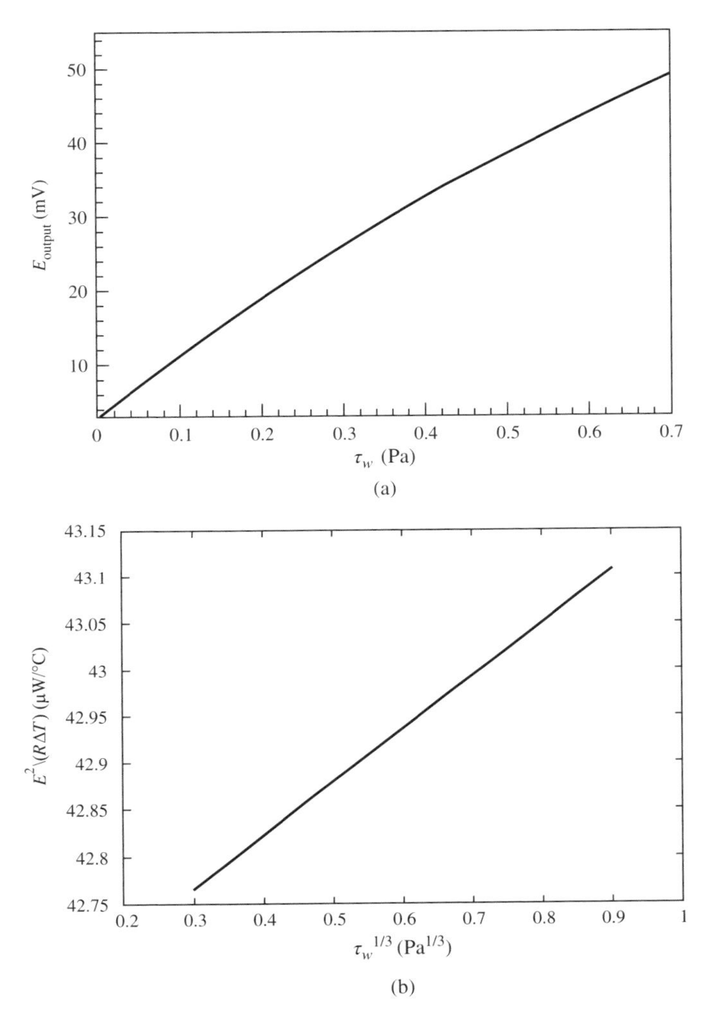

Figure 12.19 Static calibration curve: (a) the bridge output (E) versus shear stress (τ_w) and (b) the input power versus (shear stress)$^{1/3}$ of thermal shear stress sensor

12.6.2 Dynamic Calibration

The shear stress sensor for turbulent flow needs to accurately capture the complete turbulent fluctuation spectrum. Therefore, the shear stress sensor should possess a large bandwidth with flat and minimum frequency–phase relationship. For direct measurement by floating point sensors, the resonant frequency of the floating element and the fluidic damping determines the usable bandwidth. For the thermal sensor, the thermal inertia of the sensor element and the

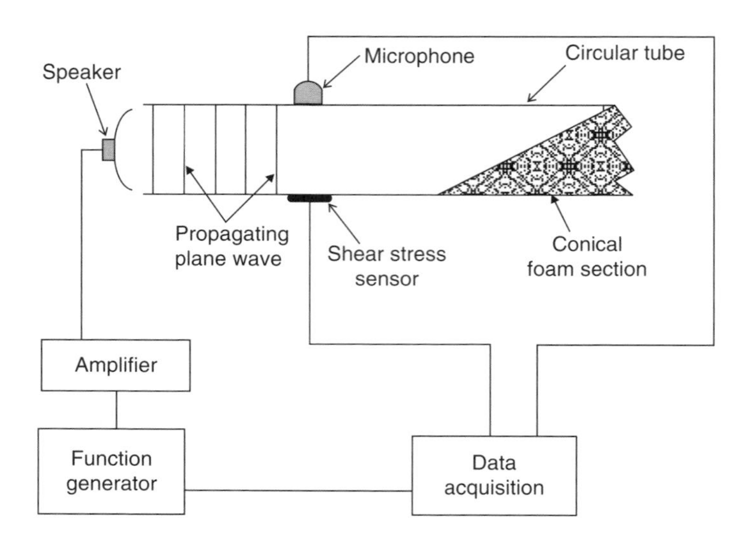

Figure 12.20 The schematic of a dynamic calibration apparatus for shear sensor

frequency-dependent heat conduction to the substrate influence the usable bandwidth. It is complicated to analytically predict the frequency response of the thermal sensor. Therefore, dynamic calibration is essential to characterize the frequency response of the sensor.

The Stokes-layer excitation in a cylindrical duct is one of the effective techniques for dynamic calibration of shear stress sensors (Sheplak *et al.* 2001). The flow inside the duct is driven by an oscillating pressure gradient generated by a loud speaker (see Figure 12.20). The loud speaker driven by an amplifier generates an acoustic wave. The amplifier receives sinusoidal input from a function generator. The microphone and shear stress sensor are mounted at opposite location of the tube. The data acquisition system records the signal from the microphone and shear stress sensor.

The solution of compressible momentum equation provides the relationship between the shear stress, pressure gradient, and excitation frequency as

$$\tau(z,t) = -\frac{p'\exp\{j(2\pi ft - kz)\}}{c}\sqrt{\frac{j2\pi f\mu}{\rho}}\frac{I_1(R\sqrt{j2\pi f/V})}{I_o(R\sqrt{j2\pi f/V})} \tag{12.27}$$

where p' is the magnitude of pressure fluctuation, c is the speed of sound, f is the frequency of acoustic wave, $k = 2\pi f/c$ is the wave number, z is the axial direction, μ is the fluid viscosity, ρ is the density, R is the radius of tube, υ is the kinematic viscosity, and I is the Bessel function. The frequency response function $H(f)$ from equation (12.27) is used to determine the frequency response of the shear stress sensor:

$$H(f) = \frac{V(f)}{\tau(f)}\frac{\delta\tau}{\delta V}$$

Here, $V(f)$ is the sensor voltage, $\tau(f)$ is the shear stress calculated from equation (12.27) as a function of the pressure fluctuation magnitude, and $\frac{\delta\tau}{\delta V}$ is the static sensitivity from the static calibration of the shear stress sensor. For an ideal sensor, the static and dynamic sensitivity should be equal, leading to a normalized frequency transfer function ($H(f)$) of 0 dB at all measurable frequency range.

12.7 Uncertainty and Noise

For proper design and practical implementation of shear stress measurement procedure, it is important to be aware of different sources of noise and uncertainty during the measurement. When using shear stress sensor based on MEMS technology, all circuit elements are located on the substrate surface. Hence, electronics is one of the primary sources for drift or noise during shear stress measurement. The other important contribution to noise is the physical sources, that is, during low flow velocity situations; the free convection effect disturbs the signal. The packaging and mounting of the sensor element can generate offset signals, which may change in time due to the change in heat dissipation of the sensor element. This influences the frequency response of the shear stress sensor. The MEMS-based shear stress sensor can have the following noise sources: (1) thermal or white noise, (2) shot noise, (3) flicker noise, and (4) photon noise. The thermal noise is the movement of electrons due to the finite temperature difference leading to noise in the output voltage or current. Shot noise is due to the flow of charge carriers past different potential barrier sources in the electronic circuits. Flicker noise or $1/f$ noise is the low-frequency noise due to flow of charge carriers in discontinuous medium, that is, resistors, transistors, thin films, and so on. Photon noise is due to the fluctuation in the number of quanta or power reaching the photodiode or CCD element. Photon noise is strongly temperature dependent and can be reduced by effective cooling.

References

Fujisawa N, Aoyama A, and Kosaka S 2003 Measurement of shear stress distribution over a surface by liquid-crystal coating. *Meas. Sci. Technol.*, **14**, pp. 1655–1661.

Gad-el-Hak M 2000 *Flow Control, Passive, Active and Reactive Flow Management*. Cambridge University Press.

Hanratty TJ and Campbell JA 1996 Chapter 9, Measurement of wall shear stress. In: *Fluid Mechanics Measurements*, 2nd Edition, Editor: Goldstein RJ, Taylor and Francis, Washington, DC, pp. 575–648.

Holman R, Utturkar V, Mittal R, Smith BL, and Cattafesta L 2005 Formation criterion for synthetic jets. *AIAA J.*, **43**, pp. 2110–2116.

Lee C, Hong G, Ha QP, and Mallinson SG 2003 A piezoelectrically actuated micro synthetic jet for active flow control. *Sens. Actuators, A*, **108**, pp. 168–174.

Liu C, Tsao T, Lee GB, Leu TS, Yong WV, Tai VC, and Ho CM 1999 Out-of-plane magnetic actuators with electroplated permalloy for fluid dynamics control. *Sens. Actuators*, **78**, pp. 190–197.

Naughton JW and Sheplak M 2002 Modern developments in shear-stress measurement. *Prog. Aerosp. Sci.*, **38**, pp. 515–570.

Nguyen N and Wereley ST 2006 *Fundamentals and Applications of Microfluidics*. Artech House.

Panigrahi PK, Schroeder A, and Kompenhans J 2005 PIV investigation of flow behind surface mounted permeable ribs. *Exp. Fluids*, **40**, pp. 277–300.

Schober M, Obermeier S, Pirskawetz S, and Fernholz HH 2004 A MEMS skin-friction sensor for time resolved measurements in separated flows. *Exp. Fluids*, **36**, pp. 593–599.

Sheplak M, Padmanabhan A, Schmidt MA, and Breuer KS 2001 Dynamic calibration of a shear-stress sensor using Stokes-layer excitation. *AIAA J.*, **39**(5), pp. 819–823.

Tung S, Maines B, Jiang F, and Tsao T 2004 Development of a MEMS-based control system for compressible flow separation. *J. Microelectromech. Syst.*, **13**, pp. 91–99.
Xu Y, Jiang F, Newbern S, Huang A, and Ho CM 2003 Flexible shear-stress sensor skin and its application to unmanned aerial vehicles. *Sens. Actuators, A*, **105**, pp. 321–329.

Supplemental Reading

Fernholz HH, Janke G, Schober M, Wagner PM, and Warnack D 1996 New developments and applications of skin-friction measuring techniques. *Meas. Sci. Technol.*, **7**, pp. 1396–1409.
Huang A, Lew J, Xu V, Tai VC, and Ho CM 2004 Microsensors and actuators for microfluidic control. *IEEE Sens. J.*, **4**, pp. 494–502.
Khoo M and Liu C 2001 Micro magnetic silicone elastomer membrane actuator. *Sens. Actuators*, **89**, pp. 259–266.
Löfdahl L and Gad-el-Hak M 1999 MEMS-based pressure and shear stress sensors for turbulent flows. *Meas. Sci. Technol.*, **10**, pp. 665–686.
Panigrahi PK 2008a Shear stress sensors. *Encyclopedia of Micro and Nano Fluidics* Chief Editor: Li D, Section Editor: Kandlikar SG, Springer Publications.
Panigrahi PK 2008b Turbulence control (Microflaps, Microballoon, Microsynthetic jet). *Encyclopedia of Micro and Nano Fluidics* Chief Editor: Li D, Section Editor: Steve W., Springer Publications.
Rathnasingham R and Breuer K 2003 Active control of turbulent boundary layers. *J. Fluid Mech.*, **495**, pp. 209–233.

13

Heat Pipe

13.1 Introduction

The electronic equipments are becoming smaller and faster during the recent technological boon. There has been lot of emphasis on heat dissipation as these devices generate large quantity of heat. The main aim in electronical cooling applications is to provide minimal resistance to thermal path. The microheat pipe is being looked at as one of the emerging technologies in thermal cooling. One of the primary advantages of microheat pipe is its size. It has dimensions in microns and can be fabricated on the substrate itself. This helps in removing heat directly from the hotspot. The other important advantage is that it is a passive heat transfer device. The mechanism of energy transport in heat pipe is very similar to that of microheat pipe. The role of porous wick in a heat pipe is accomplished by the corner region of microheat pipe. Both heat pipe and microheat pipe have been discussed in this chapter considering the small-scale phenomena occurring in both cases.

13.2 Applications of Heat Pipe

Heat pipe has many practical applications. Some of the possible applications are listed as follows:

1. The removal of heat from laser diodes or other small localized heat generating devices.
2. Thermal control of photovoltaic cells.
3. Removal or dissipation of heat from the leading edge of a hypersonic aircraft.
4. Heat pipes are embedded in silicon radiator panel of the spacecraft to dissipate large amount of generated heat.
5. Heat pipes are used for nonsurgical treatment of cancerous tissue. There is no bleeding and little pain during the operation. The probe is inserted into the tumor, and the surgery takes place by freezing.
6. In solar collector, the heat pipe is used to collect solar energy from a large area and transferred to a small area where the storage liquid is heated. There is no need to circulate

Transport Phenomena in Microfluidic Systems, First Edition. Pradipta Kumar Panigrahi.

Companion Website: www.wiley.com/go/panigrahi/microfluidic

liquid in large collector plates. It also works as diodes, which cut off the loss of heat from the storage liquid to the atmosphere when the collector plate is at a lower temperature.

7. Laptop computer commonly uses heat pipe for dissipating the heat generation from electronic components.

13.3 Advantages of Heat Pipe

The unique features of a heat pipe for various practical applications are due to the following advantages:

1. *Very high thermal conductivity*: Heat pipe requires less temperature difference to transport heat than traditional materials (thermal conductivity up to 90 times greater than copper for the same size), resulting in low thermal resistance.
2. *Temperature control*: The evaporator and condenser temperature can remain nearly constant (at saturation temperature, T_{sat}), while heat flux into the evaporator may vary.
3. *Geometry control*: The condenser and evaporator can have different areas to fit variable area spaces. High heat flux inputs can be dissipated with low heat flux outputs by only using natural or forced convection.
4. *Passive device*: It is a passive device with no moving parts.
5. *Maintenance*: It has no wear, has long life, and is maintenance-free.
6. *Noise*: Heat pipe has noiseless operation.

13.4 Heat Pipe Operation

The schematic of a conventional heat pipe has been shown in Figure 13.1. It consists of three sections: (1) evaporator, (2) condenser, and (3) adiabatic section. It is a surface tension-driven liquid–vapor phase-change device. The movement of condensate flow takes place through the capillary structure. It also has large heat transport capability at small temperature drop. It has

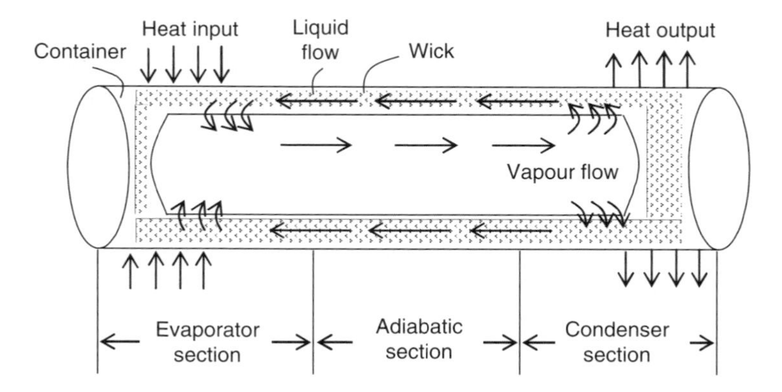

Figure 13.1 A schematic showing the structure and operation of a conventional heat pipe

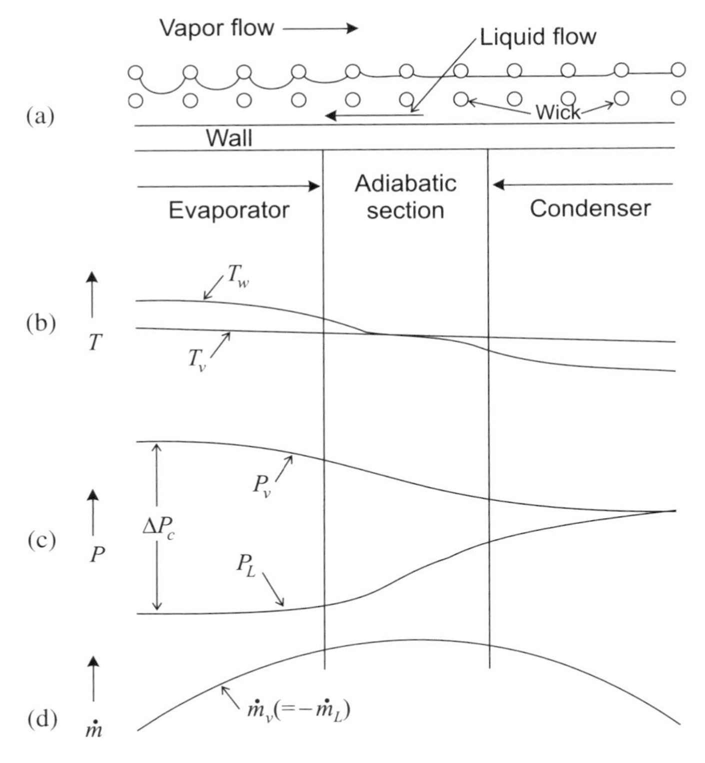

Figure 13.2 (a) Liquid and vapor flow, (b) wall and vapor temperature, (c) vapor and liquid pressure distribution, and (d) vapor and liquid flow rates

a large operating temperature range (4–2200 K). The heat source and sink can be decoupled from each other. The wick material near the wall is responsible for the transport of liquid from the condenser section to the evaporator section by capillary force.

Figure 13.2 shows the distribution of different operating parameters in a heat pipe. Figure 13.2(a) shows the wick structure and the interface of the liquid–vapor film. Figure 13.2(b) shows the wall temperature of the heat pipe and the corresponding vapor temperature distribution. Figure 13.2(c) shows the vapor and liquid pressure distribution. The liquid pressure drop from the condenser side to the evaporator side is due to Darcy pressure drop. Figure 13.2(d) indicates that the liquid and vapor flow rates are equal at any cross section as there is no inflow and outflow due to the closed end of the heat pipe.

13.5 Wick Structure

The wick material is the most critical component of a heat pipe. The primary functions of a wick are:

1. To provide necessary flow passage to the return of condensed liquid.

2. Surface pores at the liquid–vapor interface help in the development of the required capillary pumping pressure.
3. Heat flow path between the inner wall of the capillary and the liquid–vapor interface.

Figure 13.3 shows few samples of wick structures currently being used in heat pipes. The wicks in a heat pipe are used to pump the condensate from the condenser to the evaporator. The grooved and sintered particles are the most commonly used wicks of heat pipe. The grooved wicks are widely used in laptop computers. The groove dimensions significantly affect the effective thermal conductivity. There exists an optimized groove configuration for the best heat transfer performance of a given application. Sintered metal wick also has high effective thermal conductivity. The heat transfer performance in a heat pipe with sintered wicks largely depends on the wick thickness, particle size, and porosity.

The following properties affect the wick design:

1. *High pumping pressure*—A small capillary pore radius (channels through which the liquid travels in the wick) results in a large pumping (capillary) pressure.
2. *Permeability*—A large pore radius results in low liquid pressure drops and low flow resistance.

Design choice should be made that balances large capillary pressure with low liquid pressure drop.

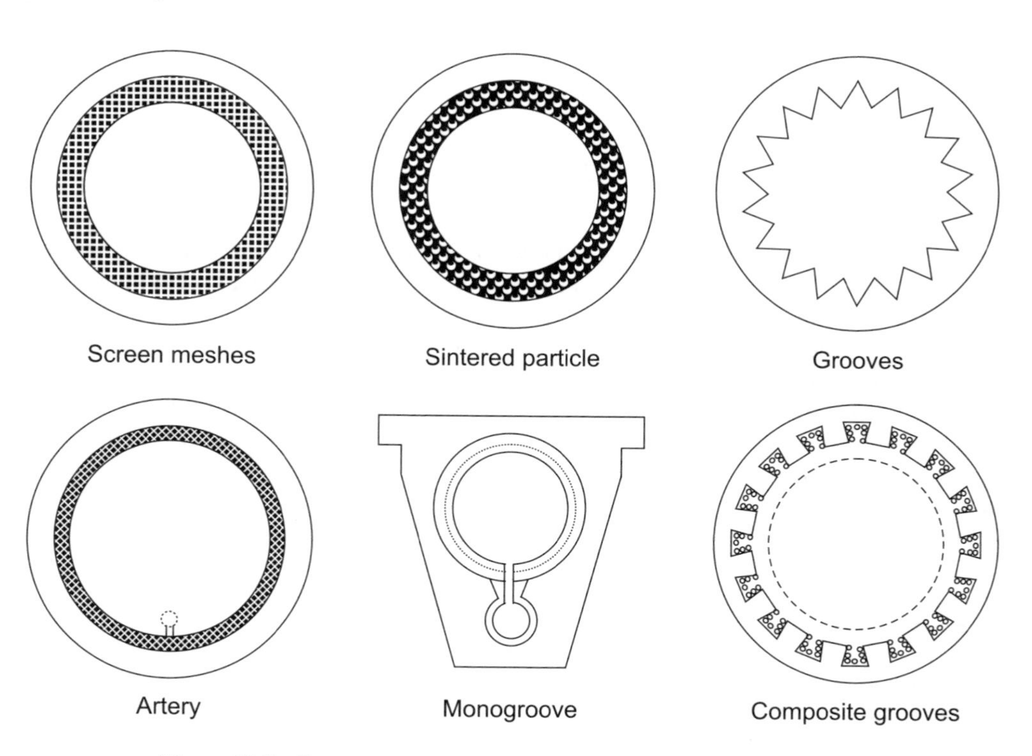

Figure 13.3 Some common wick structures used in traditional heat pipe

13.6 Working Fluids and Structural Material of Heat Pipe

Table 13.1 shows the recommended working fluids corresponding to different operating temperatures of the heat pipe. Typical working fluids for cryogenic heat pipes include helium, argon, oxygen, and krypton. Ammonia, acetone, and water are commonly employed for the most common low-temperature heat pipes ranging from 200 to 550 K. The typical working fluids being used in high-temperature heat pipes are sodium, lithium, silver, and potassium.

Table 13.1 Working fluids and operating temperatures of heat pipe

Working fluids	Melting point, K at 1 atm	Boiling point, K at 1 atm	Useful range, K
Helium	1.0	4.21	2–4
Hydrogen	13.8	20.38	14–31
Neon	24.4	27.09	27–37
Nitrogen	63.1	77.35	70–103
Argon	83.9	87.29	84–116
Oxygen	54.7	90.18	73–119
Methane	90.6	111.4	91–150
Krypton	115.8	119.7	116–160
Ethane	89.9	184.6	150–240
Freon 22	113.1	232.2	193–297
Ammonia	195.5	239.9	213–373
Freon 21	138.1	282.0	233–360
Freon 11	162.1	296.8	233-393
Pentane	143.1	309.2	253–393
Freon 113	236.5	320.8	263–373
Acetone	180.0	329.4	273–393
Methanol	175.1	337.8	283–403
Flutec PP2	223.1	349.1	283–433
Ethanol	158.7	351.5	273–403
Heptane	182.5	371.5	273–423
Water	273.1	373.1	303–473
Toluene	178.1	383.7	323–473
Flutec PP9	203.1	433.1	273–498
Naphthalene	353.4	490	408–478
Dowtherm	285.1	527	423–668
Mercury	234.2	630.1	523–923
Sulfur	385.9	717.8	530–947
Cesium	301.6	943.0	723–1173
Rubidium	312.7	959.2	800–1275
Potassium	336.4	1032	773–1273
Sodium	371	1151	873–1473
Lithium	453.7	1615	1273–2073
Calcium	1112	1762	1400–2100
Lead	600.6	2013	1670–2200
Indium	429.7	2353	2000–3000
Silver	1234	2485	2073–2573

Table 13.2 Experimental compatibility tests

	Aluminum	Brass	Copper	Iron	Nickel	Silica	Stainless steel	Titanium	Tungsten
Acetone	C	C	C			C	C		
Ammonia	C		I	C	C		C		
Cesium								C	
Dowtherm			C			C	C		
Freon-11	C								
Heptane	C								
Lead					I		I	I	C
Lithium					I		I	I	C
Mercury				I	I		C	I	
Methanol		C	C	C		C	C		
Silver					I		I	I	C
Sodium					C		C	I	
Water	I		C		C	C	C	C	

The structural materials need to be selected such that the material can withstand the temperature and there is no chemical reaction between the wall material and the working fluid. Noncondensable gas generation and corrosion are the two most important factors for selecting heat pipe wicks/material and working fluids. Table 13.2 shows the compatibility test results for various working fluid and structure material combinations. The other factors are wettability of fluid–wick combination, strength to weight ratio, thermal conductivity, and manufacturability. Al alloys and steels are used as structural material at low operating temperature. Copper and alloys of copper are used at medium operating temperature. Stainless steel and Ni alloys are used as structural material at high operating temperatures. The more important factor in the selection of working fluid is its compatibility with the case and wick materials.

13.7 Operating Temperature of Heat Pipe

Heat pipes are designed for various operating temperatures depending on the nature of applications. The heat pipe can operate only in the temperature range between the thermodynamic triple point and the critical point of the working fluid as it involves liquid–vapor phase transition process. Figure 13.4 shows the pressure–temperature diagram of a pure substance. These two end points (triple point and critical point) represent the theoretical lower and upper bounds of the operating temperature. Operating near the low end is hampered by adverse vapor dynamics (such as sonic limit, entrainment limit, or excessive vapor pressure drop, which are discussed later) due to low vapor densities and correspondingly high vapor velocities. The high vapor pressure near the critical point is a major concern in the mechanical design of the pipe structure. The high end, that is, the steeply sloped region of the vapor pressure curve, is the more desirable operating region, since pressure drops in vapor flow result in smaller temperature differences in the heat pipe.

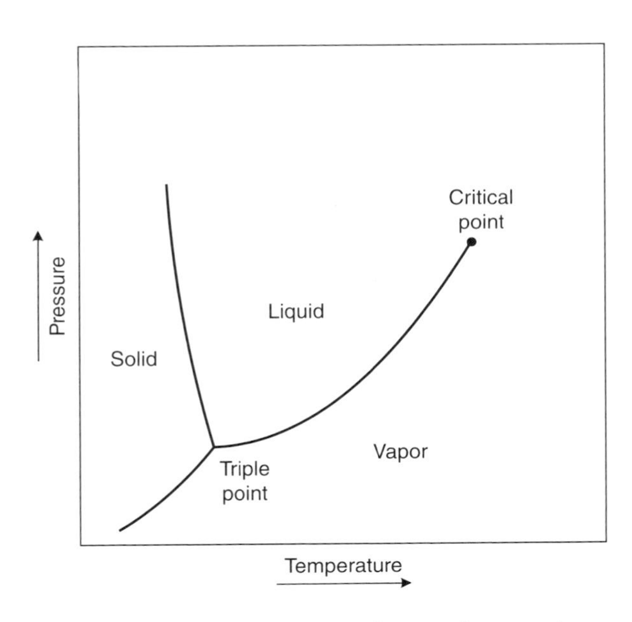

Figure 13.4 Pressure–temperature diagram of a pure substance

13.8 Ideal Thermodynamic Cycle of Heat Pipe

The ideal thermodynamic cycle describing the operation of a heat pipe is shown in Figure 13.5. The process 1–2 corresponds to the evaporator section of the heat pipe. Heat applied to the evaporator through external sources vaporizes the working fluid to a saturated (2) or superheated (2) vapor state. The process 2–3 corresponds to the adiabatic section, when the vapor pressure drives vapor through the adiabatic section to the condenser. The process 3–4 corresponds to the condenser section, when the vapor condenses, releasing heat to a heat sink. In process 4–1, capillary pressure created by menisci in wick pumps condensed fluid into the evaporator section. Thus, the working fluid undergoes repetition of the process.

13.9 Microheat Pipe

Cotter (1984) first introduced the concept of a very small microheat pipe. The fundamental principles of microheat pipes are essentially the same as those of larger, more conventional heat pipes. The cornered region is used to pump the condensate from the condenser to the evaporator. A schematic of a microheat pipe has been shown in Figure 13.6. Heat applied to one end of the heat pipe vaporizes the liquid in that region and forces it to move to the cooler end where it condenses and gives up the latent heat of vaporization. This vaporization and condensation process causes the liquid–vapor interface in the corner regions to change continually along the pipe, resulting in a capillary pressure difference between the evaporator and condenser regions. This capillary pressure difference promotes the flow of the working fluid from the condenser back to the evaporator through the triangular-shaped corner regions. These corner regions in microheat pipes serve as the liquid arteries, and no additional wicking structure is required.

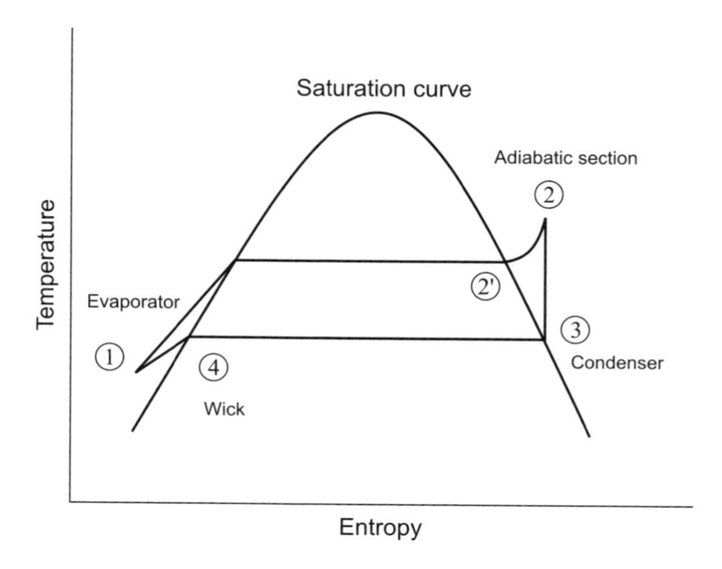

Figure 13.5 Temperature entropy diagram showing the operation of a heat pipe

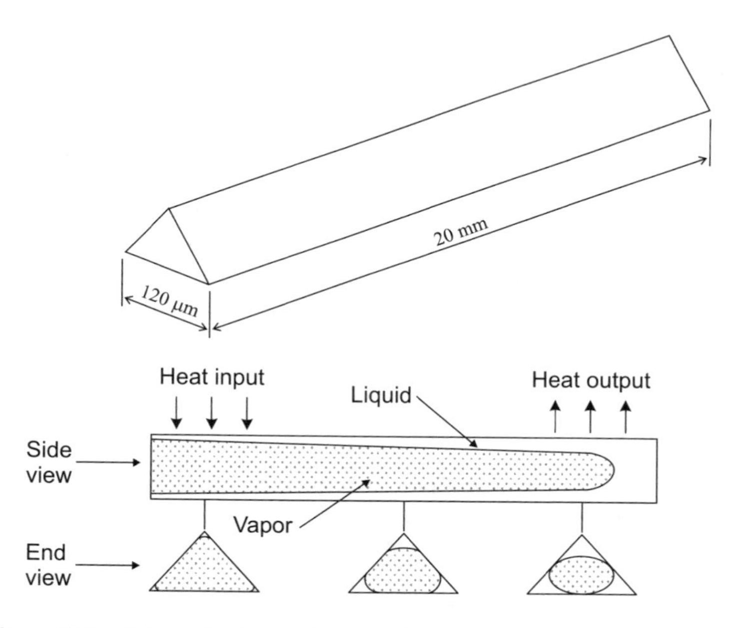

Figure 13.6 Schematic of the microheat pipe showing its operational principle

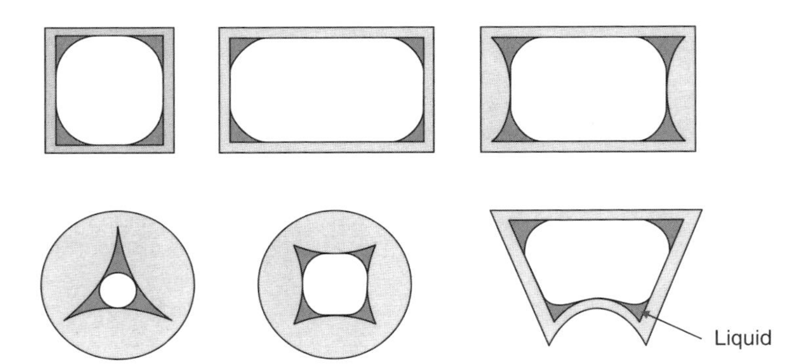

Figure 13.7 Various microheat pipe profiles (square, rectangular, modified rectangular, circular outer and triangular inner, circular outer and square inner, tapered)

Microheat pipes are subjected to the same operating limits as the conventional heat pipe. The capillary limitation is the most important operating limit of the microheat pipe.

Figure 13.7 shows various shapes of microheat pipe profiles used in literature. The hydraulic radius/diameter is used to describe the size of an internal duct. Hydraulic radius is defined as $r_h = \frac{2\times A}{P}$, where A is the cross-sectional area and P is the perimeter. Similarly, the capillary radius is used to describe the liquid–vapor interface as r_c = capillary radius = reciprocal of the mean curvature of the liquid–vapor interface. For microheat pipe, $\frac{r_c}{r_h} \geq 1$.

13.10 Effective Thermal Conductivity

Effective thermal conductivity, K_{eff}, is used to describe the characteristic of a heat pipe. It is defined on the basis of Fourier's law as

$$Q = \frac{K_{\text{eff}} A_{\text{eff}} \Delta T}{L_{\text{eff}}} \tag{13.1}$$

where Q is the input power, A_{eff} is the cross-sectional area of the device, ΔT is the difference between the evaporator and condenser average temperature, and L_{eff} is the effective heat pipe length given as

$$L_{\text{eff}} = 0.5L_e + L_a + 0.5L_c \tag{13.2}$$

where L_e is the length of evaporator, L_a is the length of adiabatic region, and L_c is the length of the condenser region.

13.11 Operating Limits

Heat pipe has limits on the maximum heat transfer that it can achieve. These limits originate from fluid mechanic consideration. Figure 13.8 shows the schematic diagram of various limits in a heat pipe. The working fluids have a narrower operating temperature range for room temperature and low-temperature heat pipes where only two or three limits, that is, the capillary and boiling limits, are usually in effect. The following section discusses various limits of heat pipes.

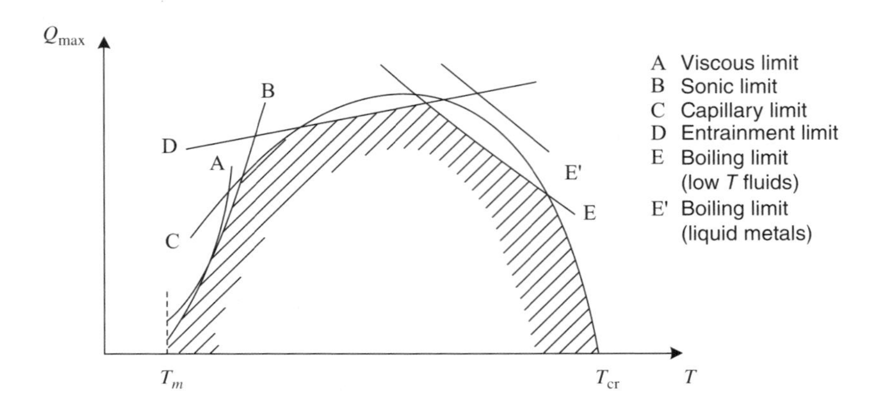

Figure 13.8 Possible limiting range of a heat pipe as a function of operating temperature

13.11.1 Capillary Limitation

The capillary pressure at position x along the heat pipe is the difference between vapor pressure and liquid pressure denoted as $P_c(x) = P_v(x) - P_l(x)$, where v is the vapor and l the liquid. The capillary pressure is established by the menisci that form at the interface. At the liquid–vapor interface, the Young–Laplace equation can be written as

$$P_c = \sigma\left(\frac{1}{R_1} + \frac{1}{R_2}\right) \tag{13.3}$$

The maximum capillary pressure is

$$P_{\text{cm}} = \frac{2\sigma}{r_c} \tag{13.4}$$

where $\frac{2}{r_c}$ is the maximum possible value of $\left(\frac{1}{R_1} + \frac{1}{R_2}\right)$. If the heat pipe is operating in a gravitational field and circumferential communication of liquid within the liquid is possible, the maximum effective capillary pressure ($P_{cm,e}$) available for axial transport of fluid decreases due to the effect of the gravitational force in the direction perpendicular to the heat pipe axis:

$$P_{\text{cm}}, e = \frac{2\sigma}{r_c} - \rho_l g d_v \cos\psi \tag{13.5}$$

where ρ_l is the density of liquid, d_v is the diameter of the vapor portion of the pipe, and ψ is the heat pipe inclination with respect to the horizontal.

13.11.1.1 Capillary Pressure

The capillary pressure difference at a liquid–vapor interface ΔP_c for most heat pipe applications is

$$\Delta P_{c,m} = \left(\frac{2\sigma}{r_{c,e}}\right) - \left(\frac{2\sigma}{r_{c,c}}\right) \tag{13.6}$$

where $r_{c,e}$ is the radius of curvature in evaporator and $r_{c,c}$ is the radius of curvature in the condenser region. During heat pipe operation, the vaporization occurs in the evaporator, and the $r_{c,e}$ decreases while the $r_{c,c}$ increases. This difference in curvature pumps the liquid from the condenser to the evaporator. During steady-state operation, it is generally assumed that the $r_{c,c}$ approaches infinity. Hence, the maximum capillary pressure is

$$\Delta P_{c,m} = \left(\frac{2\sigma}{r_{c,e}}\right) \tag{13.7}$$

The values for $r_{c,e}$ can be found theoretically for simple geometries and experimentally for other geometries:

Cylindrical pore: The capillary pressure for cylindrical pore is given as

$$P_c = \frac{2\sigma \cos\theta}{r} \tag{13.8}$$

because $R_1 = R_2 = r/\cos\theta$, where r is the radius of the cylindrical pore and θ is the contact angle. The maximum capillary pressure exists when $\theta = 0$. Hence, maximum capillary pressure is equal to $P_{c,m} = \frac{2\sigma}{r}$, with radius of curvature $r_c = r$.

Rectangular groove: One radius of curvature is infinity, and the other is equal to zero for the rectangular groove. Hence, for maximum capillary pressure, the critical radius is equal to half of the groove width, $r_c = w$, where w is the groove width.

Triangular groove: At zero wetting angle of triangular groove, the radius of curvature is

$$R = \frac{w}{2\cos\beta} \tag{13.9}$$

where w is the perpendicular distance from the interface to the groove centerline and β is the half-included angle of the triangular groove (see Figure 13.9).
Hence,

$$\frac{2}{r_c} = \frac{1}{R} = \frac{2\cos\beta}{w} \tag{13.10}$$

$$\text{Therefore,} \quad r_c = \frac{w}{\cos\beta} \tag{13.11}$$

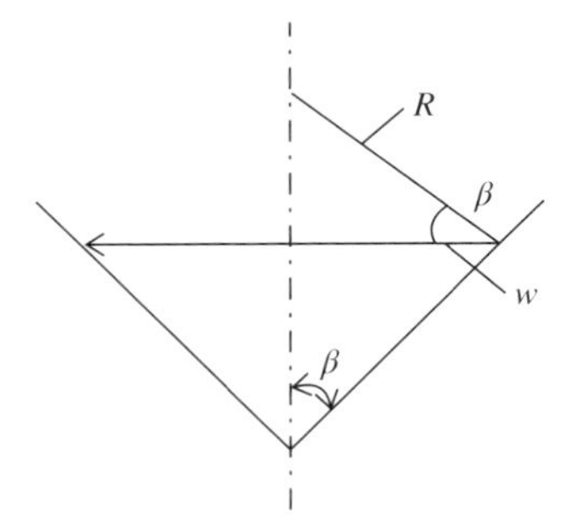

Figure 13.9 Schematic showing the nomenclature of the triangular groove for calculation of capillary pressure

13.11.1.2 Hydrostatic Pressure Drop

The normal and axial hydrostatic pressure drops ΔP_+ and ΔP_- are the result of the local gravitational force and can be expressed as

$$\Delta P_+ = \rho_l g d_v \cos \psi \tag{13.12}$$

$$\Delta P_- = \rho_l g L \sin \psi \tag{13.13}$$

where ρ_l is the density of liquid, d_v is the diameter of vapor portion of the pipe, L is the length of the heat pipe, and ψ is the angle of the heat pipe with respect to the horizontal. The axial hydrostatic pressure term may either assist or hinder the capillary pumping forces depending on whether the tilt of the heat pipe promotes or hinders the flow of liquid back to the evaporator, that is, the evaporator either lies below or above the condenser.

13.11.1.3 Capillary Limitation

Capillary limit occurs when the capillary pressure is too low to provide enough liquid to the evaporator from the condenser. It leads to dryout in the evaporator. The dryout prevents continuation of thermodynamic cycle, and the heat pipe cannot function properly any more.

The difference in capillary pressure across the liquid–vapor interface in the evaporator and condenser must exceed the sum of all the pressure losses throughout the liquid and vapor flow paths, that is,

$$\Delta P_c \geq \Delta P_+ + \Delta P_- + \Delta P_l + \Delta P_v \tag{13.14}$$

where ΔP_l is the liquid pressure drop and ΔP_v is the vapor pressure drop. In microheat pipes, where the surface tension forces dominate, ΔP_+ and ΔP_- terms can be neglected.

13.11.1.4 Liquid Pressure Drop

As the liquid returns from the condenser to the evaporator, it experiences a viscous pressure drop (ΔP_l), which can be written as

$$\frac{dP_l}{dx} = -\frac{2\tau_l}{r_{h,l}} \tag{13.15}$$

where τ_l is the frictional shear stress at the liquid–solid interface and $r_{h,l}$ is the hydraulic radius ($\frac{2A}{P}$, where A is the cross-sectional area and P is the wetted perimeter).

The above expression is analogous to the balance between shear force and pressure force. For a pipe flow of radius R and length L, the force balance gives

$$\tau \times 2\pi RL = \Delta P \times \pi R^2$$

$$\frac{\Delta P}{L} = \frac{\tau}{R/2} \tag{13.16}$$

The Reynolds number of the liquid flow can be written as

$$Re_l = \frac{2(r_{h,l})\rho_l v_l}{\mu_l} \tag{13.17}$$

The coefficient of friction can be written as

$$f_l = \frac{2\tau_l}{\rho_l v_l^2} \tag{13.18}$$

where v_l is the local liquid velocity related to the local heat flow.

We can write the local heat flux as

$$Q = v_l A_l \tag{13.19}$$

where A_l is defined as

$$A_l = \varepsilon A_w \rho_l h_{fg} \tag{13.20}$$

where ε is the wick porosity, A_w is the wick cross-sectional area, and h_{fg} is the latent heat of vaporization. Substituting equations (13.17)–(13.20) in equation (13.15), we have

$$\begin{aligned} \frac{dp_l}{dx} &= -\left(\frac{(f_l Re_l)\mu_l}{2\varepsilon A_w (r_{h,l})^2 h_{fg} \rho_l}\right) Q \\ &= -\left(\frac{\mu_l}{K A_w h_{fg} \rho_l}\right) Q \end{aligned} \tag{13.21}$$

where the permeability K is expressed as

$$K = \frac{2\varepsilon \gamma_{h,l}^2}{f_l \cdot Re_l} \tag{13.22}$$

For steady-state operation, with constant heat addition and removal, equation (13.21) can be integrated to yield

$$\Delta P_l = -\frac{\mu_l}{K A_w h_{fg} \rho_l} L_{\text{eff}}\, Q \tag{13.23}$$

13.11.1.5 Vapor Pressure Drop

In the steady state, the mass flow rate for the vapor is equal to that of the liquid at the same axial position. However, because of the low density for the vapor in comparison with that of the liquid, the vapor velocity can be considered much larger than the liquid velocity. Hence, the flow of vapor can be laminar or turbulent, and the compressibility of the vapor may also be important. Similar to the liquid flow case (equation (13.21)), we can write

$$\frac{dp_v}{dx} = -\left(\frac{f_v Re_v \mu_v}{2 r_{h,v}^2 A_v \rho_v h_{fg}}\right) Q \tag{13.24}$$

where the Reynolds number for vapor flow is

$$Re_v = \frac{2 r_{h,v} Q}{A_v \mu_v h_{fg}} \tag{13.25}$$

Using

$$v_v = \frac{Q}{A_v \rho_v h_{fg}} \tag{13.26}$$

the Mach number can be defined as

$$M_v = \frac{Q}{A_v \rho_v h_{fg} \sqrt{\gamma_v R_v T_v}} \tag{13.27}$$

Here, γ_v is the specific heat ratio.

Case-I

The vapor pressure drop expression can be simplified depending on the flow conditions as mentioned here. When $Re_v < 2300$ and $M_v < 0.2$, the flow is assumed to be laminar and incompressible:

$$f_v Re_v = 16 \tag{13.28}$$

Case-II

When $Re_v < 2300$ and $M_v > 0.2$, the compressibility effect is important, and we write for the laminar flow

$$\frac{f_{v,c}}{f_{v,i}} = \left(1 + \frac{\gamma_v - 1}{2} M_v^2\right)^{-(1/2)} \tag{13.29}$$

where $f_{v,c}$ and $f_{v,i}$ are the friction factor for compressible and incompressible flow condition, respectively. The vapor pressure drop expression (equation (13.24)) is written as

$$\frac{dp_v}{dx} = -\left(\frac{f_v Re_v \mu_v}{2r_{h,v}^2 A_v \rho_v h_{fg}}\right) Q \left(1 + \frac{\gamma_v - 1}{2} M_v^2\right)^{-(1/2)} \tag{13.30}$$

Case-III

When $Re_v > 2300$ and $M_v < 0.2$, the flow is considered to be incompressible and turbulent. We have the friction factor expression for turbulent flow as

$$f_v = \frac{0.038}{Re_v^{0.25}} \tag{13.31}$$

Hence,

$$\frac{dp_v}{dx} = -\left(\frac{0.019\mu_v}{A_v r_{h,v}^2 \rho_v h_{fg}}\right) Q(Re_v)^{3/4} \tag{13.32}$$

Case-IV

When $Re_v > 2300$ and $M_v > 0.2$, the flow is classified as compressible turbulent flow. We have

$$\frac{f_{v,c}}{f_{v,i}} = \left(1 + \frac{\gamma_v - 1}{2} M_v^2\right)^{-(3/4)} \tag{13.33}$$

$$\frac{dp_v}{dx} = -\left(\frac{0.019\mu_v}{A_v r_{hv}^2 \rho_v h_{fg}}\right) Q(Re_v)^{3/4} \left(1 + \frac{\gamma_v - 1}{2} M_v^2\right)^{-(3/4)} \tag{13.34}$$

13.11.2 Viscous Limit

The vapor pressure difference between the condenser and the evaporator may not be enough to overcome the viscous forces at low temperature. The vapor from the evaporator does not move to the condenser, and the thermodynamic cycle does not occur.

At low temperatures, viscous forces are dominant in the vapor flow down the heat pipe. At very low operating temperature, the vapor pressure difference between the closed ends of the evaporator (the high-pressure region) may be extremely small. Because of the small pressure difference, the viscous forces within the vapor region may prove to be dominant and hence limit the heat pipe operation. From the 2-D analysis by Busse,

$$Q = \frac{r_v^2 L \rho_v P_v}{16 \mu_v l_{\text{eff}}} \tag{13.35}$$

where P_v and ρ_v refer to the pressure and density at the evaporator end of the pipe.

This limit is a concern when the following flow condition is met:

$$\frac{\Delta P_v}{P_v} > 0.1 \tag{13.36}$$

13.11.3 Sonic Limit

The vapor flow in a heat pipe is analogous to the compressible fluid flow in a converging–diverging nozzle (see Figure 13.10). In curve A of Figure 13.10, at back pressure (P_a), the pressure decreases in the converging section with an increase in velocity up to the throat. In the divergent section, a pressure recovery occurs with a decrease in velocity. In curve B, at back pressure (P_b), the velocity becomes sonic at the throat, and the maximum mass flow rate

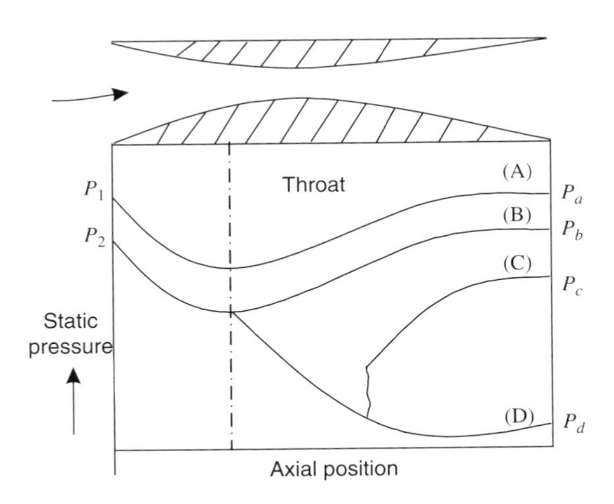

Figure 13.10 Compressible flow characteristic in a converging–diverging nozzle

is attained (choked flow). Further reduction in back pressure will not increase the flow rate. In curve C, at further reduced back pressure (P_c), the velocity in the divergent section becomes supersonic, and pressure recovery often takes the form of a shock front. In curve D, at back pressure (P_d), there is a continuous acceleration of the gas over the length of the divergent section. Decreasing the back pressure below this value has no effect upon the conditions in the converging section.

The vapor flow in the heat pipe vapor case is quite similar to the flow characteristics encountered in a converging–diverging nozzle. Very high velocity, choked flow, and pressure recovery are evident in the operation of heat pipes, which are functions of the heat input and rejection rates.

The temperature distribution is plotted in Figure 13.11 rather than the pressure in the sodium heat pipe experiment due to the two-phase system. The temperature and pressure profiles are identical between Figures 13.10 and 13.11. The following observations can be made about Figure 13.11:

Curve A: It demonstrates a subsonic condition with slight temperature recovery in the condenser. The temperature decreases along the evaporator section as the vapor stream is accelerated due to mass addition caused by evaporation.

Curve B: When the condenser temperature is lowered by increasing the heat rejection rate, the evaporator temperature is also lowered. The vapor velocity at the exit becomes sonic and critical, and choked flow condition exists.

Curves C and D: Further increasing the heat rejection rate only lowers the condenser temperature because the rate of heat transfer could not be increased due to the existence of choked flow. The change in condenser temperature has no effect upon the evaporation

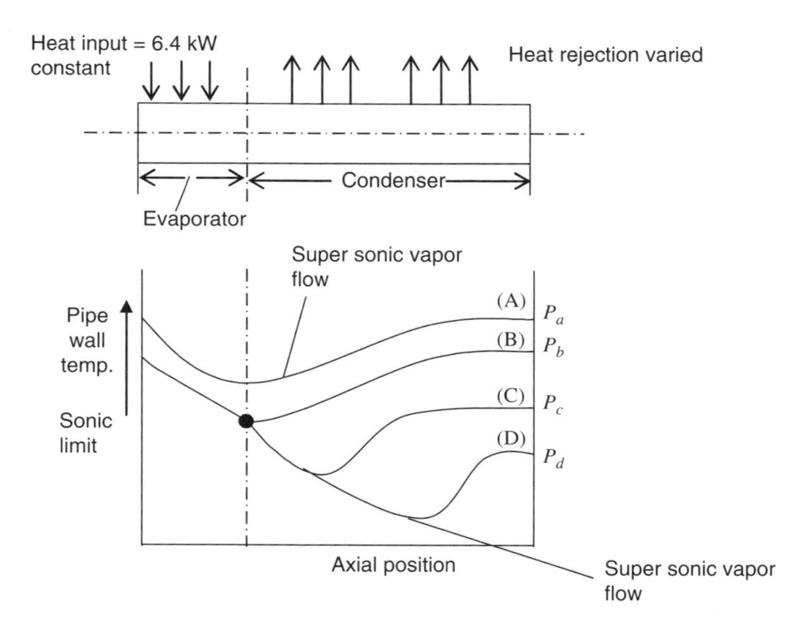

Figure 13.11 Temperature–axial location plot of a typical sodium heat pipe to explain the sonic limit

temperatures because the vapor is moving at the speed of sound at the evaporator exit and changes in condenser conditions cannot be transmitted to the upstream section. At this condition, there is a maximal axial heat transport rate due to the choked flow and a fixed axial temperature drop along the evaporator. Increasing the heat rejection rate beyond the sonic limit lowers the condenser temperature, induces supersonic flow, and creates very large axial temperature gradients along the pipe.

In a heat pipe of constant vapor core diameter, the vapor stream accelerates and decelerates because of the vapor addition in the evaporator and vapor removal in the condenser. Velocity variations in a converging–diverging nozzle result from a constant mass flow through a variable area, whereas in a heat pipe velocity, variations result from a variable mass flow through a constant area.

Maximum heat transfer: We assume 1-D vapor flow theory, that is, (1) ideal gas law for vapor flow, (2) dominance of inertial effects, and (3) negligible frictional effects. We can write the ideal gas law as

$$\frac{P_0}{\rho_0 T_0} = \frac{P_v}{\rho_v T_v} \tag{13.37}$$

Using conservation of energy and momentum with negligible frictional effects, we have

$$\frac{T_0}{T_v} = 1 + \frac{V_v^2}{2CpT_v} \tag{13.38}$$

$$\frac{P_0}{P_v} = 1 + \frac{\rho_v V_v^2}{P_v} \tag{13.39}$$

$$Q = \rho_v V_v A_v h_{fg} \tag{13.40}$$

where Q is the axial heat flux. In terms of local Mach number (M_v), using sonic velocity equal to $\sqrt{\gamma_v R_v T_v}$, we can write the above three equations as

$$\frac{T_0}{T_v} = 1 + \frac{\gamma_v - 1}{2} M_v^2 \tag{13.41}$$

$$\frac{P_0}{P_v} = 1 + \gamma_v M_v^2 \tag{13.42}$$

$$Q = \rho_v M_v A_v h_{fg} \sqrt{\gamma_v R_v T_v} \tag{13.43}$$

By combining the above equations for T_0 and P_0 with the gas law relationship equation (13.37), we get

$$\frac{\rho_0}{\rho_v} = \frac{1 + \gamma_v M_v^2}{1 + \frac{\gamma_v - 1}{2} M_v^2} \tag{13.44}$$

By combining equations (13.41)–(13.44), we can write

$$Q = \frac{A_v \rho_0 h_{fg} (\gamma_v R_v T_0)^{1/2} M_v \left(1 + \frac{\gamma_v - 1}{2} M_v^2\right)^{1/2}}{1 + \gamma_v M_v^2} \tag{13.45}$$

The sonic limit occurs when the Mach number M_v at the evaporator exit is unity.

Hence,

$$Q_{\max} = A_v \rho_0 h_{fg} \left(\frac{\gamma_v R_v T_0}{2(\gamma_v + 1)} \right)^{1/2} \tag{13.46}$$

13.11.4 Entrainment Limit

Droplets of liquid in the wick are torn from the wick at higher velocities and sent into the vapor resulting in dryout, which is known as entrainment limit. As the vapor and the liquid move in opposite directions, a shear force exists at the vapor–liquid interface. If the vapor velocity is high, a limit can be reached when the liquid is torn from the surface and entrained in the vapor. There is a sudden increase in the fluid circulation, and the liquid return system cannot accommodate the increased flow. There are excess liquid accumulation in the condenser and dryout in the evaporator. The shear force at the liquid–vapor interface is proportional to the dynamic pressure of the moving vapor $(\rho_v V_v^2)/2$ and the area (A_s):

$$F_s = K_1 \frac{\rho_v V_v^2 A_s}{2} \tag{13.47}$$

where K_1 is the constant of proportionality. The surface force which holds the liquid is proportional to the product of the surface tension coefficient σ and the wetted parameter P:

$$F_t = K_2 \sigma P \tag{13.48}$$

where K_2 is the constant of proportionality.

$$\text{Weber number}, We = \frac{F_S}{F_t} = \frac{K_1 \rho_v V_v^2 A_s}{2 K_2 \sigma P} \tag{13.49}$$

For entrainment limitation, $We \simeq 1$.

We know that

$$r_{h,s} = \frac{2A_s}{P} \tag{13.50}$$

We may assume from available data

$$\frac{K1}{K2} \simeq 8 \tag{13.51}$$

Thus, we have entrainment limit expressed as

$$\frac{2 r_{h,s} \rho_v V_v^2}{\sigma} = 1 \tag{13.52}$$

$$\text{Vapor velocity} = V_v = \frac{Q}{A_v \rho_v h_{fg}} \tag{13.53}$$

Substituting in the entrainment limit, we get the heat transport limit:

$$Q_{e,\max} = A_v h_{fg} \left(\frac{\sigma \rho_v}{2 r_{h,s}} \right)^{1/2} \tag{13.54}$$

13.11.5 Boiling Limit

All the limitations discussed earlier depend upon the axial heat transfer. The boiling limit, however, depends upon the evaporator heat flux (radial). Boiling limit occurs when the radial heat flux into the heat pipe causes the liquid in the wick to boil and evaporate causing dryout. It also occurs when the nucleate boiling in the evaporator creates vapor bubbles that partially block the return of fluid. The presence of vapor bubbles requires both (1) the formation of bubbles and also (2) the subsequent growth of these bubbles. Let us imagine a spherical vapor bubble that is very close to the heat pipe surface. At equilibrium, we have

$$\pi r_b^2(P_{pw} - P_l) = 2\pi r_b \sigma \tag{13.55}$$

where P_{pw} is the saturation vapor pressure at the pipe surface temperature, P_l is the liquid pressure, r_b is the bubble radius, and σ is the coefficient of surface tension.

The liquid pressure is equal to the difference between the liquid–vapor interface pressure, P_v, and the capillary pressure, P_c, that is,

$$P_l = P_v - P_c \tag{13.56}$$

Substituting equation (13.56) in equation (13.55), we get

$$\pi r_b^2(P_{pw} - P_v + P_c) = 2\pi r_b \sigma \tag{13.57}$$

$$P_{pw} - P_v = \frac{2\sigma}{r_b} - P_c \tag{13.58}$$

Clausius–Clapeyron equation relating T and P along the saturation line is

$$\frac{dP}{dT} = \frac{h_{fg}\rho_v}{T_v} \tag{13.59}$$

We can approximate the left-hand side of the above equation as

$$\frac{dP}{dT} \simeq \frac{P_{pw} - P_v}{T_{pw} - T_{wv}} \tag{13.60}$$

Combining equations (13.58) and (13.60), we have

$$T_{pw} - T_{wv} = \frac{T_v}{h_{fg}\rho_v}\left(\frac{2\sigma}{r_b} - P_c\right) \tag{13.61}$$

Here, $T_{pw} - T_{wv}$ is the temperature drop of the liquid at the evaporator due to conduction. Total heat transfer at the evaporator section can be written based on 1-D conduction in a pipe as

$$Q = \frac{2\pi L_e K_e (T_{pw} - T_{wv})}{\ln\,(r_i/r_v)} \tag{13.62}$$

where L_e is the evaporator length, K_e is the effective thermal conductivity of the liquid, r_i is the inner radius of the pipe, and r_v is the vapor core radius. Using $T_{pw} - T_{wv}$ from equation (13.61), we get

$$Q = \frac{2\pi L_e K_e T_v}{h_{fg}\rho_v \ln\left(\frac{r_i}{r_v}\right)}\left(\frac{2\sigma}{r_b} - P_c\right) \tag{13.63}$$

The above heat transfer is required to maintain equilibrium vapor bubbles of radius r_b. If the heat transfer (Q) is less than the above value, then the vapor bubbles will grow. Hence, the boiling heat transport limit is

$$Q_{b,\max} = \frac{2\pi L_e K_e T_v}{h_{fg}\rho_v l_n \frac{r_i}{r_v}} \left(\frac{2\sigma}{r_n} - P_c \right) \tag{13.64}$$

Here, r_n is the initial radius of the vapor bubbles at its formation dependent on the surface conditions and affected by the presence of dissolved gases in the liquid.

Note: The type of limitation restricting the heat transport capability of a heat pipe is the heat transport mechanism leading to lowest heat transfer value at the temperature under consideration. The magnitude of these different limitations depends on various properties of the working fluids, for example, dimensions and structure of the heat pipe. The boiling limit can be increased by using liquid of large effective conductance. The entrainment limit can be increased by using small-size pipes or channels.

13.12 Cleaning and Charging

Maintaining cleanliness of all materials used in heat pipe is essential to achieve two important objectives: to ensure (1) that the working fluid should wet the materials and (2) that no foreign matter is present that could hinder capillary action. Oils from machining and oxides formed on the wall can decrease the wettability of the wick. Solid particles can physically block the structure, decreasing the liquid flow rate and increasing the likelihood of encountering the capillary limit. Solvent cleaning, vapor degreasing, acid cleaning, and ultrasonic cleaning are needed to properly clean the heat pipe.

Working fluid should be most highly pure, and further purification may be necessary after purchase. This may be carried out by distillation. Working fluids such as acetone, methanol, and ammonia in the presence of water can lead to incompatibilities, and the minimum possible water content should be achieved.

The amount of working fluid required for a heat pipe can be estimated by calculating the approximate volume occupied inside the wick structure or corner region of the microheat pipe. The amount of working fluid significantly influences the heat transfer performance of the heat pipe. The optimal liquid charge also depends on the heat transfer rate. The working fluid can be introduced into the heat pipe by evacuation and backfilling technique. The proper charging procedure prevents introduction of noncondensable gas into the heat pipe.

Reference

Cotter TP 1984 Micro heat pipes. *Proceedings of the 5th International Heat Pipe Conference*, Tsukuba, Japan, pp. 328–335.

Supplemental Reading

Faghri A 1995 *Heat Pipe Science and Technology*. Taylor & Francis, US.
Peterson GP 1994 *An Introduction to Heat Pipes*. John Wiley & Sons, Inc., Canada.

Index

Transport Phenomena in Microfluidic Systems, First Edition. Pradipta Kumar Panigrahi.

Companion Website: www.wiley.com/go/panigrahi/microfluidic